Biotechnology
Second Edition

Volume 5b
Genomics and Bioinformatics

Biotechnology

Second Edition

Fundamentals

Volume 1
Biological Fundamentals

Volume 2
Genetic Fundamentals and
Genetic Engineering

Volume 3
Bioprocessing

Volume 4
Measuring, Modelling and Control

Products

Volume 5a
Recombinant Proteins, Monoclonal
Antibodies and Therapeutic Genes

Volume 5b
Genomics and Bioinformatics

Volume 6
Products of Primary Metabolism

Volume 7
Products of Secondary Metabolism

Volumes 8a and b
Biotransformations I and II

Special Topics

Volume 9
Enzymes, Biomass, Food and Feed

Volume 10
Special Processes

Volumes 11a–c
Environmental Processes I–III

Volume 12
Legal, Economic and
Ethical Dimensions

All volumes are also displayed on our Biotech Website:
http://www.wiley-vch.de/home/biotech

A Multi-Volume Comprehensive Treatise

Biotechnology

Second, Completely Revised Edition

Edited by
H.-J. Rehm and G. Reed
in cooperation with
A. Pühler and P. Stadler

Volume 5b

Genomics and Bioinformatics

Edited by
C. W. Sensen

WILEY-VCH

Weinheim · New York · Chichester · Brisbane · Singapore · Toronto

Series Editors:
Prof. Dr. H.-J. Rehm
Institut für Mikrobiologie
Universität Münster
Corrensstraße 3
D-48149 Münster
FRG

Prof. Dr. A. Pühler
Biologie VI (Genetik)
Universität Bielefeld
P.O. Box 100131
D-33501 Bielefeld
FRG

Dr. G. Reed
1029 N. Jackson St. #501-A
Milwaukee, WI 53202-3226
USA

Prof. Dr. P. J. W. Stadler
Artemis Pharmaceuticals
Geschäftsführung
Pharmazentrum Köln
Neurather Ring
D-51063 Köln
FRG

Volume Editor:
Dr. C. W. Sensen
University of Calgary
Faculty of Medicine
Department of Biochemistry and
Molecular Biology
Calgary, Alberta
Canada, T2N 4N1

This book was carefully produced. Nevertheless, authors, editors and publisher do not warrant the information contained therein to be free of errors. Readers are advised to keep in mind that statements, data, illustrations, procedural details or other items may inadvertently be inaccurate.

Library of Congress Card No.: applied for

British Library Cataloguing-in-Publication Data:
A catalogue record for this book is available from the British Library

Die Deutsche Bibliothek – CIP-Cataloguing-in-Publication Data:

A catalogue record for this book
is available from Die Deutsche Bibliothek
    ISBN 3-527-28328-5

© WILEY-VCH Verlag GmbH, D-69469 Weinheim (Federal Republic of Germany), 2001

Printed on acid-free paper.

Composition and Printing: Zechner, Datenservice und Druck, D-67346 Speyer.
Bookbinding: J. Schäffer, D-67269 Grünstadt.
Printed in the Federal Republic of Germany

Preface

In recognition of the enormous advances in biotechnology in recent years, we are pleased to present this Second Edition of "Biotechnology" relatively soon after the introduction of the First Edition of this multi-volume comprehensive treatise. Since this series was extremely well accepted by the scientific community, we have maintained the overall goal of creating a number of volumes, each devoted to a certain topic, which provide scientists in academia, industry, and public institutions with a well-balanced and comprehensive overview of this growing field. We have fully revised the Second Edition and expanded it from ten to twelve volumes in order to take all recent developments into account.

These twelve volumes are organized into three sections. The first four volumes consider the fundamentals of biotechnology from biological, biochemical, molecular biological, and chemical engineering perspectives. The next four volumes are devoted to products of industrial relevance. Special attention is given here to products derived from genetically engineered microorganisms and mammalian cells. The last four volumes are dedicated to the description of special topics.

The new "Biotechnology" is a reference work, a comprehensive description of the state-of-the-art, and a guide to the original literature. It is specifically directed to microbiologists, biochemists, molecular biologists, bioengineers, chemical engineers, and food and pharmaceutical chemists working in industry, at universities or at public institutions.

A carefully selected and distinguished Scientific Advisory Board stands behind the series. Its members come from key institutions representing scientific input from about twenty countries.

The volume editors and the authors of the individual chapters have been chosen for their recognized expertise and their contributions to the various fields of biotechnology. Their willingness to impart this knowledge to their colleagues forms the basis of "Biotechnology" and is gratefully acknowledged. Moreover, this work could not have been brought to fruition without the foresight and the constant and diligent support of the publisher. We are grateful to VCH for publishing "Biotechnology" with their customary excellence. Special thanks are due to Dr. Hans-Joachim Kraus and Karin Dembowsky, without whose constant efforts the series could not be published. Finally, the editors wish to thank the members of the Scientific Advisory Board for their encouragement, their helpful suggestions, and their constructive criticism.

H.-J. Rehm
G. Reed
A. Pühler
P. Stadler

Scientific Advisory Board

Contents

Ethical, Legal and Social Issues

Contributors

Dr. Lothar Altschmied
AG Expressionskartierung
Institut für Pflanzengenetik und
Kulturpflanzenforschung (IPK)
Corrensstraße 3
D-06466 Gatersleben
Germany
Chapter 1

Dr. Rolf Apweiler
EMBL Outstation
The European Bioinformatics Institute
Hinxton Hall, Hinxton
Cambridge, CB10 1SD
United Kingdom
Chapter 12

Dr. Gary D. Bader
Samuel Lunenfeld Research Institute
Mount Sinai Hospital
600 University Avenue
Toronto, Ontario, M5G 1X5
Canada
Chapter 18

Dr. Roland Brousseau
National Research Council of Canada
6100 Royalmount Avenue
Montreal, Quebec, H4P 2R2
Canada
Chapter 9

Dr. Detlev Buttgereit
Zell- und Entwicklungsbiologie
FB Biologie
Universität Marburg
Karl-von-Frisch-Straße
D-35032 Marburg
Germany
Chapter 1

Dr. Miroslaw Cygler
Biotechnology Research Institute
6100 Royalmont Avenue
Montreal, Quebec, H4P 2R2
Canada
Chapter 15

Prof. Dr. Antoine Danchin
Hong Kong University,
Pasteur Research Centre
Dexter HC Man Building
8, Sassoon Road, Pokfulam
Hong Kong
and
Genetics of Bacterial Genomes,
CNRS URA 2171
Institut Pasteur
28, rue du Docteur Roux
75724 Paris Cedex 15
France
Chapter 1

Prof. Dr. Horst Feldmann
Adolf-Butenandt-Institut
Schillerstraße 44
D-80336 München
Germany
Chapter 1

Dr. Daniel B. Davison
Director of Bioinformatics
Bristol Myers Squibb
Bioinformatics-853
5 Research Parkway
Wallingford, CT 06492-7660
USA
Chapter 5

Dr. Daniel Figeys
MSD Ocata Inc.
480 University Avenue
Suite 401
Toronto, Ontario, M5G 1V2
Canada
Chapter 10

Dr. Graham Dellaire
MRC – Human Genetics Unit
Crewe Road
Edinburgh, EH4 2XU
United Kingdom
Chapter 3

Dr. Terry Gaasterland
The Rockefeller University
1230 York Avenue
New York, NY 10021-6399
USA
Chapters 16 and 17

Dr. Norman Dovichi
Department of Chemistry
University of Washington
Seattle, WA 98195-1700
USA
Chapter 11

Dr. Shuba Gopal
The Rockefeller University
1230 York Avenue
New York, NY 10021-6399
USA
Chapter 16

Dr. Thure Etzold
EMBL Outstation
The European Bioinformatics Institute
Hinxton Hall, Hinxton
Cambridge, CB10 1SD
United Kingdom
Chapter 12

Dr. Paul Gordon
National Research Council of Canada
1411 Oxford Street
Halifax, Nova Scotia, B3H 3Z1
Canada
Chapter 17

Dr. Roger Green
Memorial University of Newfoundland
Health Service Center
St. John's, Newfoundland, A1B 3V6
Canada
Chapter 4

Dr. Dwayne Hegedus
AAFC, Saskatoon Research Center
Molecular Genetics Section
107 Science Place
Saskatoon, Saskatchewan, S7N 0X2
Canada
Chapter 6

Dr. Christopher Hogue
Samuel Lunenfeld Research Institute
Mount Sinai Hospital
600 University Avenue
Toronto, Ontario, M5G 1X5
Canada
Chapter 18

Dr. Shen Hu
Department of Chemistry
University of Washington
Seattle, WA 98195-1700
USA
Chapter 11

Dr. Patrick G. Humphrey
Advanced Research and Development
Li-Cor, Inc.
4308 Progressive Avenue
P.O. Box 4000
Lincoln, NE 68504
USA
Chapter 8

Dr. Doris Jording
Lehrstuhl für Genetik
Universität Bielefeld
D-33594 Bielefeld
Germany
Chapter 1

Dr. Jörn Kalinowski
Lehrstuhl für Genetik
Universität Bielefeld
D-33594 Bielefeld
Germany
Chapter 1

Dr. Hans-Peter Klenk
EPIDAUROS Biotechnologie AG
D-82347 Bernried
Germany
Chapter 1

Prof. Bartha Maria Knoppers
Université de Montréal
Faculté de Droit
3101 chemin de la Tour
C.P. 6128, succursale A
Montreal, Quebec, H3C 3J7
Canada
Chapter 19

Prof. Dr. Manfred Kröger
Institut für Mikro- und Molekularbiologie
Universität Gießen
Heinrich-Buff-Ring 26–32
D-35392 Gießen
Germany
Chapter 1

Dr. Sergej N. Krylov
Department of Chemistry
York University
Toronto, Ontario, M3J 1P3
Canada
Chapter 11

Dr. Rodrigo Lopez
EMBL Outstation
The European Bioinformatics Institute
Hinxton Hall, Hinxton
Cambridge, CB10 1SD
United Kingdom
Chapter 12

Dr. Derek Lydiate
AAFC, Saskatoon Research Center
Molecular Genetics Section
107 Science Place
Saskatoon, Saskatchewan, S7N 0X2
Canada
Chapter 6

Dr. Alan Matte
Biotechnology Research Institute
6100 Royalmont Avenue
Montreal, Quebec, H4P 2R2
Canada
Chapter 15

Dr. David Michels
Department of Chemistry
University of Washington
Seattle, WA 98195-1700
USA
Chapter 11

Dr. Lyle R. Middendorf
Advanced Research and Development
Li-Cor, Inc.
4308 Progressive Avenue
P.O. Box 4000
Lincoln, NE 68504
USA
Chapter 8

Dr. Narasimhachari Narayanan
Advanced Research and Development
Li-Cor, Inc.
4308 Progressive Avenue
P.O. Box 4000
Lincoln, NE 68504
USA
Chapter 8

Dr. Isobel A. P. Parkin
AAFC, Saskatoon Research Center
Molecular Genetics Section
107 Science Place
Saskatoon, Saskatchewan, S7N 0X2
Canada
Chapter 6

Prof. Dr. Alfred Pühler
Lehrstuhl für Genetik
Universität Bielefeld
D-33594 Bielefeld
Germany
Chapter 1

Dr. Chandra S. Ramanathan
Bristol Myers Squibb
Bioinformatics-853
5 Research Parkway
Wallingford, CT 06492-7660
USA
Chapter 5

Prof. Dr. Renate Renkawitz-Pohl
Zell- und Entwicklungsbiologie
FB Biologie
Universität Marburg
Karl-von-Frisch-Straße
D-35032 Marburg
Germany
Chapter 1

Dr. Peter Rice
Sanger Centre
Wellcome Trust Genome Campus
Hinxton
Cambridge, CB10 1SA
United Kingdom
Chapter 13

Dr. Stephen J. Robinson
AAFC, Saskatoon Research Center
Molecular Genetics Section
107 Science Place
Saskatoon, Saskatchewan, S7N 0X2
Canada
Chapter 6

Dr. Stephen C. Roemer
Advanced Research and Development
Li-Cor, Inc.
4308 Progressive Avenue
P.O. Box 4000
Lincoln, NE 68504
USA
Chapter 8

Dr. Kevin Rozwadowski
AAFC, Saskatoon Research Center
Molecular Genetics Section
107 Science Place
Saskatoon, Saskatchewan, S7N 0X2
Canada
Chapter 6

Dr. Steve W. Scherer
The Hospital for Sick Children
HSC Research Institute
555 University Avenue
Toronto, Ontario, M5G 1X8
Canada
Chapter 2

Dr. Joseph D. Schrag
Biotechnology Research Institute
6100 Royalmont Avenue
Montreal, Quebec, H4P 2R2
Canada
Chapter 15

Dr. Christoph W. Sensen
University of Calgary
Faculty of Medicine
Department of Biochemistry and Molecular
Biology
Calgary, Alberta, T2N 4N1
Canada
Chapters 17 and 20

Dr. Andrew G. Sharpe
AAFC, Saskatoon Research Center
Molecular Genetics Section
107 Science Place
Saskatoon, Saskatchewan, S7N 0X2
Canada
Chapter 6

Dr. Daniel C. Tessier
National Research Council of Canada
6100 Royalmount Avenue
Montreal, Quebec, H4P 2R2
Canada
Chapter 9

Dr. David Y. Thomas
National Research Council of Canada
6100 Royalmount Avenue
Montreal, Quebec, H4P 2R2
Canada
Chapter 9

Dr. Theerayut Toojinda
DNA Fingerprinting Unit
Biotechnology Center
Kasetsart University
Kampangsaen Campus
Nakorn Pathom 73140
Thailand
Chapter 7

Dr. Somvong Tragoonrung
DNA Fingerprinting Unit
Biotechnology Center
Kasetsart University
Kampangsaen Campus
Nakorn Pathom 73140
Thailand
Chapter 7

Prof. Lap Chee Tsui
The Hospital for Sick Children
HSC Research Institute
555 University Avenue
Toronto, Ontario, M5G 1X8
Canada
Chapter 2

Dr. Apichart Vanavichit
DNA Fingerprinting Unit
Biotechnology Center
Kasetsart University
Kampangsaen Campus
Nakorn Pathom 73140
Thailand
Chapter 7

Dr. David Wishart
University of Alberta
Faculty of Pharmacy and Pharmaceutical
Sciences
Dentistry/Pharmacy Centre 2123
Edmonton, Alberta, T6G 2N8
Canada
Chapter 14

Dr. Evgeni M. Zdobnov
EMBL Outstation
The European Bioinformatics Institute
Hinxton Hall, Hinxton
Cambridge, CB10 1SD
United Kingdom
Chapter 12

Dr. Zheru Zhang
Analytical R & D
Pharmaceutical Research Institute
Bristol – Myers Squibb Company
Syracuse, NY 13221-4755
USA
Chapter 11

Introduction

CHRISTOPH W. SENSEN

Calgary, Canada

Genomics has revolutionized biological and medical research and development over the last fifteen years. The speed and magnitude by which Genomics has outgrown the disciplines from which it originally developed have taken many by surprise. One indication for this is that the Second Edition of Biotechnology added a Volume 5b: *Genomics and Bioinformatics* to the original concept and volume layout. The rapid development of the field has left much of the early history of Genomics behind and many key events have not been recorded properly.

It may be even all but forgotten how the term "genomics" originated. According to the first editorial of *Genomics* (1987, **1**, 1–2), the term was coined by T. H. RODERICK from the Jackson Laboratories in Bar Harbor, MN, some time around 1987 in discussions with editors VICTOR A. MCKUSICK and FRANK H. RUDDLE, who were looking for suggestions to name their new journal.

There is no all-encompassing definition for genomics, the word is used with many meanings. At the time when MCKUSICK and RUDDLE wrote their editorial, they understood genomics to be mapping and sequencing to analyze the structure and organization of genomes. When the Genomics journal was founded, only three years had passed since the invention of automated DNA sequencers, which dominated the first phase of the development of genomics as a science. Thus a definition such as MCKUSICK and RUDDLE's of the word genomics can be understood in the context of that time.

Today, genomics is very often subdivided into "structural genomics", which deals with the determination of the complete sequence of genomes (DNA sequencing), or the complete set of proteins (proteome) in an organism (proteomics), and "functional genomics", which studies the functioning of genes and metabolic pathways (metabolomics) or the gene expression patterns in an organism (chip technologies). To complicate matters, X-ray crystallographers have adopted the term structural genomics to refer to protein 3-D structure determination.

For the purpose of this book, genomics has the broader meaning of "genome research", including bioinformatics and other studies of the genome and proteome to understand the blueprint and function of organisms. Many of the technologies that are part of today's genomics toolkit were developed previously and then automated in an attempt to apply them in large-scale, high-throughput environments.

Some people, including the late Canadian Nobel laureate MICHAEL SMITH (UBC), have claimed that they were doing genomics all along, which is true to a certain degree when using a broad definition of genomics, considering its strong roots in molecular biology, biophysics, and biochemistry. With such a definition, we may say that genomics really started when WATSON and CRICK discovered the structure of DNA.

Without a doubt, the introduction of computers into molecular biology laboratories was one of the key factors in the development of genomics. Laboratory automation led to the production of large amounts of data, and the need to analyze, combine, and understand these resulted in the development of "bioinformatics", a new discipline at the interface of several traditional disciplines. Bioinformatics is the glue that integrates all the diverse aspects of genomics. Of similar importance to the development of the field is the development of laser-based technologies. The use of laser-based systems, which can be coupled to computerized detection systems, has replaced most of the radioactive techniques in genomics laboratories, allowing the complete automation of many types of experiments.

Considering the rapid pace of development, it is quite difficult to organize a book that reflects all aspects of genomics. Chapters about model organisms are followed by overviews of the key technologies. Because of the importance to the field, several chapters are dedicated to bioinformatics. Genomics is a science with huge impact on society, thus ethical and legal issues that need to be dealt with arise daily. One of the book chapters is devoted to ethical and legal aspects of genome research. The book closes with an outlook to future developments in genomics.

With the completion of the human genome, the true tasks for genomics are only starting to emerge. We are far from understanding how organisms with small genomes function, let alone how the human genome is organized. As more and more scientific disciplines get "genomicized", the field will undergo continual transformation, thus a book like this one can only capture a flavor and a moment in time. This may be frustrating to some, but this book is intended to summarize the essence of the first fifteen years of research and development in genomics, during which the cornerstone for a very exiting future was laid.

Calgary, February 2001   Christoph W. Sensen

Application Domains

1 Genome Projects of Model Organisms

ALFRED PÜHLER,
DORIS JORDING,
JÖRN KALINOWSKI

Bielefeld, Germany

DETLEV BUTTGEREIT
RENATE RENKAWITZ-POHL

Marburg, Germany

LOTHAR ALTSCHMIED

Gatersleben, Germany

ANTOIN E. DANCHIN

Hong Kong

HORST FELDMANN

Munich, Germany

HANS-PETER KLENK

Bernried, Germany

MANFRED KRÖGER

Giessen, Germany

1 Introduction

Genome research allows the establishment of the complete genetic information of organisms. The first complete genome sequences established were those of prokaryotic and eukaryotic microorganisms followed by plants and animals (see, e.g., the TIGR web page at *http://www.tigr.org/*). The organisms selected for genome research were mostly those which already played an important role in scientific analysis, thus they can be considered model organisms. In general, organisms are defined as model organisms when a large amount of scientific knowledge has been accumulated in the past. For this chapter on genome projects of model organisms, a number of experts in genome research have been asked to give an overview on specific genome projects and to report on the respective organism from their specific point of view. The organisms selected include prokaryotic and eukaryotic microorganisms as well as plants and animals.

We have chosen the prokaryotes *Escherichia coli*, *Bacillus subtilis*, and *Archaeoglobus fulgidus* as representative model organisms. The *E. coli* genome project is described by M. KRÖGER (Giessen, Germany). He gives a historical outline about the intensive research on microbiology and genetics of this organism, which cumulated in the *E. coli* genome project. Many of the technological tools presently available have been developed during the course of the *E. coli* genome project. *E. coli* is without doubt the best analyzed microorganism of all. The knowledge of the complete sequence of *E. coli* has confirmed its reputation to represent the leading model organism of gram-negative eubacteria.

A. DANCHIN (Hong Kong) reports on the genome project of the environmentally and biotechnologically relevant gram-positive eubacterium *B. subtilis*. The contribution focuses on the results and analysis of the sequencing effort and gives a number of examples for specific and sometimes unexpected findings of this project. Special emphasis is given to genomic data which support the understanding of general features such as translation, specific traits relevant for living in its general habitat or its usefulness for industrial processes.

A. fulgidus is the subject of the contribution by H.-P. KLENK (Bernried, Germany). This genome project was started before the genetic properties of the organism had been extensively studied. However, its unique lifestyle as a hyperthermophilic and sulfate-reducing organism makes it a model for a large number of environmentally important microorganisms and species with a high biotechnological potential. The structure and results of the genome project are described in the contribution.

The yeast *Saccharomyces cerevisiae* has been selected as a representative of eukaryotic microorganisms. The yeast project is presented by H. FELDMANN (Munich, Germany). *S. cerevisiae* has a long tradition in biotechnology as well as a long-term research history as the eukaryotic model organism *per se*. It was the first eukaryote to be completely sequenced and has led the way to sequencing other eukaryotic genomes. The wealth of the yeast sequence information as a useful reference for plant, animal or human sequence comparisons is outlined in this contribution.

Among plants, the small crucifere *Arabidopsis thaliana* was identified as the classical model plant due to simple cultivation and short generation time. Its genome was originally considered to be the smallest in the plant kingdom and was, therefore, selected for the first plant genome project which is described here by L. ALTSCHMIED (Gatersleben, Germany). The sequence of *A. thaliana* helped to identify that part of the genetic information unique to plants. In the meantime, other plant genome sequencing projects were started, many of which focus on specific problems of crop cultivation and nutrition.

The roundworm *Caenorhabditis elegans* and the fruitfly *Drosophila melanogaster* have been selected as animal models due to their specific model character for higher animals and also for humans. The genome project of *C. elegans* is summarized by D. JORDING (Bielefeld, Germany). The contribution describes how the worm – in spite of its simple appearance – became an interesting model organism for features such as neuronal growth, apoptosis, or signaling pathways. This genome project has also provided a number of bioinformatic tools which are widely used for other genome projects.

The genome project concerning the fruitly *D. melanogaster* is described by D. BUTTGE-REIT and R. RENKAWITZ-POHL (Marburg, Germany). *D. melanogaster* being the best analyzed multicellular organism up to now, is capable of serving as a model system for features such as the development of limbs, the nervous system, circadian rythms, and even for complex human diseases. The contribution gives examples for the genetic homology and similarities between *Drosophila* and humans and outlines perspectives for studying features of human diseases using the fly as a model.

2 Genome Projects of Selected Prokaryotic Model Organisms

2.1 The Gram-Negative Enterobacterium *Escherichia coli*

2.1.1 The Organism

The development of the most recent field of molecular genetics is directly connected to one of the best described model organisms, the eubacterium *Escherichia coli*. There is no textbook in biochemistry, genetics or microbiology without extensive sections describing the numerous basic observations, which have been noted first in *E. coli* cells or the respective bacteriophages, or using *E. coli* enzymes as a tool, respectively. Consequently, several monographs solely regarding *E. coli* have been published. Although it seems to be impossible to name or count the number of scientists involved in the characterization of *E. coli*, Tab. 1 is an attempt to name some of the most relevant people in chronological order.

The scientific career of *E. coli* (Fig. 1) started in 1885, when the German paediatrician T. ESCHERICH described the first strain from the faeces of newborn babies. As late as 1958, this is internationally recognized by using his name to classify the respective group of bacterial strains. In 1921 the very first report on virus formation was published for *E. coli*. Today we call the respective observation "lysis by bacteriophages". In 1935 these bacteriophages became the most powerful tool in defining the characters of individual genes. Because of their small size, they turned out to be ideal tools for statistical calculations performed by the former theoretical physicist M. DELBRÜCK. His very intensive and successful work has attracted many others to this area of research. In addition, DELBRÜCK's extraordinary capability to catalyze the exchange of ideas and methods yielded the legendary Cold Spring Harbor Phage course. Everybody interested in basic genetics once attended the famous summer course or at least came to the respective annual phage meeting. This course, which was an optimal combination of joy and work, became an ideal source to spread practical methods. For many decades it was the most important exchange forum for results and ideas, as well as strains and mutants. Soon, the so-called phage family was formed, which interacted almost like one big laboratory, e.g., results were communicated by preprints preferentially. Finally, 15 Nobel prize laureates have their roots in this summer school (see Tab. 1).

The substrain *E. coli* K12 was first used by E. TATUM as a prototrophic strain. It was chosen more or less by chance from the strain collection of Stanford Medical School. Since it

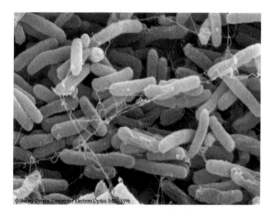

Fig. 1. Scanning electron micrograph (SEM) of *Escherichia coli* cells (image courtesy of: SHIRLEY OWENS, Center for Electron Optics, MSU; found at *http://commtechlab.msu.edu/sites/dlc-me/zoo/ zah0700.html#top#top*).

Tab. 1. Chronology of the Most Important Primary Detections and Method Applications with *E. coli*

1885	"bacterium coli commune" by T. ESCHERICH
1921	Lysogeny and prophages by D'HERELLE
1939	Growth kinetics for a bacteriophage by M. DELBRÜCK (Nobel prize 1969)
1943	Statistical interpretation of phage growth curve (game theorie) by S. LURIA (Nobel prize 1969)
1946	Konjugation by E. TATUM and J. LEDERBERG (Nobel prize 1958)
	Repair of UV damage by A. KELNER and R. DULBECCO (Nobel prize for tumor virology)
1952	DNA as the carrier of genetic information, proven via radioisotopes by M. CHASE and A. HERSHEY (Nobel prize 1969)
1953	Phage immunity as the first example for gene regulation by A. LWOFF (Nobel prize 1965)
	Transduction of *gal* genes (first isolated gene) by E. and J. LEDERBERG
	Host-controlled modification of phage DNA by G. BERTANI and J. J. WEIGLE
1956	DNA polymerase I by A. KORNBERG (Nobel prize 1959)
	Polynucleotide phosphorylase (RNA synthesis) by M. GRUNBERG-MANAGO and S. OCHOA (Nobel prize 1959)
1958	Semiconservative duplication of DNA by M. MESELSON and F. STAHL
1959	Operon theory and induced fit by F. JACOB and J. MONOD (Nobel prize 1965)
1962	Restriction enzymes by W. ARBER (Nobel prize 1978)
1963	Physical genetic map with 99 genes by A. L. TAYLOR and M. S. THOMAN
	Strain collection by B. BACHMANN
1967	DNA ligase by several groups at the same time
1972	DNA hybrids by P. LOBBAN and D. KAISER
1973	Recombinant DNA from *E. coli* and SV40 by P. BERG (Nobel prize 1980)
	Patent on genetic engineering by H. BOYER and S. COHEN
1974	Sequencing techniques using *lac* operator by W. GILBERT and *E. coli* polymerase by F. SANGER (Nobel prize 1980)
1975	Promoter sequence by H. SCHALLER
	Attenuation by C. YANOWSKY
	General ribosome structure by H. G. WITTMANN
1977	Rat insulin expressed in *E. coli* by H. GOODMANN
	Synthetic gene expressed by K. ITAKURA and H. BOYER
1978	Site directed mutagenesis by M. SMITH (Nobel prize 1993)
1984	Polymerase chain reaction by K. B. MULLIS (Nobel prize 1993)
1987	Restriction map of the complete genome by Y. KOHARA and K. ISONO
1989	Organism specific sequence data base by M. KRÖGER
1995	Total sequence of *Haemophilus influenzae* using an *E. coli* comparison
1996	Systematic sequence finished by a Japanese consortium under leadership of H. MORI
1997	Systematic sequence finished by F. BLATTNER
1999	Three dimensional structure of ribosome by four groups at the same time

was especially easy to cultivate and since it is, as an inhabitant of our gut, a nontoxic organism by definition, the strain became very popular. Because of the already acquired vast knowledge and because of its lack to form fimbriae, in 1975 *E. coli* K12 was chosen as the only organism to allow early cloning experiments in the famous Asilomar Conference on bio-safety (BERG et al., 1975). No wonder that almost each of the following basic observations in life sciences was either done with or within *E. coli*. However, what started as the "phage family", dramatically split into hundreds of individual groups working in tough competion. As one of the most important outcomes, sequencing of *E. coli* was performed more than once. Because of the separate efforts, the genome finished only as number seven (BLATTNER et al., 1997; YAMAMOTO et al., 1997; KRÖGER and WAHL, 1998). However, the amount of knowledge acquired is certainly second to none, and the way how this knowledge has been acquired is interesting, both for the history of sequencing methods and bioinformatics, and for the influence of national and individual pride.

The work on *E. coli* is not finished with the completion of the DNA sequence. Data will be continuously acquired to fully characterize the genome in terms of genetic function and protein structures (THOMAS, 1999). This is very important, since a number of toxic *E. coli* strains is known. Thus research on *E. coli* has turned from basic science into applied medical research. Consequently, the human toxic strain O157 has been also completely sequenced, again more than once (PERNA et al., 2001 and unpublished data).

2.1.2 Characterization of the Genome and Early Sequencing Efforts

With its history in mind and realizing the impact of the respective data, it is obvious that an ever growing number of colleagues worldwide worked with or on *E. coli*. Consequently, there was an early need for the organization of data. This led to the first physical genetic map of any living organism comprising 99 genes, published by TAYLOR and THOMAN (1964). This map was improved and refined for several decades by BACHMANN (1983), and M. BERLYN (see Neidhard, 1996). These researchers still maintain a very useful collection of strains and mutants at Yale University. A number of 1,027 loci had been mapped in 1983 (BACHMANN, 1983), these were used as a basis for the very first sequence database specific to a single organism (KRÖGER and WAHL, 1998). As shown in Fig. 2 of KRÖGER and WAHL (1998), sequencing of *E. coli* started as early as 1967, with one of the first ever characterized tRNA sequences. Immediately after DNA sequencing had been established, numerous laboratories started to determine sequences of their personal interest.

2.1.3 Structure of the Genome Project

In 1987 the group of K. ISONO published a very informative and incredibly exact restriction map of the entire genome (KOHARA et al., 1987). With the help of K. RUDD, it was possible to locate sequences quite precisely (see NEIDHARD, 1996; ROBERTS, 2000). But only very few saw any advantage in closing the sometimes very small gaps, thus a worldwide joint sequencing approach could not be established. Two groups, one in Kobe, Japan (YAMAMOTO et al., 1997) and one in Madison, WI (BLATTNER et al., 1997) started systematic genome sequencing in parallel. In addition, another laboratory at Harvard University used *E. coli* as a target to develop a new sequencing technology. Several meetings, organized especially on *E. coli*, did not result in any unified systematic approach, thus many genes have been sequenced two or three times. Specific databases have been maintained to bring some order into the increasing chaos. However, even this type of tool has been developed several times in parallel (KRÖGER and WAHL, 1998; Roberts, 2000). Whenever a new contiguous sequence was published, about 75% had already previously been submitted to the international databases by other laboratories. The progress of data acquisition followed a classical e-curve, as shown in Fig. 2 of KRÖGER and WAHL (1998). Thus in 1992 it was possible to predict the completeness of the sequence for 1997, without knowing about the enormous technical innovations in between (KRÖGER and WAHL, 1998).

Both the Japanese consortium and the group of F. BLATTNER started early; people say they started too early. They subcloned the DNA first, and they used manual sequencing and older informatic systems. Sequencing has been performed semiautomatically, and many students were employed to read and control the X-ray films. When the first genome sequence of *Haemophilus influenzae* appeared in 1995, the science foundations wanted to discontinue the support of the *E. coli* projects, which received their grant support mainly because of the model character of the sequencing techniques developed.

Three facts and the truly international protest, convinced the juries to continue financial support: First, in contrast to other completely sequenced organisms, *E. coli* is an autonomously living organism. Second, when the first complete very small genome sequence was released, even the longest contiguous sequence for *E. coli* was already longer. Third, the other laboratories could only finish their sequences because the *E. coli* sequences were already publicly available. Consequently, the two competing main laboratories were allowed to purchase several of the meanwhile developed sequencing machines and use the shutgun approach to complete their efforts. Finally, they finished almost at the same time: H. MORI and his colleagues included already published sequences from other laboratories into their sequence data and sent them to the international databases on December 28, 1996 (YAMAMOTO et al., 1997) while F. BLATTNER turned in an entirely new sequence on January

16, 1997 (BLATTNER et al., 1997). They added last changes and additions as late as October 1998. Very sadly, at the end *E. coli* had been sequenced almost three times (KRÖGER and WAHL, 1998). However, most people nowadays forget about all other sources and refer to the BLATTNER sequence.

2.1.4 Results of the Genome Project

When the sequences finally were finished, most of the features of the genome had already been known. Consequently, people did no longer celebrate the *E. coli* sequence as a major breakthrough. At that time, everybody knew that the genome is almost completely covered with genes, although less than half of them have been genetically characterized. Tab. 2 illustrates this and shows the differences

Tab. 2. Some Statistical Features of the *E. coli* Genome

Total size		4,639,221 bp[a]	acc. to Regulon[d]	acc. to BLATTNER[e]
Transcription units	proven	528		
	predicted	2,328		
Genes	total found	4,408	4,403	
	regulatory	85		
	essential	200		
	nonessential[b]	2,363	1,897	
	unknown[c]	1,761	2,376	
	tRNAS	84	84	
	rRNA	29	29	
Promoters	proven	624		
	predicted	4,643		
Sites (?)		469		
Regulatory interactions	found	642		
	predicted	275		
Terminators	found	96		
RBSs		98		
Gene products	regulatory proteins	85		
	RNAs	115	115	
	other peptides	4,190	4,201	

[a] Additional 63 bp compared to the original sequence.
[b] Genes with known or predicted function.
[c] Yet no other data available than the existence of an open reading frame with a start sequence and more than 100 codons.
[d] Data from *http://tula.cifn.unam.mx8850/regulondb/regulon_doc/summary.*
[e] Data from *http://www.genome.wisc.edu.*

in counting. Because of this high density of genes, F. BLATTNER and coworkers defined "grey holes" whenever they found more than 2 kb of noncoding region (BLATTNER et al., 1997). It turned out that the termination of replication is almost exactly opposite to the origin of replication. No special differences have been found for either direction of replication. Approximately 40 formerly described genetic features could not yet be located or supported by the sequence (KRÖGER and WAHL, 1998; NEIDHARD, 1996). On the other hand, we have several examples for multiple functions encoded by the same gene. It turned out that the multifunctional genes are mostly involved in gene expression and used as general control factor. M. RILEY determined the number of gene duplications, which is also not unexpectedly low when neglecting the ribosomal operons (see ROBERTS, 2000).

Everybody is convinced that the real work is starting only now. A number of strain differences may be the cause of deviations between the different sequences available. Thus the number of genes as well as nucleotides differ slightly (see Tab. 2). Everybody would like to know the function of each of the open reading frames (THOMAS, 1999), but nobody has received the grant money to work on this important problem. Seemingly, the other model organisms are of more public interest, thus, it may well be that research on other organisms will now help to understand *E. coli*, just the way how *E. coli* provided information to understand them. In contrast to yeast, it is very hard to produce knock-out mutants. Thus, we may have the same situation in the postgenomic era as we had before the genome was finished. A number of laboratories will continue to work with *E. coli*, they will constantly characterize one or the other open reading frame, but there will not be any mutual effort (THOMAS, 1999).

2.1.5 Follow up Research in the Postgenomic Era

Today, it seems to be more attractive to work with toxic *E. coli* strains like O157, rather than with *E. coli* K12. This strain has recently been completely sequenced, the data are available via Internet. The comparison between toxic and nontoxic strains will certainly help to understand the toxic mechanisms. On the other hand, it turned out that it was correct to use *E. coli* K12 as the most intensively used strain for biological safety regulations (BERG et al., 1975). No additional features turned out to change this.

Surprisingly, the colleagues from mathematics or informatics are those who showed most interest in the bacterial sequences. They did all kind of statistical analyses and tried to find some evolutionary roots. Here another fear of the public is already formulated: People are afraid of the attempts to reconstruct the first living cell. So there are at least some attempts to find the minimal set of genes for the most basic needs of a cell. We have to ask again the very old question: Do we really want to "play God"? If we want, *E. coli* indeed could serve as an important milestone.

2.2 The Gram-Positive Spore Forming *Bacillus subtilis*

2.2.1 The Organism

Self-taught ideas have a long life: articles about *Bacillus subtilis* (Fig. 2) almost invariably begin with words such as: "*B. subtilis*, a soil bacterium …", nobody taking the elementary care to check whether this is based on experimental observations. *Bacillus subtilis*, first identified in 1885, is named *ko so kin* in japanese and *laseczka sienna* in polish, "hay bacterium", and this refers to the real biotope of the organism, the surface of grass or low lying plants (SEKOWSKA, 1999). Interestingly, it required its genome to be sequenced to conquer again its right biotope. Of course, plant leaves fall on the soil surface, and one must find *B. subtilis* there, but its normal niche is the surface of leaves, the phylloplane. Hence, if one wishes to use this bacterium in industrial processes, to engineer its genome, or simply to understand the functions coded by its genes, it is of fundamental importance to understand where it normally thrives, and what the environmental parameters are controlling its life

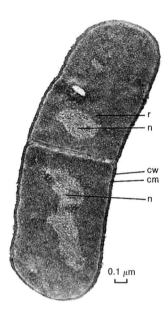

Fig. 2. Electron micrograph of a thin section of *Bacillus subtilis*. The dividing cell is surrounded by a relatively dense wall (CW), enclosing the cell membrane (cm). Within the cell, the nucleoplasm (n) is distinguishable by its fibrillar structure from the cytoplasm, densely filled with 70S ribosomes (r).

nome corresponds to several gene families which have been probably expanded by gene duplication. The largest family contains 77 known and putative ATP-binding cassette (ABC) permeases indicating that, despite its large metabolism gene number, *B. subtilis* has to extract a variety of compounds from its environment (KUNST et al., 1997). In general, the permeating substrates are unchanged during permeation. Group transfer, where substrates are modified during transport, however, plays an important role in *B. subtilis*. Its genome codes for a variety of phosphoenolpyruvate-dependent systems (PTS) which transport carbohydrates and regulate general metabolism as a function of the nature of the supplied carbon source. A functionally-related catabolite repression control, mediated by a unique system (not cyclic AMP), exists in this organism (SAIER, 1998). Remarkably, apart from the expected presence of glucose-mediated regulation, it appears that carbon sources related to sucrose play a major role, via a very complicated set of highly regulated pathways, indicating that this carbon supply is often encountered by the bacteria. In the same way, *B. subtilis* can grow on many of the carbohydrates synthesized by grass-related plants.

In addition to carbon, oxygen, nitrogen, hydrogen, sulfur, and phosphorus are the core atoms of life. Some knowledge about other metabolisms in *B. subtilis* has accumulated, but significantly less than in its *E. coli* counterpart. However, knowledge of its genome sequence is rapidly changing the situation, making *B. subtilis* a model of similar general use as *E. coli*. A frameshift mutation is present in an essential gene for surfactin synthesis in strain 168, but it has been found that including a small amount of a detergent into plates allowed these bacteria to swarm and glide extremely efficiently (C.-K. WUN and A. SEKOWSKA, unpublished observations). The first lesson of the genome text analysis is thus that *B. subtilis* must be tightly associated with the plant kingdom, with grasses in particular (KUNST et al., 1997). This should be considered a priority when devising growth media for this bacterium, in particular in industrial processes.

cycle and the corresponding gene expression. Among other important ancillary functions, *B. subtilis* has thus to explore, colonize, and exploit the local resources, while at the same time it must maintain itself, dealing with congeners and with other organisms: Understanding *B. subtilis* requires understanding the general properties of its normal habitat (SEKOWSKA, 1999).

2.2.2 A Lesson from Genome Analysis: The *Bacillus subtilis* Biotope

The genome of *B. subtilis* (strain 168), sequenced by a team of European and Japanese laboratories, is about 4,214,780 bp long. Of the more than 4,100 protein-coding genes, 53% are represented once. One quarter of the ge-

Another aspect of the *B. subtilis* life cycle consistent with a plant-associated life, is that it can grow over a wide range of different temperatures, up to 54–55 °C – an interesting feature for large-scale industrial processes. This indicates that its biosynthetic machinery comprises control elements and molecular chaperones that permit this versatility. Gene duplication may permit adaptation to high temperature, with isozymes having low and high temperature optima. Because the ecological niche of *B. subtilis* is linked to the plant kingdom it is subjected to rapid alternating drying and wetting. Accordingly, this organism is very resistant to osmotic stress, and can grow well in media containing 1 M NaCl. Also, the high levels of oxygen concentration reached during daytime are met with protective systems: *B. subtilis* appears to have as many as six catalase genes, both of the heme-containing type (*katA*, *katB*, and *katX* in spores) and of the manganese-containing type (*ydbD*, *yjqC*, and *cotJC* in spores).

The obvious conclusion of these observations is that the normal *B. subtilis* niche is the surface of leaves (ARIAS et al., 1999). This is consistent with the old observation that *B. subtilis* makes up the major population of bacteria of rotting hay. Furthermore, consistent with the extreme variety of conditions prevailing on plants, *B. subtilis* is an endospore-forming bacterium, making spores highly resistant to the lethal effects of heat, drying, many chemicals, and radiation.

2.2.3 General Organization of the Genome: A First Law of Genomics

Analysis for repeated sequences in the *B. subtilis* genome discovered an unexpected feature: strain 168 does not contain insertion sequences. A strict constraint on the spatial distribution of repeats longer than 25 bp was found in the genome, in contrast to the situation in *E. coli*. The correlation of the spatial distribution of repeats and the absence of insertion sequences in the genome suggests that mechanisms aiming at their avoidance and/or elimination have been developed (ROCHA et al., 1999a). This observation is particularly relevant for biotechnological processes where one has multiplied the copy number of genes in order to improve production. Although

there is generally no predictable link between the structure and function of biological objects (DANCHIN, 1999), the pressure of natural selection has adapted together gene and gene products. Biases in features of predictably unbiased processes is evidence for prior selective pressure. In the case of *B. subtilis* one observes a strong bias in the polarity of transcription with respect to replication: 70% of the genes are transcribed in the direction of the replication fork movement (KUNST et al., 1997). Global analysis of oligonucleotides in the genome demonstrated that there is a significant bias not only in the base or codon composition of one DNA strand with respect to the other, but, quite surprisingly, there is a strong bias at the level of the amino acid content of the proteins. The proteins coded by the leading strand are valine-rich, and those coded by the lagging strand are threonine + isoleucine-rich. This first law of genomics seems to extend to many bacterial genomes (ROCHA et al., 1999b). It must result from a strong selection pressure of a yet unknown nature, demonstrating that, contrary to an opinion frequently held, genomes are not, at a global scale, plastic structures.

2.2.4 Translation: Codon Usage and the Organization of the Cell's Cytoplasm

Exploiting the redundancy of the genetic code, coding sequences exhibit highly variable biases of codon usage. The genes of *B. subtilis* are split into three classes on the basis of their codon usage bias. One class comprises the bulk of proteins, another is made up of genes that are expressed at a high level during exponential growth, and a third class, with A + T-rich codons, corresponds to portions of the genome that have been horizontally exchanged (KUNST et al., 1997).

When mRNA threads are emerging from DNA they become engaged by the lattice of ribosomes, and ratchet from one ribosome to the next, like a thread in a wiredrawing machine (DANCHIN et al., 2000). In this process, nascent proteins are synthesized on each ribosome, spread throughout the cytoplasm by the linear diffusion of the mRNA molecule from ribosome to ribosome. However, upon a sudden shift in environmental conditions, the transcription complex must often break up. Truncated mRNA is likely to be a dangerous molecule because, if translated, it would produce a truncated protein. Such protein fragments are often toxic, because they can disrupt the architecture of multisubunit complexes. A process copes with this kind of accidents in *B. subtilis*. When a truncated mRNA molecule reaches its end, the ribosome stops translating, and waits. A specialized RNA, tmRNA, that is folded and processed at its 3′ end like a tRNA and charged with alanine, comes in, inserts its alanine at the C-terminus of the nascent polypeptide, then replaces the mRNA within a ribosome, where it is translated as ASFNQNVALAA. This tail is a protein tag that is then used to direct the truncated tagged protein to a proteolytic complex (ClpA, ClpX), where it is degraded (GOTTESMAN et al., 1998).

2.2.5 *Bacillus subtilis* Genome Evolution

Three principal modes of transfer of genetic material, namely transformation, conjugation, and transduction naturally occur in prokaryotes. In *B. subtilis*, transformation is an efficient process (at least in some *B. subtilis* species such as the strain 168) and transduction with the appropriate carrier phages is well understood.

The presence in the *B. subtilis* genome of uniquely local repeats, suggesting Campbell-like integration of foreign DNA, is consistent with a strong involvement of recombination processes in its evolution (ROCHA et al., 1999a). In addition, recombination must be involved in mutation correction. In *B. subtilis*, MutS and MutL homologs exist, presumably for the purpose of recognizing mismatched base pairs. However, no MutH activity was identified, which would allow the daughter strand to be distinguished from its parent. It is, therefore, not known how the long-patch mismatch repair system corrects mutations in the

newly synthesized strand: the nicks caused in the daughter strands by excision of newly misincorporated uracil instead of thymine during replication might provide the appropriate signal.

The recently sequenced genome of the pathogen *Listeria monocytogenes* has many features in common with the genome of *B. subtilis*. Preliminary analysis suggests that the genome may be organized around the genes of core metabolic pathways, such as that of sulfur metabolism, consistent with a strong correlation between the organization of the genome and the architecture of the cell (ROCHA et al., 2000).

2.2.6 Industrial Processes

Bacillus subtilis is Generally Recognized as Safe (GRAS). It is much used at the industrial level for both enzyme production or for food supply fermentation. Riboflavin is derived from genetically modified *B. subtilis* using fermentation techniques. For some time, high levels of heterologous gene expression in *B. subtilis* was difficult to achieve. In contrast to gram-negatives, A + T-rich gram-positive bacteria have optimized transcription and translation signals: although *B. subtilis* has a counterpart of the *rpsA* gene, this organism lacks the function of the corresponding ribosomal S1 protein, which permits recognition of the ribosome binding site upstream of the translation start codons (DANCHIN, 1997). Traditional techniques (e.g., random mutagenesis followed by screening; *ad hoc* optimization of poorly defined culture media) are important and will continue to be utilized in the food industry, but biotechnology must now include genomics to target artificial genes that follow the sequence rules of the genome at precise positions, adapted to the genome structure, as well as to modify intermediary metabolism while complying with the adapted niche of the organism, as revealed by its genome. As a complement to standard genetic engineering and transgenic technology, knowing the genome text has opened a whole new range of possibilities in food product development, in particular allowing "humanization" of the content of food products (adaptation to the human metabolism, and even adaptation to sick or healthy conditions). These techniques provide an attractive method to produce healthier food ingredients and products that are presently not available or very expensive. *B. subtilis* will remain a tool of choice in this respect.

2.2.7 Open Questions

The complete genome sequence of *B. subtilis* contains information that remains underutilized in the current prediction methods applied to gene functions, most of which are based on similarity searches of individual genes. In particular it is now clear that the order of the genes in the chromosome is not random, and that some hot spots allow gene insertion without much damage, while other regions are forbidden. For the production of small molecules, one must utilize higher-level information on metabolic pathways to reconstruct a complete functional unit from a set of genes. The reconstruction *in silico* of selected portions of metabolism using the existing biochemical knowledge of similar gene products has been undertaken. The core biosynthetic pathways of all 20 amino acids have been completely reconstructed in *B. subtilis*. However, many satellite or recycling pathways have not been identified yet, in particular in the synthesis of pyrimidines, in sulfur and in short carbon chain acids metabolism. Finally, there remain some 800 genes of completely unknown function in the genome of strain 168 and many more in the genome of related species. Functional genomics is aiming at identifying their role.

2.3 The Archaeon *Archaeoglobus fulgidus*

2.3.1 The Organism

Archaeoglobus fulgidus is a strictly anaerobic, hyperthermophilic, sulfate-reducing archaeon. It is the first sulfate-reducing organism from which the complete genome sequence has been determined and published (KLENK et al., 1997). Sulfate-reducing organisms are essential to the biosphere, because biological sulfate reduction is part of the global sulfur cycle. The ability to grow by sulfate reduction is restricted to only a few groups of prokaryotes. The Archaeoglobales are in two ways unique within this group: (1) they are members of the Archaea and, therefore, unrelated to all other sulfate reducers, which belong to the Bacteria; (2) the Archaeoglobales are the only hyperthermophiles within the sulfate reducers, a feature which allows them to occupy extreme environments, e.g., hydrothermal fields and subsurface oil fields. The production of iron sulfide as an end product of high-temperature sulfate reduction by *Archaeoglobus* species contributes to oil-well "souring", which causes corrosion of iron and steel in submarine oil and gas processing systems. *A. fulgidus* is also a model for hyperthermopilic organisms and for the Archaea, because it is only the second hyperthermophile whose genome has been completely deciphered (after *Methanococcus jannaschii*), and it is the third species of Archaea (after *M. jannaschii* and *Methanobacterium thermoautotrophicum*) whose genome has been completely sequenced and published.

A. fulgidus DSM4304 (Fig. 3) is the type strain of the Archaeoglobales (STETTER, 1988). Its glycoprotein-covered cells are irregular spheres (diameter 2 µm) with four distinct monopolar flagellae. It is able to grow not only organoheterotrophically using a variety of carbon and energy sources, but also lithoautotrophically on hydrogen, thiosulfate, and car-

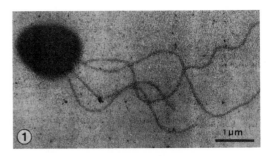

Fig. 3. Electron micrograph of *A. fulgidus* DSM4303 (strain VC-16), kindly provided by O. STETTER, University of Regensburg, Germany. The bar in the lower right corner represents 1 µm.

bon dioxide. Within a range of 60–95 °C it grows best at 83 °C.

Prior to genome sequencing very little was known about the genomic organization of *A. fulgidus*. The first estimation of its genome size using pulsed field gel electrophoresis was published after the final assembly of the genome sequences was already performed. Since extra-chromosomal elements are absent from *A. fulgidus*, it was clear that the genome consists of only one circular chromosome. Data about genetic or physical mapping of the genome were unknown before the sequencing project, however, a small-scale approach to physical mapping was performed late in the project for confirmation of the genome assembly. The sequences of only eleven genes from *A. fulgidus* had been published before the sequencing project started, altogether covering less than 0.7% of the genome.

2.3.2 Structure of the Genome Project

The whole-genome random sequencing procedure was chosen as the sequencing strategy for the *A. fulgidus* genome project. This procedure has previously been applied to four microbial genomes sequenced at The Institute for Genomic Research (TIGR): *Haemophilus influenzae*, *Mycoplasma genitalium*, *M. jannaschii*, and *Helicobacter pylori* (TOMB et al., 1997). Chromosomal DNA for the construction of libraries was prepared from a culture derived from a single cell isolated by optical tweezers and provided by K. O. STETTER. Three libraries were used for sequencing: two plasmid libraries (1.42 kbp and 2.94 kbp insert size) for mass sequence production and one large insert λ-library (16.38 kbp insert size) for the genome scaffold. The initial random sequencing phase was performed with these libraries until 6.7-fold sequence coverage was achieved. At this stage the genome was assembled into 152 contigs separated by sequence gaps and five groups of contigs separated by physical gaps. Sequence gaps were closed by a combined approach of editing the ends of sequence traces and by primer walking on plasmid- and λ–clones spanning the gaps. Physical

gaps were closed by direct sequencing of PCR fragments generated by combinatorial PCR reactions. Only 0.33% of the genome (90 regions) was covered by only one single sequence after the gap closure phase. These regions were confirmed by additional sequencing reactions to ensure a minimum sequence coverage of two for the whole genome. The final assembly consisted of 29,642 sequencing runs which cover the genome sequence 6.8-fold.

The *A. fulgidus* genome project was financed by the US Department of Energy (DOE) within the Microbial Genome Program. This program financed several of the early microbial genome sequencing projects performed at various genome centers, e.g., *M. jannaschii* (TIGR), *M. thermoautotrophicum* (Genome Therapeutics), *Aquifex aeolicus* (Recombinat BioCatalysis, now DIVERSA), *Pyrobaculum aerophilum* (California Institute of Technology), *Pyrococcus furiosus* (University of Utah), and *Deinococcus radiodurans* (Uniformed Services University of the Health Sciences). Like the *M. jannaschii* project which was started one year earlier, the *A. fulgidus* genome was sequenced and analyzed by a collaboration between researchers at TIGR and the Department of Microbiology at the University of Illinois, Champaign-Urbana. Construction of the plasmid libraries was done in Urbana, whereas the λ-library was constructed at TIGR. Sequencing and assembly was performed at TIGR using automated ABI sequencers and TIGR assembler, respectively. Confirmation of the assembly by mapping with large-size restriction fragments was done in Urbana. ORF prediction and identification of functions, as well as the data mining and interpretation of the genome content was done in a joint effort by both teams.

Coding regions in the final genome sequence were identified with a combination of two sets of ORFs generated by programs developed by members of the two teams: GeneSmith by H. O. SMITH at TIGR, and CRITICA by G. J. OLSEN and J. H. BADGER in Urbana. The two sets of ORFs identified by GeneSmith and CRITICA were merged into one consensus set containing all members of both initial sets. The amino acid sequences derived from the consensus set were compared to a non-redundant protein database using BLASTX. ORFs shorter than 30 codons were carefully inspected for database hits and eliminated in cases without significant database match. The results of the database comparisons were first inspected and categorized by TIGR's microbial annotation team. This initial annotation database was then further analyzed and refined by a team of experts for all major biological categories.

The sequencing strategy chosen for the *A. fulgidus* genome project has some advantages compared to alternative strategies applied in genome research:

(1) Given the relatively large set of automated sequencers available at TIGR, the whole-genome random sequencing procedure is much faster than any strategy that includes a mapping step prior to the sequencing phase;

(2) within the DOE Microbial Genome Program the TIGR strategy and the sequencing technology applied for *M. jannaschii* and *A. fulgidus* genome projects proved to be clearly superior in competition with projects based on multiplex sequencing (*M. thermoautotrophicum* and *P. furiosus*), by finishing two genomes in less time than the competing laboratories needed for one genome each;

(3) the interactive annotation with a team of experts for the organism and for each biological category ensured a more sophisticated final annotation than any automated system can ever achieve.

2.3.3 Results of the Genome Project

Although the initial characterization of the genome revealed all its basic features, annotation of biological functions for the ORFs will continue to be updated for new functions identified either in *A. fulgidus* or for homologous genes characterized in other organisms. The size of the *A. fulgidus* genome was determined to be 2,178,400 bp, with an average G + C-con-

tent of 48.5%. Three regions with low G + C content (< 39%) were identified, two of them encoding enzymes for lipopolysaccharide biosynthesis. The two regions with the highest G + C content (> 53%) contain the ribosomal RNAs and proteins involved in heme biosynthesis. With the bioinformatics tools available at completion of the genome no origin of replication could be identified. The genome contains only one set of genes for ribosomal RNAs. Other RNAs encoded in *A. fulgidus* are 46 species of tRNA, five of them with introns 15–62 bp long, no significant tRNA clusters, 7S RNA and RNase P. Alltogether 0.4% of the genome is covered by genes for stable RNAs. Three regions with short (< 49 bp) non-coding repeats (42–60 copies) were identified. All three repeated sequences are similar to short repeated sequences found in *M. jannaschii* (BULT et al., 1996). Nine classes of long, coding repeats (> 95% sequence identity) were identified within the genome, three of them may represent IS elements, and three other repeats encode conserved hypothetical proteins found previously in other genomes. The consensus set of ORFs contains 2,436 members with an average length of 822 bp, similar to *M. jannaschii* (856 bp), but shorter than in most bacterial genomes (average 949 bp). With 1.1 ORFs per kb, the gene density appears to be slightly higher than in other microbial genomes, although the fraction of the genome covered by protein coding genes (92.2%) is comparable to other genomes. The elevated number of ORFs per kbp might be artificial, due to a lack in stop codons in high G + C organisms. Predicted start codons are 76% ATG, 22% GTG, and 2% TTG. No inteins were identified in the genome. The isoelectric point of the predicted proteins in *A. fulgidus* is rather low (median pI is 6.3), other prokaryotes show distributions with peaks between 5.5 and 10.5. Putative functions could be assigned to about half of the predicted ORFs (47%) by significant matches in database searches. One quarter (26.7%) of all ORFs are homologous to ORFs previously identified in other genomes ("conserved hypotheticals"), whereas the remaining quarter (26.2%) of ORFs in *A. fulgidus* appears to be unique, without any significant database match. *A. fulgidus* contains an unusually high number of paralogous gene families:

242 families with 719 members (30% of all ORFs). This may explain why the genome is larger than most other archaeal genomes (average about 1.7 Mbp). Interestingly, one third of the identified families (85 out of 242) have no single member for which a biological function could be predicted. The largest families contain genes assigned to "energy metabolism", "transporters", and "fatty acid metabolism".

The genome of *A. fulgidus* is neither the first archaeal genome to be sequenced completely nor is it the first genome of a hyperthermophilic organism. The novelties for both features were already reported together with the genome of *M. jannaschii* (BULT et al., 1996). *A. fulgidus* is, however, the first sulfate-reducing organism whose genome was completely deciphered. Model findings in respect to sulfur and sulfate metabolism were not expected from the genome, because sulfate metabolism was already heavily studied in *A. fulgidus* prior to the genome project. The genes for most enzymes involved in sulfate reduction were already published, and new information from the genome confirmed only that the sulfur oxide reduction systems in Archaea and Bacteria are highly similar. The single most exciting finding in the genome of *A. fulgidus* was the identification of multiple genes for acetyl-CoA synthase and the presence of 57 β-oxidation enzymes. It has been reported that the organism is incapable of growth on acetate (VORHOLT et al., 1995), and no system for β-oxidation has previously been described in the Archaea. It appears now that *A. fulgidus* can gain energy by degradation of a variety of hydrocarbons and organic acids, because genes for a least five types of ferredoxin-dependent oxidoreductases and at least one lipase were also identified. Interestingly, at about the same time when the unexpected genes for metabolizing enzymes were identified, it was also reported that a close relative of *A. fulgidus* is able to grow on olive oil (K. O. STETTER, personal communication), a feature that would require the presence of the genes just identified in *A. fulgidus*. On the other hand, not all genes necessary for the pathway described could be identified. Glucose has been described as a carbon source for *A. fulgidus* (STETTER et al., 1987), but neither an uptake transporter nor a catabolic pathway could be identified in the genome. There is still a chance that the required genes are hidden in the pool of functionally uncharacterized ORFs. Other interesting findings with respect to the biology of *A. fulgidus* concern the sensory functions and the regulation of gene expression. *A. fulgidus* appears to have complex sensory and regulatory networks – a major difference to what has been described for *M. jannaschii* – consistent with its extensive energy-producing metabolism and its versatile system for carbon utilization. These networks contain more than 55 proteins with presumed regulatory functions, as well as several iron-dependent repressor proteins. At least 15 signal-transducing histidine kinases were identified, but only nine response regulators.

2.3.4 Follow-Up Research

Almost 30 papers about *A. fulgidus* were published between the initial description of the organism in 1987 and the genome sequence ten years after. Twice as many papers were published about the organism within the three years since the genome has been finished. Although functional genomics is a hot topic at many scientific meetings and discussions, *A. fulgidus* appears to be no prime candidate for such studies. So far, not a single publication deals with proteomics, transcriptomics, or serial mutagenesis in this organism. Almost every second publication (44%) that refers to the genome of *A. fulgidus* describes the functional or structural characteristics of novel proteins, mostly enzymes. The other major fraction of publications on *A. fulgidus* (42%) deals with comparative genomics, about 10% of the publications deal with structure–function relationships. The most interesting follow-up story is probably that on the flap endonucleases. In October 1999 HOSFIELD et al. described newly discovered archaeal flap endonucleases (FENs) from *A. fulgidus, M. jannaschii*, and *P. furiosus*, that show a structure-specific mechanism for DNA substrate binding and catalysis resembling human flap endonuclease. In the spring of 1999, LYAMICHEV et al. (1999) showed how FENs could be used for polymorphism identification and quantitative detection of genomic DNA by invasive cleavage of oligonucleotide probes. More recently, COOKSEY et al. (2000) described an invader assay based on *A. fulgidus* FEN that allows a linear signal amplification for identification of mutatons. This procedure could eventually become important as a non-PCR based procedure for SNP detection.

3 Genome Projects of Selected Eukaryotic Model Organisms

3.1 The Budding Yeast *Saccharomyces cerevisiae*

3.1.1 Yeast as a Model Organism

The budding yeast, *Saccharomyces cerevisiae* (Fig. 4), can be viewed to be one of the most important fungal organisms used in biotechnological processes. It owes its name to the fact that it can ferment saccharose and has served mankind for several thousands of years in making bread and alcoholic beverages. The introduction of yeast as an experimental system dates back to the 1930s and has since

attracted increasing attention. Unlike more complex eukaryotes, yeast cells can be grown on defined media giving the investigator complete control over environmental parameters. The elegance of yeast genetics and the ease of manipulation of yeast have substantially contributed to the explosive growth in yeast molecular biology. This success is also due to the recent notion that the extent to which basic biological processes have been conserved throughout eukaryotic life is remarkable and makes yeast a unique unicellular model organism, in which cell architecture and fundamental cellular mechanisms can be successfully investigated. No wonder then that yeast had again reached the forefront in experimental molecular biology by being the first eukaryotic organism of which the entire genome sequence became available (GOFFEAU et al., 1996; Anonymous, 1997). The wealth of sequence information obtained in the yeast genome project turned out to be extremely useful as a reference to which sequences of human, animal, or plant genes could be compared.

The first genetic map of *S. cerevisiae* was published by LINDEGREN in 1949; many revisions and refinements have appeared since. At the outset of the sequencing project, about 1,200 genes had been mapped, and detailed biochemical knowledge about a similar number of genes encoding either RNA or protein products had accumulated (MORTIMER et al., 1992). The existence of 16 chromosomes ranging in size between 250 kb and ~2,500 kb was firmly established when it became feasible to separate all chromosomes in pulsed field gel electrophoresis (PFGE). This also provided definition of "electrophoretic karyotypes" of strains by sizing chromosomes (CARLE and OLSON, 1985). Not only do laboratory strains possess different karyotypes, because of chromosome length polymorphisms and chromosomal rearrangements, but so do industrial strains. Therefore, a defined laboratory strain (αS288C) was chosen for the yeast sequencing project.

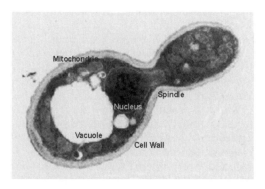

Fig. 4. Micrograph of the budding yeast *Saccharomyces cerevisiae* during spore formation. Cell wall, nucleus, vacuole, mitochondria, and the spindle are indicated.

3.1.2 The Yeast Genome Sequencing Project

The yeast sequencing project was initiated in 1989 within the framework of the EU biotechnology programs. It was based on a network approach in which initially 35 European laboratories became involved, and chromosome III – the first eukaryotic chromosome ever to be sequenced – was completed in 1992. In the following years and engaging many more laboratories, sequencing of further complete chromosomes was tackled by the European network. Soon after its beginning, laboratories in other parts of the world joined the project to sequence other chromosomes or parts thereof, ending up in a coordinated international enterprise. Finally, more than 600 scientists in Europe, North America, and Japan became involved in this effort. The sequence of the entire yeast genome was completed in early 1996 and released to public databases in April 1996.

Although the sequencing of chromosome III started from a collection of overlapping plasmid or phage lambda clones, it was anticipated that in the following cosmid libraries had to be used to aid large-scale sequencing (STUCKA and FELDMANN, 1994). Assuming an average insert length of 35–40 kb, a cosmid library containing 4,600 random clones would represent the yeast genome at about twelve times the genome equivalent. The advantages of cloning DNA segments in cosmids were at hand: clones turned out to be stable for many years and the low number of clones was advantageous in setting up ordered yeast cosmid libraries or sorting out and mapping chromosome-specific sublibraries. High-resolution physical maps of the chromosomes to be sequenced were constructed by application of classical mapping methods (fingerprints, cross-hybridzation) or by novel methods developed for this program, such as site-specific chromosome fragmentation (THIERRY and DUJON, 1992) or a high resolution cross-hybridization matrix, to facilitate sequencing and assembly of the sequences.

In the European network, chromosome-specific clones were distributed to the collaborating laboratories according to a scheme worked out by the DNA coordinators. Each contracting laboratory was free to apply sequencing strategies and techniques of its own provided that the sequences were entirely determined on both strands and unambiguous readings were obtained. Two principal approaches were used to prepare subclones for sequencing:

(1) generation of sublibraries by the use of a series of appropriate restriction enzymes or from nested deletions of appropriate subfragments made by exonuclease III;

(2) generation of shotgun libraries from whole cosmids or subcloned fragments by random shearing of the DNA.

Sequencing by the Sanger technique was either done manually, labeling with [^{35}S]dATP being the preferred method of monitoring or by automated devices (on-line detection with fluorescence labeling or direct blotting electrophoresis system) following the various established protocols. Similar procedures were applied to the sequencing of the chromosomes contributed by the Sanger laboratory and the laboratories in the USA, Canada, and Japan. The American laboratories largely relied on machine-based sequencing.

Due to their repetitive substructure and the lack of appropriate restriction sites, the yeast chromosome telomeres presented a particular problem. Conventional cloning procedures were successful only with a few exceptions. Largely, telomeres were physically mapped relative to the terminal-most cosmid inserts using the chromosome fragmentation procedure (THIERRY and DUJON, 1992). The sequences were then determined from specific plasmid clones obtained by "telomere trap cloning", an elegant strategy developed by LOUIS and BORTS (1995).

Within the European network, all original sequences were submitted by the collaborating laboratories to the Martinsried Institute of Protein Sequences (MIPS) who acted as an informatics center. They were kept in a data library, assembled into progressively growing contigs, and in collaboration with the DNA coordinators, the final chromosome sequences were derived. Quality control was performed by anonymous resequencing of selected regions and suspected or difficult zones (total of 15–20% per chromosome). Similar procedures were employed for sequence assembly and quality control in the other laboratories. During the past years, further quality control was carried out that resulted in a nearly absolute accuracy of the total sequence.

The sequences of the chromosomes were subjected to analysis by computer algorithms, identifying open reading frames (ORFs) and other genetic entities, and monitoring compositional characteristics of the chromosomes (base composition, nucleotide pattern frequencies, GC profiles, ORF distribution profiles, etc.). As the intron splice site/branchpoint pairs in yeast are highly conserved, they could be detected by using defined search patterns. It finally turned out that only 4% of the yeast genes do contain (mostly short) introns. Centromere and telomere regions, as well as tRNA genes, sRNA genes, or the retrotransposons were sought by comparison with previously characterized datasets or appropriate search programs. All putative proteins were annotated by using previously established yeast data and evaluating searches for similarity to entries in the databases or protein signatures detected by using the PROSITE dictionary.

3.1.3 Life with some 6,000 Genes

With its 12.8 Mb, the yeast genome is about 250 times smaller than the human genome. The complete genome sequence now defines 6,240 open reading frames (ORFs) which are likely to encode specific proteins in the yeast cell. A protein-encoding gene is found every two Kb in the yeast genome, with nearly 70% of the total sequence consisting of ORFs. This leaves only limited space for the intergenic regions which can be thought to harbor the major regulatory elements involved in chromosome maintenance, DNA replication, and transcription. Generally, the genes are rather evenly distributed among the two strands of the single chromosomes, although arrays longer than eight genes that are transcriptionally oriented into the same direction can be found eventually. With a few exceptions, transcribed genes on complementing strands are not overlapping,

and no "genes-in-genes" are observed. Although the intergenic regions between two consecutive ORFs sometimes are extremely short, they are normally maintained as separate units and not coupled for transcription. In "head-to-head" gene arrangements, the intervals between the divergently transcribed genes might be interpreted to mean that their expression is regulated in a concerted fashion involving the common promoter region. This, however, seems not to hold for the majority of the genes and appears to be a principle reserved for a few cases. The sizes of the ORFs vary between 100 to more than 4,000 codons; less than 1% is estimated to be below 100 codons. Additionally, the yeast genome contains some 120 ribosomal RNA genes in a large tandem array on chromosome XII, 40 genes encoding small nuclear RNAs (sRNAs), and 275 tRNA genes (belonging to 43 families) which are scattered throughout the genome. Overall, the yeast genome is remarkably poor in repeated sequences, except for the transposable elements (Tys) which account for approximately 2% of the genome, which due to their genetic plasticity are the major source of polymorphisms between different strains. Finally, the sequences of non-chromosomal elements, such as the 6 kb of the 2 μ plasmid DNA, the killer plasmids present in some strains, and the yeast mitochondrial genome (ca.75 kb) have to be considered.

With the completion of the yeast genome sequence it became possible to define the proteome of a eukaryotic cell, for the first time: detailed information was deposited in inventory databases and the majority of the proteins could be classified according to function (*http://www.mips.mpg.biochem.de; http://www.quest7.proteome.com*). It was seen that nearly 40% of the proteome consist of membrane proteins, and that an estimated 8–10% of the nuclear genes encode mitochondrial functions. It came as an initial surprise that no function could be attributed to some 40% of the yeast genes. However, even with the exponential growth of entries in protein databases and the refinement of *in silico* analyses, this figure could not be reduced decisively. The same notion turned out to be valid for all other genomes that have been sequenced since then. As an explanation, we have to envisage that a considerable portion of every genome is reserved for species- or genus-specific functions.

An interesting observation made for the first time was the occurrence of regional variations of base composition with similar amplitudes along the chromosomes. The analysis of chromosomes III and XI revealed an almost regular periodicity of the GC content, with a succession of GC-rich and GC-poor segments of ~ 50 kb each, a further interesting observation being that the compositional periodicity correlated with local gene density. Profiles obtained from similar analyses of chromosomes II and VIII again showed these phenomena, albeit with less pronounced regularity. Other chromosomes also show compositional variation of similar range along their arms, with pericentromeric and subtelomeric regions being AT-rich, though spacing between GC-rich peaks is not always regular. In most cases, however, there is a broad correlation between high GC content and high gene density.

A comparison of all yeast sequences revealed that there is a considerable degree of internal genetic redundancy in the yeast genome, which at the protein level is approximately 40%. Whereas an estimate of sequence similarity (both at the nucleotide and the amino acid level) is highly predictive, it still remains difficult to correlate these values to functional redundancy. Interestingly, it turned out that all other genomes sequenced since then showed the same phenomenon. Gene duplications in yeast are of a different type. In many instances, the duplicated sequences are confined to nearly the entire coding region of these genes and do not extend into the intergenic regions. Thus, the corresponding gene products share high similarity in terms of amino acid sequence or sometimes are even identical and, therefore, may be functionally redundant. However, as suggested by sequence differences within the promoter regions or demonstrated experimentally, expression varies. It may well be that one gene copy is highly expressed while another one is lowly expressed. Turning on or off expression of a particular copy within a gene family may depend on the differentiated status of the cell (such as mating type, sporulation, etc.). Biochemical studies also revealed that in particular cases "redundant" proteins can substitute for each other, thus accounting for the fact that a large portion of single gene disruptions in yeast do not impair growth or cause "abnormal" phenotypes. This does not imply, however, that these "redundant" genes were *a priori* dispensable. Rather they may have arisen through the need to help adapt yeast cells to particular environmental conditions.

Subtelomeric regions in yeast are rich in duplicated genes, which are of functional importance for carbohydrate metabolism or cell wall integrity, but there is also a great variety of (single) genes internal to chromosomes that appear to have arisen from duplications. An even more surprising phenomenon became apparent when the sequences of complete chromosomes were compared to each other, revealing that there are 55 large chromosome segments (up to 170 kb) in which homologous genes are arranged in the same order, with the same relative transcriptional orientations, on two or more chromosomes (WOLFE and SHIELDS, 1997). The genome has continued to evolve since this ancient duplication occurred: genes have been inserted or deleted, Ty elements and introns have been lost and gained between two sets of sequences. If optimized for maximum coverage, up to 40% of the yeast genome is found to be duplicated in clusters, not including Ty elements and subtelomeric regions. All observed clusters do not overlap and intra- and interchromosomal cluster duplications have similar probabilities.

The availability of the complete yeast genome sequence not only provided further insight into genome organization and evolution in yeast, but also offered a reference to search for orthologs in other organisms. Of particular interest were those genes that are homologs to genes that perform differentiated functions in multicellular organisms or that might be of relevance to malignancy. Comparing the catalog of human sequences available in the databases with the yeast ORFs reveals that more than 30% of the yeast genes have homologs among the human genes of known function. Approximately 100 yeast genes were seen to exhibit significant similarity to human disease genes (BASSETT et al., 1996), and some of the latter could even be predicted from comparisons with the yeast genes.

3.1.4 The Yeast Postgenome Era

It was evident to anyone engaged in the project that the determination of the entire sequence of the yeast genome should only be considered a prerequisite for functional studies of the many novel genes to be detected. Thus, a European functional analysis network (EUROFAN) was initiated in 1995 and similar activities were started in the international collaborating laboratories in 1996. The general goal is to systematically investigate the yeast genes of unknown function by the following approaches:

(1) Improved data analysis by computer (*in silico* analysis);
(2) systematic gene disruptions and gene overexpression;
(3) analysis of phenotypes under different growth conditions, such as temperature, pH, nutrients, stress;
(4) systematic transcription analysis by conventional methods; gene expression under varying conditions;
(5) *in situ* cellular localization and movement of proteins by the use of tagged proteins (GFP-fusions);
(6) analysis of gene expression under varying conditions by 2D gel electrophoresis of proteins;
(7) complementation tests with genes from other organisms.

In this context, a most compelling approach is the genome-wide analysis of gene expression profiles by chip technology. High-density microarrays of all yeast ORFs were the first to be successfully used in studying various aspects of a transcriptome (e.g., DERISI et al., 1997).

Now that the entire sequence of a laboratory strain of *S. cerevisiae* is available, the complete sequences of other yeasts of industrial or medical importance are within our reach. Such knowledge would considerably accelerate the development of productive strains needed in other areas (e.g., *Kluyveromyces*, *Yarrowia*) or the search for novel anti-fungal drugs. It may even be unnecessary to finish the entire genomes whenever a yeast or fungal genome displays considerable synteny with that of *S. ce-revisiae*. A special program devoted to this problem, the analysis of hemiascomycetes yeast genomes by tag sequence studies for the approach of speciation mechanisms and preparation of tools for functional genomics, has been finalized recently by a French consortium.

In all, the yeast genome project has demonstrated that an enterprise like this could successfully be carried out in "small steps" and in team work. Clearly, the wealth of fresh and biologically relevant information collected from the yeast sequences and the functional analyses had an impact on the other large-scale sequencing projects.

3.2 The Plant *Arabidopsis thaliana*

3.2.1 The Organism

Arabidopsis thaliana is a small cruciferous plant of the mustard family (Fig. 5) which was first described by the German physician JOHANNES THAL in 1577 in his Flora Book of the Harz Mountains, Germany, and named after him later. In 1907, *A. thaliana* was recognized to be a versatile tool for genetic studies by LAIBACH (1907) when he was a student with E. STRASBURGER in Bonn, Germany. More than 30 years later in 1943 – then Professor of Botany in Frankfurt – he published an influential paper (LAIBACH, 1943) clearly describing the favorable features making this plant a true model organism:

(1) short generation time of only two months,
(2) high seed yield,
(3) small size,
(4) simple cultivation,
(5) self-fertilization, but
(6) easy crossing yielding fully fertile hybrids,
(7) only five chromosomes in the haploid genome, and
(8) the possibility to isolate spontaneous and induced mutants.

An attempt by RÉDEI in the 1960s to convince funding agencies to develop *Arabidopsis*

Fig. 5. The model plant *Arabidopsis thaliana*. (**a**) Adult plant, approx. height 20 cm, (from LAIBACH, 1951, (**b**) flower, approx. height 4 mm, (from MÜLLER, 1961), (**c**) chromosome plate showing the five chromosomes of *Arabidopsis* (from LAIBACH, 1907).

as a plant model system was unsuccessful, mainly because geneticists at that time had no access to genes at the molecular level and, therefore, no reason to work with a plant irrelevant for agriculture.

With the development of molecular biology two additional properties of the *A. thaliana* genome made this little weed the superior choice as an experimental system. LAIBACH had already noted in 1907 that *A. thaliana* contained only one third of the chromatin of related *Brassica* species. Much later it became clear that this weed possesses (1) the smallest genome of any higher plant (ARUMUGANATHAN and EARLE, 1991) with (2) a low amount of repetitive DNA. Within the plant kingdom characterized by its large variation of genome sizes (see e.g., *http://www.rbgkew.org.uk/cval/database1.html*) caused mainly by a differing content of repetitive DNA these features support efficient map-based cloning of genes for a detailed elucidation of their function at the molecular level. A first set of tools for that purpose became established during the 1980s with a comprehensive genetic map containing 76 phenotypic markers obtained by mutagenesis, RFLP maps, *Agrobacterium*-mediated transformation, and cosmid and YAC libraries covering the genome several fold with only a few thousand clones.

Soon it was realized that projects involving resources shared by many laboratories would profit from a centralized collection and distribution of stocks and related information. "The Multinational Coordinated *Arabidopsis thaliana* Genome Research Project" was launched in 1990 and as a result, two stock centers at Nottingham, UK and Ohio, USA as well as the *Arabidopsis* database at Boston, USA (CHERRY et al., 1992) were created in 1991. With respect to seed stocks these centers succeeded an effort already started by LAIBACH in 1951 and continued by RÖBBELEN and KRANZ. The new, additional collections of clones and clone libraries provided the basis for the genome sequencing project later on. With the increased use of the Internet at that time the database soon became a central tool for data storage and distribution and has ever since served the community as a central one-stop shopping point for information, despite its move to Stanford and its restructuring to become "The *Arabidopsis* Information Resource" (TAIR; *http://www.arabidopsis.org*).

During the following years many of the research tools were improved and new ones were added: mutant lines based on insertions of T-DNA and transposable elements were created in large numbers, random cDNA clones were sequenced partially, maps became available for different types of molecular markers such as RAPDs, CAPS, microsatellites, AFLPs, and SNPs which were integrated with each other, recombinant inbred lines were established to facilitate the mapping process, a YAC library with large inserts, as well as BAC and P1 libraries were constructed, physical maps based on cosmids, YACs and BACs were built and tools for their display developed.

3.2.2 Structure of the Genome Project

In August 1996 the stage was prepared for large-scale genome sequencing. At a meeting in Washington, DC representatives of six research consortia from North America, Europe, and Japan launched the "*Arabidopsis* Genome Initiative". They set the goal to sequence the complete genome by the year 2004, agreed about the strategy, the distribution of tasks and guidelines for the creation and publication of sequence data. The genome of the ecotype Columbia was chosen for sequencing, because all large insert libraries had been prepared from this line and it was one of the most prominent ecotypes for all kinds of experiments worldwide besides Landsberg erecta (Ler). Since Ler is actually a mutant isolated from an inhomogeneous sample of the Landsberg ecotype after X-ray irradiation (RÉDEI, 1992), it was not suitable to serve as a model genome. The sequencing strategy rested on BAC and P1 clones from which DNA can be isolated more efficiently than from YACs and which on average contain larger inserts than cosmids. This strategy was chosen despite the fact that most attempts to create physical maps had been based on cosmid and YAC clones at that time and that initial sequencing efforts had employed mostly cosmids. BAC and P1 clones for

sequencing were chosen via hybridization to YACs and molecular markers of known and well separated map positions. Later on, information from BAC end sequences as well as fingerprint and hybridization data, created while genome sequencing was already in progress, were used to minimize redundant sequencing caused by clone overlap. This multinational effort has been very fruitful and led to the complete sequences of two of the five *Arabidopsis* chromosomes, namely chromosomes 2 (LIN et al., 1999) and 4 (MAYER et al., 1999), with exception of their rDNA repeats and the heterochromatic regions around their centromeres. More than three years ahead of the original time table the genomic sequence was completed by the end of the year 2000 (Anonymous 2000b). Sequences of the mitochondrial (UNSELD et al., 1997) and the plastid genome (SATO et al., 1999) have been determined as well so that the complete genetic information of *Arabidopsis* is now available.

3.2.3 Results of the Genome Project

The two sequenced chromosomes have yielded no surprises with respect to their structural organization. With the exception of one sequenced marker more than hundred have been observed in the expected order. Repetitive elements and repeats of transposable elements are concentrated in the heterochromatic regions around the centromers where gene density and recombination frequency are below their averages (22 genes per 100 kbp, 1 cM/50–250 kbp). With a few minor exceptions these average values do not vary strongly in other regions of the chromosomes, which is in sharp contrast to larger plant genomes (GILL et al., 1996a; 1996b; KÜNZEL et al., 2000). In addition, genomes such as that of maize do contain repetitive elements and transposons interspersed with genes (SANMIGUEL et al., 1996) so that *Arabidopsis* is certainly not a model for the structure of large plant genomes.

From the sequences available it has been calculated that the 120 Mbp gene containing part of the nuclear genome of *Arabidopsis* contains approximately 26,000 genes (MEYE-

ROWITZ 2000; chr. 2: 4,037, chr. 4: 3,744 annotated genes), whereas the mitochondrial and the plastid genomes carry only 57 and 78 genes on 366,924 and 154,478 bp of DNA, respectively. Therefore, most of the organellar proteins have to be encoded in the nucleus and are targeted to their final destinations via N-terminal transit peptides. Recently, it has been estimated that 10 or 14% of the nuclear genes encode proteins localized in mitochondria or plastids, respectively (EMANUELSSON et al., 2000). Only for a fraction of the predicted plastid proteins homologous cyanobacterial proteins could be identified (LIN et al., 1999; MAYER et al., 1999), even though lateral gene transfer from the endosymbiotic organelle to the nucleus has been assumed to be the main source of organellar proteins. These data indicate that either the large evolutionary distance between plants and cyanobacteria prevents the recognition of orthologs and/or that many proteins from other sources in the eukaryotic cell have acquired plastid transit peptides as suggested earlier based on Calvin cycle enzymes (MARTIN and SCHNARRENBERGER, 1997). Furthermore, a substantial number of proteins without predicted transit peptides, but with higher homology to proteins from cyanobacteria than to any other organism have been recognized (LIN et al., 1999; MAYER et al., 1999), indicating that at least plastids have contributed a significant part of the protein complement of other compartments. These data clearly show that plant cells have become highly integrated genetic systems during evolution assembled from the genetic material of the eukaryotic host and two endosymbionts (HERRMANN, 2000). That this system integration is an ongoing process is revealed by the many small fragments of plastid origin in the nuclear genome (BEVAN et al., 1998) and the unexpected discovery of a recent gene transfer event from the mitochondrial to the nuclear genome. Within the genetically defined centromere of chromosome 2 a 270 kbp fragment with 99% identity to the mitochondrial genome has been identified and its location confirmed via PCR across the junctions with unique nuclear DNA (LIN et al., 1999). For a comprehensive list of references concerning *Arabidopsis thaliana* readers are referred to SCHMIDT (2000). In this chapter only articles were cited which are not listed there or which are of utmost importance to the matters discussed here.

3.2.4 Follow-Up Research in the Postgenome Era

Potential functional assignments can be made for up to 60% of the predicted proteins based on sequence comparisons, the identification of functional domains, and motifs as well as structural predictions. Interestingly, 65% of the proteins show no significant homology to proteins of the completely sequenced genomes of bacteria, yeast, *C. elegans*, and *D. melanogaster* (LIN et al., 1999), clearly reflecting the large evolutionary distance of the plant and other kingdoms and the independent development of multicellularity accompanied by a large increase in gene number. The discovery of protein classes and domains specific for plants, e.g., Ca^{2+}-dependent protein kinases containing four EF-hand domains (MAYER et al., 1999) or the B3 domain of *ABI3*, *VP1* and *FUS3* (SUZUKI et al., 1997), and the significantly different abundance of several proteins or protein domains when compared with *C. elegans* or *D. melanogaster*, e.g., myb-like transcription factors (JIN and MARTIN, 1999), C3HC4 ring finger domains and P450 enzymes (MAYER et al., 1999), further support this notion. Already the larger number of genes in the *Arabidopsis* genome (approx. 25,000) as compared to the genomes of *C. elegans* (approx. 19,000), and *D. melanogaster* (approx. 14,000) seems to indicate different ways to evolve organisms of comparable complexity. Currently, the underlying reason for this large difference is unclear. It may simply reflect a larger proportion of duplicated genes, as it is the case for *C. elegans* vs. *D. melanogaster* (RUBIN et al., 2000). The large number of observed tandem repeats (LIN et al., 1999; MAYER et al., 1999) and the large duplicated segments between chromosome 2 and 4, altogether 2.5 Mbp, and between chromosomes 4 and 5, a segment containing 37 genes (MAYER et al., 1999), seem to favor this explanation. But other specific properties of plants may as well contribute to their large number of genes, such as autotrophic, non-mobile life, rigid body structure, continuous organ development, successive gameto-phytic and sporophytic generations, other ways as animals to process information and to respond to environmental stimuli, a lesser extent of combinatorial use of proteins or any combination of those factors. At present, the functional assignments of proteins are not advanced enough to distinguish between all these possibilities.

The data currently available indicate that the function of many plant genes is different from animals and fungi. Besides the basic eukaryotic machinery which was present in the last common ancestor before endosymbiosis with cyanobacteria created the plant kingdom and which can be delineated via identification of orthologs in eukaryotic genomes (MAYER et al., 1999), all the plant-specific functions have to be elucidated. Since more than 40% of all predicted proteins from the genome of *Arabidopsis* have no assigned function and many others have not been investigated thoroughly, this will require an enormous effort using various approaches. New techniques from chip-based expression analysis and genotyping to high-throughput proteomics and protein–ligand interaction studies as well as metabolite profiling have to be applied in conjunction with the identification of diversity present in nature or created via mutagenesis, transformation, gene knock-out, etc. Since multicellularity was established independently in all kingdoms it might be wise to sequence the genome of a unicellular plant. Its gene content would help to identify all the proteins required for cell-to-cell communication and transport of signals and metabolites. To collect, store, evaluate, and access all the relevant data from these various approaches, many of which are parallelized and designed for high-throughput analyses, new bioinformatic tools have to be created. To coordinate such efforts, a first meeting for "The Multinational Coordinated *Arabidopsis* 2010 Project" (*http://www.arabidopsis.org/workshop1.html*) aiming at functional genomics and the creation of a virtual plant within the next decade was held in January 2000 at the Salk Institute in San Diego, USA. From just two examples it is evident that the rapid progress in *Arabidopsis* research will continue in the future. A centralized facility for chip-based expression analysis

has been set up already in Stanford, USA, and a company provides access to 39,000 potential single nucleotide polymorphisms which will speed up mapping and genotyping considerably. At the end the question remains the same as in the beginning: Why should all these efforts be directed towards a model plant irrelevant to agriculture? The answer can be given again as a question: Which other plant system would provide better tools to tacle all the basic questions of plant development and adaptation than *Arabidopsis thaliana*?

Acknowledgement
The author likes to thank HELMUT BÄUMLEIN and ULRICH WOBUS for helpful discussions and critical reading of the manuscript.

3.3 The Roundworm *Caenorhabditis elegans*

3.3.1 The Organism

The free-living nematode *Caenorhabditis elegans* (Fig. 6) is the first multicellular animal whose genome has been completely sequenced. This worm, although often viewed as a featureless tube of cells, has been studied as a model organism for more than 20 years. In the 1970s and 1980s, the complete cell lineage of the worm from fertilized egg to adult was determined by microscopy (SULSTON, 1988), and later on the entire nervous system was reconstructed (CHALFIE and WHITE, 1988). It has proved to have several advantages as a biological study object such as simple growth conditions, rapid generation time with an invariant cell lineage, and well-developed genetic and molecular tools for its manipulation. Many of the discoveries made with *C. elegans* became in particular relevant to the study of higher organisms since it shares many of the essential biological features such as neuronal growth, apoptosis, intra- and intercellular signaling pathways, food digestion, etc. that are in the focus of interest of, e.g., human biology.

A special review by the *C. elegans* Genome Consortium (WILSON, 1999) gives an interesting summary of how "the worm was won" and

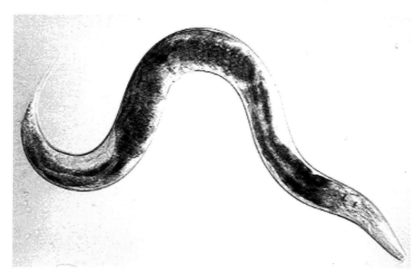

Fig. 6. The free-living bacteriovorous soil nematode *Caenorhabditis elegans* which is a member of the Rhabditidae (found at *http://www.nematodes.org*).

examines some of the preliminary findings from the near-complete sequence data.

The *C. elegans* genome was estimated at 100 Mb deduced from a clone-based physical map. This map was initially based on cosmid clones using a fingerprinting approach. Later, yeast artificial chromosome (YAC) clones were incorporated to bridge the gaps between cosmid contigs and provided coverage of regions that were not represented in the cosmid libraries. By 1990, the physical map consisted of less than 20 contigs and was useful for rescue experiments that were able to localize a phenotype of interest to a few kilobases of DNA (COULSON et al., 1988). Alignment of the existing genetic and physical maps into the *C. elegans* genome map was greatly facilitated through the cooperation of the entire "worm community". After the physical map had been nearly completed, the effort to sequence the entire genome became both feasible and desirable. By that time, this attempt was significantly larger than any sequencing project before and was nearly two orders of magnitude more expensive than the mapping effort.

3.3.2 The Structure of the Genome Project

In 1990 a three-year pilot project for sequencing 3 Mb of the genome was initiated as a collaboration between the Genome Sequencing Center in St. Louis, USA, and the Sanger Centre in Hinxton, UK. Funding was obtained from the NIH and the UK MRC. The genome sequencing effort initially focused on the central regions of the five autosomes which were well represented in cosmid clones, because the majority of genes known at that time were contained in these regions. At the beginning of the project, sequencing was still based on standard radioisotopic methods using "walking" primers and cosmid clones as templates for the sequencing reactions. However, a serious problem of this primer-directed sequencing approach on cosmids were multiple priming events due to repetitive sequences and efficient preparation of sufficient template DNA. To address these problems the strategy was changed to a more classic shotgun sequencing strategy based on cosmid subclones generally sequenced from universal priming sites in the subcloning vectors. Further developments in automation of the sequencing reactions, fluorescent sequencing methods, improvements of the dye-terminator chemistry (LEE et al., 1992), and the generation of assembly and contig editing programs, led to the phasing out of the instrument in favor of four-color, single-lane sequencing. The finishing phase then used a more ordered, directed sequencing strategy as well as the walking approach, to close specific remaining gaps and resolve ambiguities. Hence, the worm project grew into a collaboration among *C. elegans* Sequencing Consortium members and the entire international community of *C. elegans* researchers. In addition to the nuclear genome sequencing effort, other researchers sequenced its 15 kb mitochondrial genome and carried out extensive cDNA analyses that facilitated gene identification.

The implementation of high-throughput devices and semi-automated methods for DNA purification and sequencing reactions led to an overwhelming success in scaling up of the sequencing. The first 1 Mb threshold of *C. elegans* finished genome sequences was reached in May 1993. In August 1993, the total had already increased to over 2 Mb (WILSON et al., 1994), and by December 1994, over 10 Mb of the *C. elegans* genome had been finished. Bioinformatics played an increasing role in the genome project. Software developments made the processing, analysis, and editing of thousands of data files per day a manageable task. Indeed, many of the software tools developed in the *C. elegans* project, e.g., ACeDB, PHRED, and PHRAP, have become key components in the current approach to sequencing the human genome.

The 50 Mb mark was passed in August 1996. At this point, it became obvious that 20% of the genome was not covered by cosmid clones, and a closure strategy has been implemented. For gaps in the central regions, either long-range PCR was used or a fosmid library was probed in search of a gridging clone. For remaining gaps in the central regions and for regions of chromosomes contained only in YACs, purified YAC DNA was used as the starting material for shotgun sequencing. All of these final regions have been essentially completed with the exception of several repetitive elements. Hence, the final genome sequence of the worm is a composite from cosmids, fosmids, YACs, and PCR products. The exact genome size is still approximate, mainly because of repetitive sequences that cannot be sequenced in their entity, although most of the extensive tandem repeats have been accurately sized by restriction digestion. It is expected that the repeats will be resolved after more effective methods will have been developed for sequencing them to completion. These methods, which will be of great value for sequencing difficult regions of other genomes, include short insert libraries (MCMURRAY et al., 1998), the use of alternative enzymes (TABOR and RICHARDSON, 1995), and "overgo" probing for specific subclones. Some tandem repeats in the larger YACs are of unknown size. Telomeres were sequenced from plasmid clones provided by WICKY et al. (1996). Of twelve chromosome ends, nine have been linked to the outermost YACs on the physical map.

3.3.3 Results of the Genome Project

Analysis of the 97 Mb of the total *C. elegans* genome revealed 19,099 predicted genes, considerably more than expected before sequencing began (HERMAN, 1988; WATERSTON and SULSTON, 1995), with an average density of one predicted gene per 5 kb. Each gene has an average of five introns and 27% of the genome resides in exons. The gene number is about three times the number found in yeast (CHERVITZ et al., 1998) and is about 1/5 to 1/3 the number predicted for humans (WILSON, 1999).

The interruption of the coding sequence by introns and the relatively low gene density make accurate gene prediction more challenging than in microbial genomes. Valuable bioinformatics tools have been developed and used to identify putative coding regions and to provide an initial overview of the gene structure. In order to refine the computer-generated gene structure predictions, the available EST and protein similarities as well as the genomic sequence data from the related nematode *Caenorhabditis briggsae* were used for verification. About 40% of the predicted genes were confirmed by EST matches, but as ESTs match only a portion of the gene, only about 15% of the total coding sequence is presently confirmed. The analysis of several ESTs provided direct evidence for alternative splicing (WILSON, 1999).

Similarities to known proteins provide an inital glimpse into the possible function of many of the predicted genes. WILSON (1999) outlined that approximately 42% of predicted protein products have cross-phylum matches, most of which provide putative functional information (GREEN et al., 1993). Another 34% of predicted proteins match only other nematode proteins, a few of which have been functionally characterized. The fraction of genes with informative similarities is far less than the 70% observed for microbial genomes. This may reflect the smaller proportion of nematode genes devoted to core cellular functions (CHERVITZ et al., 1998), the comparative lack of knowledge of functions involved in building an animal, and the evolutionary divergence of nematodes from other animals extensively studied so far at the molecular level. Interestingly, genes encoding proteins with cross-phylum matches were more likely to have a matching EST (60%) than those without cross-phylum matches (20%). This observation suggested that conserved genes are more likely to be highly expressed, perhaps reflecting a bias for "housekeeping" genes among the conserved set. Alternatively, genes lacking confirmatory matches may be more likely to be false predictions, although the analyses did not suggest this.

In addition to the protein-coding genes, the genome contains several hundred genes for noncoding RNAs. 659 widely dispersed transfer RNA genes have been identified, and at least 29 tRNA-derived pseudogenes are present. 44% of the tRNA genes were found on the X chromosome, which represents only 20% of the total genome. Several other noncoding RNA genes, such as those for spliceosomal RNAs, occur in dispersed multigene families. Several RNA genes are located in the introns of protein coding genes, which may indicate RNA gene transposition (WILSON, 1999). Other noncoding RNA genes are located in long tandem repeat regions; the ribosomal RNA genes were found solely in such a region at the end of chromosome I, and the 5S RNA genes occur in a tandem array on chromosome V.

Extended regions of the entire genome do neither code for proteins nor RNA. Among these are regions that are involved in gene regulation, replication, maintenance and movement of chromosomes. Furthermore, a significant fraction of the *C. elegans* genome is repetitive and can be classified as either local repeats (e.g., tandem, inverted, and simple sequence repeats) or dispersed repeats. Tandem repeats amount to 2.7% and inverted repeats to 3.6% of the genome. These local repeats are distributed non-uniformly throughout the genome, with respect to genes. Not surprisingly, only a small percentage of tandem repeats are found within the 27% of the protein coding genes. Conversely, the density of inverted repeats is higher in regions predicted as intergenic (WILSON, 1999). Although local repeat structures are often unique in the genome, other repeats are members of families. Some repeat families show a chromosome-specific bias in representation. Altogether 38 dispersed repeat families have been recognized. Most of these dispersed repeats are associated with transposition in some form (SMIT, 1996), and include the previously described transposons of *C. elegans*. In addition to multiple-copy repeat families, a significant number of simple duplications have been observed involving segments that range from hundreds of bases to tens of kilobases. In one case, a segment of 108 kb containing six genes was duplicated tandemly with only ten nucleotide differences observed between the two copies. In another example, immediately adjacent to the telomere at the left end of chromosome IV, an inverted repeat of 23.5 kb was present, with only eight differences found between the two copies. There are many instances of smaller duplications, often separated by tens of kilobases or more that may contain coding sequences. This could provide a mechanism for copy divergence and the subsequent formation of new genes (WILSON, 1999).

The GC content is more or less uniform throughout all chromosomes at 36%, unlike human chromosomes that have different isochores (BERNARDI, 1995). There are no localized centromers as are found in most other metazoa. Instead, the extensive highly repetitive sequences that are involved in spindle attachment in other organisms may be represented by some of the many tandem repeats found scattered among the genes, particularly on the chromosome arms. Gene density is also uniform across the chromosomes, although some differences are apparent, particularly between the centers of the autosomes, the autosome arms, and the X chromosome.

More striking differences become evident upon examination of other features. Both inverted and tandem repeat sequences are more frequent on the autosome arms than in the central regions or on the X chromosome. This abundance of repeats on the arms is likely the reason for the difficulties in cosmid cloning and sequence completion in these regions. The fraction of genes with cross-phylum similarities tends to be lower on the arms as does the fraction of genes with EST matches. Local clusters of genes also appear to be more abundant on the arms.

3.3.4 Follow-Up Research in the Postgenome Era

While sequencing of the *C. elegans* genome has essentially been completed, analysis and annotation will continue for years and will – hopefully – facilitated by further information and better technologies to become available. However, it is now possible to describe some interesting features of the *C. elegans* genome, based on the analysis of completed genome sequences.

The observations and findings of the *C. elegans* genome project provide a preliminary glimpse of the biology of metazoan development. There is much left to be uncovered and understood in the sequence. Of primary interest, all of the genes necessary to build a multicellular organism are now essentially in hand, although their exact boundaries, relationships, and functional roles have to be elucidated more precisely. The basis for a better understanding of the regulation of these genes is also now within grasp. Furthermore, many of the discoveries made with *C. elegans* are relevant to the study of higher organisms. This ex-

tends beyond fundamental cellular processes such as transcription, translation, DNA replication and cellular metabolism. For these reasons, and because of its intrinsic practical advantages, *C. elegans* has proved to be an invaluable tool for understanding features such as vertebrate neuronal growth and pathfinding, apoptosis and intra- and intercellular signaling pathways.

3.4 The Fruitfly *Drosophila melanogaster*

3.4.1 The Organism

In 1910, T. H. MORGAN started his analysis of the fruitfly *Drosophila melanogaster* and identified the first white-eyed mutant fly strain (RUBIN and LEWIS, 2000). Now, less than 100 years later, nearly the complete genome of the insect has been sequenced offering the ultimate opportunity to elucidate processes ranging from the development of an organism up to its daily performances.

Drosophila (Fig. 7), as a small insect with a short life span and a very rapid, well-characterized development represents one of the best analyzed multicellular organism up to now. During the last century, more than 1,300 genes – mostly based on mutant phenotypes – were genetically identified, cloned, and sequenced. Surprisingly, most of them were found to have counterparts in other metazoa including even humans. Soon it became evident, that not only the genes are conserved among species, but also the functions of the encoded proteins. Processes like the development of limbs, the nervous system, the eyes and the heart, or the presence of circadian rhythms and innate immunity are highly conserved, even though that genes that were taken from human sources can supplement the function of genes in the fly (RUBIN et al., 2000). Early in the 1980s identification of the HOX genes lead to the discovery that the anterior-postererior body patterning genes are conserved from fly to man (VERAKSA et al., 2000). Furthermore, genes essential for the development and the differentiation of muscles such as twist, MyoD, and MEF2 act in most of the organisms analyzed

Fig. 7. The fruitfly *Drosophila melanogaster*. (**a**) Adult fly, (**b**) stage 16 embryo. In dark-brown, the somatic muscles are visualized using an *β*3-tubulin specific antibody; in red-brown, expression of *β*1-tubulin in the attachment sites of the somatic muscles is shown.

so far (ZHANG et al., 1999). The most striking example is the capability of a certain class of genes, the PAX-6 family, to induce ectopic eye development in *Drosophila*, regardless from which animal source it was taken (HALDER et al., 1995). Besides these developmental processes, in addition human disease networks involved in replication, repair, translation, metabolism of drugs and toxins, neural disorders like Alzheimer's desease and also higher-order functions like memory and signal transduction cascades have shown to be highly conserved (ADAMS et al., 2000; MAYFORD and KANDEL, 1999; VERAKSA et al., 2000). As a quite surprising conclusion, apart from being simply an insect, *Drosophila* is capable of serving as a model system even for complex human diseases.

The genomic organization of *Drosophila melanogaster* has been known for many years. The fly has one small and three large chromosomes with an early estimate of 1.1×10^8 bp. Using polytene chromosomes as tools, in 1935 and 1938, BRIDGES published maps of such an

accuracy that they are still used today (cited in RUBIN and LEWIS, 2000). Making extensive use of chromosomal rearrangements, he constructed cytogenetic maps that assigned genes to specific sections and even specific bands. With the development of techniques like *in situ* hybridization to polytene chromosomes, genes could be mapped with a resolution of less than 10 kb. Another major advantage of *Drosophila* is given by the presence of randomly shuffled chromosomes, the so-called balancers, which allow easy monitoring and pursuit of given mutations on the homologuous chromosome, as well as to guarantee the persistence of such a mutation at the chromosomal place where it has initially occurred by suppression of meiotic recombination.

3.4.2 Structure of the Genome Project

The strategy chosen for the sequencing was the whole-genome shotgun sequencing (WGS) technique. For this technique, the whole genome is broken into small pieces, subcloned into suitable vectors and sequenced. For the *Drosphila* genome project, libraries containing 2 kb, 10 kb, and 130 kb inserts were chosen as templates. They were sequenced from the ends, and assembled by pairs of reads, called mates, from the ends of the 2 kb and 16 kb inserts (MYERS et al., 2000). The assembly of the sequences was facilitated by the absence of interspersed repetitive elements like the human ALU-repeat family. The ends of the large BAC clones were taken to unequivocally localize the sequences into large contigs and scaffolds. Three million reads of ~500 bp were used to assemble the *Drosophila* genome. The detailed strategies, techniques and algorithms used are described in a *Science* issue (Anonymous, 2000a) which was published in March 2000. The work was organized by the Berkeley *Drosophila* Genome Project (BDGP) and performed at the BDGP, the European DGP, the Canadian DGP, and at Celera Genomics Inc. under the auspices of the federally funded Human Genome Project. The combined DGPs produced all the genomic resources and finished 29 Mbp of sequences. Altogether, the

Drosophila genome is ~180 Mbp in size, a third of which is heterochromatin. From the ~120 Mbp of euchromatin, 98% are sequenced with an accuracy of at least 99.5%. Due to the structure of the heterochromatin, which is mainly built up by repetitive elements, retrotransposons, rRNA clusters as well as some unique sequences, most of it could not be cloned or propagated in YACs, which is in contrast to the *C. elegans* genome project (ADAMS et al., 2000).

3.4.3 Results of the Genome Project

As a major result, the number of genes was calculated with ~13,600 using a *Drosophila*-optimized version of the program GENIE (ADAMS et al., 2000). Over the last 20 years, 2,500 of them have already been characterized by the fly community, and their sequences were continously made available during the ongoing of the fly project. The average density is one gene in 9 kb, however, it ranges from none to nearly 30 genes per 50 kbp, and in contrast to *C. elegans* gene-rich regions are not clustered. Regions of high gene density correlate with G + C-rich sequences. Computational comparisons with known vertebrate genes have presented both expected findings as well as surprises. So genes encoding the basic DNA replication machinery are conserved among eukaryotes; especially all of the proteins known to be involved in recognition of the replication start point are present as single-copy genes, and the ORC3 and ORC6 proteins share close similarity with vertebrate proteins, but are highly divergent relative to yeast and apparently much further diverged in the worm. Focusing on chromosomal proteins, the fly seems to lack orthologs to most of the mammalian proteins associated with centromeric DNA such as the CENP-C/MIF-2 family. Furthermore, as *Drosophila* telomeres lack the simple repeats characteristic of most eukaryotic telomeres, consequently the known telomerase components are absent. Concerning gene regulation, RNA polymerase subunits and cofactors are more closely related to their mammalian counterparts than to yeast; for example, the promoter interacting factors UBF and TIF-IA are present in *Drosophila* but not in yeast. The overall set of transcription factors in the fly seems to comprise about 700 members, about half of which are zinc-finger proteins, while in the worm out of 500 factors only one third belong to this family. Nuclear hormone receptors appear to be more rare, as only four new members werde detected, bringing the total up to 20 compared to more than 200 in *C. elegans*. As an example for metabolic processes iron pathway components were analyzed. A third ferritin gene has been found that probably encodes a subunit belonging to cytosolic ferritin, the predominant type in vertebrates. Two newly discoverd transferrins are homologs of the human melanotransferrin p97, which is of special interest, as the analyzed iron transporters in the fly so far are involved in antibotic response rather than in iron transport. Otherwise, proteins homologous to transferrin receptors appear to be absent from the fly, so that the melanotransferrin homologs may mediate the main insect pathway for iron uptake (all data taken from ADAMS et al., 2000). The sequences and the data compiled are freely available to the scientific community on servers in several countries (see: *http://www.fruitfly.org*). In addition, large collections of ESTs (expressed sequence tags), cDNA libraries, and genomic resources are available, as well as databases presenting expression patterns of identified genes or enhancertrap lines, e.g., flyview at the University of Münster, Germany (*http://pbio07.uni-muenster.de*), started by the group of W. JANNING.

3.4.4 Follow-Up Research in the Postgenome Era

A comparative analysis between the genomes of *Drosophila melanogaster*, *Caenorhabditis elegans* and *Saccharomyces cerevisiae* was performed (RUBIN et al., 2000). It showed that the core proteome of *Drosophila* consists of 9,453 proteins, which is only twice the number of *S. cerevisiae* (4,383 proteins). Using stringent criteria, the fly genome is much closer related to the human than to the *C. elegans* genome. Interestingly, the fly has orthologs to

177 of 289 human disease genes examined so far and provides an excellent basis for the rapid analysis of some basic processes involved in human disease. Furthermore, hitherto unknown counterparts for human genes involved in other disorders were found: for example, *menin, tau, limb girdle muscular dystrophy type 2B, Friedrich ataxia*, and *parkin*. Of the cancer genes surveyed, at least 68% appear to have *Drosophila* orthologs, and even a p53-like tumor suppressor protein was detected. All of these fly genes are present as single copy and can be genetically analyzed without uncertainty about additional functionally redundant copies. Hence, all the powerful *Drosophila* tools such as genetic analysis, exploitation of developmental expression patterns, loss-of-function as well as gain-of-function phenotypes may be used to elucidate the function of human genes that are not yet understood.

4 Conclusions

This chapter summarizes the genome projects of selected model organisms with completed or almost finished genomic sequences. The organisms represent members of the major phylogenetic lineages, the eubacteria, archaea, fungi, plants, and animals, thus covering unicellular prokaryotes with singular, circular chromosomes, as well as uni- and multicellular eukaryotes with multiple linear chromosomes.

Their genome sizes range from ~2 to ~180 Mbp with estimated gene numbers ranging from 2,000 to 25,000 (Tab. 3).

Concerning the genome organization, each of the organisms presented exhibits its own specific and often unexpected characteristics. In *B. subtilis*, for example, a strong bias in the polarity of transcription of the genes with respect to the replication fork was observed, whereas in *E. coli* and *S. cerevisiae* the genes are more or less equally distributed on both strands. Furthermore, although insertion sequences are widely distributed in bacteria, none were found in *B. subtilis*. For *A. thaliana*, an unexpectedly high percentage of proteins showed no significant homology to proteins of organisms outside the plant kingdom and thus are obviously specific to plants. Another interesting observation resulting from the genome projects concern gene density. In prokaryotic organisms, i.e., eubacteria and archaea, the genome sizes vary considerably. However, their gene density is relatively constant at about one gene per kbp. During the evolution of eukaryotic organisms, genome sizes grew, but gene density decreased from one gene in 2 kbp in *Saccharomyces* to one gene in 10 kbp in *Drosophila*. This led to the surprising observation that some bacterial species can have more genes than lower eukaryotes and that the number of genes in *Drosophila* is only about three times higher than the number of genes in *E. coli*.

Another interesting question concerns the general homology of genes or gene products

Tab. 3. Summarized Information on the Genomes of Model Organims Described in this Book Chapter (Status September 2000)

Organism	Genome Structure	Genome Size (kb)	Estimated Numbers of Genes/ORFs
Escherichia coli	1 chromosome, circular	4,600	4,400
Bacillus subtilis	1 chromosome, circular	4,200	4,100
Archaeoglobus fulgidus	1 chromosome, circular	2,200	2,400
Saccharomyces cerevisiae	16 chromosomes, linear	12,800	6,200
Arabidopsis thaliana	5 chromosomes, linear	130,000	25,000
Caenorhabditis elegans	6 chromosomes, linear	97,000	19,000
Drosophila melanogaster	4 chromosomes, linear	180,000	13,600

between model organisms. Comparisons of protein sequences have shown that certain gene products can be found in a wide variety of organisms.

Thus, comparative analysis of predicted protein sequences encoded by the genomes of *C. elegans* and *S. cerevisiae* suggested that most of the core biological functions are carried out by orthologous proteins that occur in comparable numbers in these organisms. Furthermore, comparing the yeast genome with the catalog of human sequences available in the databases revealed that a significant number of yeast genes have homologs among the human genes of unknown function. *Drosophila* has proven to be of special importance in this respect since many human disease networks have shown to be highly conserved in the fruitfly. Hence, in an outstanding manner the insect is capable of serving as a model system even for complex human diseases. Since the respective genes in the fly were found to be present as single copies, they can be genetically analyzed much easier.

Genome research of model organisms has just begun. At the time when this chapter was written, more than 50 genome sequences mainly from prokaryotes have been published and more than 300 genomes are currently being sequenced worldwide. The full extent of this broad and rapidly expanding field of genome research on model organisms cannot be covered by a single book chapter. However, the World-Wide Web represents an ideal platform for making available the outcome of large genome projects in an adequate and timely manner. Beside several specialized entry points, two web pages are recommended to start with a broad overview on model organism research, the WWW Virtual Library: Model Organisms (*http://ceolas.org/VL/mo*) and the WWW Resources for Model Organisms (*http://genome.cbs.dtu.dk/gorm/model-organism.html*). Both links represent excellent start sites for detailed information on model organism research.

5 References

ADAMS, M. D., CELNIKER, S. E., HOLT, R. A., EVANS, C. A., GOCAYNE, J. D. et al. (2000), The genome sequence of *Drosophila melanogaster*, *Science* **287**, 2185–2195.

Anonymous (2000a), The *Drosophila* Genome, *Science* **287** (issue 5461), 2181–2224.

Anonymous (2000b), *Arabidopsis thaliana* genome, *Nature* **408**, 781–826.

Anonymous (1997), Dictionary of the yeast genome, *Nature* **387** (Suppl.), 3–105.

ARIAS, R. S., SAGARDOY, M. A., VAN VUURDE, J. W. (1999), Spatio-temporal distribution of naturally occurring *Bacillus* spp. and other bacteria on the phylloplane of soybean under field conditions, *J. Basic Microbiol.* **39**, 283–292.

ARUMUGANATHAN, K., EARLE E. D. (1991), Nuclear DNA content of some important plant species, *Plant. Mol. Biol. Rep.* **9**, 208–218.

BACHMANN, B. J. (1983), Linkage map of *Escherichia coli* K-12, edition 7, *Microbiol. Rev.* **47**, 180–230.

BASSETT, D. E., BOGUSKI, M. S., HIETER, P. (1996), Yeast genes and human disease, *Nature* **379**, 589–590.

BERG, P., BALTIMORE, D., BRENNER, S., ROBLIN, R. O. III, SINGER, M. F. (1975), Asilomar conference on recombinant DNA molecules, *Science* **188**, 991–994.

BERNARDI, G. (1995), The human genome: organization and evolutionary history, *Annu. Rev. Genet.* **29**, 445–476.

BEVAN, M., BENNETZEN, J. L., MARTIENSSEN, R. (1998), Genome studies and molecular evolution. Commonalities, contrasts, continuity and change in plant genomes, *Curr. Opin. Plant Biol.* **1**, 101–102.

BLATTNER, F. R., PLUNKETT, G. III, BLOCH, C. A., PERNA, N. T., BURLAND, V. et al.(1997), The complete genome sequence of *Escherichia coli* K-12, *Science* **277**, 1453–1474.

BULT, C. J., WHITE, O., OLSEN, G. J., ZHOU, L., FLEISCHMANN, R. D. et al. (1996), Complete genome sequence of the methanogenic archaeon *Methanococcus jannaschii*, *Science* **273**, 1058–1073.

CARLE, G. F., OLSON, M. V. (1985), An electrophoretic karyotype for yeast, *Proc. Natl. Acad. Sci. USA* **82**, 3756.

CHALFIE, M., WHITE, J. (1988), The Nervous system, in: *The Nematode Caenorhabditis elegans* (Wood, W. B., Ed.), pp. 337–391. Cold Spring Harbor, N.Y.: CSHL Press.

CHERRY, J. M., CARTINHOUR, S. W., GOODMANN, H. M. (1992), AAtDB, an *Arabidopsis thaliana* database, *Plant Mol. Biol. Rep.* **10**, 308–309, 409–410.

CHERVITZ, S. A., ARAVIND, L., SHERLOCK, G., BALL, C. A., KOONIN, E. V. et al. (1998), Comparison of the complete protein sets of worm and yeast: orthology and divergence, *Science* **282**, 2022–2027.

COOKSEY, R. C., HOLLOWAY, B. P., OLDENBURG, M. C., LISTENBEE, S., MILLER, C. W. (2000), Evaluation of the invader assay, a linear signal amplification method, fro identification of mutations associated with resistance of rifampin and isoniazid in *Mycobacterium tuberculosis, Antimicrob. Agents Chemother.* **44**, 1296–1301.

COULSON, A., WATERSTON, R., KIFF, J., SULSTON, J., KOHARA, Y. (1988), Genome linking with yeast artificial chromosomes, *Nature* **335**, 184–186.

DANCHIN, A. (1997), Comparison between the *Escherichia coli* and *Bacillus subtilis* genomes suggests that a major function of polynucleotide phosphorylase is to synthesize CDP, *DNA Res.* **4**, 9–18.

DANCHIN, A. (1999), From protein sequence to function, *Curr. Opin. Struct. Biol.* **9**, 363–367.

DANCHIN, A., GUERDOUX-JAMET, P., MOSZER, I., NITSCHKE, P. (2000), Mapping the bacterial cell architecture into the chromosome, *Philos. Trans. R. Soc. Lond. B Biol. Sci.* **355**, 179–190.

DERISI, J. L., IYER, V. R., BROWN, P. O. (1997), Exploring the metabolic and genetic control of gene expression on a genomic scale, *Science* **278**, 680–686.

EMANUELSSON, O., NIELSEN, H., BRUNAK, S., VON HEIJNE, G. (2000), Predicting subcellular localization of proteins based on their N-terminal amino acid sequence, *J. Mol. Biol.* **300**, 1005–1016.

GILL, K. S., GILL, B. S., ENDO, T. R., TAYLOR, T. (1996a), Identification and high density mapping of gene-rich regions in the chromosome group 5 of wheat, *Genetics* **143**, 1001–1012.

GILL, K. S., GILL, B. S., ENDO, T. R., TAYLOR, T. (1996b), Identification and high density mapping of gene-rich regions in the chromosome group 1 of wheat, *Genetics* **144**, 1883–1891.

GOFFEAU, A. BARRELL, B. G., BUSSEY, H., DAVIS, R. W., DUJON, B. et al. (1996), Life with 6000 genes, *Science* **274**, 546–567.

GOTTESMAN, S., ROCHE, E., ZHOU, Y., SAUER, R. T. (1998), The ClpXP and ClpAP proteases degrade proteins with carboxy-terminal peptide tails added by the SsrA-tagging system, *Genes Dev.* **12**, 1338–1347.

GREEN, P., LIPMAN, D., HILLIER, L., WATERSTON, R., STATES, D., CLAVERIE, J. M. (1993), Ancient conserved regions in new gene sequences and the protein database, *Science* **259**, 1711–1716.

HALDER, G., CALLAERTS, P., GEHRING, W. J. (1995), Induction of ectopic eyes by targeted expression of the eyeless gene in *Drosophila, Science* **267**, 1788–1792.

HERMAN, R. K. (1988), Genetics, in: *The Nematode Caenorhabditis elegans* (WOOD, W. B., Ed.), pp. 17–45. Cold Spring Harbor, N.Y.: CSHL Press.

HERRMANN, R. G. (2000), Organelle genetics – part of the integrated plant genome, *Vortr. Pflanzenzüchtg.* **48**, 279–296.

HOSFIELD, D. J., FRANK, G., WENG, Y., TAINER, J. A., SHEN, B. (1999), Newly discovered archaebacterial flap endonucleases show a structure-specific mechanism for DNA substrate binding and catalysis resembling human flap endonuclease-1, *J. Biol. Chem.* **273**, 27154–27161.

JIN, H., MARTIN, C. (1999), Multifunctionality and diversity within the plant MYB-gene family, *Plant Mol. Biol.* **41**, 577–585.

KLENK, H.-P., CLAYTON, R. A., TOMB, J.-F., WHITE, O., NELSON, K. E. et al. (1997), The complete genome of the hyperthermophilic, sulfate-reducing archaeon *Archaeoglobus fulgidus, Nature* **390**, 364–370.

KOHARA, Y., AKIYAMA, K., ISONO, K. (1987), The physical map of the whole *E. coli* chromosome: Application of a new strategy for rapid analysis and sorting of a large genomic library, *Cell* **50**, 495–508.

KRÖGER, M., WAHL, R. (1998), Compilation of DNA sequences of *Escherichia coli* K12: description of the interactive databases ECD and ECDC, *Nucl. Acids Res.* **26**, 46–49.

KUNST, F., OGASAWARA, N., MOSZER, I., ALBERTINI, A. M., ALLONI, G., et al. (1997), The complete genome sequence of the gram-positive bacterium *Bacillus subtilis, Nature* **390**, 249–256.

KÜNZEL, G., KORZUN, L., MEISTER, A. (2000), Cytologically integrated physical restriction fragment length polymorphism maps for the barley genome based on translocation breakpoints, *Genetics* **154**, 397–412.

LAIBACH, F. (1907), Zur Frage nach der Individualität der Chromosomen im Pflanzenreich, *Beih. Bot. Centralblatt, Abt. I* **22**, 191–210.

LAIBACH, F. (1943), *Arabidopsis Thaliana* (L.) Heynh. als Objekt für genetische und entwicklungsphysiologische Untersuchungen, *Bot. Arch.* **44**, 439–455.

LAIBACH, F. (1951) Über sommer- und winterannuelle Rassen von *Arabidopsis thaliana* (L.) Heynh. Ein Beitrag zur Atiologie der Blütenbildung, *Beitr. Biol. Pflanzen* **28**, 173–210.

LEE, L. G., CONNELL, C. R., WOO, S. L., CHENG, R. D., MCARDLE, B. F. et al. (1992), DNA sequencing with dye-labeled terminators and T7 DNA polymerase: Effect of dyes and dNTPs on incorporation of dye-terminators, and probability analysis of termination fragments, *Nucleic Acids Res.* **20**, 2471–2483.

LIN, X., KAUL, S., ROUNSLEY, S., SHEA, T. P., BENITO, M.I., et al. (1999), Sequence and analysis of chromosome 2 of the plant *Arabidopsis thaliana, Nature* **402**, 761–768.

LINDEGREN, C. C. (1949), *The Yeast Cell, its Genetics and Cytology.* St. Louis, MI: Educational Publishers.

LOUIS, E. J., BORTS, R. H. (1995), A complete set of marked telomeres in *Saccharomyces cerevisiae* for physical mapping and cloning, *Genetics* **139**, 125–136.

LYAMICHEV, V., MAST, A. L., HALL, J. G., PRUDENT, J. R., KAISER, M. W. et al. (1999), Polymorphism identification and quantitative detection of genomic DNA by invasive cleavage of oligonucleotide probes, *J. Biol. Chem.* **273**, 27154–27161.

MARTIN, W., SCHNARRENBERGER, C. (1997), The evolution of the Calvin cycle from prokaryotic to eukaryotic chromosomes: a case study of functional redundancy in ancient pathways through endosymbiosis, *Curr. Genet.* **32**, 1–8.

MAYER, K., SCHULLER, C., WAMBUTT, R., MURPHY, G., VOLCKAERT, G. et al. (1999), Sequence and analysis of chromosome 4 of the plant *Arabidopsis thaliana*, *Nature* **402**, 769–777.

MAYFORD, M., KANDEL, E. R. (1999), Gentic approaches to memory storage, *Trends in Genetics* **15**, 463–470.

MCMURRAY, A. A., SULSTON, J. E., QUAIL, M. A. (1998), Short-insert libraries as a method of problem solving in genome sequencing, *Genome Res.* **8**, 562–566.

MEYEROWITZ, E. M. (2000), Today we have the naming of the parts, *Nature* **402**, 731–732.

MORTIMER, R. K., CONTOPOULOU, R., KING, J. S. (1992), Genetic and physical maps of *Saccharomyces cerevisiae*, Edition 11, *Yeast* **8**, 817–902.

MÜLLER, A. (1961), Zur Charakterisierung der Blüten und Infloreszenzen von *Arabidopsis thaliana* (L.) Heynh, *Die Kulturpflanze* **9**, 364–393.

MYERS, E. W., SUTTON, G. G., DELCHER, A. L., DEW, I. M., FASULO, D. P. et al. (2000), A whole-genome assembly of *Drosophila*, *Science* **287**, 2196–2204.

NEIDHARD, F. C. (1996), *Escherichia coli* and *Salmonella typhimurium – Cellular and Molecular Biology,* 2. Ed. Washington DC: American Society for Microbiology.

PERNA, N. T., PLUNKETT, G., BURLAND, V., MAU, B., GLASNER, J. D. et al. (2001), Genome sequence of enterohaemorrhagic *Escherichia coli* O157:H7, *Nature* **409**, 529–533.

RÉDEI, G. P. (1992), A heuristic glance at the past of *Arabidopsis genetics*, in: *Methods in Arabidopsis Research* (KONCZ, C., CHUA, N.-H., SCHELL, J., Eds.), pp. 1–15. Singapore: World Scientific Publishing.

ROBERTS, R. (2000), Database issue of nucleic acids research, *Nucl. Acids Res.* **28**, 1–382.

ROCHA, E. P., DANCHIN, A., VIARI, A. (1999a), Analysis of long repeats in bacterial genomes reveals alternative evolutionary mechanisms in *Bacillus subtilis* and other competent prokaryotes, *Mol. Biol. Evol.* **16**, 1219–1230.

ROCHA, E. P., DANCHIN, A., VIARI, A. (1999b), Universal replication biases in bacteria, *Mol. Microbiol.* **32**, 11–16.

ROCHA, E. P. C., GUERDOUX-JAMET, P., MOSZER, I., VIARI, A., DANCHIN, A. (2000), Implication of gene distribution in the bacterial chromosome for the bacterial cell factory, *J. Biotechnol.* **78**, 209–219.

RUBIN, G. M., LEWIS, E. B. (2000), A brief history of Drosophila's contributions to genome research, *Science* **287**, 2216–2218.

RUBIN, G. M., YANDELL, M. D., WORTMAN, J. R., GABOR MIKLOS, G. L., NELSON, C. R. et al. (2000), Comparative genomics of eukaryotes, *Science* **287**, 2204–2215.

SAIER, M. H., Jr. (1998), Multiple mechanisms controlling carbon metabolism in bacteria, *Biotechnol. Bioeng.* **58**, 170–174.

SANMIGUEL, P., TIKHONOV, A., JIN, Y. K., MOTCHOULSKAIA, N., ZAKHAROV, D. et al. (1996), Nested retrotransposons in the intergenic regions of the maize genome, *Science* **274**, 765–768.

SATO, S., NAKAMURA, Y., KANEKO, T., ASAMIZU, E., TABATA, S. (1999), Complete structure of the chloroplast genome of *Arabidopsis thaliana*, *DNA Res.* **6**, 283–290.

SCHMIDT, R. (2000), The *Arabidopsis* genome, *Vortr. Pflanzenzüchtg.* **48**, 228–237.

SEKOWSKA, A. (1999), *Thesis.* Université de Versailles-Saint-Quentin-en-Yvelines.

SMIT, A. F. (1996), The origin of interspersed repeats in the human genome, *Curr. Opin. Genet. Dev.* **6**, 743–748.

STETTER, K. O., LAUERER, G., THOMM, M., NEUNER, A. (1987), Isolation of extremely thermophilic sulfate reducers: evidence for a novel branch of archaebacteria, *Science* **236**, 822–824.

STETTER, K. O. (1988), Archaeoglobus fulgidus gen. nov., sp. nov.: a new taxon of extremely thermophilic archaebacteria, *Syst. Appl. Microbiol.* **10**, 172–173.

STUCKA, R., FELDMANN, H. (1994), Cosmid cloning of Yeast DNA, in: *Molecular Genetics of Yeast – A Practical Approach* (JOHNSTON, J., Ed.), pp. 49–64. Oxford: Oxford University Press.

SULSTON, J. (1988), Cell lineage, in: *The Nematode Caenorhabditis elegans* (WOOD, W. B., Ed.), pp. 123-155. Cold Spring Harbor, N.Y.:CSHL Press.

SUZUKI, M., KAO, C. Y., MCCARTY, D. R. (1997), The conserved B3 domain of VIVIPAROUS1 has a cooperative DNA binding activity, *Plant Cell* **9**, 799–807.

TABOR, S., RICHARDSON, C. C. (1995), A single residue in DNA polymerases of the *Escherichia coli* DNA polymerase I family is critical for distinguishing between deoxy- and dideoxynucleotides, *Proc. Natl. Acad. Sci. USA* **92**, 6339–6343.

TAYLOR, A. L., THOMAN, M. S. (1964), The genetic map of *Escherichia coli* K-12, *Genetics* **50**, 659–677.

THIERRY, A., DUJON, B. (1992), Nested chromosomal fragmentation in yeast using the meganuclease *I-Sce* I: a new method for physical mapping of eukaryotic genomes, *Nucl. Acids Res.* **20**, 5625–5631.

THOMAS, G. H. (1999), Completing the *E. coli* proteome: a database of gene products characterised since completion of the genome sequence, *Bioinformatics* **15**, 860–861.

TOMB, J.-F., WHITE, O., KERLAVAGE, A. R., CLAYTON, R. A., SUTTON, G. G. et al. (1997), The complete genome sequence of the gastric pathogen *Heliobacter pylori*, *Nature* **388**, 539–547.

UNSELD, M., MARIENFELD, J. R., BRANDT, P., BRENNICKE, A. (1997), The mitochondrial genome of *Arabidopsis thaliana* contains 57 genes in 366,924 nucleotides, *Nature Genet.* **15**, 57–61.

VERAKSA, A., DEL CAMPO, M., MCGINNIS, W. (2000), Developmental patterning genes and their conserved functions: From model organisms to humans, *Mol. Genet. Metabolism* **69**, 85–100.

VORHOLT, J., KUNOW, J., STETTER, K. O., THAUER, R. K. (1995), Enzymes and coenzymes of the carbon monoxide dehaydrogenase pathway for autotrophic CO_2 fixation in *Archaeoglobus lithotrophicus* and the lack of carbom monoxide dehydrogenase in the heterotrophic *A. profundus*, *Arch. Microbiol.* **163**, 112–118.

WATERSTON, R., SULSTON, J. (1995), The genome of *Caenorhabditis elegans*, *Proc. Natl. Acad. Sci. USA* **92**, 10836–10840.

WICKY, C, VILLENEUVE, A. M., LAUPER, N., CODOUREY, L., TOBLER, H., MULLER, F. (1996), Telomeric repeats (TTAGGC) are sufficient for chromosome capping in *Caenorhabditis elegans*. *Proc. Natl. Acad. Sci. USA* **93**, 8983–8988.

WILSON, R., AINSCOUGH, R., ANDERSON, K., BAYNES, C., BERKS, M. et al. (1994), 2.2 Mb of contiguous nucleotide sequence from chromosome III of *C. elegans*, *Nature* **368**, 32–38.

WILSON, R. K. (1999), How the worm was won. The *C. elegans* genome sequencing project, *Trends Genet.* **15**, 51–58.

WOLFE, K. H., SHIELDS, D. C. (1997), Molecular evidence for an ancient duplication of the entire yeast genome, *Nature* **387**, 708–713.

YAMAMOTO, Y., AIBA, H., BABA, T., HAYASHI, K., INADA, T. et al. (1997), Construction of a contiguous 874-kb sequence of the *Escherichia coli* – K12 genome corresponding to 50.0–68.8 min on the linkage map and analysis of its sequence features, *DNA Res.* **4**, 91–113.

ZHANG, J. M., CHEN, L., KRAUSE, M., FIRE, A., PATERSON, B. M. (1999), Evolutionary conservation of MyoD function and differential utilization of E proteins, *Dev. Biol.* **208**, 465–472.

2 The Human Genome Project

Lap Chee Tsui

Steve W. Scherer

Toronto, Canada

1 Introduction

The Human Genome Project (HGP) is a worldwide research initiative with the goal of analyzing the complete sequence of human DNA to identify all of the genes. Further understanding of the structure and organization of genes will allow for a systematic analysis of their normal function and regulation in an organism. A comprehensive description of the human genome is thus the foundation of human biology and the essential prerequisite for an in-depth understanding of disease mechanisms. As such, information generated by the HGP will represent a source book for biomedical science in the 21st century. It will help scientists and clinicians alike to understand, diagnose, and eventually treat many of the 5,000 genetic diseases that afflict humankind, including the multifactorial diseases in which genetic predisposition plays an important role.

With the introduction of somatic cell technology, recombinant DNA, and polymorphic DNA markers, human geneticists began to dissect the human genome at the molecular level in the early 1980s. Due to the limited resources and knowledge base available, most efforts were concentrated only in scattered regions of chromosomes with known biological significance or disease relevance. Every project had to invest an enormous amount of effort in setting up the technology and recruiting the appropriate resources before any reasonable progress could be made. Also, it was realized that detailed characterization of chromosomes and isolation of disease genes were just the first steps required to approach the biological problems. It soon became apparent that a large-scale effort would be more efficient and cost-saving, if the ultimate goal was to identify all the molecular defects in diseases, as well as the structure and function of the genes in the human genome (COOK-DEEGAN, 1989; SINSHEIMER, 1989; DULBECCO, 1986). It was also clear that such a large-scale study would necessitate collaboration of research laboratories at the international level (CANTOR, 1990).

The international HGP was officially launched in the early 1990s. Its progress and general strategy can be described in three phases (outlined in Fig. 1). These include (1) the generation of chromosome maps, (2) large-scale DNA sequencing, and (3) annotating the DNA sequence (GUYER and COLLINS, 1995; COLLINS et al., 1998). Prior to the initiation of each of these phases as well as the HGP itself, significant advances in technology were necessary (Fig. 1). Moreover, due to these technological advances the international HGP has experienced changes in strategy. There were several genome-wide efforts to complete each of these stages, but to monitor the true progress it was often easiest to measure the level of completeness in chromosome units. Presently, almost all regions of chromosomes have high resolution maps, and a "working draft" DNA sequence of the human genome has been assembled. The DNA sequence of chromosomes 21 and 22 is complete (HATTORI et al., 2000; DUNHAM et al., 2000). The next five years of the HGP will involve annotating the DNA sequence. This will include completing the DNA sequence, characterizing all of the genes, identifying DNA sequence variations and mutations associated with disease, and collating all of the resulting data as a reference for future biomedical studies.

2 Chromosome Maps

There are two types of chromosome maps: physical maps and genetic maps. Physical mapping uses a variety of methods to assign genes and DNA markers to particular locations along a chromosome, so the actual distances between the genes (measured in nucleotide base pairs) are known (as discussed below, the finished DNA sequence is the highest resolution physical map, but to achieve an ordered sequence map lower resolution maps are first required). Genetic mapping describes the arrangement of genes based on the relationship of their linkage. DNA markers or probes can also be used in the construction of genetic maps, if they detect sequence changes (polymorphism) among different individuals. The tendency of two genes or DNA markers to segregate together through meiosis in family studies gives a description of genetic linkage, but

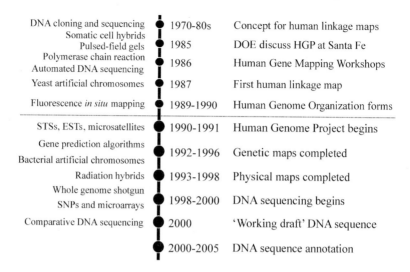

DNA cloning and sequencing	1970-80s	Concept for human linkage maps
Somatic cell hybrids		
Pulsed-field gels	1985	DOE discuss HGP at Santa Fe
Polymerase chain reaction	1986	Human Gene Mapping Workshops
Automated DNA sequencing		
Yeast artificial chromosomes	1987	First human linkage map
Fluorescence *in situ* mapping	1989-1990	Human Genome Organization forms
STSs, ESTs, microsatellites	1990-1991	Human Genome Project begins
Gene prediction algorithms		
Bacterial artificial chromosomes	1992-1996	Genetic maps completed
Radiation hybrids	1993-1998	Physical maps completed
Whole genome shotgun		
SNPs and microarrays	1998-2000	DNA sequencing begins
Comparative DNA sequencing	2000	'Working draft' DNA sequence
	2000-2005	DNA sequence annotation

Fig. 1. Timeline of the Human Genome Project (HGP). A historical summary of the enabling technologies (on the left) and the achievements (on the right) leading up to and including the HGP are shown. The formal international HGP began around 1990. Other aspects of the HGP are the study of model organisms, analysis of genome variation, and the development of bioinformatics. Establishing training and public education programs, as well as the study of the ethical, legal, and social issues of genetics research in society are also important priorities.

not their physical location. The order of genes on a chromosome measured by genetic linkage is the same as the order in physical maps, but there is no constant scale factor that relates physical and genetic distance. The variation in scale occurs because recombination does not occur at equal frequencies for different intervals along a chromosome. Since most of the genes and DNA markers used in the construction of genetic maps exist as cloned and sequenced DNA fragments, they can also be readily placed on a physical map.

2.1 Genetic Maps

Due to the initial interest in disease gene cloning, most of the early efforts (pre-HGP) in generating genetic maps were focussed on the isolation of genetic markers that could be used in linkage analysis for mapping of disease loci. Throughout the course of the HGP, the choice of genetic markers evolved from restriction fragment length polymorphisms (RFLPs) (KAN and DOZY, 1980; BOTSTEIN et al., 1980; WHITE et al., 1985; DONIS-KELLER et al., 1987), to variable number tandem repeats (VNTRs) (NAKAMURA et al., 1987), to microsatellites (length variation of simple di-, tri-, and tetranucleotide repeats) (WEBER and MAY, 1989; LITT and LUTY, 1989). By following the inheritance of marker alleles, linkage groups were constructed on each chromosome. The first example of a genome-wide effort was the establishment of the Centre d'Etude du Polymorphisme Humain (CEPH) in 1984 to produce reference maps with a set of well-defined pedigrees (DAUSSET, 1986). Since microsatellites are highly polymorphic (ideal for linkage analysis) and abundant in the genome, extensive efforts were devoted to the isolation of these markers and their use in the generation of genetic maps. Studies from Genethon and CEPH maps (WEISSENBACH et al., 1992; GYAPAY et al., 1994; DIB et al., 1996), the Collaborative (DONIS-KELLER et al., 1992) and Co-operative

Human Linkage Centers (BUETOW et al., 1994), and the Utah Marker Development Group (The Utah Marker Development Group, 1995), generated human genetic maps at the 1 centiMorgan (cM) resolution level. These genetic maps provided essential markers for constructing accurate physical maps of the chromosomes. Single nucleotide polymorphisms (SNPs) represent the most common polymorphism in the human genome occurring on average, once in every 1,000 base pairs (bp) (COLLINS et al., 1997). Their abundance and amenability for automated analysis by high-throughput technologies will allow genome-wide association studies to be conducted in projects aimed at identifying genes involved in multifactorial diseases. High-density SNP maps of human chromosomes are currently being constructed (KRUGLYAK, 1997; WANG et al., 1998).

2.2 Physical Maps

Early physical mapping studies were limited to the use of *in situ* hybridization of DNA probes to metaphase chromosome spreads and gel-blot hybridization analysis of DNA digested with various combinations of restriction enzymes. The introduction of pulsed field gel electrophoresis (PFGE) (SCHWARTZ and CANTOR, 1984) and fluorescence *in situ* hybridization (FISH) (TRASK et al., 1989; LAWRENCE, 1990; LICHTER et al., 1990; LAWRENCE et al., 1990; Pinkel et al. 1986) were major technology advances, but significant improvement of the mapping strategy only came after the availability of the yeast artificial chromosome (YAC) (BURKE et al., 1987) cloning system. YAC vectors are capable of propagating large fragments of human DNA insert 500,000 bp (500 kb) to 1 million bp (1 Mb) in size in yeast. Any genomic DNA fragments and cDNA (copies of messenger RNA) clones could be used to assemble overlapping clones (contigs) by hybridization screening. The introduction of sequence tagged sites (STSs), which are short DNA segments defined by their unique sequences, allowed the use of polymerase chain reaction (PCR) in contig assembly (known as anchor loci mapping or STS-content mapping; OLSON et al., 1989). STSs that marked genes

were called expressed sequence tags (ESTs) (ADAMS et al., 1991; OKUBO et al., 1992). DNA fingerprinting is another method of assembling overlapping cloned DNA molecules based on restriction fragment analysis or Southern blot hybridization patterns (KOHARA et al., 1987; COULSON et al., 1986; OLSON et al., 1986; BELLANNÉ-CHANTELOT, 1992).

Another powerful large-scale physical mapping technique is the use of radiation-hybrid (RH) mapping (COX et al., 1990). Human chromosome fragments were generated by X-ray irradiation of somatic cells and maintained in rodent cell backgrounds through the use of cell fusion techniques. Since only a portion of the human chromosomes is retained in each hybrid cell line, statistical methods could be used to assess physical linkage relationships between DNA markers and build linkage groups. With the inclusion of anchor loci (genes, genetic markers, STSs) in the mapping, it was possible to rapidly position unknown genes or DNA markers to specific chromosome regions resulting in the generation of useful RH maps of the human genome (WALTER et al., 1994).

Much of the early success of YAC-based contig mapping came from defined chromosome regions and some of the smaller chromosomes (GREEN and OLSON, 1990; CHUMAKOV et al., 1992; FOOTE et al., 1992). Considerable excitement was generated from the whole genome mapping work at CEPH-Genethon and subsequently the Whitehead Genome Center (COHEN et al., 1993; CHUMAKOV et al., 1995; HUDSON et al., 1995). The two groups generated YAC contig coverage initially estimated to be 75% and 95% complete for the entire genome, respectively. It became clear, however, that the level of coverage reported by both groups was probably overestimated. The overestimation was primarily caused by a paucity of DNA markers for some chromosomal regions and the high percentage (40–50%) of YAC clones in the libraries containing noncontiguous DNA sequences (chimeric clones) which were not readily detectable by the algorithms when used without additional analysis. In addition, there exist regions of human chromosomes not intrinsically amenable to cloning in yeast. Nevertheless, the data generated in both efforts was extremely valuable for disease gene identification studies and it provided an ordered scaffold of DNA markers which could be used in the DNA sequencing stage of the HGP (Fig. 2). The development of bacterial-based cloning vectors (BACs) (IOAN-NOU et al., 1994; SHIZUYA et al., 1992; see below) was important since the majority of DNA sequences that could not be cloned or mapped using YACs, could be analyzed using this system. Bacterial-based DNA molecules also provided the substrate for the clone-by-clone approach to sequence the human genome (Fig. 2).

Therefore, the genetic and physical mapping resources generated by the many different experimental strategies, from both chromosome-specific and whole-genome efforts, were complimentary and essential for the success of the HGP. The establishment of practical, accurate, and detailed maps of ordered DNA markers along each chromosome provided the framework for DNA sequence maps (see Sect. 3). Moreover, since many of these DNA markers and clones were used in biological studies, they remain useful for phase III of the HGP; annotating the DNA sequence. The earliest examples of integrated physical and genetic maps were for the smallest chromosomes, 21 (CHUMAKOV et al., 1995) and 22 (COLLINS et al., 1995). Beginning in 1973 and throughout the HGP, single chromosome data, maps, and DNA sequences were collated at the International Gene Mapping Workshops and Single Chromosome Workshops often sponsored by the Human Genome Organization (HUGO) (MCKUSICK, 1989). This data was submitted to relevant scientific databases such as the Genome Database and GenBank (see Tab. 1).

3 DNA Sequencing

The haploid human genome is estimated to contain approximately 3 billion bp of DNA sequence with the smallest chromosomes (21, 22, and Y) being about 40–50 Mb and the larger chromosomes (1 and 2) being 260 Mb; the average size chromosome is 130 Mb, approximately equal in size to the entire genome of the fruitfly *Drosophila melanogaster* (MYERS et al., 2000). If unwound and tied together, the strands of adenines, thymines, cytosines, and guanines (A,C,G,Ts) comprising human DNA, would stretch more than 5 feet but would be only 50 trillionths of an inch wide.

Due to the obvious enormity of the task at hand, it became clear that large-scale genomic DNA sequencing would be most economical when conducted in organized and automated facilities with substantial computerization and instrumentation (OLSON, 1995). There were 20 such centers involved in the international human sequencing consortium with the largest contributions coming from the Sanger Center in Cambridge, UK, Washington University in St. Louis, USA, and the Whitehead Institute in

Fig. 2. Systematic mapping and sequencing of human chromosomes using a clone-by-clone approach. DNA markers from a chromosomal region are positioned and simultaneously used to order large cloned DNA molecules (YACs and BACs). BACs are broken down in "sub-clones" for DNA sequencing. Obtaining partial sequence of the BACs represents "working draft" sequence. The DNA sequence is the highest resolution physical map.

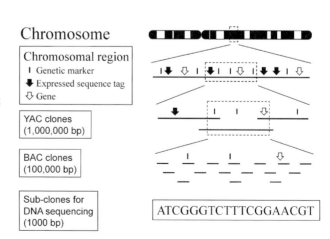

Boston, USA. While a number of sequencing technologies were being developed, one of the original methodologies described in the 1970s by SANGER and colleagues (SANGER et al., 1977) provided the basic chemistry (flourescence-based dideoxy sequencing) (HUNKAPILLER et al., 1991) performed in the HGP. The principles behind the Sanger method include using (1) an enzymatic procedure to synthesize DNA chains of varying lengths that terminate at either the A, T, C, or G nucleotides, and (2) separating the fragments on gels by electrophoresis to determine the identity and order of nucleotides based on the size of the fragment.

Due to limitations of DNA isolation and sequencing technologies, however, the ordered DNA molecules cloned into YAC contigs (which comprised the first-generation physical maps of the human genome) had to first be converted to smaller-size clones to prepare for genomic DNA sequence determination (see Fig. 2). Therefore, the sequencing protocols currently used are almost exclusively based on

using clones from existing bacterial-based cloning libraries (see below). If necessary, YAC clones could be subcloned directly in plasmid vectors for shotgun sequencing of regions, as was demonstrated in the *C. elegans* genome sequencing project (The *C. elegans* Sequencing Consortium, 1998). For the human genome project BAC cloning systems were primarily used. BACs are capable of carrying inserts in the 100–500 kb size range, with the libraries used for the HGP having inserts of about 150 kb.

Two general strategies were followed to yield a "working draft" DNA sequence of the human genome. These include using a systematic clone-by-clone approach (Fig. 2) and the whole-genome shotgun strategy (WEBER and MYERS, 1997; VENTER et al., 1998; Fig. 3). In addition to genomic DNA sequencing, many cDNA sequencing projects were initiated to capture information about protein coding genes.

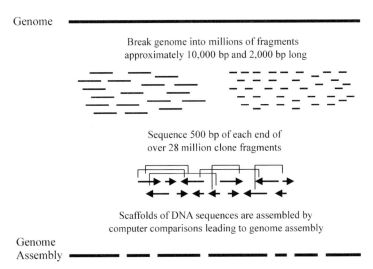

Fig. 3. Whole Genome Shotgun (WGS) DNA sequencing strategy of the human genome. The genome is broken into random DNA fragments that are cloned and sequenced from each end. Large scaffolds of DNA sequences can be assembled by identifying overlapping stretches of DNA sequence and these can be ordered and extended because both ends of the clone are known. This strategy bypasses the physical mapping stage used in the clone-by-clone approach (Fig. 2). However, for the human genome the assembled DNA sequences need to be positioned on chromosomes by comparison to DNA markers derived from well-characterized genetic and physical maps.

3.1 cDNA Sequencing

Prior to starting large-scale genomic sequencing projects, many initiatives were undertaken to capture (through DNA sequencing) those portions of the genome that are transcribed into genes. The reason for this is that only 5% of the DNA contained on a chromosome encodes for genes. These DNA sequences are called exons and the intervening genomic sequences are called introns (see Fig. 5, for example). By isolating messenger RNA (mRNA) from human tissues, the exon coding sequences were captured as cDNA allowing rapid isolation of putative gene sequences. The resulting cDNA clones were sequenced and ESTs for genes could be made. The first major public cDNA sequencing effort was initiated at the Institute for Genomic Research (TIGR) (ADAMS et al., 1995). Subsequently, the Integrated Molecular Analysis of Gene Expression (IMAGE) Consortium and the Merck–Washington University (WU) Initiative were formed (HILLIER et al., 1996). Since the introduction of the EST concept in 1991 the number of EST entries in public databases grew to over 2,800,000. These clones are derived from hundreds of different cDNA libraries constructed from most human tissues. Obviously, there is a substantial redundancy in these ESTs and they can be grouped into consensus groups or "Unigenes" based on sequence overlap. A consortium of investigators was formed and in one effort over 30,000 ESTs were mapped to defined chromosomal regions providing gene-based DNA markers for constructing YAC and BAC contigs (DELOUKAS et al., 1998). Despite these efforts, comprehensive gene identification, and in particular identifying cDNA covering the entire gene, almost always requires additional experimentation and verification. As part of the HGP, experiments for full-length gene identification and characterization through DNA sequencing have been undertaken the results of which will contribute substantially in annotating the genome sequence (STRAUSBERG et al., 1999).

3.2 Systematic Mapping and DNA Sequencing of Human Chromosomes

To provide the starting material for DNA sequencing of complex genomes, the standard approach relies on first building sequence-ready clone maps over regions ranging in size from hundreds of kilobases to entire chromosomes (see Fig. 2). As described in Sect. 2, the HGP BAC clone maps are assembled and ordered based on their DNA marker content as well as using fingerprinting techniques (GREGORY et al., 1997; MARRA et al., 1997) and by determining the DNA sequences of the ends of clones (VENTER et al., 1996). Additional mapping information on the chromosome location of the BACs along the chromosome could be determined using the clones directly as probes for hybridization analysis of chromosomes (see Fig. 5, for example). The BAC map of the human genome consists of >330,000 clones grouped into 1,743 contigs covering upwards of 95% of the genome.

For sequencing, 32,221 clones forming the minimal set covering the maximum region (a "tiling path") of each human chromosome was selected. Each clone then needs to broken down into smaller subclones for sequencing. Initially, the plan was to put the fragments to be sequenced in order, followed by complete sequence determination of each fragment in a systematic manner so that the entire human DNA sequence of the BAC (and, therefore, the corresponding region on the chromosome) was known. This method produces highly accurate sequence with few gaps. However, the upfront process of building the sequence-ready maps, subclone library construction, and directed gap filling is costly, time consuming, and, therefore, often rate-limiting.

To accelerate progress, in 1998 a major deviation from this plan was the decision to collect only partial data from each DNA fragment, hence, a "working" or "rough" draft (GUYER and COLLINS, 1995) (the change of strategy was partly due to the launching of a privately funded company, Celera, who decided to determine the DNA sequence of small pieces of human DNA totally at random using the whole genome shotgun approach described below). Working draft DNA sequence usually covers 95% of the BAC (maintaining 99% accuracy), but it is divided into 10–100 largely unordered segments. Additional sequencing is required to generate the finished DNA sequence such that there are no gaps or ambiguities and the final product is greater than 99.99% accurate. At the time of writing this chapter, close to 90% of the euchromatic human genome sequence was determined (approximately 75% of this was working draft and 25% in finished form). Results from the finished DNA sequence of chromosomes 21 and 22 are likely to be instructive for what a finished product of the human genome might look like. Chromosome 22q (an acrocentric chromosome) was determined to contain 33.5 Mb represented in 10 contigs. Although usually small, the gaps in the DNA sequence map could not be closed even after exhaustive attempts using multiple cloning systems (DUNHAM et al., 2000). For chromosome 21, 33.8 Mb of DNA sequence could be assembled and represented in three contigs with only a few detectable small gaps (HATTORI et al., 2000).

3.3 Whole Genome Shotgun

The whole genome shotgun strategy (WGS) involves shearing all of the DNA of an organism into segments of a defined length which are cloned into a plasmid vector for DNA sequencing (WEBER and MYERS, 1997; VENTER et al., 1998). Sufficient DNA sequencing is performed so that each nucleotide of DNA in the genome is covered numerous times in fragments of about 500 bp. After sequencing, the fragments are assembled to reconstruct the complete genome. WGS was used by scientists at The Institute for Genomic Research to generate the first complete sequence of a self-replicating organism called *Haemophilus influenzae* (FLEISCHMANN et al., 1995) as well as many other prokaryotic organisms. The advantage of WGS is that the upfront steps of constructing a physical map are not completed. For organisms with much larger and more complex genomes, such as *Drosophila melanogaster* and human, assemblies of sequences are expected to be complicated by the presence of a vast number of repetitive elements (approximately 50% of human DNA is repeats). Notwithstanding, VENTER and colleagues at Celera Corporation initiated a WGS project of the *Drosophila* genome and in doing so, 3.2 million sequence reads were completed (giving a 12.8X coverage of the 120 Mb genome). Based on this data 115 Mb of DNA sequence could be assembled and although it was quite comprehensive, the genome was still divided by 1,630 gaps (MYERS et al., 2000). Closure of the gaps is being completed using the scaffold of BACs generated from physical mapping projects.

Following the experience gained from the *Drosophila melanogaster* project, Celera Corporation planned a whole-genome assembly of human DNA (VENTER et al., 1998). The aim of the project was to produce highly accurate, ordered sequence spanning more than 99.9% of the human genome. Based on the size of the human genome and the results from the *Drosophila* experiment it was predicted that over 70 million sequencing reactions would need to be completed. This would be divided into sequencing both ends of 30 million 2 kb clones, 5 million 10 kb clones, and 300,000 150 kb clones. The alignment of the resulting sequence assemblies along the chromosomes would be accomplished using the large number of DNA markers and physical maps generated by the ongoing HGP (see Sect. 2). Since efforts were escalated by the publicly funded HGP using the clone-by-clone "working draft" approach, Celera could also easily integrate this data into their assemblies, thereby greatly reducing the amount of sequencing required. Moreover, the sequencing of the ends of the 150 kb clones (which were BACs) was completed by publicly funded efforts. In the end, to assemble sequence contigs, Celera completed approximately 28 million reads and merged this data with DNA sequence in public databases (Fig. 4). The final sequence contigs were ordered using the DNA markers and maps from the HGP (see Sect. 2).

The completeness and accuracy of the first draft of the human genome sequence will be tested by many types of experimentation over the next decade. Based on the results from chromosome 21 and 22 it is expected that there will be chromosome regions that will not be represented, since some DNA cannot be cloned

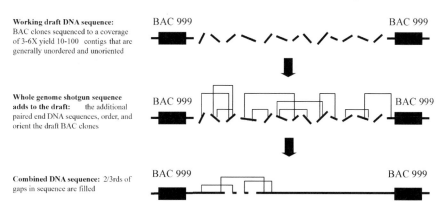

Working draft DNA sequence:
BAC clones sequenced to a coverage of 3-6X yield 10-100 contigs that are generally unordered and unoriented

Whole genome shotgun sequence adds to the draft: the additional paired end DNA sequences, order, and orient the draft BAC clones

Combined DNA sequence: 2/3rds of gaps in sequence are filled

Fig. 4. Working draft DNA sequence combined with the whole genome shotgun (WGS) strategy. The strategy of the publicly funded HGP using mapped clones (Fig. 2) occurs in two phases, (1) the "shotgun" stage which involves determining most of the sequence from a clone that is assembled into a product ("working draft") that contains gaps and ambiguities and (2) the "finishing" stage which involves additional directed sequencing for gap closure and resolution of ambiguities. DNA sequence data from the WGS strategy (see Fig. 3) can expedite the "finishing" stage since it can often extend contigs, fill gaps, and resolve orders of sequence scaffolds. Some WGS sequences could also reside in regions of the human difficult to clone using the existing vectors.

using currently available vectors. Also, incorrect assemblies of sequence will occur due to the presence of repeats and duplications. It will be interesting to determine if data derived from the WGS approach can close some of the gaps that could not be completed using directed chromosome walking and sequencing. An example of the genomic representation of working draft DNA sequence alone compared to the equivalent sequence combined with Celera WGS data for the q35 region of human chromosome 7 is shown in Fig. 5.

4 Annotation of the Genome DNA Sequence

A complete understanding of the biology and function of the genome will be the ultimate goal of the HGP. The DNA sequence will undergo continual refinement as new sequences and new types of biological data are added. With proper annotation of the DNA sequence, a full description of all the genes and other important biological information stored within the DNA will be known. Given this information it will be possible to investigate the roles of all of the gene products, how they are controlled and interact, and their possible involvement in disease. The process of annotating the human DNA sequence will take several forms including (1) cataloguing all of the genes, (2) identifying the genes and DNA sequence variations that either directly cause or are associated with disease, (3) studying genetic variation, and (4) establishing integrated and curated databases containing all DNA sequence annotation. A full description of the databases is beyond the context of this chapter. We have summarized and provided links to relevant www sites where initiatives are ongoing to establish integrated databases (Tab. 1). A vital next step will be to organize experts in different fields of genomics for proper curation of this information.

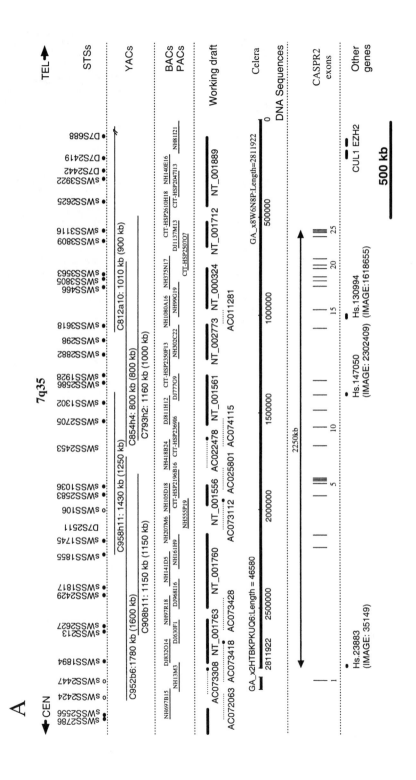

B

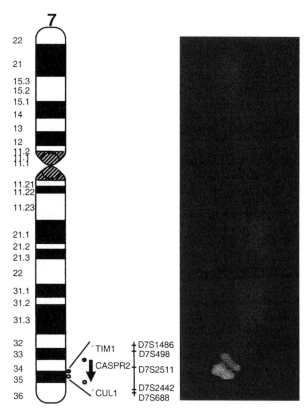

Fig. 5. Integrated physical, genetic and DNA sequence map of human chromosome 7q35. (**A**) The physical order of DNA markers (microsatellites, STSs, ESTs, and genes) as determined by their presence or absence in YAC and BAC clones, is shown. CASPR2 was determined to be one of the largest genes in the human genome spanning 2.3 Mb (K. NAKABAYASHI and S. W. SCHERER, unpublished data). The representation of "working draft" DNA sequence alone compared and combined with "whole genome shotgun" (Celera contigs) data is shown. (**B**) A BAC from exon 1 (pink) and another from exon 25 (green) of CASPR2 were hybridized using FISH to map the gene to chromosome 7q35 (see color plates p. 455).

4.1 Cataloguing the Genes

Just as difficult as constructing maps or determining the DNA sequence of the human genome will be to identify all of the genes it encodes. At the present there are two general approaches to gene finding. The first involves detecting, isolating, and analyzing transcripts of genes from genomic DNA. This can be accomplished by studying cDNAs, which are cloned copies of messenger RNA. An alternative approach involves computer-based analysis of DNA sequence to search for structural features of genes such as start sites for gene transcription, protein coding regions (coding exons), and transcription stop sites (UBERBACHER and MURAL, 1991; FICKETT, 1996; CLAVERIE, 1997). Part of this strategy can include comparing the DNA sequence of two organisms to search for evolutionarily conserved sequences to identify gene coding regions (O'BRIEN et al., 1999). A combination of both approaches and a finished human genome sequence will likely be necessary for the construction of an accurate gene map.

Tab. 1. Web Resources for the Human Genome Project (HGP)

Topic	Site	Description
General information on the Human Genome Project	*http://www.ornl.gov/TechResources/Human_Genome/home.html*	excellent general site maintained by the U.S. Department of Energy providing information and links on the history, general science, technical strategies, ethics, and applications of the HGP
	http://www.hgmp.mrc.ac.uk/GenomeWeb/	established by the United Kingdom HGP to provide many useful genome links
	http://www.kumc.edu/gec/	relevant links for educators, counselors, lawmakers, and lay people interested in genetics and the HGP
General and *scientific* genome news	*http://genomics.phrma.org/today/index.html*	daily links to wire stories from newspaper and broadcast www-sites and press releases
	http://www.geneletter.org/index.epl	"GeneLetter" analyzes topics about the ethical, legal, and social issues of genetics and the HGP
Scientific mapping, DNA sequence, and annotation data	*http://www.gdb.org/hugo/*	maintained by the Human Genome Organization chromosome editors. Excellent links for all chromosome-specific genome/disease information
	http://www.ncbi.nlm.nih.gov/ *http://www.ncbi.nlm.nih.gov/LocusLink/* *http://www.ncbi.nlm.nih.gov/omim/*	NCBI hosts GenBank and many other online databases of DNA and protein sequences with bibliographic information. "LocusLink" provides information on genes and "OMIM" on diseases
	http://bioinfo.weizmann.ac.il/cards/index.html *http://www.gdb.org/*	GeneCards and Genome Database catalog genes, their products and their involvement in diseases
Scientific bioinformatic tools	*http://www.hgsc.bcm.tmc.edu/SearchLauncher/*	The Baylor College of Medicine Launcher is an ongoing project to organize molecular biology-related search and analysis software
	http://www.embl-heidelberg.de/Services/index.html	European Molecular Biology Laboratory hosted software and sequence analysis; also see NCBI

It is difficult to predict with accuracy what the final gene map of the human genome might look like because our definition of a gene continues to change as our knowledge grows. The current definition used by the HUGO Gene Nomenclature Committee is deliberately vague: "A gene is a segment of DNA that contributes to phenotype/function. In the absence of demonstrated function a gene may be characterized by sequence, transcription, or homology to another gene" (WHITE et al., 1997). Most genes can be easily recognized by identifying the known characteristic features, but a growing number exhibit properties that add to the current definition. For example:

(1) Overlapping genes have been observed which share RNA transcripts that are in different reading frames (STOTT et al., 1998).

(2) Fusion genes have been found that are comprised of an mRNA producing one protein product, but are derived from coding regions on different chromosomes. Fusion genes can be due to translocation or they can arise from genes that may be widely separated on the same chromosome (MAGRANGEAS et al., 1998).

(3) Bicistronic genes can be formed, which arise from one segment of chromosomal DNA (locus), but produce two protein products from one stretch of mRNA (REISS et al., 1998, 1999).

(4) "Antisense" genes arise from one locus, but produce two mRNAs (and perhaps two products). In this case, one gene product is encoded on the "sense" strand of DNA and the other on the "anti-sense" DNA strand which is often involved in control of the "sense" gene (SEROUSSI et al., 1999).

(5) Genes that do not appear to encode protein products (pseudogenes) (MIGHELL et al., 1999).

(6) Imprinted genes which yield different products depending on the parental origin of the chromosome they reside on (some genes are only expressed off the maternally inherited chromosome and others off the paternally inherited chromosome) (TILGHMAN, 1999).

Interestingly, many unconventional RNA products (bicistronic, antisense, or non-coding) are imprinted. Further annotation will surely reveal other phenomena that will add to the repertoire of genes in the human genome.

The estimates of the number of genes in the human genome have ranged from 28,000 to over 150,000. Early estimates using EST sampling (FIELDS et al., 1994; BOGUSKI and SCHULER, 1995) and counting DNA rich in CpG sequences (that often demarcate genes) suggested 60,000–80,000 genes (ANTEQUERA and BIRD, 1994). Subsequent estimates using much larger EST datasets ranged from 35,000 to 120,000 genes (EWING and GREEN, 2000; CROLLIUS et al., 2000; LIANG et al., 2000). From more recent tallies, as determined by identifying evolutionarily conserved regions between species and by extrapolating the number of genes found on chromosomes 21 and 22 to the rest of the genome, a number of 30,000–40,000 genes arises (HATTORI et al., 2000; DUNHAM et al., 2000). The latter estimates seem to be lower than what might be expected based on what has been observed in other species. However, one of the most interesting observations to come out of genome sequencing is that gene number does not necessarily correlate with organism complexity or its evolutionary hierarchy. For example, the fruitfly *Drosophila melanogaster* has 13,000 genes (MYERS et al., 2000) and is comprised of 10 times more cells than the worm *C. elegans*, which has 19,000 genes (The *C. elegans* Sequencing Consortium, 1998).

4.2 Disease Gene Identification

The immediate medical application of human genome information is in identification and annotation of genes associated with disease, the pursuit of new diagnostics, and treatments for these diseases (Fig. 6). About 5,000 human diseases are known to have a genetic component and 1,000 disease-associated markers or genes have already been isolated over the years (ANTONARAKIS and MCKUSICK, 2000), some even prior to the HGP. Until the 1980s, the primary strategy for identifying human genetic disease genes was to focus on biochemical and physiological differences be-

Disease Gene Identification

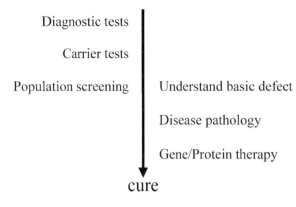

Diagnostic tests

Carrier tests

Population screening   Understand basic defect

Disease pathology

Gene/Protein therapy

cure

Fig. 6. Applications of the Human Genome Project. The immediate application of the human genome information are identification of genes associated with disease, and development of diagnostic and predictive tests. Based on this information the pursuit of new treatments for these diseases can be pursued.

tween normal and affected individuals. However, for the vast majority of inherited single gene disorders in human adequate biochemical information was either insufficient or too complicated to give insight into the basic biological defect. In the early 1980s an alternative strategy was introduced that suggested disease genes could be isolated solely on their chromosome location (ORKIN, 1986). This approach is now called "positional cloning" (TSUI and ESTIVILL, 1991; COLLINS, 1992; KING, 1997).

A disease gene can be localized to a particular chromosome by (1) identifying gross chromosome rearrangements, such as translocations, deletions, and duplications, which serve as markers for the position of the gene, or (2) by collecting pedigrees in which the responsible gene is segregating and performing linkage analysis or association studies with polymorphic markers. Prior to the initiation of the HGP, with the exception of the cystic fibrosis gene (ROMMENS et al., 1989; RIORDAN et al., 1989; KEREM et al., 1989), all disease discoveries relied on the identification of detectable rearrangements at the cytogenetic or DNA blot level. Notable targets of positional cloning of loci for heritable disorders identified by a cytogenetic rearrangement included; chronic granulomatous disease (ROYER-POKORA et al., 1986), Duchenne/Becker muscular dystrophy (MONACO et al., 1986; BURGHES et al., 1987),

and familial adenomatous polyposis (JOSLYN et al., 1991; GRODEN et al., 1991; KINZLER et al., 1991).

With the development of new mapping resources and technologies generated by the HGP, the ability to clone rearrangement breakpoints and map disease loci in pedigrees was greatly simplified. This accelerated the pace of discovery of new disease loci and the underlying mutational mechanisms. For example, trinucleotide repeats located within the genes for X-linked spinal and bulbular muscular atrophy (LASPADA et al., 1991), fragile X syndrome (VERKERK et al., 1991; FU et al., 1991; KREMER et al., 1991), and Huntington disease (The Huntington's Disease Collaborative Group, 1993) were found expanded in copy number in individuals with the disease. In developmental diseases such as aneuridia (FANTES et al., 1995) and holoprosencephaly (BELLONI et al., 1996) it was observed that mutations or chromosome rearrangement breakpoints distant to the coding regions could alter expression of the disease causing genes via position effect mutations. Other examples of medically relevant genes cloned using HGP information included those involved in inherited breast cancer (MIKI et al., 1994), early onset Alzheimer disease (SHERRINGTON et al., 1995), and the hereditary non-polyposis colorectal cancer (HNPCC) genes. For HNPCC, the observation that HNPCC tumors exhibited widespread alterations of microsatellites distributed throughout the genome suggested that mutations in genes involved in DNA mismatch repair may be involved in the disease. In agreement with this prediction, the human versions of the bacterial and yeast DNA repair genes *mut*S and *mut*L mapped to the same region where HNPCC had been localized in family studies, chromosomes 2p16 (LEACH et al., 1993; FISHEL et al., 1993) and 3p21 (PAPADOPOULOS et al., 1994), respectively.

After a decade of experience in positional cloning and the HGP sequence now well advanced, it becomes possible to dissect the molecular genetics of multifactorial diseases such as cancer and cardiovascular disease, which involve multiple combinations of genes and strong environmental components. The study of these common diseases will require careful patient and family data collection, thorough clinical examination, systematic DNA analysis, and complex statistical modeling as demonstrated in the description of the calpain-10 gene as a candidate for involvement in type II diabetes (HORIKAWA et al., 2000).

4.3 Human Genome Sequence Variation

Genetic variation is the basis of evolution. The haploid genomes inherited from each parent demonstrate differences between each other and it is this genetic variation that contributes, in part, to health or disease. There are many types of variation in the human genome with single nucleotide changes (polymorphisms) or SNPs being the most frequent occurring one in 1,000 nucleotides (COLLINS et al., 1997). Other DNA sequence variations include copy number changes, insertions, deletions, duplications, translocations, inversions, and more complex rearrangements, but these are not as frequent. When DNA sequence variants occur within genes or affect their expression disease can sometimes result. Technological advances, availability of genetic markers, and higher-resolution SNP maps generated by the HGP, have revolutionized the study of human genetic variation. The biomedical applications include identifying variations within specific genes that cause or predis-

pose to disease, finding gene–environment interactions that might have pharmacogenetic and toxologic implications, and identifying variations in immune response genes which will have implications for transplantation and vaccine development (ROSES, 2000). The study of genomic variability has also greatly improved our grasp of human history and migration, evolution of the human gene pool, and our basic understanding of genome evolution (OWENS and KING, 1999). For example, studies of DNA variation have revealed that the vast majority of genetic diversity (>80%) occurs between individuals in the same population even in small or geographically isolated groups. Moreover, most variation seems to predate the time when humans migrated out of Africa. Therefore, the concept of homogeneous groups (or races) having major biological differences is not consistent with genetic evidence.

5 Conclusion

From conceptualization through to chromosome mapping and DNA sequencing, each success of the HGP followed close behind advances in technology and implementation of new strategies, one building on the other. The result has been the generation of massive amounts of data describing DNA, genes, and chromosomes and how they influence our development in health and disease. Scientists will continue to work on the HGP with an emphasis on annotation of the DNA sequence to find new genes, determining the function of the gene products (functional genomics), and to apply all of this information to the study of common diseases. Due to significant advances in the international HGP over the last decade, we are also experiencing a knowledge explosion in human biology. We have observed unprecedented developments in disease research, accompanied with genetic diagnosis, and predictive testing. As a result, novel drugs, disease treatment, and therapies are being developed (Fig. 6). The HGP has also shed light on historical puzzles such as the roots and migrations of ancient peoples, historical demography of cultures, and human genetic diversity bringing anthropologists, geneticists, historians, and theologians to converse on the same topics. Perhaps more difficult than the genetic science will be the social implication of this knowledge. There are issues of genetic privacy and ownership of our genome. It will be the responsibility of all in society to ensure that there is proper education, adequate discussion of these topics, and sufficient consensus for the lawmakers to safeguard acceptable standards. Therefore, programs studying ethical, legal, and societal implications of genetic information will be just as important to continued success of the HGP as the science.

6 References

ADAMS, M. D., KELLEY, J. M., GOCAYNE, J. D., DUB-
NICK, M., POLYMEROPOULOS, M. H. et al. (1991),
Complementary DNA sequencing: Expressed sequence tags and human genome project, *Science*
252, 1651–1656.

ADAMS, M. D., KERLAVAGE, A. R., FLEISCHMANN, R.
D., FULDNER, R. A., BULT, C. J. et al. (1995), Initial
assessment of human gene diversity and expression patterns based upon 83 million nucleotides
of cDNA sequence, *Nature* **28**, 3–174.

ANTEQUERA, F., BIRD, A. (1994), Predicting the total
number of human genes, *Nature Genet.* **8**, 114.

ANTONARAKIS, S. E., MCKUSICK, V. A. (2000), OMIM
passes the 1,000-disease-gene mark, *Nature Genet.* **1**, 11.

BELLANNÉ-CHANTELOT, C., LACROIX, B., OUGEN, P.,
BILLAULT, A., BEAUFILS, S. et al. (1992), Mapping
the whole human genome by fingerprinting yeast
artificial chromosomes, *Cell* **70**, 1059–1068.

BELLONI, E., MUENKE, M., ROESSLER, E., TRAVERSO,
G., SIEGEL-BARTELT, J. et al. (1996), Identification
of Sonic hedgehog as a candidate gene responsible for holoprosencephaly, *Nature Genet.* **14**,
353–356.

BOGUSKI, M. S., SCHULER, G. D. (1995), ESTablishing
a human transcript map, *Nature Genet.* **10**,
369–371.

BOTSTEIN, D., WHITE, R. L., SKOLNICK, M., DAVIS, R.
W. (1980), Construction of a genetic linkage map
in man using restriction fragment length polymorphisms, *Am. J. Hum. Genet.* **32**, 314–331.

BUETOW, K., WEBER, J. L., LUDWIGSEN, S., SCHERP-
BIER-HEDDEMA, T., DUYK, G. M. et al. (1994), Integrated human genome-wide maps constructed

using the CEPH reference panel, *Nature Genet.* **6**,
391–393.

BURGHES, A. H. M., LOGAN, C., HU, X., BELFALL, B.,
WORTON, R. G., RAY, P. N. (1987), A cDNA from
the Duchenne/Becker muscular dystrophy gene,
Nature **328**, 434–437.

BURKE, D. T., CARLE, G. F., OLSON, M. V. (1987),
Cloning of large segments of exogenous DNA into yeast by means of yeast artificial chromosome
vectors, *Science* **236**, 806–812.

CANTOR, C. (1990), Orchestrating the human genome project, *Science* **248**, 49–51.

CHUMAKOV, I., RIGAULT, P., GUILLOU, S., OUGEN, P.,
BILLAUT, A. et al. (1992), Continuum of overlapping clones spanning the entire human chromosome 21q, *Nature* **359**, 380–387.

CHUMAKOV, I. M., RIGAULT, P., LE GALL, I., BALLAN-
NE-CHANTELOT, C., BILLAULT, A. et al. (1995), A
YAC contig map of the human genome, *Nature*
377 (Suppl.), 175–297.

CLAVERIE, J.-M. (1997), Computational methods for
the identification of genes in vertebrate genomic
sequences, *Hum. Mol. Genet.* **6**, 1735–1744.

COHEN, D., CHUMAKOV, I., WEISSENBACH, J. (1993), A
first generation physical map of the human genome, *Nature* **366**, 698–701.

COLLINS, F. S., PATRINOS, A., CHAKRAVARTI, A.,
GESTELAND, R., WALTERS, L. (1998), New goals
for the U.S. human genome project: 1998–2003,
Science **282**, 682–689.

COLLINS, F. S., GUYER, M. S., CHARKRAVARTI, A.
(1997), Variations on a theme: cataloging human
DNA sequence variation, *Science* **278**, 1580–1581.

COLLINS, J. E., COLE, C. G., SMINK, L. J., GARRETT, C.
L., LEVERSHA, M. A. et al. (1995), A high-density
YAC contig map of human chromosome 22, *Nature* **377**, 367–379.

COLLINS, F. S. (1992), Positional cloning: Let's not call
it reverse anymore, *Nature Genet.* **1**, 3–6.

COOK-DEEGAN, R. M. (1989), The Alta Summit, *Genomics* **5**, 661–663.

COULSON, A., SULSTON, J., BRENNER, S., KARN, J.
(1986), Towards a physical map of the nematode
genome *Caenorhabditis elegans, Proc. Natl. Acad.
Sci. USA* **83**, 7821–7825.

COX, D. R., BURMEISTER, M., PRICE, E. R., KIM, S.,
MYERS, R. M. (1990), Radiation hybrid mapping:
A somatic cell genetic method for constructing
high-resolution maps of mammalian chromosomes, *Science* **250**, 245–250.

CROLLIUS, H. R., JAILLON, O., BERNOT, A., DASILVA,
C., BOUNEAU, L. et al. (2000), Estimate of human
gene number provided by genome-wide analysis
using *Tetraodon nigroviridis* DNA sequence, *Nature Genetics* **25**, 235–238.

DAUSSET, J. (1986), Le centre d'étude du polymorphisme humain, *Press Med.* **15**, 1801–1802.

DELOUKAS, P., SCHULER, G. D., GYAPAY, G., BEASLEY,

E. M., SODERLUND, C. et al. (1998), A physical map of 30,000 human genes, *Science* **282**, 744–746.

DIB, C., FAURE, S., FIZAMES, C., SAMSON, D., DROUOT, N. et al. (1996), A comprehensive genetic map of the human genome based on 5,264 microsatellites, *Nature* **380**, 152–154.

DONIS-KELLER, H., GREEN, P., HELMS, C., CARTIN-HOUR, S., WEIFENBACH, B. et al. (1987), A genetic linkage map of the human genome, *Cell* **51**, 319–337.

DONIS-KELLER, H., GREEN, P., HELMS, C., CARTIN-HOUR, S., WEIFFENBACH, B. et al. (1992), A genetic linkage map of the human genome, *Cell* **51**, 319; NIH/CEPH Collaborative Mapping Group (1992), A comprehensive genetic linkage map of the human genome, *Science* **258**, 67–86.

DULBECCO, R. (1986), A turning point in cancer research: Sequencing the human genome, *Science* **231**, 1055–1056.

DUNHAM, I. N., SHIMIZU, N., ROE, B., CHISSOE, S. et al. (2000), The DNA sequence of human chromosome 22, *Nature* **404**, 904–920.

EWING, B., GREEN, P. (2000), Analysis of expressed sequence tags indicates 35,000 human genes, *Nature Genet.* **25**, 232–234.

FANTES, J., REDEKER, B., BREEN, M., BOYLE, S., BROWN, J. et al. (1995), Aniridia-associated cytogenetic rearrangements suggest that a position effect may cause the mutant phenotype, *Hum. Mol. Genet.* **4**, 415–422.

FICKETT, J. W. (1996), Finding genes by computer: the state of the art, *Trends Genet.* **12**, 316–320.

FIELDS, C., ADAMS, M. D., WHITE, O., VENTER, J. C. (1994), How many genes in the genome? *Nature Genet.* **7**, 345–346.

FISHEL, R., LESCOE, M., RAO, M. R. S., COPELAND, N. G., JENKINS, N. A. et al. (1993), The human mutator gene homolog *MSH2* and its association with hereditary nonpolyposis colon cancer, *Cell* **75**, 1027–1038.

FLEISCHMANN, R. D., ADAMS, M. D., WHITE, O., CLAYTON, R. A., KIRKNESS, E. F. et al. (1995), Whole-genome random sequencing and assembly of *Haemophilus influenzae* Rd, *Science* **269**, 496–512.

FOOTE, S., VOLLRATH, D., HILTON, A., PAGE, D. (1992), The human Y chromosome overlapping DNA clones spanning the euchromatic region, *Science* **258**, 60–66.

FU, Y.-H., KUHL, D. P. A., PIZZUTI, A., PIERETTI, M., SUTCLIFFE, J. S. et al. (1991), Variation of the CGG repeat at the fragile X site results in genetic instability: resolution of the Sherman paradox, *Cell* **67**, 1047–1058.

GREEN, E. D., OLSON, M. V. (1990), Chromosomal region of the cystic fibrosis gene in yeast artificial chromosomes: A model for human genome mapping, *Science* **250**, 94–98.

GREGORY, S. G., HOWELL, G. R., BENTLEY, D. R. (1997), Genome mapping by fluorescent fingerprinting, *Genome Res.* **12**, 1162–1168.

GRODEN, J., THLIVERIS, A., SAMOWITZ, W., CARLSON, M., GELBERT, L. et al. (1991), Identification and characterization of the familial adenomatous polyposis coli gene, *Cell* **66**, 589–600.

GUYER, M. S., COLLINS, F. S. (1995), How is the Human Genome Project doing, and what have we learned so far? *Proc. Natl. Acad. Sci. USA* **92**, 10841–10848.

GYAPAY, G., MORISSETTE, J., VIGNAL, A., DIB, C., FIZAMES, C. et al. (1994), The 1993–94 Généthon human genetic linkage map, *Nature Genet.* Special Issue. **7**, 246–339.

HATTORI, M., FUJIYAMA, A., TAYLOR, T. D. et al. (2000), The DNA sequence of human chromosome 21, *Nature* **405**, 311–319.

HILLIER, L. D., LENNON, G., BECKER, M., BONALDO, M. F., CHIAPELLI, B. et al. (1996), Generation and analysis of 280000 human expressed sequence tags, *Genome Res.* **6**, 807–828.

HORIKAWA, Y., ODA, N., COX, N. J., LI, X., ORHO-MELANDER, M. et al. (2000), Genetic variation in the gene encoding calpain-10 is associated with type 2 diabetes mellitus, *Nature Genet.* **2**, 163–175.

HUDSON, T. J., STEIN, L. D., GERETY, S. S., MA, J., CASTLE, A. B., SILVA, J. et al. (1995), An STS-based map of the human genome, *Science* **270**, 1945–1954.

HUNKAPILLER, T., KAISER, R. J., KOOP, B. F., HOOD, L. (1991), Large scale and automated DNA sequence determination, *Science* **254**, 59–67.

IOANNOU, P. A., AMEMIYA, C. T., GARNES, J., KROISEL, P. M., SHIZUYA, H. et al. (1994), A new bacteriophage P1-derived vector for the propagation of large human DNA fragments, *Nature Genet.* **6**, 84–89.

JOSLYN, G., CARLSON, M., THLIVERIS, A., ALBERTSEN, H., GELBERT, L. et al. (1991), Identification of deletion mutations and three new genes at the familial polyposis locus, *Cell* **66**, 601–609.

KAN, Y. W., DOZY, A. M. (1980), Evolution of the hemoglobin S and C genes in the world populations, *Science* **209**, 388–391.

KEREM, B., ROMMENS, J. M., BUCHANAN, J. A., MARKIEWICZ, D., COX, T. et al. (1989), Identification of the cystic fibrosis gene: genetic analysis, *Science* **245**, 1073–1080.

KING, M. C. (1997), Leaving Kansas … finding genes in 1997, *Nature Genet.* **15**, 8–10.

KINZLER, K. W., NILBERT, M. C., SU, L.-K., VOGELSTEIN, B., BRYAN, D. B. et al. (1991), Identification of the FAP locus gene from chromosome 5q21, *Science* **245**, 661–665.

KOHARA, Y., AKIYAMA, K., ISONO, K. (1987), The physical map of the whole *E. coli* chromosome: Application of a new strategy for rapid analysis and sorting of a large genomic library, *Cell* **50**, 495–508.

KREMER, E. J., PRITCHARD, M., LYNCH, M., YU, S., HOLMAN, K. et al. (1991), Mapping of DNA instability at the fragile X to a trinucleotide repeat sequence p(CCG)n, *Science* **252**, 1711–1174.

KRUGLYAK, L. (1997), The use of a genetic map of bi-allelic markers in linkage studies, *Nature Genet.* **17**, 21–24.

LASPADA, A. R., WILSON, E. M., LUBAHN, D. B., HARDING, A. E., FISHBECK, H. (1991), Androgen receptor gene mutations in X-linked spinal and bulbar muscular atrophy, *Nature* **352**, 77–79.

LAWRENCE, J. B. (1990), A fluorescence *in situ* hybridization approach for gene mapping and the study of nuclear organization, in: *Genome Analysis* Vol. 1, pp. 1–38. Cold Spring Harbor Laboratory Press.

LAWRENCE, J. B., SINGER, R. H., MCNEIL, J. A. (1990), Interphase and metaphase resolution of different distances within the human dystrophin gene, *Science* **249**, 928–932.

LEACH, F., NICOLAIDES, N. C., PAPADOPOULOS, N., LIU, B., JEN, J. et al. (1993), Mutation of a *mut*S homolog in hereditary nonpolyposis colorectal cancer, *Cell* **75**, 1215–1225.

LIANG, F., HOLT, I., PERTEA, G., KARAMYCHEVA, S., SALZBERG, S. L., QUACKENBUSH, J. (2000), Gene Index analysis of the human genome estimates approximately 120,000 genes, *Nature Genet.* **25**, 239–240.

LICHTER, P., LEDBETTER, S. A., LEDBETTER, D. H., WARD, D. C. (1990), Fluorescence *in situ* hybridization with Alu and L1 polymerase chain reaction probes for rapid characterization of human chromosomes in hybrid cell lines, *Proc. Natl. Acad. Sci. USA* **87**, 6634–6638.

LITT, M., LUTY, J. A. (1989), A hypervariable microsatellite revealed by *in vitro* amplification of a dinucleotide repeat within the cardiac muscle actin gene, *Am. J. Hum. Genet.* **44**, 397–401.

MAGRANGEAS, F., PITIOT, G., DUBOIS, S., BRAGADO-NILSSON, E., CHEREL, M. et al. (1998), Cotranscription and intergenic splicing of human galactose-1-phosphate uridylyltransferase and interleukin-11 receptor alpha-chain genes generate a fusion mRNA in normal cells. Implication for the production of multidomain proteins during evolution, *J. Biol. Chem.* **26**, 16005–16010.

MARRA, M. A., KUCABA, T. A., DIETRICH, N. L., GREEN, E. D., BROWNSTEIN, B. et al. (1997), High throughput fingerprint analysis of large-insert clones, *Genome Res.* **11**, 1072–1084.

MCKUSICK, V. A., (1989), HUGO news. The Human Genome Organisation: history, purposes, and membership, *Genomics* **5**, 385–387.

MIGHELL, A. J., SMITH, N. R., ROBINSON, P. A., MARKHAM, A. F. (1999), Vertebrate pseudogenes, *FEBS Lett.* **468**, 109–114.

MIKI, Y., SWENSEN, J., SHATTUCK-EIDENS, D., FUTREAL, P. A., HARSHMAN, K. et al. (1994), A strong candidate for the breast and ovarian cancer susceptibility gene BRCA1, *Science* **266**, 66–71.

MONACO, A. P., NEVE, R. L., COLLETTI-FEENER, C., BERTELSON, C. L., KUNIT, D. M., KUNKEL, L. M. (1986), Isolation of candidate cDNA's for portions of the Duchenne muscular dystrophy gene, *Nature* **323**, 646–650.

MYERS, E. W., SUTTON, G. G., DELCHER, A. L., DEW, I. M., FASULO, D. P. et al. (2000), A whole-genome assembly of *Drosophila*, *Science* **287**, 2196–2204.

NAKAMURA, Y., LEPPERT, M., O'CONNELL, P., WOLFF, R., HOLM, T. et al. (1987), Variable number of tandem repeat (VNTR) markers for human gene mapping, *Science* **235**, 1616–1622.

O'BRIEN, S. J., MENOTTI-RAYMOND, M., MURPHY, W. J., NASH, W. G., WIENBERG, J. et al. (1999), The promise of comparative genomics in mammals, *Science* **286**, 458–462.

OKUBO, K., HORI, N., MATOBA, R., NIIYAMA, T., FUKUSHIMA, A. et al. (1992), Large Scale cDNA sequencing for analysis of quantitative and qualitative aspects of gene expression, *Nature Genet.* **2**, 173–179.

OLSON, M. V. (1995), A time to sequence, *Science* **270**, 394–396.

OLSON, M. V., DUTCHIK, J. E., GRAHAM, M. Y., GARRETT, M., BRODEUR, G. et al. (1986), Random-clone strategy for genomic restriction mapping in yeast, *Proc. Natl. Acad. Sci. USA* **83**, 7826–7830.

OLSON, M. V., HOOD, L., CANTOR, C., BOTSTEIN, D. (1989), A common language for physical mapping of the human genome, *Science* **245**, 1434–1435.

ORKIN, S. H. (1986), Reverse genetics and human disease, *Cell* **47**, 845–850.

OWENS, K., KING, M.-C. (1999), Genomic views on human history, *Science* **286**, 451–455.

PAPADOPOULOS, N., NICOLAIDES, N. C., WEI, Y.-F., RUBEN, S. M., CARTER, K. C. et al. (1994), Mutation of a *mut*L homolog in hereditary colon cancer, *Science* **263**, 1625–1629.

PINKEL, D., STRAUME, T., GRAY, J. W. (1986), Cytogenetic analysis using quantitative, high-sensitivity, fluorescence hybridization, *Proc. Natl. Acad. Sci. USA* **83**, 2934–2938.

REISS, J., COHEN, N., DORCHE, C., MANDEL, H., MENDEL, R. R. et al. (1998), Mutations in a polycistronic nuclear gene associated with molybdenum cofactor deficiency, *Nature Genet.* **1**, 51–53.

REISS, J., DORCHE, C., STALLMEYER, B., MENDEL, R. R., COHEN, N., ZABOT, M. T. (1999), Human molybdopterin synthase gene: genomic structure and mutations in molybdenum cofactor deficiency type B, *Am. J. Hum. Genet.* **3**, 706–711.

RIORDAN, J. R., ROMMENS, J., KEREM, B., ALON, N., ROZMAHEL, R. et al. (1989), Identification of the cystic fibrosis gene: cloning and characterization of complementary DNA, *Science* **245**, 1066–1073.

ROMMENS, J. M., IANNUZZI, M. C., KEREM, B., DRUMM, M. L., MELMER, G. et al. (1989), Identification of the cystic fibrosis gene: Chromosome walking and jumping, *Science* **245**, 1059–1065.

ROSES, A. D. (2000), Pharmacogenetics and the practice of medicine, *Nature* **405**, 857–865.

ROYER-POKORA, B., KUNKEL, L. M., MONACO, A. P., GOFF, S. C., NEWBURGER, P. E. et al. (1986), Cloning the gene for an inherited human disorder – chronic granulomatous disease – on the basis of its chromosomal location, *Nature* **322**, 32–38.

SANGER, F., NICKLEN, S., COULSON, A. R. (1977), DNA sequencing with chain-terminating inhibitors, *Proc. Natl. Acad. Sci. USA* **74**, 5463–5467.

SCHWARTZ, D. C., CANTOR, C. R. (1984), Separation of yeast chromosome-sized DNA's by pulsed-field gradient electrophoresis, *Cell* **37**, 67–75.

SEROUSSI, E., KEDRA, D., PAN, H. Q., PEYRARD, M., SCHWARTZ, C. et al. (1999), Duplications on human chromosome 22 reveal a novel Ret Finger Protein-like gene family with sense and endogenous antisense transcripts, *Genome Res.* **9**, 803–814.

SHERRINGTON, R., ROGAEV, E. I., LIANG, Y., ROGAEVA, E. V., LEVESQUE, G. et al. (1995), Cloning of a gene bearing a missense mutations in early-onset familial Alzheimer's disease, *Nature* **375**, 754–760.

SHIZUYA, H., BIRREN, B., KIM, U.-J., MANCINO, V., SLEPAK, T. et al. (1992), Cloning and stable maintenance of 300-kilobase-pair fragments of human DNA in *Escherichia coli* using an F-factor-based vector, *Proc. Natl. Acad. Sci. USA* **89**, 8794–8797.

SINSHEIMER, R. L. (1989), The Santa Cruz Workshop, *Genomics* **5**, 954–956.

STOTT, F. J., BATES, S., JAMES, M. C., McCONNELL, B. B., STARBORG, M. et al. (1998), The alternative product from the human CDKN2A locus, p14(ARF), participates in a regulatory feedback loop with p53 and MDM2, *EMBO J.* **17**, 5001–5014.

STRAUSBERG, R. L., FEINGOLD, E. A., KLAUSNER, R. D., COLLINS, F. S. (1999), The mammalian gene collection, *Science* **286**, 455–457.

The *C. elegans* Sequencing Consortium (1998), Genome sequence of the nematode *C. elegans*: a platform for investigating biology, *Science* **282**, 2012–2018.

The Huntington's Disease Research Collaborative Group (1993), A novel gene containing a trinucleotide repeat is expanded and unstable on Huntington's disease chromosomes, *Cell* **72**, 971–983.

The Utah Marker Development Group (1995), A collection of ordered tetranucleotide repeat markers rom the human genome, *Am. J. Hum. Genet.*

57, 619–628.

TILGHMAN, S. M. (1999), The sins of the fathers and mothers: genomic imprinting in mammalian development, *Cell* **2**, 185–193.

TRASK, B., PINKEL, D., VANDEN ENGH, G. (1989), The proximity of DNA sequences in interphase cell nuclei is correlated to genomic distance and permits ordering of cosmids spanning 250 kilobase pairs, *Genomics* **5**, 710–717.

TSUI, L.-C., ESTIVILL, X. (1991), Identification of disease genes on the basis of chromosomal localization, *Genome Anal.* **3**, 1–36.

UBERBACHER, E. E., MURAL, R. J. (1991), Locating protein-coding regions in human DNA sequence by a multiple sensor-neural network approach, *Proc. Natl. Acad. Sci. USA* **88**, 11261–11265.

VENTER, J. C., ADAMS, M. D., SUTTON, G. G., KERLAVAGE, A. R., SMITH, H. O., HUNKAPILLER, M. (1998), Shotgun sequencing of the human genome. *Science* **280**, 1540–1542.

VENTER, J. C., SMITH, H. O., HOOD, L. (1996), A new strategy for genome sequencing, *Nature* **381**, 364–366.

VERKERK, A. J. M. H., PIERETTI, M., SUTCLIFFE, J. S., FU, Y.-H., KUHL, D. P. A. et al. (1991), Identification of a gene (*FMR-1*) containing a CGG repeat coincident with a breakpoint cluster region exhibiting length variation in fragile X syndrome, *Cell* **65**, 905–914.

WALTER, M. A., SPILLETT, D. J., THOMAS, P., WEISSENBACH, J., GOODFELLOW, P. N. (1994), A method for constructing radiation hybrid maps of whole genomes, *Nature Genetics* **7**, 22–28.

WANG, D. G., FAN, J. B., SIAO, C. J., BERNO, A., YOUNG, P. et al. (1998), Large-scale identification, mapping, and genotyping of single-nucleotide polymorphisms in the human genome, *Science* **15**, 1077–1082.

WEBER, J. L., MAY, P. E. (1989), Abundant class of human DNA polymorphisms which can be typed using the polymerase chain reaction, *Am. J. Hum. Genet.* **44**, 388–396.

WEBER, J. L., MYERS, E. W. (1997), Human whole-genome shotgun sequencing, *Genome Res.* **7**, 401–409.

WEISSENBACH, J., GYAPAY, G., DIB, C., VIGNAL, A., MORISSETTE, J. et al. (1992), A second generation linkage map of the human genome, *Nature* **359**, 794–801.

WHITE, R., LEPPERT, M., BISHOP, D. T., BARKER, D., BERKOWITZ, J. et al. (1985), Construction of linkage maps with DNA markers for human chromosomes, *Nature* **313**, 101–105.

WHITE, J. A., McALPINE, P. J., ANTONARAKIS, S., CANN, H., EPPIG, J. T. et al. (1997), Guidelines for human gene nomenclature (1997). HUGO Nomenclature Committee, *Genomics* **15**, 468–471.

3 Genetic Disease

GRAHAM DELLAIRE

Edinburgh, UK

1 Introduction

Genes are units of heredity which provide the blueprint for our physical body, determining not only how long we may live, but also the quality of that life. The extent and quality of life can be drastically altered by disease, and genetic disease is perhaps the purest illustration of the relationship between our genes and our health. In the case of a rare genetic disease such as cystic fibrosis, the inheritance of two mutated copies of a single gene, the cystic fibrosis transmembrane conductance regulator (CFTR) (ROMMENS et al., 1989), directly predisposes the individual to the disease. For common diseases, such as heart disease or cancer the situation is less clear.

Environment and lifestyle play essential roles in disease susceptibility, yet a genetic component may underpin most, if not all disease. Infectious disease for example, requires the host to be susceptible to infection by the disease agent. This susceptibility can be affected by diet, and the presence of the disease agent is most certainly environmental. As well, the expression of this susceptibility may rely on the presence or absence of certain alleles for genes that either predispose an individual to infection or provide some degree of resistance to a particular infectious agent. For example, mutant alleles of the CFTR gene cause cystic fibrosis and predispose an individual to lung infections when present in two copies, but actually provide resistance to diarrheal diseases such as typhoid fever (JOSEFSON, 1998), when present in only one copy. This concept can also be extended to the genetic susceptibility to carcinogens in regard to cancer or to a predisposition for high-serum cholesterol in regard to coronary heart disease. Thus, all disease may have some genetic basis and the penetrance of that disease phenotype is modulated by a potent combination of environment, lifestyle, and the particular compliment of genes an individual has inherited.

The Human Genome Project promises to revolutionize medicine by providing the sequence of every human gene. It is hoped that this information will help determine the cause of many common diseases whose genetic components have been previously recalcitrant to dissection. The following chapter provides an overview of genetic disease and aims to aquaint the reader with the various classes of disease, the methods of determining their aetiology and common strategies for their treatment or management. The chapter concludes with observations on the current and future roles of academia and industry in genetic disease research.

1.1 The Spectrum of Genetic Disease

Usually, when one thinks of genetic disease, simple Mendelian disorders come to mind such as cystic fibrosis or sickle cell anemia (Tab. 1). These disorders are considered monogenic, in which predisposition for the disease is directly associated with the presence of a single gene allele. Unfortunately, such disorders are more the rarity than the rule. The reality is that most common genetic diseases either have many genes that together predispose the individual to disease (i.e., polygenic) or result from the action of a small number of genes, each contributing moderately to the overall phenotype (i.e., oligogenic). It is these polygenic, complex diseases that have the greatest impact on the human population. The leading causes of death in the Western world are cancer and heart disease, and the various subclasses within these disease categories have oligogenic and polygenic components. In addition, chromosomal disorders caused by changes in chromosome number or by chromosomal rearrangements, provide a further class of genetic disease that often resemble complex disorders (i.e., Down syndrome). The last classification of genetic disease, the mitochondrial disorders, result from mutations and rearrangements in the mitochondrial genome, and are perhaps the rarest of all. Clearly, the treatment and dissection of genetic disorders, in the search for therapies, must reflect the diversity of the various classes of genetic disease. The promise of the Human Genome Project is that the lessons learned from the study of monogenic and chromosomal disorders when combined with the sequencing of the entire human genome, will one day aid in the elucidation of the full spectrum of genetic disease mechanisms.

Tab. 1. Important Monogenic Disorders (adapted from Jorde et al., 1999)

Disorder	Gene(s)	Mode of Inheritance	Population Incidence
Ademomatous polyposis coli	APC	AD	1/6,000
Adult polycistic kidney disease	PKD1 and 2	AD	1/1,000
α-1-Antitrypsin deficiency	AAT	AD	1/1,000
Cystic fibrosis	CFTR	AR	1/2,500 to 1 /10,000 Caucasians
Duchenne muscular dystrophy	Dystrophin	XLR	1/3,500 males
Familial hypercholesterolemia	LDLR	AD	1/500
Familial polyposis coli	APC	AD	1/8,000
Fragile X syndrome	FMR1	XLD reduced penetrance	1/1,500 males; 1/2,500 females
Friedreich's ataxia	Frataxin, X25	AR	1/50,000
Hemophilia A	Factor XIII	XLR	1/10,000 males
Hereditary nonpolyosis colorectal cancer	MLH1, MSH2, MSH6, PMS1 and 2	AD	1/200
Huntington disease	Huntingtin	AD reduced penetrance	1/20,000 Caucasians
Marfan syndrome	Fibrillin-	AD	1/10,000 to 1/20,000
Myotonic dystrophy	DMPK	AD	1/7,000 to 1/20,000 Caucasians
Neurofibromatosis type 1	NF1	AD	1/3,000 to 1/5,000
Osteogenesis imperfecta	COL1A1 and COL1A2	AD	1/5,000 to 1/10,000
Phenylketonuria	PAH	AR	1/10,000 to 1/15,000 Caucasians
Retinoblastoma	Retinblastoma	AD reduced penetrance	1/20,000
Sickle cell anaemia	Beta globin	AR	1/400 to 1/600 African-Americans (up to 1/50 in Central Africa)
Tay–Sachs disease	HEX-A	AR	1/3,000 Ashkenazi Jews

AD: autosomal dominant; AR: autosomal recessive; XLD: X-linked dominant; XLR: X-linked recessive.

1.1.1 Rare Genetic Diseases are often Mendelian and Monogenic

Mendelian traits adhere to two main principles outlined by Gregor Mendel, the 19th century Augustinian monk and father of modern genetics. These two principles are: the principle of segregation, which states that sexually reproducing organisms have genes in pairs and transmit only one member of this pair to their offspring; and the principle of independent assortment, which states that genes at different loci are transmitted independently (reviewed in Jorde et al., 1999). Mendel also provided geneticists with the useful definitions of dominant and recessive to describe the link between genotype and phenotype. Dominant alleles produce a phenotype when present as a single copy (i.e., an individual is heterozygous for that allele), whereas recessive alleles produce a phenotype only when an individual carries two copies of that allele (i.e., the individual is homozygous for that allele). As well, genes can reside on either a sex chromosome (i.e., the X or Y, as in X- or Y-linked genes) or on one of the other 22 non-sex chromosomes, which are referred to as autosomes. Thus, monogenic traits are the result of mutations in single genes and they are said to be Mendelian, if those alleles assort independently of other genes and are either dominant or recessive.

The 2000 on-line edition of Mendelian Inheritance in Man lists 11,293 entries for different genetic traits, of which 10,573 are autosomal and 622 are X-linked (McKusick, 2000). More than 4,500 of these gene traits are autosomal dominant and the majority are rare in populations, occurring at frequencies far below 1/1,000. These single-gene traits have been the primary focus of medical genetics in the past. It is hoped that the insights gained from their analysis will eventually aid in the determination of the genetic components of complex common diseases such as heart disease and cancer, which combined have an incidence rate of 1 in 3.

Cystic fibrosis (CF) is one of the most common single-gene disorders affecting 1/2,000 to 1/4,000 Caucasian newborns (see Tab. 1) (Jorde, et al., 1999). The disease is characterized by fibrotic lesions in the pancreas and insufficiency in the levels of pancreatic enzymes, high levels of chlorine in the perspiration, absence or obstruction of the vas deferens in males, and obstruction of the lungs by thick mucus. The lung problems are the most serious for patients with CF. Chronic obstruction of the airways and high susceptibility to infection by bacteria such as *Staphylococcus aureus* and *Pseudomonas aeruginosa* leads to pulmonary disease and the eventual death of more than 90% of CF patients before the age of 30 (Jorde et al., 1999). The multifaceted nature of the CF phenotype is also a good example of pleiotropy, a term used to describe the multiple and seemingly unrelated phenotypic effects of a single gene. The CF gene was cloned in 1989 (Riordan et al., 1989) and encodes a cyclic-AMP-regulated chloride ion channel that spans the membrane of specialized epithelial cells found in the bowel and lung. This gene, known as the cystic fibrosis transmembrane conductance regulator (CFTR), spans 250 kb of genomic DNA on chromosome 7q31.2 and includes 27 exons (Rommens et al., 1989). Defective ion transport associated with mutations in the CFTR gene can account for the pancreatic insufficiency, the abnormal accumulation of chloride in the sweat secretions, and the depletion of water in the airways of CF patients that leads to the accumulation of mucus in their lungs. Thus, mutations in the CFTR gene locus directly predispose an individual to cystic fibrosis.

Another excellent example of the connection between a single gene mutation and a disease phenotype is sickle cell anemia, which results from a missense mutation in the β-globin gene and leads to anemia, tissue infarction, and infection (see Tab. 1) (Jorde et al., 1999). The disorder is autosomal recessive and very rare among Caucasians, although, among African-Americans it can affect 1/400 to 1/600 newborns. The disorder is very common in some parts of Africa where it can affect up to one in 50 children born. Approximately 15% of children born homozygous for the sickle cell allele die before the age of 5 (Jorde, 1999). The majority of sickle cell anemia cases are caused by the same mutation, a single nucleotide change leading to the substitution of valine for glutamic acid at position 6 in the β-globin protein (Jorde et al., 1999). Similarly, although more than 700 different mutations in the CFTR gene have been described, nearly 70% of the known cases of cystic fibrosis are a result of the deletion of phenylalanine 508 (ΔF508) (Schroeder et al., 1995).

Sometimes certain disease alleles reach a balance in a population due to some selective advantage bestowed on carriers of the disease who are heterozygous for that particular allele. Such alleles are called balanced polymorphisms, and they can also arise when populations occupy heterozygous environments where each environment favors a different allele. It is believed that the sickle cell mutation became a balanced polymorphism in some parts of Africa, because of the reduced susceptibility to malaria seen in individuals who are heterozygous for the sickle cell allele. The CFTR allele, ΔF508, may have reached a balance in Caucasian populations, because of an enhanced resistance heterozygous individuals may have had against diarrheal diseases such as typhoid fever (Josefson, 1998) and lung diseases such as bronchial asthma (Schroeder et al., 1995).

In contrast, some monogenic diseases exhibit allelic heterogeneity where multiple mutations for the underlying disease gene are found in a population. One such disease is Duchenne muscular dystrophy (DMD), which is one of the most common muscle wasting diseases occurring in 1 of 3,500 males (Jorde et al., 1999). DMD is an X-linked disorder characterized by skeletal muscle degeneration and impaired cardiac and diaphragm muscle function. DMD patients usually die before the age of 25 from respiratory and cardiac failure (JORDE et al., 1999). The disease is caused by mutations in the dystrophin gene isolated in 1986 (MONACO et al., 1986). The majority of mutations involve gene duplications or deletions that result in a truncated dystrophin protein. This truncated dystrophin can severely disrupt the cytoskeletal integrity of the muscle fiber by altering the dystrophin–glycoprotein complex that is responsible for linking the cytoskeleton of the cell to the extracellular matrix (MICHALAK and OPAS, 1997). The dystrophin gene is the largest human gene known, spanning 2.3 mio base pairs of DNA and containing 79 exons (JORDE et al., 1999). Such a large gene provides an immense target for genetic mutations. Therefore, many different DMD alleles exist within the population making genetic diagnosis and therapy much more difficult. For example, a clinically related, but milder form of muscular dystrophy called Becker muscular dystrophy (BMD), was thought to be caused by another X-linked gene until it was shown to result from different mutations in the dystrophin gene than found in DMD (GILLARD et al., 1989). Rather than deletions leading to frameshifts in the dystrophin gene, BMD mutations often lead to in-frame deletions, and this may account for the reduced severity of the disease when compared to DMD.

Thus, although monogenic disorders are much easier to dissect genetically, their successful diagnosis and treatment can be affected by the nature of the specific mutations involved and the prevalence of those alleles within a population. Locus heterogeneity and the presence of balanced polymorphisms provide interesting challenges and opportunities for medical genetics to devise therapies that are effective on an individual or population-wide basis (respectively). Another important concept to consider is the penetrance of a disease, which describes the frequency of the disease phenotype in a population possessing a disease-causing genotype. This aspect of genetic disease will be further explored in Sect. 1.1.3.

1.1.2 Common Genetic Diseases are often Polygenic and Multifactorial

As discussed in the previous section, diseases that are revealed to be Mendelian by pedigree analysis, directly indicate the involvement of genes in those disorders. These simple, monogenic disorders make up only a small portion of the spectrum of human diseases. Most congenital malformations and many common adult-onset diseases, such as cancer, diabetes, and heart disease, appear to involve multiple genes coupled with environmental factors (Tab. 2). These traits are often referred to as multifactorial, and the term polygenic is used to describe their genetic component. Thus complex, polygenic diseases are the chief concern of most health care professionals as taken together they have the greatest impact on public health.

1.1.2.1 Polygenic Traits can be Continuous or Discontinuous

Non-Mendelian, complex traits can be broadly split into:

(1) continuous (quantitative) traits, which everyone has, but to varying degrees and can be measured on a continuous numerical scale (e.g., height or weight), or
(2) discontinuous (dichotomous) traits, such as club foot, which some people have but others do not (JORDE et al., 1999).

Many quantitative traits are also multifactorial. For example, if we look at blood pressure, the additive effects of genetic, environmental, and dietary factors leads to a range of blood pressure values in a population. A genetic component of blood pressure is indicated by the correlation between the blood pressure of parents and their children, but dietary intake of salt and stress can also cause variations in this trait. Fig. 1 depicts the expected blood pressure values in a population taking into account the effect of one, two, or multiple genet-

Tab. 2. Common Multifactorial Disorders (adapted from JORDE et al., 1999)

Disorder	Population Incidence
Congenital	
Cleft lip/palate	1/500 to 1/1,000
Club foot	1/1,000
Congenital heart defects	1/200 to 1/500
Neural tub defects	1/200 to 1/1,000
Pyloric stenosis	1/200 in males; 1/1,000 in females
Adult disease	
Alcoholism	1/10 to 1/20
Alzheimer disease	1/10 (American over 65)
Bipolar affective disorder	1/100 to 1/200
Cancer	1/3
Diabetes (type I and II)	1/10
Cardiovascular disease	1/3 to 1/5
Schizophrenia	1/100

ic factors. The distribution begins to take on a bell-shape as we extend our model from one or two genes with a moderate affect on blood pressure, to multiple genes, each contributing an incremental effect on blood pressure. Thus, continuous polygenic traits exhibit a normal distribution within populations.

Some complex diseases do not follow a normal distribution; instead, they are discontinuous and seem to be either present or absent on an individual basis. These traits are of particular medical interest as many common diseases have discontinuous characters including heart disease, diabetes, and some cancers. Commonly, the threshold model is invoked to explain these diseases (FALCONER, 1981). In this model, an underlying liability distribution is envisaged for a population which takes into account multiple genetic and environmental factors (Fig. 2). Individuals who have little chance of developing a particular disease are found on the lower end of the distribution. Such individuals would have very few genetic and environmental factors that would lead to the disease. Other individuals with many genetic and environmental factors would be at the high end of the distribution. The phenotypic expression of a discontinuous, multifactorial disease would then occur as an individual passes some empirical threshold of liability, beyond which the disease is manifested. In addition, the threshold of liability may differ between the sexes or between particular populations. For example, pyloric stenosis is much more common in males than females, affecting 1/200 males and 1/1,000 females (JORDE et al., 1999). This disorder results from a narrowing or obstruction of the pylorus (the area between the stomach and the intestine) which causes chronic vomiting, constipation, and weight loss. It is believed there are two thresholds of liability, one for males and one for females (see Fig. 2). The lower threshold for men would indicate fewer risk factors are required to cause the disorder in males. A number of other congenital malformations fit the threshold model, including: cleft lip or palate, neural tube defects (anencephaly and spina bifida), club foot, congenital heart defects, and infantile autism (JORDE et al., 1999). Tab. 2 summarizes the most common congenital and adult multifactorial diseases.

Statistics such as the recurrence risk are often used to distinguish discontinuous polygenic disorders from each other and from monogenic disorders with reduced penetrance. Recurrence risk is defined as the probability that another affected offspring will be produced in a family in which one or more affected offspring has already been born (JORDE et al., 1999). In cases, such as cleft lip, which is not lethal or severely debilitating, the recurrence

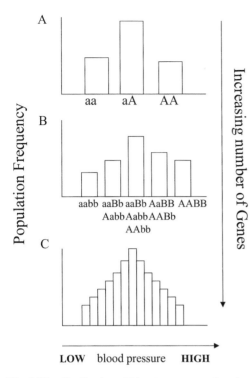

Fig. 1. The distribution of blood pressure values in a population assuming monogenic, oligogenic, or polygenic inheritance. (**A**) The distribution of blood pressure assuming only one locus controls blood pressure (locus A, i.e., monogenic); (**B**) The distribution of blood pressure assuming only two loci control blood pressure (locus A and B, i.e., oligogenic); (**C**) The distribution of blood pressure assuming multiple loci control blood pressure (i.e., polygenic).

risk can also be estimated for the children of affected parents. The occurrence risk, in contrast, is the risk of having the first affected child or proband. The recurrence risk of monogenic disorders is typically 50%, for autosomal dominant, or 25%, for autosomal recessive traits. In contrast, the recurrence risk of polygenic disorders can vary over a wide range and may differ substantially between populations. Recurrence risk can also vary between genders. For example, in pyloric stenosis, where males are more likely to have the disorder than females, the recurrence risk is much higher for siblings of a female proband (7%) then for a male (3.5%) (JORDE et al., 1999) . Since more risk factors would have to be present for the female child to manifest the disease it follows that siblings of this child would be more likely to have the disease, especially male siblings.

When an empirically derived recurrence risk for a multifactorial disease is not available, the square root of the population incidence is a good rule of thumb for the recurrence risk for first-degree relatives (JORDE et al., 1999). For example, if the occurrence of disease in a population is 1/1,000 then the recurrence risk for the sibling, parent, or child of a proband would be about 1 in 30. This connection between population incidence and recurrence rates in families only holds true for multifactorial disease as monogenic traits show recurrence rates independent of population incidence rates. Finally, the difference between the

Fig. 2. The liability distribution for a multifactorial disease within a population. For an individual to be affected by many multifactorial diseases, they must surpass a threshold on an underlying liability distribution that takes into account multiple genetic and environmental factors. As shown, this liability threshold can be different depending on the sex of the individual, e.g., lower in males than in females as is seen in pyloric stenosis (Sect. 1.1.2.1).

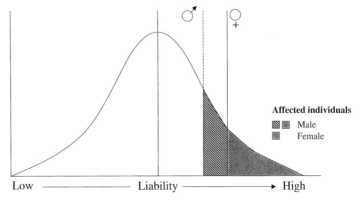

empirical recurrence risk and the theoretical recurrence risk can be used to calculate the heritability of a condition, which is the proportion of the cause of disease that can be attributed to genetic factors. A necessary caveat is that shared environment rather than shared genetic factors can also lead to a discrepancy between empirical and theoretical recurrence risk. Thus, recurrence risk is a powerful statistical tool in determining, if a disease is polygenic and multifactorial versus a monogenic disorder with low penetrance. Tab. 3 summarizes the criteria used to define multifactorial inheritance.

1.1.2.2 Oligogenic Traits, Locus Heterogeneity, and Phenocopy

Purely monogenic and multifactorial, polygenic disorders are two extremes within the spectrum of genetic disease. Somewhere in between are oligogenic traits that are governed by a small number of genes acting together, often superimposed on a polygenic background or severely affected by environmental factors. The factors contributing to an oligogenic disorder are sometimes called susceptibility loci, and segregation analysis can help provide evidence for these loci. Segregation analysis uses the segregation ratio, or proportion of affected children to determine the mode of inheritance (i.e., sporadic, polygenic, dominant, or recessive) (JORDE et al., 1999). The segregation ratio must be corrected for any difference between the sexes, as well as, any bias in determining affected individuals (i.e., a bias of ascertainment) to produce a true segregation ratio. Computer simulations are then used to perform a maximum likelihood analysis in which a general model, including a mix of monogenic, polygenic, and random

environmental influences, is compared to the ability of each specific model of inheritance (i.e., sporadic, polygenic, etc.) to explain the observed segregation ratio. The results of the segregation analysis will let the researcher determine which mode of inheritance is most likely, and hence whether major susceptibility loci may exist. Alleles, that when mutated, provide a dramatic increase in risk for an oligogenic or polygenic disorder are sometimes referred to as major genes. One such major gene is apolipoprotein E (apoE), for which the presence of certain allelic polymorphisms greatly increase an individual's risk for Alzheimer disease, stroke, and coronary heart disease (JARVIK, 1997). Several common multifactorial diseases believed to have oligogenic components include: bipolar disorder, schizophrenia, Alzheimer disease, multiple sclerosis, polycystic ovary syndrome, type 1 diabetes, and idiopathic epilepsies.

Oligogenic disorders must also be distinguished from single-gene disorders that show locus heterogeneity. That is to say, disorders such as adult polycystic kidney disease (APKD), in which a mutated gene at more than one chromosomal location can cause the disease. APKD is an autosomal dominant disorder that results in the progressive accumulation of renal cysts and is sometimes associated with hypertension, cerebral aneurysms, and cardiac valvular defects (JORDE et al., 1999). The genes responsible for APKD are located on chromosome 16 (PKD1; The European Polycystic Kidney Disease Consortium, 1994) and chromosome 4 (PKD2; MOCHIZUKI et al., 1996). Both proteins encode membrane-spanning glycoproteins that are believed to have roles in cellular signaling. Mutation in either gene can cause an identical disease phenotype with both forms of APKD being clinically indistinguishable. Another example of a mono-

---

Tab. 3. Criteria for Determining Multifactorial Inheritance (JORDE et al., 1999)

1) The recurrence risk is higher if more than one family member is affected.
2) When the expression of the disease in a proband is more severe, the recurrence risk is higher.
3) The recurrence risk is higher if the proband is the less commonly affected sex.
4) The recurrence risk for the disease usually decreases rapidly in more remotely related relatives.
5) When the prevalence of the disease in a population is X, the risk for offspring and siblings of probands (first degree relatives) is approximately $\sqrt{X}$.

genic disorder with locus heterogeneity is osteogenesis imperfecta (OI), in which mutation of either of the two genes that make up type 1 collagen results in brittle or malformed bones (POPE et al., 1985). Type 1 collagen consists of a triple helix of three polypeptide chains. Two of the polypeptide chains are encoded by the COL1A1 gene on chromosome 17 and the other polypeptide is encoded by the COL1A2 gene on chromosome 7. To complicate matters further, both APKD and OI are subject to allelic heterogeneity that can drastically affect the severity of the disease.

Lastly, sometimes the observed phenotype of one genetic disorder can be recapitulated either by purely environmental factors or by a completely different disease pathway involving different genes. Such a resemblance between disease phenotypes is referred to as phenocopy (JORDE et al., 1999). One example is Reye syndrome, a fatal acute encephalopathy that is thought to be caused by viral challenge (GLICK et al., 1970). A phenotypically identical encephalopathy to Reye syndrome is associated with some urea cycle defects that produce hyperammonemia (e.g., ornithine transcarbamylase deficiency) (KRIEGER et al., 1979). Phenocopy can also arise in some developmental disorders from purely environmental factors. One example is polydactyly, a genetically heterogeneous disorder resulting in multiple digits of the hand or foot (WINTER and TICKLE, 1993). Polydactyly can sometimes occur sporadically due to environmental factors that disrupt the normal development of the fetus. The sporadic occurrence of what appears to be a genetic disorder, in an individual who does not have affected relatives, is sometimes a good indication that the disease phenotype may be explained by environmental phenocopy. Caution must be taken in such cases, as *de novo* mutation in a major susceptibility gene cannot be ruled out.

1.1.3 Phenotypic Variation in Genetic Disease

In determining the mode of inheritance, geneticists look at several parameters that can be determined from epidemiological data and pedigree analysis. These parameters include the frequency of occurrence in the population and the recurrence risk of a disorder for the relatives of an affected individual, as discussed in Sect. 1.1.2.1. Sometimes these ratios can deviate from the expected values, indicating variable transmission of the disease trait. Often, not only is transmission of the disorder variable, but so is the severity of the disease among affected individuals. The severity of a disease may also increase progressively with each generation in a phenotypic pattern known as anticipation. In addition, both the age and sex of an individual can determine the expression and severity of a genetic disease. Lastly, the phenotypic expression of some disorders relies on a mechanism known as imprinting, which describes the dependence of disease expression on the sex of the parent transmitting the trait. The various anomalies in disease expression are collectively known as the phenotypic variation of a trait.

Nearly all disease traits vary somewhat from person to person. A good example of a wide range in the severity of a monogenic autosomal disease is neurofibromatosis type 1 (NF1), which shows variation between and within families as well as a variable disease progression among affected individuals (JORDE et al., 1999). NF1, or von Recklinghausen disease, is an autosomal dominant disorder arising from mutations in neurofibromin (NF1 gene), a tumor suppressor gene located at chromosome 17q11.2 (RASMUSSEN and FRIEDMAN, 2000). The disorder affects approximately 1 in 3,000 individuals. Nearly half of all cases of NF1 are the result of new mutations making the disorder one of the most common genetic diseases to arise *de novo*. The mild form of the disease is characterized by small discolored patches of skin (*café-au-lait* spots), small lumps (Lisch nodules) in the iris of the eye, and the presence of a few skin neurofibromas (JORDE et al., 1999). Individuals with more severe NF1 have thousands of neurofibromas and often succumb to cancer. Lastly, patients with identical NF1 gene mutations can exhibit different phenotypic expression suggesting that multiple factors may modulate the severity of the disease.

1.1.3.1 Penetrance and the Age of Onset of Genetic Disease

As discussed above, the phenotype of individuals with type 1 neurofibromatosis can be highly variable, yet the disease demonstrates 100% penetrance (JORDE, 1999). Penetrance refers to the proportion of individuals possessing a disease-causing genotype that actually express the disease phenotype within a population. When the proportion is less than 100%, the trait is said to have reduced or incomplete penetrance. Retinoblastoma, a malignant eye tumor, is a classic example of a disease with reduced penetrance. About 1 in 20,000 children develop retinoblastoma, an autosomal dominant disorder caused by mutations in the retinoblastoma gene (Rb) (JORDE, 1999). The protein product of the Rb gene is a key regulator of cell proliferation, and when mutated, can result in the uncontrolled growth of retinal cells in one or both eyes. Familial retinoblastoma accounts for 40% of all retinoblastoma cases and usually occurs in both eyes (bilateral) (JORDE, 1999), whereas novel cases of retinoblastoma require the mutation of both Rb alleles and usually result in a tumor in only one eye (unilateral) although the disease is autosomal dominant, approximately 10% of the individuals who carry a mutated Rb gene never actually develop the disease (JORDE, 1999). Thus, retinoblastoma has a penetrance of 90%. This reduced penetrance can partly be explained by the fact that individuals who carry one mutated Rb gene must develop a mutation in the other allele within the retinal cells of one or both eyes. Since the second mutation in the unaffected Rb allele occurs by chance, some individuals never develop a second mutation. These individuals can still pass on the trait to their children who are very likely to develop retinoblastoma. Another dominant autosomal trait, polydactyly (see Sect. 1.1.2.2), manifests incomplete penetrance as "skipped generations" within an affected family pedigree. Several other genetic diseases show incomplete penetrance, including renal adysplasia, Huntington disease, acute intermittent porphyria, and type 1 diabetes.

In most cases, the reasons for incomplete penetrance are not clear, but sometimes it is found that a specific allele at one locus masks the effect of the disease causing allele at another locus. This phenomenon is called epistasis, not to be confused with dominance, a situation describing the masking of one allele by another allele at the same locus (BROOKER, 1999). Epistasis is also used to describe the incremental effect of genes, at different loci, to the overall expression of a phenotype. For example, the development of high blood pressure can result from two or more genes, acting in independent physiologic pathways, which contribute independently to several millimeters of elevated blood pressure. Together, these factors have the overall additive effect of greatly elevating an individual's blood pressure.

Environmental factors such as diet or exposure to infectious agents may also affect the expression of a disease phenotype and thus the penetrance of that trait. For example, children with phenylketonuria (PKU) are not able to metabolize the amino acid phenylalanine because of mutations in the phenylalanine hydroxylase (PAH) gene (SCRIVER and WATERS, 1999). Deficiency in the PAH enzyme translates to dangerous levels of phenylalanine in the affected individuals' blood and this hyperphenylalaninemia is thought to cause the mental retardation associated with PKU. Diets low in phenylalanine can prevent hyperphenylalaninemia and thus allows normal cognitive development in children affected by PKU. In addition, environmental impact on phenotypic variation can extend to the social transmission of habits or customs (e.g., the consumption of alcohol or smoking). This is particularly the case for people with alpha 1-antitrypsin deficiency (alpha1AD). Alpha-1-antitrypsin is a protease inhibitor, and deficiency in this enzyme predisposes an individual to chronic obstructive pulmonary diseases (COPDs) such as emphysema and chronic bronchitis, in addition to cirrhosis of the liver (WIEDEMANN and STOLLER, 1996; YANG et al., 1999). The severity of the disease is greatly exacerbated by smoking tobacco. COPD and smoking are both strong risk factors in the development of pulmonary cancers, therefore, individuals who have alpha1AD and smoke, have a greatly increased risk for developing lung cancer (YANG et al., 1999).

Incomplete penetrance may also be explained by a delayed onset of the disease phenotype. In such cases, the individual may die before the full expression of the disease. Late onset disease may arise from the accumulation of toxic substances, slow and progressive tissue death or the inability to repair environmental damage to tissues as we age. Many late-onset diseases can be 100% penetrant, if the affected individual lives long enough. Important adult-onset diseases include Alzheimer disease, cancer (all types), diabetes (I and II), coronary disease, and schizophrenia. In many of these diseases, both the risk of developing the disease and the severity of the phenotype increases with age. For geneticists, this can complicate the determination of the heritable component of a disorder because, as an individual ages, the contribution of environmental and other non-genetic factors becomes increasingly important in the expression of a disease phenotype. Conversely, the early onset of a disease can strongly suggest a major genetic component is responsible for the expression of disease in that individual. In familial breast cancer, e.g., two loci were identified by linkage analysis in families with early onset and highly penetrant breast cancer. The genes at these two loci, BRCA1 and BRCA2, encode transcription factors that can interact with the protein Rad51 (SCULLY et al., 1997) as well as a number of other proteins involved in DNA repair (WANG et al., 2000). Mutations in these genes may lead to genomic instability that in turn can result in further mutations and cancer. Mutations in BRCA1 account for approximately 3% of all cases of breast cancers and also predispose female carriers to ovarian cancer (JORDE et al., 1999). Other examples of early onset disease include familial hypercholesterolemia, type I diabetes, and familial colon cancer.

Huntington disease, which affects 1 in 20,000 Caucasians, is a late onset disorder that is manifested in most individuals between the ages of 30 and 50 (JORDE et al., 1999). This autosomal dominant neurological disorder is characterized by a difficulty of swallowing, progressive dementia, and the uncontrolled movement of the patients limbs. The brain of individuals with Huntington disease can often show a marked atrophy resulting in the loss of up to 25% or more of brain mass. Individuals can expect to live approximately 15 years from the time of the first clinical diagnosis of the disorder. Death usually arises from aspiration pneumonia or cardiorespiratory failure. Mutations in the trinucleotide repeat, CAG, repeated several times within the coding region of the huntingtin gene, leads to the disease (REDDY et al., 1999). The repeat CAG codes for the amino acid glutamine, and large glutamine tracts within the huntingtin gene may lead to the formation of toxic protein aggregates in the brain. Unaffected individuals typically have 11–35 copies of the CAG repeat. These repeated sequences are expanded to anywhere from 36–100 copies in individuals who have Huntington disease, with incomplete penetrance of the disorder occurring in individuals with 36–38 copies of the repeat (JORDE et al., 1999). Curiously, the onset of the disease is earlier, and the phenotype of the disease becomes steadily more severe in progressive generations. This phenomenon is called anticipation. Usually, longer repeats are correlated with earlier onset and more severe Huntington disease. Above the threshold number of 35 repeats, the CAG trinucleotides progressively expand from generation to generation, leading to a more severe disease with each expansion. This trinucleotide repeat expansion (TRNE) is common among several muscle and neurological disorders including fragile X syndrome (CGG and GCC repeat), myotonic muscular dystrophy (CTG repeat), spinal and bulbar muscular dystrophy (CAG repeat), and spinocerebellar ataxia (CAG repeat) (BROOKER, 1999). The severity of Huntington disease like some other TRNE disorders depends on the parental origin of the mutated huntingtin allele. Generally, the disease is more severe, if you inherit the Huntington disease allele from your father. Up to 80% of all juvenile or early-onset Huntington disease is due to paternal transmission (JORDE et al., 1999). This paternal transmission bias was later shown to involve the preferential expansion of the CAG repeats when affected males transmit the allele. One plausible reason for male-specific expansion of trinucleotide repeats is the fact that male sperm production involves many more cell divisions than the generation of female oocytes. With each cell division, the risk of replication errors such as slipping or polymerase stuttering increase and, therefore, male gametes, which are the product of many cell divisions, are more likely to exhibit TRNE. Cytoplasmic (mitochondrial) factors and secondary modifying loci on one of the sex chromosomes (X or Y) are two other commonly invoked mechanisms used to explain such inheritance patterns (JORDE et al., 1999). Sex specific effects on disease expression and another mode of phenotypic variation called imprinting, which is dependent on the parental origin of an allele, will be discussed in the following section.

1.1.3.2 Sex-Specific Phenotypic Variation and Imprinting

Several modes of inheritance depend on either the sex of the individual or the sex of the parent transmitting the trait. One of the first mechanisms of sex-specific variation in disease to be characterized is X-linked inheritance. Human females carry two copies of the X chromosome and males carry one X and one Y. Male children receive their X chromosome from their mother and the Y from their father. Therefore, males have a 0% chance of receiving any particular X-linked allele from their father and a 50% chance of receiving the same allele from their mother, if she is heterozygous for that particular mutation. Females on the other hand will always receive one X from each parent. Thus, females will always receive the X-linked mutation from their fathers and will receive an X-linked allele from their mother 50% of the time when the mother is heterozygous for that allele. As expected, Y-linked inheritance is strictly father–son and will not be discussed here. By looking for such sex-specific ratios of inheritance, it is possible to determine, if traits are X-linked. For example, a disease is most likely X-linked recessive, if

(1) the disease occurs predominantly in males,
(2) there is no male–male transmission and affected males rarely have affected female progeny, and
(3) 100% of the male and none of the female progeny of an affected female exhibit the disorder.

In contrast, a disease is probably X-linked dominant, if

(1) twice as many females are affected as males in the population,
(2) the disease is seen in every generation, and
(3) 50% of both male and female progeny of an affected female exhibit the disease.

The majority of common X-linked diseases are recessive such as Duchenne muscular dystrophy, which predominately affects males as discussed in Sect. 1.1.1. Probably the most well known X-linked recessive disease is hemophilia A, a severe bleeding disorder that affects 1/10,000 males world-wide (JORDE, 1999). Historically, the disease has been associated with the royal families of Europe, and Queen Victoria of England was a carrier of the disease. Hemophilia A is caused by a deficiency in or a defect in the factor VIII protein, an essential component in the blood-clotting cascade. Inversions in the factor VIII gene are the most common mutations accounting for 45% of severe hemophilia A cases (JORDE et al., 1999). Without sufficient levels of factor VIII, individuals can develop prolonged and severe bleeding episodes from external wounds and internal hemorrhaging in the joints and muscles. These individuals often bruise easily and develop hemarthroses or bleeding into the joints, which can eventually destroy or severely diminish joint function. Platelet function is normal in hemophilia patients and consequently, superficial cuts do not lead to excessive bleeding. Before the advent of purified factor VIII supplementation, hemophiliacs generally died in their early twenties from a bleeding episode. With factor VIII supplementation the average life expectance of an individual with hemophilia A is about 68 years (JORDE et al., 1999). Another bleeding disease, hemophilia B is also X-linked. Hemophilia B, caused by a deficiency in factor IX, is less severe than hemophilia A, and is treated with donor-derived or recombinant factor IX supplementation (JORDE et al., 1999).

X-linked dominant diseases are less frequent in the population, but twice as frequent in females than males. Fragile X is an important X-linked dominant disorder that accounts for approximately 40% of all X-linked mental retardation occurring in 1/1,250 males and in 1/2,500 females (DE VRIES et al., 1998; JORDE et al., 1999). In addition to mental retardation, the disease is characterized by a distinct facial appearance with large ears and an elongated face, as well as hypermobile joints. The disease is highly penetrant in males (80%) as compared to females (30%) who tend to exhibit less severe mental retardation (JORDE, 1999).

The term "fragile X" is derived from the peculiar observation that fibroblasts cultured from patients with this disorder exhibit chromosomal breaks when grown in medium lacking folic acid. Unlike most X-linked dominant disorders, fragile X exhibits anticipation, and affected males rarely have affected female progeny. Anticipation is a hallmark of diseases involving trinucleotide repeat expansion (TNRE). Indeed, when the gene responsible for fragile X syndrome, FMR1, was finally cloned, it was found that the 5′ untranslated region of the gene contains several repeats of the trinucleotide (DE VRIES et al., 1998). In normal individuals, there are 6–50 copies of the CGG repeat. Those with fragile X syndrome have 230–1,000 copies of this repeat. An intermediate or pre-mutation level of 50–230 repeats is often seen in normal transmitting males and their female offspring (JORDE et al., 1999). When normal females transmit the allele to their offspring, an expansion can occur in the number of repeats leading to a full mutation level of more than 230 CGG repeats. This maternal expansion of repeats helps explain the anticipation seen in fragile X pedigrees. Further analysis of the FMR1 region led to the discovery of a second fragile site that is subject to TNRE of a GCC repeat. In this case, the repeat shows expansion when transmitted by both sexes (JORDE et al., 1999). The expression level of the FMR1 gene is adversely affected by expansion in the number of CGG repeats. In individuals with a low repeat number, the FMR1 gene is expressed and in affected individuals with a high number of repeats, there is a complete loss of FMR1 expression. Because the CGG repeats are highly methylated in affected individuals, it is believed that this methylation is responsible for the silencing of the FMR1 gene (JORDE et al., 1999).

Methylation also plays a role in genomic imprinting, which is a term used to describe a means of differential gene expression based on the parental origin of an allele (reviewed in BRANNAN and BARTOLOMEI, 1999; TILGHMAN, 1999). The expression of imprinted genes is regulated by the differential DNA methylation of maternal and paternal alleles, which provides a mechanism of alleles-specific transcriptional silencing (reviewed in LI et al., 1993). Thus, imprinting can be seen as another type of sex-specific phenotypic expression, which relies on the sex of the parent transmitting the trait rather than the sex of the affected individual.

Two major clusters of imprinted genes have been identified in the human genome: a 1 Mb region on chromosome 11p15, containing seven imprinted genes (LEE et al., 1999) and a 2.3 Mb cluster at 15q11-13 which also contains at least seven genes (SCHWEIZER et al., 1999) . The imprinted region on chromosome 15q11-13 is associated with a pair of syndromes known as Angelman syndrome (AS) and Prader–Willi syndrome (PWS). PWS is characterized by an inverted upper lip, small hands, mild obesity, hypogonadism, and mild to moderate mental retardation. The disease is often caused by a deletion of the 15q11-13 region that is inherited from the father. In contrast, when the same deletion is inherited from the mother the child will have AS, or "happy-puppet" syndrome. AS is characterized by severe mental retardation, seizures, and ataxic gait. The two disorders are seen in 1/15,000 individuals, and most cases arise from deletions in the 15q11-13 region which encodes several genes including SNRPN and UBE3A, both of which are preferentially expressed in the brain (JORDE et al., 1999; TILGHMAN, 1999). Although PWS is caused by the loss of expression of multiple paternal-specific genes in the 15q11-13 region, there is evidence that AS results from the loss of function of a single gene, UBE3A, a brain-specific ubiquitin ligase (ROUGEULLE et al., 1997; VU and HOFFMAN, 1997). A minority of PWS and AS cases arise from uniparental disomy of the maternal or paternal chromosome 15 (respectively), where two copies of the chromosome are inherited from one parent. Uniparental disomy is also seen in another imprinted disorder, Beckwith–Weidemann syndrome (BWS).

Beckwith–Weidemann syndrome is associated with a cluster of imprinted genes on chromosomes 11p15, including H19 and the insulin-like growth factor 2 (IGF2). The disorder is characterized by a large tongue, a predisposition for Wilms tumor of the kidney, and abdominal wall defects. Some cases of BWS arise from uniparental disomy of the paternal chromosomes 11. In contrast to Angelman and Prader–Willi syndrome, which result from the loss of gene expression, it is believed that BWS is primarily caused by the over-expression of a gene or genes on chromosome 11 (JORDE et al., 1999). When two copies of a paternal chromosome 11 are inherited, genes such as IGF2 are expressed at twice the normal level. Increased levels of IGF2 can explain some of the over-growth related symptoms of BWS, such as the enlarged tongue and the rapid growth of Wilms tumors. Thus, both the gain and loss of function (or expression) of imprinted genes can lead to human disease.

As discussed so far, a wide range of phenotypic variation exists for most disease traits. This variation can take the form of reduced or incomplete penetrance, as explained by a late age of onset or the interaction of multiple genetic and environmental factors in disease development. In other cases, sex-specific variation such as X-linked inheritance and imprinting are responsible for altering the expression of a disease phenotype. Some other types of sex-related variation exist such as sex limitation, in which the trait is expressed in only one gender and may require a male- or female-specific anatomy to develop the disease (e.g., ovarian cancer), and sex-influenced inheritance, where the disease allele has reduced penetrance or is recessive in one gender and dominant in another (e.g., male baldness and pyloric stenosis, see Sect. 1.1.2.1) (JORDE et al., 1999). Lastly, mitochondrial inheritance provides yet another layer of complexity to sex-specific phenotypic variation. Because almost all mitochondria are maternally derived, mitochondrial disease such as Kearns–Sayre disease, characterized by progressive muscle weakness, cerebellar damage, and heart failure, can only be transmitted by an affected female to her offspring (JORDE et al., 1999). Thus, like imprinting, the sex of the parent transmitting the disease trait will effect the phenotypic expression of a mitochondrial disease.

1.1.4 Chromosome Disorders: The Smoking Gun of Genetic Disease

Chromosome abnormalities are responsible for a significant proportion of genetic disorders that appear to arise *de novo*. On average chromosome abnormalities occur in 1/150 live births and are the leading cause of mental retardation, as well as spontaneous abortion in the first and second-trimester of pregnancy (in 50% and 20% of cases, respectively) (JORDE et al., 1999). These abnormalities may involve several genes on one or more chromosomes and consequently, the phenotypic characteristics of a chromosomal disorder often resemble polygenic or pleiotropic diseases. For example, many chromosome disorders are associated with some form of mental retardation, in addition to congenital malformations (especially heart defects), growth delay, and very characteristic facial morphogenesis. The diagnosis of most chromosomal abnormalities can now be made during pregnancy through amniocentesis and cytogenetic analysis. Using chromosome banding techniques and fluorescent *in situ* hybridization analysis, cytogenetics can determine changes in chromosome structure at a resolution of 1–10 Mb. These abnormalities include rearrangements within a given chromosome (i.e., deletions, insertions, duplications, and inversions) or between chromosomes (i.e., translocations), as well as changes in chromosome number or ploidy. Chromosome abnormalities also allow the rapid assignment of a genetic component to any disorder that is consistently associated with chromosome rearrangements or ploidy changes. For example, genes located at the breakpoints in chromosomal translocations are sometimes directly implicated in causing the disease. Thus, chromosomal rearrangements are the "smoking guns" of genetics, implicating a gene or genes in the etiology of a genetic disorder.

Chromosome abnormalities can be present in every cell of an individual (constitutional abnormality), or may be present in only certain tissues (somatic or acquired abnormality) (JORDE et al., 1999). Somatic abnormalities can be separated into two broad classes: mosaicism, in which an individual possesses two or more genetically different subsets of cells derived from the same zygote, and chimerism, in which an individual possesses two or more genetically different subsets of cells that are derived from different zygotes. Often abnormalities that would be lethal in constitutional form, such as most trisomies, can result in viable embryos when present as chimeras or mosaics with genetically normal cells. Chimerism and mosaicism are often discovered during routine blood analysis, and aneuploidy mosaics are very common. Aneuploidy describes any change in chromosome number from the normal karyotype and since this class of chromosome abnormality is the most common, we will confine our discussion to only the most important forms of aneuploidy. Examples of the various chromosomal abnormalities are illustrated in Tab. 4A and 4B.

Probably the most common aneuploidy seen among live births are the sex chromosome aneuploidies, which collectively occur in 1/400 male and 1/650 female births (JORDE et al., 1999). These include Turner syndrome (monosomy X; 45, X), Klinefelter syndrome (47, XXY), Trisomy X (47, XXX) and 47, XYY syndrome. Turner syndrome is the rarest, occurring in 1/2,500 to 1/5,000 live births. Individuals with the disorder are often short of stature, exhibit sexual infantilism and ovarian dysgenesis, have a triangular face with a webbed neck, and

can have congenital heart defects. Klinefelter syndrome and trisomy X are more common, affecting about 1/1,000 male births (JORDE et al., 1999). Both disorders are associated with slightly lower intelligence and some minor physical abnormalities. Males with Klinefelter syndrome are often taller than average, may exhibit breast development at puberty, and are usually sterile. Females with trisomy X can be sterile, but otherwise may appear normal. Similarly, individuals with 47, XYY syndrome appear physically normal, but exhibit reduced intelligence. This condition is found in 1/1,000 live births, but has been detected at a frequency as high as 1/30 for male inmates within prison populations (JORDE et al., 1999). The reduced intelligence and a possible link between 47, XYY syndrome and behavioral disorders is thought to account for the prevalence of this disorder among prison inmates.

Trisomy 21 (47, XY +21 or 47, XX +21), better known as Down syndrome, is the most common autosomal aneuploid disorder, occurring in 1/800 live births (JORDE et al., 1999). Interestingly, 1–3% of all trisomy 21 live births result from mosaicism, in which a trisomic embryo has lost the extra chromosome 21 in some of its cells during development. Although the variation in clinical symptoms is quite large, the condition is generally characterized by mental retardation and distinct facial features. These features include a low nasal root, redun-

Tab. 4A. Common Numerical Chromosomal Abnormalities

Abnormality	Disorder	Clinical Features
Triploidy	e.g. 69, XXX	still born, lethal (1–3% of conceptions)
Nullisomy	e.g. 44	pre-implantation lethal
Monosomy[a]	e.g. 45, X; Turner syndrome	short stature, sexual infantilism, ovarian dysgenesis and webbed neck
Trisomy	e.g. 47, XX, + 21; Down syndrome	mental retardation, distinctive facial appearance and structural heart defects
	47, XY, + 13; Patau syndrome	oral-facial clefts, microphthalmia (small eyes), polydactyly
	47, XY, + 18; Edwards syndrome	prenatal growth deficiency, small mouth, congenital heart defects and occasionally spina bifida
Mosaicism	e.g. 47, XXX/46, XX; Trisomy X mosaic	variable but may include sterility, menstrual irregularity and mild mental retardation

[a] All monosomies other than Turner syndrome are embryonic lethal (JORDE et al., 1999).

Tab. 4B. Common Structural Chromosomal Abnormalities

Abnormality	Disorder	Clinical Features
Deletion	e.g. 46, XY del (4)(p); Wolf–Hirschorn syndrome	cleft lip, wide spaced eyes, mental retardation
	46, XY del (5)(p); *cri-du-chat* syndrome	distinctive cry, mental retardation and microcephaly
	46, XY del (15)(q11-13); Prader–Willi syndrome	mental retardation, inverted upper lip, small hands and mild obesity
	46, XY del (11)(p13); WAGR syndrome	anuridia (absence of the iris), Wilms tumor, mental retardation, genitourinary abnormalities
	46, XY del (22)(q11); Digeorge syndrome	cleft palate, heart defects
Inversion	e.g. 46, inv (X) (q28) Y; Hemophilia A	severe bleeding episodes, failure of blood to clot
Duplication	e.g. 46, XYdup(17)(p12); Charcot–Marie–Tooth disease	progressive atrophy of the distal limb muscles
Insertion	e.g. 46, ins(X;X)(q21.2;q12q13.3) Y; Disomy Xq12q13.3	mental retardation, some facial dysmorphia
Ring	e.g. 46, X r(X); X ring chromosome	can be phenotypically normal, although ring chromosomes can be lost resulting in X monosomy in some cells (i.e., mosaicism)
Translocation, reciprocal	e.g. 46, XY t(9;22) (q34;q11); Chronic myelogenous leukemia (CML)	leukemia
	46, XY t(8;14)(q24;q32); Burkitt lymphoma	lymphoma, tumors of the jaw
Translocation, Robertsonian[a]	e.g. 45, XY, -14, -21, + t(14q21q)	usually phenotypically normal 1/3 of offspring will have a monosomy or partial trisomy of chromosome 21 or 14

[a] Translocation in which the short arms of two chromosomes are lost and the long arms fuse at the centromere forming a single chromosome.

dant skin on the inner eyelid, protruding tongue, and small overfolded ears. In addition, many individuals with Down syndrome exhibit a deep crease across their palms and broad hands and feet. Males are generally sterile and approximately half of females fail to ovulate. However, the most serious complications associated with trisomy 21 are structural heart defects that occur in 40% of Down syndrome patients (JORDE et al., 1999). A critical region on chromosome 21, at 21q22, is sufficient to cause Down syndrome in individuals with a partial trisomy of only this region (PATIL et al., 1994). Two of the genes in this region, DYRK1A (a kinase) and the amyloid precursor protein (APP), may be responsible for the mental deficit and Alzheimer-like dementia seen in Down syndrome patients (IANNELLO et al., 1999; OKUI et al., 1999).

In addition to Down syndrome, only two other autosomal trisomies are viable. The first is trisomy 18, or Edward syndrome, which is the second most common autosomal trisomy and occurs in 1/6,000 live births (JORDE et al., 1999). Edward syndrome is characterized by mental retardation, congenital heart defects, prenatal growth deficiency and occasionally, spina bifida. The second disorder is trisomy 13, or Patau syndrome, which occurs in 1/10,000 births (JORDE et al., 1999). Patau syndrome is characterized by mental retardation, congenital heart defects, oral–facial clefts, microphthalmia (small eyes), and polydactyly. Individuals with trisomy 13 or 18 have a similar survival rate with 90% of children dying within one year of birth (JORDE et al., 1999). The incidence of trisomy 13, 18, and 21 seems to correlate directly with increasing maternal age.

For example, the risk of a woman under 30 giving birth to a Down syndrome child is only 1/1,000, compared to 1/400 for the same woman at 35 and 1/100 at age 40 (JORDE et al., 1999). Interestingly, chromosomes 13, 18, and 21 are the most gene poor within the human genome. Since the phenotypes associated with trisomy 13, 18, and 21 are believed to be dependent on gene dosage, the paucity of genes on these chromosomes may explain why these particular trisomies are compatible with survival.

2 Etiology, Screening and Management of Genetic Disease

Increasingly, modern medicine has begun to focus on the genetic basis of human disease rather than on the treatment of disease symptoms. Coinciding with this shift, pharmaceutical companies have transitioned to gene-based therapies, which rationally address the molecular basis of a disease once the genetic components have been elucidated. This paradigm shift in medicine has fuelled rapid growth in the biotechnology industry over the last two decades, which in turn, has provided a means of translating academic research into potential therapies (see Sect. 4). Any approach to genetic disease should address.

(1) the genetic component(s) involved in the etiology of the disease,
(2) the diagnosis and screening of affected individuals, and
(3) therapeutic intervention or the management of the disease phenotype.

Each component to this strategy can be approached independently, although, once a gene has been identified, highly specific diagnostic tests and therapies can be developed much more quickly. The following sections will compare and contrast cystic fibrosis and colon cancer to illustrate how monogenic and multifactorial disease may be approached differently with respect to their etiology, diagnosis, and management.

2.1 Finding Disease Genes for Simple, Monogenic Disorders

For monogenic disorders, like cystic fibrosis (CF), a genetic component for the disease becomes evident from pedigree analysis. Cystic fibrosis is autosomal recessive, which suggests that affected individuals would have two mutated alleles at one or perhaps a few tightly linked loci. Once a genetic component has been suggested by pedigree analysis, then a search for common alleles between affected family members can be carried out. This strategy can be referred to as a "top-down" approach, in which we progress from a disease phenotype to a candidate gene. Sometimes, the search for alleles can be narrowed down very quickly by cytogenetic analysis if chromosomal rearrangements are correlated with the disease phenotype. Thus, by determining the common chromosome or the individual chromosome bands involved in these rearrangements, the researcher can focus his search for alleles to only those candidate regions. For example, a translocation associated with chronic myelogenous leukaemia [t(9;22) (q34;q11)], led to the cloning of the BCR and ABL oncogenes (Jorde et al., 1999).

2.1.1 Finding the Cystic Fibrosis Gene: Linkage Analysis and Positional Cloning

Linkage analysis can be used to pinpoint which chromosomal regions contain disease alleles by scanning the genome of family members, within affected pedigrees, for alleles that appear to be linked to the disease phenotype. Two loci are linked, if they are found on the same chromosomal segment and do not assort independently during meiosis (i.e., have a recombination frequency of less than 50%). Linkage analysis has been quite successful at determining loci that contribute moderate to large effects to disease predisposition, as is the case for monogenic and oligogenic disorders. However, this type of analysis has been generally unsuccessful in mapping loci involved in complex, polygenic disorders, for which each locus has only an incremental effect on the expression of the disease phenotype.

For an allele to be informative for linkage studies, it must be polymorphic, existing in more than one form within the population. Consequently, the power of this approach is limited by the number of affected family pedigrees that can be examined and the availability of informative polymorphic markers. Traditionally, restriction fragment length polymorphisms (RFLPs) and variable number of tandem repeats (VNTRs) have been used for linkage analysis. RFLPs rely on the differential location of restriction sites within DNA and are less polymorphic than VNTRs, which rely on the variability in length of minisatellite DNA. More recently, polymerase chain reaction (PCR) based genotyping of DNA microsatellites and single nucleotide polymorphisms (SNPs) have been used in linkage analysis.

In the case of cystic fibrosis, extensive linkage analysis indicated that a single locus on chromosome 7, between bands q21 and q31, was linked to the disease allele (WHITE et al., 1985). Once this candidate region was identified, the DNA within this region was then scanned for genes that had the expression pattern or enzymatic activity expected for a candidate gene for cystic fibrosis. This method of gene cloning is called positional cloning. As a last confirmation step each candidate gene within the chromosomal region, identified by linkage analysis, must be screened for germline mutations that are only present in individuals with the disease phenotype. Several common methods of mutation detection are summarized in Tab. 5 (for an in-depth overview, see STRACHAN and READ, 1999). For cystic fibrosis, the candidate region at 7q21-31 was shown to contain the CFTR gene, which was found to contain a 3 base pair deletion resulting in the loss of the amino acid residue phenylalanine 508 in 70% of all cases of CF (RIORDAN et al., 1989).

2.1.2 Finding Alleles in Populations Using Linkage Disequilibrium

Linkage analysis works well for diseases with strong penetrance and is quite robust within families. However, the association of the disease and marker alleles may not be conserved between families within a population.

This is because the disease-causing mutation may have occurred multiple times within a population and may be associated with different linked alleles. Alternately, after many generations the association between the disease allele and a given marker allele may be lost after multiple rounds of meiotic recombination. If we study a large number of families and find there is no preferential association between the disease and a specific allele at a linked maker locus, the two loci are said to be in linkage equilibrium. However, this is not the case for certain populations that, for various reasons, have a reduced genetic diversity. In such populations, linkage disequilibrium can occur, if there is a non-random association of a disease with a linked marker allele between families within that population.

Linkage disequilibrium can arise because of natural selection, which fixes certain favorable allelic combinations, as well as genetic drift or founder effects that reduce the genetic diversity of a population (JORDE et al., 1999). Genetic drift is most likely to occur in small populations of a stable size, whereas founder effects arise from the expansion of a population from only a few founding members. Linkage disequilibrium provides an advantage over linkage analysis. Rather than measuring the association of alleles over a few generations within a pedigree, linkage disequilibrium indicates the association of alleles over many tens or hundreds of generations. With each generation the probability of two alleles remaining linked decreases and, after many generations, only those markers in the closest proximity to the disease allele will remain linked. Thus, a marker allele that is subject to linkage disequilibrium with a specific allele of interest can be used as a surrogate for that particular allele. Consequently, whereas linkage analysis can only map genes to intervals of one to several centimorgans (cM), linkage disequilibrium analysis can map genes to a resolution of 0.1 cM or less (JORDE et al., 1999). Thus, by choosing populations that have suffered genetic drift or founder effects (e.g., Iceland and Finland), it may be possible to find markers that map very closely to disease traits. Sect. 3 will discuss how technologies derived from the Human Genome Project, such as single nucleotide polymorphism (SNP) based genotyping, are being

Tab. 5. Common Methods Used to Scan Genes for Mutations (adapted from STRACHAN and READ, 1999)

Method	Advantages	Disadvantages
Southern blot (cDNA probe)	can detect major deletions or rearrangements	labor intensive and requires μg of DNA
Sequencing	can detect all changes and mutations are fully characterized	expensive and interpretation may be difficult
Heteroduplex gel mobility	simple and cheap	sequences of <200 bp only limited sensitivity and does not reveal position of the change
Denaturing HPLC	quick, high throughput and quantitative	expensive and does not reveal position of the change
Single-strand conformation polymorphism (SSCP) analysis	Simple, cheap	sequences of <200 bp only limited sensitivity and does not reveal position of the change
Denaturing gradient gel electrophoresis (DGGE)	highly sensitive	expensive and can be technically difficult because primer choice is crucial does not reveal position of the change
Dideoxy fingerprinting	highly sensitive	interpretation may be difficult
Mismatch cleavage chemical or	highly sensitive, indicates position of change	toxic chemicals and experimentally difficult
(enzymatic)	(lack of nasty reagents)	(poor quality results)
Protein truncation test (PTT)	highly sensitive for chain terminating mutations and shows the position of the change	detects mutations that lead to protein truncation only and requires RNA expensive and technically difficult
Oligonucleotide arrays (gene chips) hybridization arrays or minisequencing arrays	both types are: quick and high throughput able to potentially detect and define all changes	both types are: expensive and restricted to a predefined number of genes novel and still under development

used to uncover disease-associated alleles that are subject to linkage disequilibrium in large populations.

2.2 Finding Disease Genes for Complex, Multifactorial Disorders

Determining the genetic component(s) of non-Mendelian characters such as heart disease, asthma, and cancer is more difficult. As discussed in Sect. 1.1.2.1, these complex traits can be continuous or discontinuous, and their phenotypic expression often depends on the interaction of a myriad of genetic, social, and environmental factors. However, by looking at the degree of disease clustering within families, it may become apparent that a complex disease has a heritable component. The mode of inheritance can be determined from epidemiological data taken from large populations or within affected pedigrees. For instance, when looking at the siblings of affected individuals, the recurrence risk will indicate, if the disease is multifactorial and can provide a quantitative estimate of the maximum effect of genes on a trait (see Sect. 1.1.2.1).

Twin studies, involving monozygotic (identical) and dizygotic (non-identical) twins, can also help determine, if a disease has a heritable component (MACGREGOR et al., 2000). Using any genetic model, monozygotic twins who are genetically identical will be concordant for a qualitative trait, or show similarity for a quantitative trait, more often than their dizygotic counterparts, who share only 50% of their alleles in common by descent from their parents. In addition, by comparing identical twins raised apart, one can be reassured that any familial clustering is due to common genes rather than a common environment. Twin studies have been invaluable in determining to what extent complex behavioral disorders such as bipolar disorder and schizophrenia are hereditary. Twin studies, however, do not reveal the specific genes associated with a disease.

The underlying genes for complex traits are sometimes referred to as quantitative trait loci (QTLs) (JORDE et al., 1999). Traditional linkage analysis can sometimes uncover QTLs for families who appear to show a single-gene inheritance pattern for a complex disease such as breast and colon cancer. The analysis of linkage in affected sibling pairs is one useful method of determining QTLs based on the propensity of two affected siblings to share the same mutant alleles by direct descent from their parents (JORDE et al., 1999). For example, if a marker locus is not linked to an underlying the disease trait, by chance one expects 25% of sib pairs to share 2 identical alleles, 50% to share 1 allele, and 25% to share 0 alleles. However, regardless of the underlying model of disease transmission, if one finds a marker that is tightly linked to an underlying locus that influences the disease trait, siblings who are concordant for the trait or disease should share identical alleles by descent from their parents more often than expected by chance alone. Consequently, if one collects a large number of such sib pairs one will expect to find an increased number of shared alleles in proximity to the underlying trait loci. The results are then analyzed by a simple statistical test that compares the observed number of alleles shared to that expected under the null hypothesis. Sib pair studies are useful in detecting the effects of trait loci when there are a few loci that impart substantial effects on development of the disease (i.e., oligogenic). However, these studies have only limited power when there are many loci that influence the disease trait (i.e., polygenic).

For many complex diseases such as schizophrenia and bipolar disorder, traditional linkage or linkage disequilibrium analysis has failed to determine reproducible and robust linkage between these diseases and individual trait loci. This may reflect the polygenic nature of such diseases, where a myriad of genetic and environmental factors could play a role in disease development. To try to circumvent the loss of power in linkage studies of polygenic disease the transmission disequilibrium test (TDT) was devised by SPIELMAN et al. (1993). To carry out the TDT, one must first find a marker locus that is associated with a disease trait within a population. Association refers to the statistical correlation of a marker locus with the disease trait in the general population, i.e., the marker allele is found in affected individuals more often than predicted by chance. Marker alleles that are associated with a disease trait can be uncovered by linkage disequilibrium analysis (see Sect. 2.1.2). Then, using data from families with at least one affected child, the transmission of the associated marker allele from a heterozygous parent to an affected child is observed. For example, when both parents are heterozygous for the associated marker allele, one would expect the associated allele to be transmitted to an affected child 50% of the time, if the allele is not linked to the disease trait. In contrast, if the marker allele is linked to the disease it will be transmitted to the affected child more often than 50% of the time. There are two major disadvantages to the TDT. The first is that the power of the test is dramatically reduced by locus heterogeneity to a point where sample size can reach unattainable levels (SLAGER et al., 2000). Second, to use the TDT, one must first find marker alleles that associate with the disease within a population. This can be a daunting task when the entire genome must be screened. The Human Genome Project and its associated technologies hold great promise in aiding in the analysis and identification of important polymorphisms for every gene (RISCH and MERIKANGAS, 1996a). As the technology for genome-wide screening of polymorphisms becomes available, it is fortunate that the same samples currently collected for linkage studies, can also be used for TDT analysis. Thus, some synergy can be accomplished between the various methods of studying genetic disease and the burgeoning field of human genomics.

Lastly, the power of standard linkage analysis can also be increased by grouping families with similar severity or onset of disease. For example, although colon cancer is considered a multifactorial disease, strong disease clustering in some families had been reported in studies as early as 1913 (WARTHIN, 1913). Colon cancer is very common in the Western world, and it is sometimes difficult to rule out sporadic mutation as the cause of disease rather than familial inheritance. As well, environmental factors and diet have a strong effect on the penetrance of the disease. Early-onset colon cancer occurs in a minority of cases, but exhibits the strongest familial clustering. By confining linkage analysis to these families, it was possible to identify the genes responsible for two common familial forms of colon cancer: familial adenomatous polyposis and hereditary non-polyposis colorectal cancer.

2.2.1 Colon Cancer: Insights from the Familial Colorectal Cancer Syndromes

About 1 in 20 individuals in the Western world will be diagnosed with colon cancer at some time in their life. Although the majority of colon cancer cases are sporadic and involve novel mutations, it is estimated that about 13% of all cases of colon cancer are familial (HOULSTON et al., 1992). Perhaps the best understood form of hereditary colon cancer is familial adenomatous polyposis (FAP), which accounts for approximately 1% of all cases of colon cancer (LYNCH and DE LA CHAPELLE, 1999; SORAVIA et al., 1997). FAP is an autosomal dominant condition that can result in the development of hundreds of adenomatous polyps in the lower colon during early adult life. There is high probability that one or more of these polyps will eventually develop into adenocarcinoma. As well, polyps can develop in the up-

per gastrointestinal tract, and other malignancies may occur in the brain and thyroid. FAP associated with extracolonic lesions such as desmoid tumors or epidermoid cysts is sometimes separately classified as Gardner's syndrome. A candidate region for FAP on chromosome 5q had been suggested by an interstitial deletion found in an individual with FAP (HERRERA et al., 1986) and by RFLP analysis of DNA from malignant colorectal tumors (BODMER et al., 1987). This region would later be narrowed down to chromosome 5q21-22 (NAKAMURA et al., 1988). In 1991, two groups would independently clone a gene in this region that was found to be mutated in patients with FAP, the adenomatous polyposis coli gene (APC) (GRODEN et al., 1991; JOSLYN et al., 1991; KINZLER et al., 1991; NISHISHO et al., 1991). The APC gene is a tumor suppressor gene that is a component of the Wnt signaling pathway and may play a role in cell migration, apoptosis, and cell proliferation (NATHKE, 1999). The gene is large, consisting of 15 exons, with approximately 80% of the APC mutations identified occurring in exon 15 (MIYOSHI et al., 1992a). These mutations usually lead to a truncated protein product that can be detected by *in vitro* translation assays and Western analysis. A milder or attenuated form of FAP associated with fewer polyps (i.e., less than 100), designated as AFAP, is also seen in some pedigrees with classical FAP, but is not always associated with APC mutations (PEDEMONTE et al., 1998). Thus, the clinical and genetic characterization of AFAP remains to be resolved and will not be discussed here.

Although FAP is rare, occurring in only 1/8,000 individuals, mutations in the APC gene or loss of heterozygosity (LOH) for the chromosomal region containing the gene, are found in almost 80% of all cases of colon cancer (MIYOSHI et al., 1992b). In fact, mutations in the APC gene are thought to initiate the majority of sporadic colorectal tumors (POWELL et al., 1992). Many of these mutations involve deletions that prevent the interaction of β-catenin with APC, an important aspect of the tumor suppressor function of APC (MORIN, 1999). The APC gene interacts with β-catenin and a serine-threonine kinase GSK3-β, which phosphorylates β-catenin and targets the protein for degradation. APC-mediated degradation of β-catenin in turn prevents the activation of T cell factor (TCF)-mediated transcription of several genes involved in cell proliferation, including c-Myc and cyclin D. Mutations in the regulatory region of β-catenin, which is required for phosphorylation of the protein by GSK3-β, are found in 50% of all cases of FAP that do not involve mutations in APC (SPARKS et al., 1998). Thus, understanding the etiology of a relatively rare familial cancer syndrome such as FAP has led to the definition of a molecular pathway that has implications for the diagnosis and treatment of colon cancer in general. This has been a common theme in cancer research, where the analysis of genes responsible for rare cancer syndromes has provided valuable information on the various pathways of oncogenesis. Other rare cancer syndromes which have provided insight into the molecular mechanisms behind tumor development include retinoblastoma (Rb gene), Li–Fraumeni syndrome (p53), familial melanoma (p16 and CDK4), and familial breast cancer (BRCA1 and 2).

Hereditary non-polyposis colorectal cancer (HNPCC), also known as Lynch syndrome, accounts for approximately 5–7% of all cases of colorectal cancer (LYNCH and DE LA CHAPELLE, 1999; SORAVIA et al., 1997). The disease is characterized by a much smaller number of colon polyps than seen for FAP, but is highly penetrant nonetheless, and is associated with a number of other cancers including endometrial, stomach, small intestine, and brain carcinomas. Because it is often difficult to distinguish HNPCC from sporadic colon cancer, strict criteria had to be developed to allow the accurate ascertainment of affected families. Linkage studies were carried out using families that had the following criteria:

(1) 3 affected members in 2 generations,
(2) at least one family member affected before the age of 50, and
(3) in addition to colon cancer, affected individuals had to display carcinomas of the endometrium, stomach, pancreas, or urinary tract (PELTOMAKI et al., 1993; VASEN et al., 1991).

Using these criteria, a microsatellite marker D2S123 was found that mapped to a candidate region at chromosome 2 p16-p15, which predisposed an individual to HNPCC (AALTONEN et al., 1993; PELTOMAKI et al., 1993). In the same year, a secondary locus linked to HNPCC was found at 3p23-p21 (LINDBLOM et al., 1993). Thus, by limiting the choice of affected pedigrees to relatively early-onset disease with strong penetrance and severity it was possible to discern two major QTLs for this form of hereditary colon cancer.

The identity of the genes at these two loci would be discovered partly through the association of HNPCC with the instability of small CA and TA dinucleotide repeat arrays in a significant number of patients (12–13%) (AALTONEN et al., 1993; IONOV et al., 1993). The instability of these microsatellite sequences was thought to reflect an increased rate of replication errors. In addition, tumors with high levels of microsatellite instability were shown to have profound defects in mismatch repair (AALTONEN et al., 1993; PARSONS et al., 1993). The genes responsible for DNA mismatch repair in *E. coli* were already known and disruption of either the MutS or the MutL gene was shown to produce microsatellite instability in both bacteria and yeast (LEVINSON and GUTMAN, 1987; STRAND et al., 1993). Consequently, the underlying genes involved in HNPCC where believed to have a role in mismatch repair and most likely would share sequence similarity with the *E. coli* mismatch repair proteins MutS and MutL. Using these criteria, the human homolog of the *E. coli* mismatch repair protein MutS, called MSH2, was found within the candidate region on chromosome 2p16 (LEACH et al., 1993). Furthermore, germline mutations in the MSH2 gene were found in both of the affected families used to originally establish linkage to chromosome 2. Soon after, a human homolog of the bacterial MutL gene, MLH1, could be mapped to chromosome 3p21.3 by fluorescence *in situ* hybridization (PAPADOPOULOS et al., 1994), the same region that was independently identified by LINDBLOM et al. using linkage analysis (LINDBLOM et al., 1993). The same study also defined two additional genes that shared similarity to the yeast MutL homolog, designated as PMS1 and PMS2. Colon cancer resulting from mutations in MSH2 is classified as type 1 HNPCC and MLH1 mutations are found in type 2

HNPCC. PMS1 and PMS2 were later shown to also predispose an individual to colon cancer, and mutations in these genes are found in type 3 and 4 HNPCC, respectively (NICOLAIDES et al., 1994). Lastly, another MutS homolog, most similar to the yeast gene MSH6 is found very tightly linked to MSH2 on chromosome 2p16 (PAPADOPOULOS et al., 1995). Germline mutations in the human MSH6 gene have been found in atypical HNPCC families that do not fulfil the strict criteria for HNPCC linkage studies (WIJNEN et al., 1999). This form of hereditary colon cancer, associated with germline mutations in MSH6, is designated as HNPCC type 5. Thus, HNPCC exhibits locus heterogeneity and a mutation in any of the five major susceptibility genes (i.e., MSH2, MLH1, PMS1, PMS2, or MSH6) can result in hereditary non-polyposis colorectal cancer.

The cloning of the HNPCC genes reflects the need to use multiple approaches when trying to determine the etiology of a complex genetic disease. Although linkage analysis had aided researchers in determining which chromosomal region might contain a gene involved in HNPCC, the actual genes involved where cloned by correlating the satellite instability of colorectal tumors with a similar instability found in the DNA mismatch repair mutants of *E. coli*. Therefore, the previous molecular analysis of the bacterial and yeast mismatch repair genes was instrumental in the final cloning of the HNPCC genes. This example serves to illustrate why it is important to continue work in model systems such as yeast and bacteria, which provide a much simpler and genetically tractable system than is available or ethically possible with humans.

At least two other hereditary forms of colon cancer exist. The first is Peutz–Jeghers syndrome, which is caused by mutations in the STK11 a serine–threonine kinase located on chromosome 19q13.3-q13.4 (HEMMINKI et al., 1998). The second is juvenile polyposis syndrome (JPS), which may be caused by mutations in the SMAD4/DPC4 gene on chromosome 18q21.1 (HOWE et al., 1998) or may result from mutations in the PTEN tumor suppressor gene located on chromosome 10q23.3 (OLSCHWANG et al., 1998). Another variant of JPS, called hereditary mixed polyposis syndrome (HMPS), is linked to a region on chromosome 6q16 (THOMAS et al., 1996). Even with the addition of these two forms of hereditary colon cancer, the fact remains that more than 80% of colon cancer cases are not hereditary and probably arise from novel germline or somatic mutations. Interestingly, most of the sporadic cases of colorectal cancer involve either the APC gene or one of the mismatch repair genes associated with HNPCC. Therefore, the insights obtained from studying the rare hereditary forms of colon cancer can also be applied to sporadic cases of colon cancer.

2.3 The Management of Genetic Disease: Diagnosis to Therapy

Once a gene mutation, a marker allele, or a candidate region of a chromosome has been found consistently associated with a disease, diagnostic tests can be developed for use in population or prenatal screening. Screening may rely on the detection of the actual disease mutation by methods such as polymerase chain reaction (PCR), on the biochemical detection of metabolic changes caused by the disease, on the detection of chromosomal aberrations associated with the disease by fluorescent *in situ* hybridization (FISH), or by the observation of developmental abnormalities by ultrasound (JORDE et al., 1999). For example, by PCR genotyping individuals for recessive disorders such as cystic fibrosis or late-onset diseases such as Huntington disease, potential carriers or affected individuals (respectively) can be identified and provided with genetic counselling. In another example, the timely diagnosis of phenylketonuria (PKU) by the detection of high serum levels of phenylalanine, can reduce the risk of mental retardation in affected individuals, if the appropriate dietary modifications are made (see Sect. 1.1.3.1). Prenatal diagnosis can also be undertaken using genotyping, biochemical assays, FISH analysis, or ultrasound. For example, because of the increased risk of trisomies with advancing maternal age, women over the age of 35 routinely undergo prenatal diagnosis for chromosome aneuploidies by amniocentesis or chorionic villus sampling (CVS) combined with FISH analysis. Lastly, ultrasound can diagnose several developmental disorders by the second trimester of pregnancy, including: cleft lip, spina bifida, renal agenesis, anencephaly (absence of the head), and osteogenesis imperfecta.

Perhaps the most controversial form of genetic screening is preimplantation diagnosis (PID), where after *in vitro* fertilization, embryos are screened for genetic disease or chromosome abnormalities before implantation (NAGY et al., 1998). The advantage of PID over prenatal diagnosis is that both the risk associated with invasive techniques such as amniocentesis or CVS and the psychological consequences of abortion can be avoided. The technique may be particularly valuable for couples who:

(1) have faced spontaneous miscarriages without a specific genetic diagnosis,
(2) are at risk for transmitting X-linked disorders, or
(3) are experiencing infertility caused by chromosomal translocations and/or gonadal mosaicism (NAGY et al., 1998).

Currently the technique is being developed in more than 20 centers world-wide, but ethical and financial issues threaten the wide-spread adoption of the procedure. In particular, there is concern that PID will be used to choose the sex of a child or to select for specific desirable traits rather than to prevent genetic disease. An overview of genetic screening and prenatal diagnosis is presented in Tab. 6A, and a list of commonly diagnosed genetic diseases can be found in Tab. 6B.

2.3.1 Cystic Fibrosis: Screening as Prevention and Gene Therapy as a Cure

Three main criteria are usually considered before population screening is undertaken for a particular disease (JORDE et al., 1999):

(1) *Disease characteristics*: Is the condition serious enough or common enough to warrant the cost of a screening program? Is the natural history and etiology of the disease sufficiently un-

derstood to provide effective treatment or prenatal diagnosis?
(2) *Test characteristics*: Is the test acceptable to the population, simple to perform, and feasibly priced? Is the test reliable and valid?
(3) *System characteristics*: Are the resources for diagnosis and treatment of the disorder accessible? Are there strategies for the effective and efficient communication of test results?

Wide-spread population screening for phenylketonuria (PKU) and spina bifida are routinely carried out as they fit these criteria adequately. In the case of PKU, the serum phenylalanine levels of the fetus are analyzed, whereas high maternal serum levels of α-fetoprotein are a prognosticator of open neural tube defects such as spina bifida (JORDE et al., 1999). These tests are relatively inexpensive and the appropriate counselling and therapeutic interventions already exist. At first glance, cystic fibrosis (CF) also appears amenable to population screening. CF is one of the most common of the single-gene disorders, and more than 70% of all cases are caused by the same mutation in the CFTR gene. Since the mutation involved in the majority of Caucasian CF cases is a single amino acid deletion (ΔF508), a genetic test based on this deletion was feasible and cost-effective for prenatal screening. In addition to the ΔF508 mutation, several other common CF mutations are typically screened for simultaneously, allowing a detection frequency of greater than 90% (GIRODON-BOULANDET et al., 2000). The carrier frequency in Northern Europeans is quite high, approximately 1 in 25. Thus, about 1/600 couples will consist of two heterozygote carriers for CF. Typically, this subpopulation of couples at risk can be offered the option of prenatal diagnosis and counselling. Population screening for CF has been considered too costly, therefore, only couples that have a positive family history for CF, or in which one member is a known carrier, are usually screened for the disorder. However, a modified antenatal screen for CF, in which pregnant females are first tested followed by their partners, if they are positive, was found to be cost-effective and has lead to a 65% reduction in the incidence of cystic fibrosis in Edin-

Tab. 6A. Forms of Genetic Screening and Prenatal Diagnosis (partially adapted from JORDE et al., 1999)

Population Screening	Prenatal Diagnosis	Family Screening
Newborn screening Blood and genetic tests: e.g., PKU, galactosemia, hypothyroidism Cystic fibrosis and hemoglobinopathies. Urine tests: e.g., Metabolic disorders such as aminoacidopathies	*Invasive diagnostic testing* 1. Amniocentesis 2. Chorionic villus sampling 3. Percutaneous umbilical blood sampling *Fetal visualization techniques* 1. Ultrasonography 2. Radiography 3. Magnetic Resonance Imaging	*Considerations for Family Screening* 1. A family history of chromosomal rearrangements such as translocations 2. Female relatives may be screened, if the family pedigree shows X-linked inheritance (e.g., fragile X or Duchenne muscular dystrophy) 3. Heterozygote screening may be carried out within at-risk families (e.g., screening for cystic fibrosis within a family pedigree after the birth of an affected child) 4. Presymptomatic screening for diseases with late on-set (e.g., breast and colon cancer and Huntington disease)
Heterozygote screening Examples: 1. Tay–Sachs disease in the Ashkenazi Jewish population 2. Sickle-cell disease in the African-American population 3. Thalassemias in Asian populations 4. Cystic fibrosis in populations with high incidence, e.g., Scotland or in French Canadian population	*Maternal considerations for prenatal diagnosis by amniocentesis* 1. Maternal age > 35 years 2. Family history of conditions diagnosable by prenatal techniques. 3. Abnormal maternal serum α-fetoprotein[a]	*Preimplantation Diagnosis (PID)* Indications for PID[b] 1. Couples at risk for transmitting X-linked disorders for which no specific molecular diagnosis exists 2. Couples who have experienced multiple miscarriages without a specific molecular diagnosis 3. Couples who are subfertile or infertile because of chromosomal abnormalities and/or gonadal mosaicism

[a] Maternal serum α-fetoprotein (AFP) levels are good indicators of neural tube defects (high AFP) and Down syndrome (low AFP).
[b] As outlined in NAGY et al. (1998).

burgh, Scotland (CUNNINGHAM and MARSHALL, 1998).

The CFTR gene was cloned in 1989 (RIORDAN et al., 1989), a step which has led to a greater understanding of the complex phenotype of CF and to hopes of treating affected individuals through gene-based therapies. The pancreatic enzyme deficiency can be corrected by enzyme replacement therapy, but the respiratory problems associated with CF are more difficult to treat. In 1994, five years after the cloning of the CFTR gene, attempts were made to alleviate the symptoms of the disease by somatic gene therapy using adenoviral transfer of the human CFTR gene to lung epithelial cells (BOUCHER et al., 1994; CRYSTAL et al., 1994). Unfortunately, the effects were transient and most patients developed respiratory problems due to inflammation caused by an immunological response to the adenovirus. In addition, expression levels dropped after each administration as the patient developed immunity to the adenovirus. Attempts have been made to develop systems which do not produce lung inflammation such as liposomal gene transfer of non-viral DNA, but adenovirus remains the vector of choice because of its efficiency (MCELVANEY, 1996). Conse-

Tab. 6B. Commonly Diagnosed Disorders for which Diagnostic Tests are Available

Disease	Nature of the Disease
α-Thalassemia	AR severe anemia
β-Thalassemia	AR severe anemia, growth retardation, and splenomegaly
Charcot–Marie–Tooth disease	XLD neurodegenerative disorder
Cystic fibrosis	AR chloride channel disorder affecting the lung and pancreas
Duchenne–Becker muscular dystrophy	XLR progressive neuromuscular disorder
Familial breast cancer	AD breast and ovary cancer syndromes (BRCA1 and BRCA2)
Familial colon cancer (FAP and HNPCC)	AD colon cancer syndrome (APC and mismatch repair genes)
Fragile X syndrome	XLD mental retardation
Friedreich's ataxia	AR spinocerebellar degeneration and cardiomyopathy
Gaucher's disease	AR lysosomal storage disorder, splenomegaly and hepatomegaly
Hemochromatosis	AR cirrhosis of the liver with secondary diabetes and heart failure
Hemophilia A	AR bleeding disorder
Hereditary papillary renal carcinoma	AD renal cancer syndrome
Huntington disease	AD progressive neurodegenerative disorder
Li–Fraumeni syndrome (p53)	AD multiple cancer syndrome
Myotonic dystrophy	AD myotonia, muscular dystrophy and cataracts
Neurofibromatosis type I	AD cancer syndrome characterized by neurofibromas
Sickle-cell disease	AR blood disorder associated with anemia and organ damage
Spinocerebellar ataxia	AD neurodegenerative disorder
Tay–Sachs disease	AR lysosomal storage disorder that results in severe neurodegeneration
von Hippel–Lindau syndrome	AD cancer syndrome affecting the brain, spine, retina, pancreas and kidneys

AD: autosomal dominant; AR: autosomal recessive; XLD: X-linked dominant; XLR: X-linked recessive.

quently, many attempts have been made to produce adenoviral vectors that elicit a reduced immunological response, but with limited success. Model systems such as the CFTR knock-out mouse and the lungs of sheep are being used to test the next generation of liposomal and adenoviral vectors in the hopes of developing an effective gene therapy for cystic fibrosis (KITSON et al., 1999; RAMJEESINGH et al., 1998). To date the only gene therapies to provide actual clinical benefit involve *ex vivo* transfer of genes to bone marrow cells (e.g., gene therapy for human severe combined immunodeficiency, SCID) (CAVAZZANA-CALVO et al., 2000). The majority of *in vivo* gene therapy trials have been plagued by the same problems of gene delivery, efficient expression, and safety. Thus, one must consider gene therapy for cystic fibrosis with guarded optimism. Gene therapy will not be covered further in this chapter as several excellent reviews on gene therapy of genetic diseases and cancer are available (MCIVOR, 1999; SANDHU et al., 1997).

2.3.2 Colon Cancer: Early Diagnosis and Prevention

As discussed in Sect. 2.2.1, it is estimated that about 13% of all cases of colon cancer are familial (HOULSTON et al., 1992). DNA sequence-based mutation detection is possible for the genes responsible for familial adenomatous polyposis (FAP), hereditary non-polyposis colon cancer (HNPCC), Peut–Jeghers syndrome, and juvenile polyposis. It is believed that genetic screening for individuals at risk combined with clinical surveillance for polyps and dietary modification can dramatically reduce the number of colon cancer fatalities. Because the genes responsible for Peut–Jeghers syndrome and juvenile polyposis have been isolated relatively recently (see Sect. 2.2.1) and the disorders are very rare, screening and early diagnosis of FAP and HNPCC has had the most clinical impact.

In contrast to cystic fibrosis, where population screening or antenatal screening of pregnant woman has been carried out, genetic screening for colon cancer is limited to "at-risk" families that contain at least one affected member. This is partly due to the fact that about 80% of colon cancer cases are sporadic, arising from *de novo* mutations. There are two indications for genetic testing for inherited forms of colon cancer: first, to confirm the diagnosis in the affected individual, and second, to test at-risk family members, usually first degree relatives (PETERSEN et al., 1999). In the case of FAP, individuals at risk of developing colorectal cancer are first diagnosed by flexible sigmoidoscopy as having hundreds or thousands of benign adenomatous polyps lining their colon. The clinical diagnosis of FAP is then confirmed by the detection of mutations in the APC gene. Although direct DNA sequence analysis of peripheral blood can be carried out to detect mutations in the APC gene, primary screening usually involves a protein truncation test. APC gene mutations associated with FAP often involve frameshift or missense mutations that lead to a shortened protein product. Therefore, the protein truncation test can detect these shortened APC proteins by an *in vitro* protein synthesis assay (POWELL et al., 1993). Once the diagnosis of FAP has been confirmed by germline mutation in the APC gene, the individual will most often have a prophylactic colectomy to reduce their cancer risk. In addition, all first-degree relatives should be screened for the same germline mutation. If they are positive for the mutation, they will undergo sigmoidoscopic surveillance for polyps, beginning in the early teen years.

HNPCC diagnosis and management is less straightforward than FAP as the first indication of disease is usually after colon cancer has already developed. As with FAP, an individual may be referred to a specialist for gastrointestinal work-up and endoscopy after either passing polyps or having detected occult fecal blood. In addition, a striking family history of colon cancer may also indicate the need for endoscopic surveillance. Usually, there are fewer than 10 polyps, and histopathalogical analysis is carried out on the colon tissue following polypectomy. Sometimes an individual with a family history of colon cancer may elect to undergo screening for the disease before any clinical manifestation of the disease. Tumors may be restricted to colorectum (Lynch syndrome I) or may also occur at other extra-colonic sites (Lynch syndrome II) including the endometrium, small bowel, stomach, breast, ovary, and skin (LYNCH, 1999).

Novel mutations are common in many cases of colorectal cancer so unless family history clearly indicates HNPCC, colorectal tumors are usually first evaluated for microsatellite instability (MSI). Tumor DNA that shows alterations in microsatellite DNA can indicate somatic changes or suggest HNPCC, at which time mutation analysis is carried out to find germline mutations in the DNA mismatch repair genes (PETERSEN et al., 1999). Genetic screening is commercially available for the two most commonly mutated DNA mismatch repair genes, hMSH2 and hMLH1. Once a mutation is found, then at-risk family members can be screened for the mutation and endoscopic surveillance and prophylactic surgery can be undertaken. Even if a mutation in hMSH2 or hMLH1 is not found, it is still recommended that at-risk family members undergo tumor surveillance. Coloscopy for family members who are positive for the mutation should be done annually and any polyps surgically removed, commencing at age 20. In addition, because endometrial cancer is the most common extracolonic malignancy in HNPCC, transvaginal ultrasound and endometrial aspiration should begin for female family members at the age of 25–35 and repeated annually (LYNCH, 1999). As well, individuals with HNPCC may elect to have a prophylactic colectomy to reduce their lifetime cancer risks, which is approximately 80% without surgery. Female family members who are at-risk for HNPCC may also undergo a hysterectomy, with or without oophorectomy for the same reason (LYNCH, 1999).

In addition to endoscopic surveillance and prophylactic colectomy, dietary modification and chemoprevention can also have an impact on the outcome of colon cancer. For example, high dietary fat and low dietary fiber (BARTSCH et al., 1999; FERGUSON and HARRIS, 1999) are associated with an increased risk of colon cancer, whereas vitamin D and calcium may provide some level of chemoprevention (FERGUSON and HARRIS, 1999). The high fat, low fiber diet of the Western countries may partly explain the disparity in colon cancer rates between the West and Asia, where colon cancer rates are low and dietary fiber is high. Interestingly, dietary changes do not seem to have an effect on the recurrence of colon cancer in affected individuals, demonstrating the need for further studies to determine the true impact of diet on colon cancer (SCHATZKIN et al., 2000).

In contrast to dietary changes, there is clear evidence that chemoprevention may be a viable means of reducing colon cancer rates. For instance, several studies indicate that aspirin may have a protective effect on individuals who are at-risk for colon cancer (BARNES and LEE, 1998; GIOVANNUCCI et al., 1994; THUN et al., 1993). Other non-steroidal anti-inflammatory agents (NSAIDs), such as sulindac and the selective cyclooxygenase-2 (COX-2) inhibitors, have demonstrated promising results for both FAP and HNPCC (GIARDIELLO et al., 1993; SHENG et al., 1997). Thus, the earlier colon cancer can be diagnosed and at-risk family members genetically screened, the sooner prophylactic surgery or chemoprevention can be administered.

Lastly, the majority of colon cancer cases are sporadic, resulting from somatic or novel germline mutations. Nonetheless, about 10–15% of individuals with pedigrees that do not support the criteria and diagnosis of FAP or HNPCC, have other family members affected by colorectal cancer. These families often exhibit multiple forms of cancer in addition to colon cancer. This form of familial colorectal cancer syndrome may be explained by the affect of modifier genes on existing mutations of low penetrance in one or more cancer susceptibility genes. For example, a mutation in the APC gene leading to the replacement of isoleucine 1307 by lysine, has been found to cause familial colorectal cancer among a proportion of the Ashkenazic Jewish population (LAKEN et al., 1997). The mutation confers a lifetime risk of developing colorectal cancer of about

20%, which is in stark contrast to the almost 100% lifetime risk for individuals with mutations leading to truncated APC gene products (PETERSEN et al., 1999). The lower penetrance suggests that other modifying genes may be involved. Hence, a significant majority of colon cancer, like many complex diseases, may actually occur on a background of modifier alleles that are individually rare within the general population. It is hoped that the analysis of human genetic variation, made possible by the Human Genome Project, will help define these modifier genes for colon cancer and other complex genetic diseases.

3 Genetic Disease and Genomics: Impact of the Human Genome Project

At the heart of the Human Genome Project is the sequencing of the entire human genetic code. The project was launched in 1990, and the draft sequence of the human genome will be finished in 2001, with a high-quality (>99.99% accuracy) version of the sequence to be available by 2003. Even before the completion of the human genome, we have been propelled head long into the world of genomic biology through the successful sequencing of several genomes including: 20 species of bacteria, the yeast *Saccharomyces cerevisiae*, the fruit fly *Drosophila melanogaster*, and the nematode *Caenorhabditis elegans*.

At one time, each gene was regarded as a single entity coding for a single protein with usually a single enzymatic activity or function. Now with our new genomic perspective, genes are seen to exist in networks, interacting with each other in both space and time at the DNA, RNA, and protein level to carry out their function. Thus, techniques have been developed and bioinformatics tools created, which together can collect and analyze data that more accurately reflect the underlying complexity of these genetic networks (reviewed in BRENT, 2000). This complexity is demonstrated at the DNA level by multiple alleles that exist within any given population for each of the estimated 80,000 human genes. In response, DNA micro-arrays (gene chips) have been developed to help us analyze this complexity by genotyping the DNA of an individual for the major alleles of each gene using single nucleotide polymorphisms (SNPs). Even more complexity exists at the RNA level, where a gene may have multiple splice variants or promoters that are regulated in a developmental or tissue-specific manner. Gene chip technology can also be used to analyze transcription by providing a snapshot of gene expression for all the genes expressed in a given cell or tissue. This massive parallelism is also extended to protein analysis through proteomics, using techniques such as mass spectrometry and yeast two-hybrid analysis to probe the complex protein–protein interactions occurring within each cell. Consequently, this new era of genomic biology has impacted the study of genetic disease in several areas, including gene discovery, genetic screening and diagnosis, and drug development and therapy. This section outlines the impact of the Human Genome Project in each of these areas.

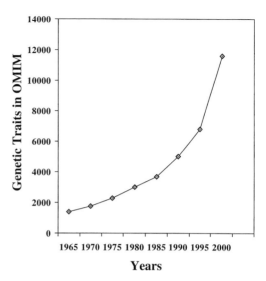

Fig. 3. Logarithmic growth of gene trait entries in the On-Line Mendelian Inheritance in Man database (OMIM). Adapted from OMIM statistics (*http://www3.ncbi.nlm.nih.gov/Omim/Stats/mimstats.html*). On May 16, 2000, there were 11,614 gene trait entries in the OMIM database.

3.1 Accelerating Gene Discovery

It is becoming increasingly evident that the Human Genome Project (HGP) has had a dramatic impact on many areas of medical genetics, particularly on gene discovery. If we use the number of genetic traits listed in the On-Line Mendelian Inheritance in Man (OMIM) database as a barometer for gene discovery, it is clear that a logarithmic increase in the number of known gene traits has occurred since the inception of the HGP in 1990 (Fig. 3) (McKusick, 2000). This pace of gene discovery is still accelerating and a quick calculation suggests that during the writing of this chapter the OMIM database has been growing by about 6 entries a day! The furious pace of gene discovery has been fuelled to some extent by the growing availability of sequence data and the bioinformatics tools to search or "mine" DNA and protein sequence databases. At first, these databases contained very little genomic sequence, being primarily repositories of expression sequence tags (ESTs), which are comple-

mentary DNA (cDNA) sequences generated from the mRNA of genes expressed in the cells of various organisms. As sequencing technologies developed, these databases facilitated the rapid cloning of novel genes and the inference of putative function by the comparison of experimental sequence to the ESTs in databases such as GenBank (Bilofsky et al., 1986). The two most common algorithms for databases searching, BLAST (Altschul et al., 1990) and FASTA (Pearson and Lipman, 1988), have been so instrumental in the cloning of novel human genes that at least one author suggests that the credit for many of these discoveries should be shared by the creators of these algorithms (Brent, 2000). When coupled with the phylogenic comparison of genes between species, data mining of sequence data can be extremely powerful at inferring gene function by sequence homology. In particular, comparative genomics in mammals provides both the opportunity to develop animal models for human disease, as well as a chance to increase our understanding of how genes and gene families have evolved (reviewed in

O'BRIEN et al., 1999). Lastly, the resolution of linkage mapping has become increasing fine as the HGP progresses and the number and density of polymorphic markers increases. The increased genetic map density, coupled with the ability to search genomic and EST databases for candidate genes, has made the positional cloning of many human disease genes possible.

3.1.1 Genetic Variation and Single Nucleotide Polymorphisms (SNPs)

Increasing emphasis is now being placed on the genetic variation of human genes and the discovery of clinically relevant allelic polymorphisms. This change in emphasis follows the realization that the genes responsible for the majority of complex human diseases are not easily mapped by traditional linkage analysis, because of their polygenic nature and/or low penetrance. Instead, to uncover disease alleles with low penetrance, researchers are turning to methods involving the association analysis of many affected and unaffected individuals in large populations (e.g., the Transmission Disequilibrium Test, TDT; see Sect. 2.2). To perform this type of analysis, with the sensitivity required to find alleles of low penetrance or frequency, requires hundreds of thousands of polymorphic markers across the entire genome (RISCH and MERIKANGAS, 1996). Such a density is not possible with the variable nucleotide repeats (VNTRs) or sequence tagged sites (STSs) used previously for linkage mapping. In contrast, the required density of markers may be provided by single nucleotide polymorphisms (SNPs), which make up 90% of the sequence variation in humans and differ from each other by single bases of DNA (COLLINS et al., 1998). It is estimated that within our genome there are about 17 million possible SNPs within the 3 billion bases of DNA in the human haploid genome. Of these SNPs, only about 5% will occur in coding regions of genes, which are the polymorphisms most likely to have consequences for gene function. Therefore, one can estimate about 500,000 informative SNPs within the human genome, which is an average of about 6 SNPs per gene (COLLINS et al., 1998). In addition, for SNPs to be informative for gene mapping, they must occur within populations at frequencies of 1% or more.

SNPs can be found by direct sequence analysis of genomic DNA or by data mining of polymorphisms in EST databases to produce cSNPs (PICOULT-NEWBERG et al., 1999). Although cSNPs by definition are found in gene coding regions, caution must be exercised as some cSNPs may arise by RNA editing. SNPs generated from genomic sequencing may also include promoter SNPs or SNPs in other DNA regulatory elements that may have an effect on gene expression (e.g., enhancers, locus control regions, splice sites, imprinting centers, etc.). Several large-scale academic and industrial SNP discovery projects have been initiated in the last 2 years (see BROOKES, 1999, and references therein). Of these projects, the most prolific is a joint academic/drug industry venture called, "The SNP Consortium (TSC)", which is committing $45 mio (U.S.) to the discovery and mapping of 300,000 SNPs. In a complimentary effort, several internet-based SNP databases have been developed to archive and annotate the tens of thousands of publicly available SNPs generated over the past decade (reviewed in a special issue of Nucleic Acids Research: 2000 Database Issue, 2000). The most comprehensive of these is the HGBASE (Human Bi-Allelic Sequences), developed by ANTHONY BROOKES and colleagues at the Center for Genomics Research at the Karolinska Insitute in Stockholm, Sweden (BROOKES et al., 2000).

A number of methods are available for SNP detection and scoring, but an exhaustive list of these techniques is beyond the scope of this chapter (for a review of these technologies, see LANDEGREN et al., 1998). Most of these procedures involve target sequence amplification by PCR and/or hybridization to DNA microarrays or chips. These microarrays where developed based on the earlier concept of gridded DNA libraries. They most often consist of glass slides or wafers on which cDNAs or oligonucleotides are adhered by spotting on the slides or by direct synthesis of the oligonucleotides on the chip (e.g., by photolithography or inkjet deposition). Detection is carried out by annealing fluorescently labeled DNA to the microarrays followed by computer-assisted scanning and scoring. More recently, high-throughput screening of SNPs has been accomplished by peptide–nucleic acid (PNA) detection of the polymorphisms coupled to MALDI-TOF mass spectrometry (GRIFFIN and SMITH, 2000). However, the intrinsic problem of processed pseudogenes plagues the genomic scoring of SNPs with any of these detection techniques. These pseudogenes can be highly similar to the gene of interest and thus will complicate the detection of SNPs for these genes, especially cSNPs based on exon sequences. It is estimated that pseudogenes may present an obstacle for approximately 20% of all cSNP assays (BROOKES, 1999). As well, problems in the detection of the most abundant mismatch, the G–T mismatch, can occur as G–T mismatches are almost as stable as true A–T base pairing (BROOKES, 1999).

The density of SNPs within the genome can facilitate increasingly fine genetic maps and provide useful markers for association analysis. In particular, if weakly penetrant cSNP alleles that produce a mild pathogenic influence are tolerated by natural selection, then some of these alleles will drift to high frequencies within certain populations. As well, SNPs that are not pathogenic themselves, but are subject to linkage disequilibrium with a pathogenic allele, thus acting as a surrogate marker for the disease mutation, can also be used for association analyses such as the Transmission Disequilibrium Test (TDT) (see Sect. 2.2). These two types of allele variants are precisely the ones that we hope association analysis will uncover, providing insight into polygenic and complex disorders such as cardiovascular disease, cancer, and diabetes. Two strategies for SNP-based association analysis are shown in Fig. 4, and a full discussion of these strategies can be found in BROOKES (1999).

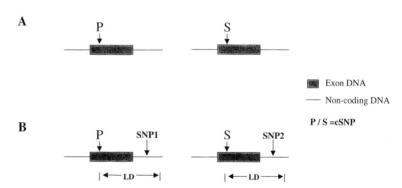

Fig. 4. Strategies for association analysis with and without linkage disequilibrium. A pathogenic allele (P, the amino acid proline) that corresponds to a coding single nucleotide polymorphism (cSNP) within an exon, and its non-pathogenic counterpart (S, the amino acid serine), are shown in a disease susceptibility gene. Other pathogenic mutations including SNPs found in non-coding DNA, such as promoters or other gene regulatory elements, could also have been envisaged. (**A**) In direct association analysis, one tests for polymorphisms that occur in candidate genes for the disease causing gene itself (i.e., for cSNP P or S). Only alleles that occur at a higher frequency in affected individuals than normal individuals are scored as having positive association with the disease phenotype. (**B**) In linkage disequilibrium (LD) based association analysis, one tests random SNP alleles distributed over chromosomal candidate regions or the entire genome (i.e., SNP1 or SNP2). The success of the analysis relies on sufficient linkage disequilibrium between one or more of the tested polymorphisms and the disease predisposing gene. As with direct association analysis, a positive association between a marker and the disease occurs, if it is more common in affected individuals than in normal individuals. Lastly, positive association may be detected in direct association analysis for polymorphisms in the disease susceptibility gene that do not predispose an individual to the disease, but are in linkage disequilibrium with the actual pathogenic mutation (adapted from BROOKES, 1999).

3.2 Screening and Diagnostics

As has been discussed in Sect. 2.2, once susceptibility genes have been identified for a disorder, specific genetic or biochemical assays can be developed for diagnostic screening. The HGP has aided screening and diagnostic efforts by accelerating gene discovery (i.e., providing gene targets) and by providing a host of new technologies for detecting mutations in susceptibility genes. DNA microarray technologies in particular have already begun to make an impact on both the screening and diagnosis of genetic disease. A number of mutation screening technologies are summarized in Tab. 5.

DNA microarrays or gene chips can be used for gene expression analysis (profiling) and for the detection of specific gene mutations or SNPs associated with a genetic disease. For example, cDNA can be prepared from a tumor biopsy and used to create an expression profile for that tumor using a gene chip array that can detect common oncogenes. Recently, the expression profile of over 8,000 genes in 60 cancer cell lines was determined in this way by researchers at Stanford University (ROSS et al., 2000). These expression profiles can also be compared to the gene expression profile of normal cells taken from matched tissue, thus providing insight into the major changes in gene expression associated with oncogenesis in that tissue. Gene expression profiles can not only help classify the molecular character of a particular cancer, but can also suggest the most appropriate therapy when coupled with pharmacogenomics (to be discussed in Sect. 3.3). One recent example of the clinical utility of expression profiling was demonstrated by the identification of two distinct forms of diffuse large B cell lymphoma, which responded differently to chemotherapy and were associated with different intrinsic survival rates (ALIZADEH et al., 2000).

Sequential Analysis of Gene Expression (SAGE) is another technology that can be used instead of microarrays for expression profiling (VELCULESCU et al., 1995). SAGE analysis is less sensitive than gene chip analysis and involves blind data collection, where cDNA sequences are obtained for all mRNA in a cell. Therefore, unlike gene chip analysis, no decision is made before data collection as to the genes to be surveyed. Thus, the main advantage of SAGE over gene chip analysis is that once sequence data is obtained it can be searched afterwards for the expression of new candidate genes when they are discovered. In addition to gene expression, chromosomal rearrangements can also be used to genetically define cancer cells. One such techniques is spectral karyotyping (SKY), which can determine chromosomal rearrangements at megabase resolution over the entire genome (SCHROCK et al., 1996). The data generated by these various techniques provides an incredible resource for the cancer biologist and recently, a publicly available annotated index of cancer genes has been developed in the United States called the Cancer Genome Anatomy Project (STRAUSBERG et al., 2000). This resource lists SNPs, SAGE data, cDNA expression profiles, as well as a directory of tumor suppressor and oncogenes.

3.3 Pharmacogenomics and the Age of Individualized Therapy

Very shortly, researchers will have a base by base account of man's entire genetic blueprint in excruciating detail. However, the greatest impact on the treatment of human disease may not come from the sequence itself, but the genetic sequence variations that occur between individuals. These differences in DNA sequence are thought to hold the key to the variation found within human populations in regard to disease susceptibility, response to environmental factors, drug response, and normal variations in development and ageing. The field of study that is primarily concerned with the affect of genetic variation on disease susceptibility and drug response is pharmacogenomics (reviewed in EVANS and RELLING, 1999). Using pharmacogenomics, the health industry is poised to enter an age of individualized treatment, where therapies are both safer and more effective.

As we know, most prescription drugs are plagued by side effects and often a percentage of patients gain no benefit at all from their therapy, or worse still, die from an adverse drug reaction (ADR). The incidence of ADRs in hospitalized patients was estimate at approximately 6–7% in the United States, which translates to about 100,000 deaths from serious ADRs per year (LAZAROU et al., 1998). In part, the problem may lie in the genetic differences between individuals, 90% of which are represented by the single nucleotide polymorphisms (SNPs). These inherited differences alter not only our susceptibility to disease, but can also affect how we respond to drugs by altering either the interaction between the drug and its target, or the way an individual metabolizes the drug itself. The study of this inherited component of drug metabolism and disposition was once termed pharmacogenetics. Today the term pharmacogenomics is more commonly used, which describes both the identification of the myriad of genes responsible for drug behavior and sensitivity, as well as the elucidation of new therapeutic targets and interventions based on the knowledge of gene polymorphisms. Therefore, the ultimate goal of pharmacogenomics is an increase in drug efficacy and safety by matching drug characteristics to the genotype of an individual.

Genetic variations that affect the efficacy of a drug can be broadly classed as polymorphisms of

(1) drug metabolism or disposition,
(2) drug transporters, or
(3) drug targets (EVANS and RELLING, 1999).

These polymorphisms were traditionally uncovered by the comparison of the phenotypic response to drugs between populations of different racial background. For example, a genetic component was suspected for the differ-

ence in frequency of isoniazid-induced neuropathies observed in Japan as compared to the United States (EVANS and RELLING, 1999). This difference in drug metabolism was later shown to occur as a consequence of a slow acetylator phenotype associated with genetic polymorphisms at the NAT1 and NAT2 locus (SPIELBERG, 1996). Polymorphisms in drug transporters may alter the absorption of drugs taken orally, which require uptake in the gastrointestinal tract, or may prevent the normal excretion of drug metabolites in urine or bile. One clinically important transporter is the P-glycoprotein, which contributes to multidrug resistance in tumor cells and is also expressed in normal tissues with excretory function such as liver, kidney, and intestine (ORLOWSKI and GARRIGOS, 1999). Most drugs carry out their function by interacting with specific target proteins such as receptors or enzymes. Therefore, polymorphisms in drug targets that alter the interaction of a drug with its target protein can severely reduce the efficiency of a medication. For instance, polymorphisms in the $\beta 2$ adrenergic receptor can alter its sensitivity to β-agonists such as albuterol, which is commonly prescribed for asthma (LIMA et al., 1999).

Although many of the known pharmacogenetic polymorphisms are monogenic, drug response is most likely a polygenic trait, in which polymorphisms in many genes contribute to the overall efficacy of a given medical treatment, as depicted in Fig. 5. Therefore, pharmaceutical companies have begun to focus on the use of SNPs to provide polymorphic markers for many hundreds or thousands of genes that have an effect on the safety and efficacy of various drug treatments. Sev-

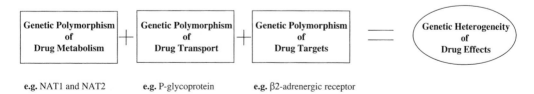

Fig. 5. The polygenic nature of drug effects. The genetic heterogeneity of a drug response arises from the sum effect of polymorphisms in many genes involved in drug metabolism, drug transport, and in the gene product being targeted by the drug. Examples of each type of polymorphism are shown (for details, see Sect. 3.3).

eral companies are now actively involved in patenting these clinically important SNPs including Celera Genomic (Rockville, USA), Genset (Paris, France), CuraGen (New Haven, USA), and Incyte Genomics (Palo Alto, USA). These SNPs can be placed on gene chips (discussed in Sect. 3.1.1), and the resulting DNA arrays can be used to genotype individuals for common polymorphisms involved in a particular drug response or indeed, for variants of susceptibility genes that interact differently with certain classes of drugs. For example, variants in the Alzheimer's disease susceptibility gene apolipoprotein E (apoE) have been correlated with a varied response to a cholinesterase inhibitor used to reduce the disease symptoms in Alzheimer's patients (POIRIER et al., 1995). Thus, the genotyping of individuals for polymorphisms involved in both disease susceptibility and drug response may be the most important legacy of the Human Genome Project by providing the basis for individualized therapy for human disease.

4 Industry and Academia: Turning Discoveries into Cures

In the last two decades biomedical research has not only led to improved concepts and tools for modern medicine, but has also played a key role in shaping and strengthening economic development. The Industrial Revolution played a dramatic role in shaping societies and economies alike through the rapid application and commercialization of technologies such as steam power and electricity. The dawn of the second millennium brings a "biological revolution" that promises an equally dramatic change in society and the possibility of unparalleled scientific discoveries, many of which could translate into cures for human genetic disease. For example, the Association of University Technology Managers (AUTM) estimates that approximately 30 billion dollars of economic activity and 250,000 jobs are created each year by the commercialization of aca-

demic discoveries world-wide (WILLEY, 1999). Therefore, the translation of research discoveries into medical cures or commercial products is vital to any economy and requires the efficient interaction of industrial, government, and academic institutions.

4.1 The Conflicting Character of Academia, Government, and Industry: Guidelines for their Collaboration

Academia has traditionally focused on the pursuit of knowledge and has been the engine of innovation that drives modern economies. Whereas government research bodies, being accountable for public money, are forced to focus on particular problems affecting society in general such as common diseases (e.g., cancer or heart disease) or key industrial and environmental problems (e.g., pollution, global warming, etc.). Industrial institutions, including pharmaceutical and biotech companies, are primarily concerned with avenues of research that may produce a marketable product or the expansion and modification of current product lines (e.g., more efficient enzymes or drugs). Not only does the mandate of these institutions differ, so does the corporate culture, which can dramatically affect collaborations and is a source of animosity and distrust between academic and non-academic scientists. Academic and government research organizations are open to collaboration and must communicate their work to the broader scientific community, not only to disseminate knowledge, but also to account for the expenditure of public funds. In contrast, industrial research institutions are generally secretive and closed because of the need to protect intellectual property, which is the first step towards commercialization of their research. Despite their differences, these institutions by necessity, have to interact during the commercialization and medical application of a scientific discovery. For instance, academic institutions may provide the initial innovation that leads to a cure for a genetic disease, but they do not have the mandate or the monetary means to carry out the required product development and clinical trials involved in commercialization of the discovery. Pharmaceutical and biotechnology companies are better suited to product development and clinical trials because they have the infrastructure and necessary funds to carry out these tasks. Biotech companies in particular have provided a popular means of obtaining the necessary capital to begin the commercialization of a scientific discovery. In addition, they can also act as a liaison with the pharmaceutical industry when clinical trials are appropriate. Usually government organizations do not have the mandate to spend public money on clinical trials, which are expensive and run the risk of failure. Cancer trials are a notable exception, where many government agencies are involved including the National Cancer Institute in the United States and Canada and the Medical Research Council in the United Kingdom. Other types of clinical trials that may be funded by the government usually involve drugs or vaccines of significant humanitarian or social value, such as treatments for Third World diseases like cholera and malaria. Such trials would not be funded by industry because they would not provide sufficient commercial value. Fig. 6 summarizes the interactions between academic, biotechnology, and pharmaceutical companies that typically occur during the commercialization of a scientific discovery.

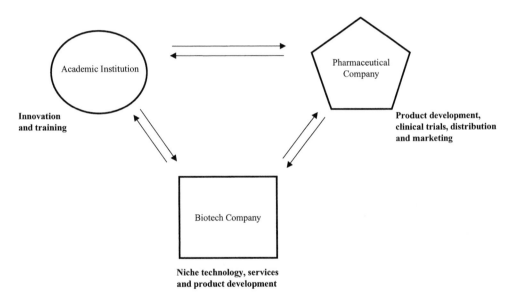

Fig. 6. Interrelationships between academic institutions, biotech, and pharmaceutical companies. As depicted in the figure, these three types of institutions play different roles in the commercialization of scientific discoveries. Academic institutions, including government sponsored laboratories, are primarily the engines of innovation that provide the basic scientific advances that can later be translated into viable products such as drugs or vaccines. In addition, academia is responsible for training scientists and providing specific expertise in a given subject area. Pharmaceutical companies on the other hand, focus on product development and clinical trials rather than the broad pursuit of science and technology. Pharmaceutical companies also provide vital distribution and marketing expertise. In between are the biotechnology companies, which can be formed to take advantage of a particular technology niche or to carry out services required by either academia or the pharmaceutical industry (e.g., high-throughput screening, sequencing, peptide production, etc.). Since product development is not the mandate of an academic institution, biotech companies are often created by "spinning-off" technology that has potential commercial value. Biotech companies not only have access to greater funds for research (e.g., venture capital), but they can also act as a liaison between the pharmaceutical industry and academia, allowing a more rapid commercialization of discoveries.

Clearly, guidelines are required to outline how productive interaction between the sometimes diametrically opposed research mandates of industry and academia can be achieved. Such guidelines have been proposed by the Canadian Institutes of Health Research (CIHR) and the National Institutes of Health (NIH) in the United States and are summarized in Tab. 7 (Department of Human Health Services, National Institutes of Health, 1999; Canadian Institutes of Health Research, 1999). Particularly important are the guidelines regarding ethical conduct and the treatment of intellectual property (IP), two areas that are likely to cause problems for any industrial/academic collaboration.

4.2 Benefits of Collaboration

There are several benefits to research collaborations between industry and academia. One important benefit is the support of excellent basic and applied research in health and biological sciences (Canadian Institutes of Health Research, 1999). This support can take the form of milestone payments provided by pharmaceutical companies to academic institutions when important advances are made, the co-funding of specific research products or research infrastructure (buildings, equipment, etc.) and the sponsoring of research chairs to provide salary and laboratory support for promising scientists. Research partnerships can also lead to the strengthening and broadening of otherwise isolated research efforts, which when combined allow the development of a critical mass of expertise in a given area of research. Bioparks or biotechnology centers such as the Roslin BioCentre in Edinburgh, provide one example of how this critical mass can be fostered. The BioCentre was built on the same site as the Roslin Institute, which is a leading institute in cloning and reproductive technologies. Because of the close proximity, collaborations between biotech companies housed by the BioCentre and the Roslin Institute are common. In addition, the expertise and facilities provided by the Roslin Institute have created a further impetus for companies to locate their operations at the BioCentre, which now employs more than 500 people.

The rate of dissemination of knowledge is also increased by partnerships, which can lead to a more efficient commercialization of research discoveries. For example, research tools developed by biotechnology companies can sometimes aid basic research in academic laboratories (e.g., gene expression technologies or cell culture techniques used for industrial protein production). As well, by commercializing scientific discoveries the profile of science in general is raised among the everyday public, this in turn can have a positive affect on the levels of government funding for research. Lastly, partnerships build bridges between the public and private sector that once formed, can encourage further collaboration and provides academic institutions with more funding options beyond the usual federal or non-profit funding agencies.

4.3 Intellectual Property, Ethics, and Conflict of Interest: Pitfalls of Collaboration

Disputes over patents and other IP issues such as the timely release of information into the public domain are good examples of how scientific collaborations break down. Academia favors the rapid dissemination of information, while industry requires nondisclosure until intellectual property can be protected by patents or the full commercial value can be realized. For example, a proposed collaboration between Celera Genomics Systems (Rockville, USA) and the HGP to complete the sequencing of the human genome in early 2001 had been thwarted by an intellectual property dispute. Celera had already completed a successful collaboration with the publicly funded *Drosophila* Genome Project in which the entire *Drosophila* genome was sequenced (BUTLER, 1999). The success of that collaboration depended on the ability of Celera to first release the data to their commercial database subscribers on a pay-per-view basis for several months before the information was publicly released in the beginning of 2000. In its negotiations with the HGP, Celera wanted a similar time lag before the public release of the completed human genome. Such a delay was unac-

Tab. 7. Guidelines for the Establishment of Research Partnerships between Industry and Academia (adapted from Canadian Institutes of Health Research, 1999)

- Definition of Partnerships: Partnerships are defined as research collaborations or agreements between academic institutions and an industrial organisation to combine resources to achieve a mutually agreed upon objective and/or to co-fund research projects.
- All partnership arrangements, programs and projects, whether for research or for capacity building, should support the overall objectives of the academic institution and must add value to the ongoing research at that institution.
- All collaborative or co-funded research partnerships must meet high quality scientific and ethical standards as determined by peer review.
- Academic institutions should endeavour to consult with more than one potential partner, preferably at the sectoral level, before creating or launching any formal partnerships programs or activities.
- Formal partnership arrangements should operate on the principle that partners are involved early and at all stages of the design of programs that will involve them.
- Before forming a formal partnership, a clear understanding should be demonstrated regarding ownership of intellectual property and an agreed dispute resolution mechanism should be in place between the researcher(s), the research institution(s) and the funder(s).
- The terms, conditions and funding ratios in co-funded partnership arrangements should reflect predetermined guidelines established set out by the academic research institution.
- When public funds are involved, all partnership arrangements should meet high standards of transparency and accountability to the public.
- As often as possible, partnership activities should actively engage all partners with the endeavor (i.e., partners should be more than just funders).
- Successes and results of partnership activity should be documented annually by written report addressed to the governing council of the academic institution.
- In seeking industrial or other partnerships, academic freedom of publication should be sought to ensure the timely disclosure of scientists research findings through publications or presentations at scientific meetings.
  Agreements should be avoided that unduly limit the freedom to publish and collaborate. Reasonable restrictions on collaboration by academic researchers involved in sponsored research agreements with an industrial partner that avoid conflicting obligations to other industrial partners, are understood and accepted. Similarly, brief delays in publication may be appropriate to permit the filing of patent applications and to ensure confidential information obtained from a sponsor or the provider of a research tool is not inadvertently disclosed. However, excessive publication delays or requirements for editorial control, approval of publications, or withholding of undermine the credibility of research data and are therefore unacceptable.[a]
- To ensure the dissemination of research resources developed through partnerships, research tools should be subject to wide distribution agreements on a non-exclusive basis (or wide distribution on reasonable terms by an exclusive distributor if appropriate), which reduces the risk that other fruitful avenues of research outside the mandate of the academic or industrial partner will not be neglected.

---

[a] Paraphrased from Department of Health and Human Services, National Institutes of Health (1999).

ceptable to the HGP whose existing guidelines stipulate the release of data within 24 h. Consequently, negotiations to finish sequencing the human genome with Celera's help were ended (MACILWAIN, 1999).

Ethical problems can also arise during scientific collaborations between academia and industry. One problem facing researchers and health professionals in Iceland is the potential for the inappropriate use of genealogy and genetic data by deCODE Genetics (Reykjavik, Iceland). The government of Iceland and deCODE had entered an agreement in which they would provide their genealogical and medical information to deCODE for the construction of an anonymous database. deCODE would then commercialize the use of this database for gene discovery. Many individuals in Iceland have since opted out of this agreement, as they feared the misuse of this data, which could potentially link individuals to their genetic information by using the genealogical data found in databases (HAUKSSON, 1999).

Conflicts of interest can also arise during industrial/academic collaborations, where ones position can be abused for personal or monetary gain. One recent example of a very serious conflict of interest involved the *New England Journal of Medicine* (WEATHERALL, 2000). Academic journals publish peer-reviewed articles that are subject to impartial treatment by viewers who should base publication solely on scientific merit. This process was severely compromised in the *New England Journal of Medicine*, where it was found that 19 out of 40 drug review papers, published between 1999 and 1997, contained reviews of drugs by authors with financial links to the manufacturers of the same drugs. Conflict of interest can also arise, if a researcher is unable to carry out his or her duties because of a prior agreement with an industrial partner. For instance, a Toronto scientist was threatened with legal action, if she disclosed her concerns about the efficacy and toxicity of a drug that she was investigating during a clinical trial (NATHAN and WEATHERALL, 1999). Both the hospital and university involved are facing allegations of conflict of interest for allowing such a situation to occur. Nevertheless, for all the problems that can arise, academia still requires industry for both core funding and the commercialization of discoveries.

4.4 Towards More Efficient Commercialization of Scientific Discoveries

The commercialization of biological and medical discoveries is still a fledgling pursuit in comparison to the commercialization of technology derived from other fields such as computer science or engineering. Many ethical and intellectual property issues still need to be addressed and their solutions and procedures streamlined to allow a more efficient transfer of biotechnology into commercial products. Ethical and IP issues can be at least prepared for by the creation of oversight committees to evaluate industrial partnerships and to establish best practices for the management of IP. Technology transfer offices (TTOs) also play an important role in the commercialization of technology developed in universities by providing guidance on IP issues and acting as a liaison between potential industrial partners and the academic institution. Unfortunately, most university TTOs are woefully funded and have problems retaining qualified staff. Realizing this problem, the CIHR has proposed a funding scheme in which TTOs will receive funding based on a percentage of the total research funds received by the academic institution (Canadian Institutes of Health Research, 1999). This is clearly a step in the right direction as is the formation of associations such as the Association of University Technology Managers (AUTM), which provides both research on best practises and training.

Another problem that poses a hurdle to the commercialization of scientific discoveries is the lack of interdisciplinary scientists who understand both the specific science behind a discovery and the chemistry or engineering skills required for product development. In particular, since the analysis of genomic data has become so essential to modern biology, their will be a great need for scientists who can feel as comfortable at a computer "mining data" as discussing signal transduction pathways in oncogenesis. As well, engineers who understand biology and biologists that understand chemistry are required to scale-up the synthesis and manufacturing of vaccines and drugs. To this end many universities have or are developing interdisciplinary graduate programs that provide the necessary flexibility to allow young scientists to undertake broad training in multiple fields including computer science, engineering, physics, and biological sciences. One example of this is the creation of the bioinformatics graduate degree at universities such as Rutgers and Stanford University in the USA, the University of Manchester in the United Kingdom, and the University of Waterloo in Canada. Another example is the formation of the Pharmaceutical Sciences and Pharmacogenomics graduate program at the University of California at San Francisco, which combines courses in bioinformatics, biochemistry, and pharmaceutical science.

In the new era of genomic biology, the pace of scientific discovery in medical genetics threatens to outstrip the capacity of biotechnology and industry to translate these discoveries into cures. Many benefits and pitfalls accompany academic/industrial partnerships but through planning and the appropriate training, these problems can be minimized. The key is to retain the flexibility and broad scientific mandate of academic institutions, but at the same time foster strong links with industry. This can be done through solid IP guidelines that both protect the industrial partners interests and maintain the intellectual freedom required by universities. Lastly, the creation of interdisciplinary graduate programs will produce scientists that can provide both the scientific expertise and technical knowledge required to translate the scientific discoveries of today into the cures of tomorrow.

5 References

2000 Database Issue, *Nucleic Acids Res.* **28**.

AALTONEN, L. A., PELTOMAKI, P., LEACH, F. S., SISTONEN, P., PYLKKANEN, L. et al. (1993), Clues to the pathogenesis of familial colorectal cancer, *Science* **260**, 812–816.

ALIZADEH, A. A., EISEN, M. B., DAVIS, R. E., MA, C., LOSSOS, I. S. et al. (2000), Distinct types of diffuse large B-cell lymphoma identified by gene expression profiling, *Nature* **403**, 503–511.

ALTSCHUL, S. F., GISH, W., MILLER, W., MYERS, E. W., LIPMAN, D. J. (1990), Basic local alignment search tool, *J. Mol. Biol.* **215**, 403–410.

BARNES, C. J., LEE, M. (1998), Chemoprevention of spontaneous intestinal adenomas in the adenomatous polyposis coli Min mouse model with aspirin, *Gastroenterology* **114**, 873–877.

BARTSCH, H., NAIR, J., OWEN, R. W. (1999), Dietary polyunsaturated fatty acids and cancers of the breast and colorectum: emerging evidence for their role as risk modifiers, *Carcinogenesis* **20**, 2209–2218.

BILOFSKY, H. S., BURKS, C., FICKETT, J. W., GOAD, W. B., LEWITTER, F. I. et al. (1986), The GenBank genetic sequence databank, *Nucleic Acids Res.* **14**, 1–4.

BODMER, W. F., BAILEY, C. J., BODMER, J., BUSSEY, H. J., ELLIS, A. et al. (1987), Localization of the gene for familial adenomatous polyposis on chromosome 5, *Nature* **328**, 614–616.

BOUCHER, R. C., KNOWLES, M. R., JOHNSON, L. G., OLSEN, J. C., PICKLES, R. et al. (1994), Gene therapy for cystic fibrosis using E1-deleted adenovirus: a phase I trial in the nasal cavity. The University of North Carolina at Chapel Hill, *Hum. Gene Ther.* **5**, 615–639.

BRANNAN, C. I., BARTOLOMEI, M. S. (1999), Mechanisms of genomic imprinting, *Curr. Opin. Genet. Dev.* **9**, 164–170.

BRENT, R. (2000), Genomic biology, *Cell* **100**, 169–183.

BROOKER, R. J. (1999), *Genetics: Analysis and Principles.* Menlo Park: Benjamin/Cummings.

BROOKES, A. J. (1999), The essence of SNPs, *Gene* **234**, 177–186.

BROOKES, A. J., LEHVASLAIHO, H., SIEGFRIED, M., BOEHM, J. G., YUAN, Y. P. et al. (2000), HGBASE: a database of SNPs and other variations in and around human genes, *Nucleic Acids Res.* **28**, 356–360.

BUTLER, D. (1999), Venter's *Drosophila* "success" set to boost human genome efforts, *Nature* **401**, 729–730.

Canadian Institutes of Health Research (1999), Canadian Institutes of Health Research. Interim Governing Council Sub-Committee on Partnership and Commercialization Working Paper. (Ref Type: Report).

CAVAZZANA-CALVO, M., HACEIN-BEY, S., DE SAINT, B. G., GROSS, F., YVON, E. et al. (2000), Gene therapy of human severe combined immunodeficiency (SCID)-X1 disease, *Science* **288**, 669–672.

COLLINS, F. S., BROOKS, L. D., CHAKRAVARTI, A. (1998), A DNA polymorphism discovery resource for research on human genetic variation, *Genome Res.* **8**, 1229–1231.

CRYSTAL, R. G., MCELVANEY, N. G., ROSENFELD, M. A., CHU, C. S., MASTRANGELI, A. et al. (1994), Administration of an adenovirus containing the human CFTR cDNA to the respiratory tract of individuals with cystic fibrosis, *Nature Genet.* **8**, 42–51.

CUNNINGHAM, S., MARSHALL, T. (1998), Influence of five years of antenatal screening on the paediatric cystic fibrosis population in one region, *Arch. Dis. Child* **78**, 345–348.

Department of Health and Human Services, National Institutes of Health (1999), Department of Health and Human Services, National Institutes of Health. Principles and Guidelines for Sharing Biomedical Research Resources: Principles and Guidelines for Recipients of NIH Research Grants and Contracts.

DE VRIES, B. B., HALLEY, D. J., OOSTRA, B. A., NIERMEIJER, M. F. (1998), The fragile X syndrome, *J. Med. Genet.* **35**, 579–589.

EVANS, W. E., RELLING, M. V. (1999), Pharmacogenomics: translating functional genomics into rational therapeutics, *Science* **286**, 487–491.

FALCONER, D. S. (1981), *Introduction to Quantitative Genetics*. London: Longman.

FERGUSON, L. R., HARRIS, P. J. (1999), Protection against cancer by wheat bran: role of dietary fiber and phytochemicals, *Eur. J Cancer Prev.* **8**, 17–25.

GIARDIELLO, F. M., HAMILTON, S. R., KRUSH, A. J., PIANTADOSI, S., HYLIND, L. M. et al. (1993), Treatment of colonic and rectal adenomas with sulindac in familial adenomatous polyposis, *N. Engl. J. Med.* **328**, 1313–1316.

GILLARD, E. F., CHAMBERLAIN, J. S., MURPHY, E. G., DUFF, C. L., SMITH, B. et al. (1989), Molecular and phenotypic analysis of patients with deletions within the deletion-rich region of the Duchenne muscular dystrophy (DMD) gene, *Am. J. Hum. Genet.* **45**, 507–520.

GIOVANNUCCI, E., RIMM, E. B., STAMPFER, M. J., COLDITZ, G. A., ASCHERIO, A., WILLETT, W. C. (1994), Aspirin use and the risk for colorectal cancer and adenoma in male health professionals, *Ann. Intern. Med.* **121**, 241–246.

GIRODON-BOULANDET, E., CAZENEUVE, C., GOOSSENS, M. (2000), Screening practices for mutations in the CFTR gene ABCC7, *Hum. Mutat.* **15**, 135–149.

GLICK, T. H., LIKOSKY, W. H., LEVITT, L. P., MELLIN, H., REYNOLDS, D. W. (1970), Reye's syndrome: an epidemiologic approach, *Pediatrics* **46**, 371–377.

GRIFFIN, T. J., SMITH, L. M. (2000), Single-nucleotide polymorphism analysis by MALDI-TOF mass spectrometry, *Trends Biotechnol.* **18**, 77–84.

GRODEN, J., THLIVERIS, A., SAMOWITZ, W., CARLSON, M., GELBERT, L. et al. (1991), Identification and characterization of the familial adenomatous polyposis coli gene, *Cell* **66**, 589–600.

HAUKSSON, P. (1999), Icelanders opt out of genetic database, *Nature* **400**, 707–708.

HEMMINKI, A., MARKIE, D., TOMLINSON, I., AVIZIENYTE, E., ROTH, S. et al. (1998), A serine/threonine kinase gene defective in Peutz–Jeghers syndrome, *Nature* **391**, 184–187.

HERRERA, L., KAKATI, S., GIBAS, L., PIETRZAK, E., SANDBERG, A. A. (1986), Gardner syndrome in a man with an interstitial deletion of 5q, *Am. J. Med. Genet.* **25**, 473–476.

HOULSTON, R. S., COLLINS, A., SLACK, J., MORTON, N. E. (1992), Dominant genes for colorectal cancer are not rare, *Ann. Hum. Genet.* **56** (Pt 2), 99–103.

HOWE, J. R., RINGOLD, J. C., SUMMERS, R. W., MITROS, F. A., NISHIMURA, D. Y., STONE, E. M. (1998), A gene for familial juvenile polyposis maps to chromosome 18q21.1, *Am. J. Hum. Genet.* **62**, 1129–1136.

IANNELLO, R. C., CRACK, P. J., DE HAAN, J. B., KOLA, I. (1999), Oxidative stress and neural dysfunction in Down syndrome, *J. Neur. Transm. Suppl.* **57**, 257–267.

IONOV, Y., PEINADO, M. A., MALKHOSYAN, S., SHIBATA, D., PERUCHO, M. (1993), Ubiquitous somatic mutations in simple repeated sequences reveal a new mechanism for colonic carcinogenesis, *Nature* **363**, 558–561.

JARVIK, G. P. (1997), Genetic predictors of common disease: apolipoprotein E genotype as a paradigm, *Ann. Epidemiol.* **7**, 357–362.

JORDE, L. B., CAREY, J. C., BAMSHAD, M. J., WHITE, R. L. (1999), *Medical Genetics*. St. Louis, MO: Mosby.

JOSEFSON, D. (1998), CF gene may protect against typhoid fever, *Brit. Med. J.* **316**, 1481.

JOSLYN, G., CARLSON, M., THLIVERIS, A., ALBERTSEN, H., GELBERT, L. et al. (1991), Identification of deletion mutations and three new genes at the familial polyposis locus, *Cell* **66**, 601–613.

KINZLER, K. W., NILBERT, M. C., SU, L. K., VOGELSTEIN, B., BRYAN, T. M. et al. (1991), Identification of FAP locus genes from chromosome 5q21, *Science* **253**, 661–665.

KITSON, C., ANGEL, B., JUDD, D., ROTHERY, S., SEVERS, N. J. et al. (1999), The extra- and intracellular barriers to lipid and adenovirus-mediated pulmonary gene transfer in native sheep airway epithelium, *Gene Ther.* **6**, 534–546.

KRIEGER, I., SNODGRASS, P. J., ROSKAMP, J. (1979), Atypical clinical course of ornithine transcarbamylase deficiency due to a new mutant (comparison with Reye's disease), *J. Clin. Endocr. Metab.* **48**, 388–392.

LAKEN, S. J., PETERSEN, G. M., GRUBER, S. B., ODDOUX, C., OSTRER, H. et al. (1997), Familial colorectal cancer in Ashkenazim due to a hypermutable tract in APC, *Nature Genet.* **17**, 79–83.

LANDEGREN, U., NILSSON, M., KWOK, P. Y. (1998), Reading bits of genetic information: methods for single-nucleotide polymorphism analysis, *Genome Res.* **8**, 769–776.

LAZAROU, J., POMERANZ, B. H., COREY, P. N. (1998), Incidence of adverse drug reactions in hospitalized patients: a meta-analysis of prospective studies, *JAMA* **279**, 1200–1205.

LEACH, F. S., NICOLAIDES, N. C., PAPADOPOULOS, N., LIU, B., JEN, J. et al. (1993), Mutations of a mutS homolog in hereditary nonpolyposis colorectal cancer, *Cell* **75**, 1215–1225.

LEE, M. P., BRANDENBURG, S., LANDES, G. M., ADAMS, M., MILLER, G., FEINBERG, A. P. (1999), Two novel genes in the center of the 11p15 imprinted domain escape genomic imprinting, *Hum. Mol. Genet.* **8**, 683–690.

LEVINSON, G., GUTMAN, G. A. (1987), High frequencies of short frameshifts in poly-CA/TG tandem repeats borne by bacteriophage M13 in *Escherichia coli* K-12, *Nucleic Acids Res.* **15**, 5323–5338.

LI, E., BEARD, C., FORSTER, A. C., BESTOR, T. H., JAENISCH, R. (1993), DNA methylation, genomic imprinting, and mammalian development, *Cold*

Spring Harbor Symp. Quant. Biol. **58**, 297–305.

LIMA, J. J., THOMASON, D. B., MOHAMED, M. H., EBERLE, L. V., SELF, T. H., JOHNSON, J. A. (1999), Impact of genetic polymorphisms of the beta2-adrenergic receptor on albuterol bronchodilator pharmacodynamics, *Clin. Pharmacol. Ther.* **65**, 519–525.

LINDBLOM, A., TANNERGARD, P., WERELIUS, B., NORDENSKJOLD, M. (1993), Genetic mapping of a second locus predisposing to hereditary non-polyposis colon cancer, *Nature Genet.* **5**, 279–282.

LYNCH, P. M. (1999), Clinical challenges in management of familial adenomatous polyposis and hereditary non-polyposis colorectal cancer, *Cancer* **86**, 2533–2539.

LYNCH, H. T., DE LA CHAPELLE, A. (1999), Genetic susceptibility to non-polyposis colorectal cancer, *J. Med. Genet.* **36**, 801–818.

MACGREGOR, A. J., SNIEDER, H., SCHORK, N. J., SPECTOR, T. D. (2000), Twins. Novel uses to study complex traits and genetic diseases, *Trends Genet.* **16**, 131–134.

MACILWAIN, C. (1999), Energy department revises terms of Venter deal after complaints, *Nature* **397**, 93.

MCELVANEY, N. G. (1996), Is gene therapy in cystic fibrosis a realistic expectation? *Curr. Opin. Pulm. Med.* **2**, 466–471.

MCIVOR, R. S. (1999), Gene therapy of genetic diseases and cancer, *Pediat. Transplant.* **3** (Suppl. 1), 116–121.

MCKUSICK, V. A. (2000), Online Mendelian Inheritance in Man, OMIM (TM), McKusick-Nathans Institute for Genetic Medicine, Johns Hopkins University (Baltimore, MD) and National Center for Biotechnology Information, National Library of Medicine (Bethesda, MD), *http://www.ncbi.nlm.nih.gov/omim/*.

MICHALAK, M., OPAS, M. (1997), Functions of dystrophin and dystrophin associated proteins, *Curr. Opin. Neurol.* **10**, 436–442.

MIYOSHI, Y., ANDO, H., NAGASE, H., NISHISHO, I., HORII, A. et al. (1992a), Germ-line mutations of the APC gene in 53 familial adenomatous polyposis patients, *Proc. Natl. Acad. Sci. USA* **89**, 4452–4456.

MIYOSHI, Y., NAGASE, H., ANDO, H., HORII, A., ICHII, S. et al. (1992b), Somatic mutations of the APC gene in colorectal tumors: mutation cluster region in the APC gene, *Hum. Mol. Genet.* **1**, 229–233.

MOCHIZUKI, T., WU, G., HAYASHI, T., XENOPHONTOS, S. L., VELDHUISEN, B. et al. (1996), PKD2, a gene for polycystic kidney disease that encodes an integral membrane protein, *Science* **272**, 1339–1342.

MONACO, A. P., NEVE, R. L., COLLETTI-FEENER, C., BERTELSON, C. J., KURNIT, D. M., KUNKEL, L. M. (1986), Isolation of candidate cDNAs for portions of the Duchenne muscular dystrophy gene, *Nature* **323**, 646–650.

MORIN, P. J. (1999), Beta-catenin signaling and cancer, *Bioassays* **21**, 1021–1030.

NAGY, A. M., DE, M. X. RUIBAL, N., LINTS, F. A. (1998), Scientific and ethical issues of preimplantation diagnosis, *Ann. Med.* **30**, 1–6.

NAKAMURA, Y., LATHROP, M., LEPPERT, M., DOBBS, M., WASMUTH, J. et al. (1988), Localization of the genetic defect in familial adenomatous polyposis within a small region of chromosome 5, *Am. J. Hum. Genet.* **43**, 638–644.

NATHAN, D. G., WEATHERALL, D. J. (1999), Academia and industry: lessons from the unfortunate events in Toronto, *Lancet* **353**, 771–772.

NATHKE, I. S. (1999), The adenomatous polyposis coli protein, *Mol. Pathol.* **52**, 169–173.

NICOLAIDES, N. C., PAPADOPOULOS, N., LIU, B., WEI, Y. F., CARTER, K. C. et al. (1994), Mutations of two PMS homologues in hereditary nonpolyposis colon cancer, *Nature* **371**, 75–80.

NISHISHO, I., NAKAMURA, Y., MIYOSHI, Y., MIKI, Y., ANDO, H. et al. (1991), Mutations of chromosome 5q21 genes in FAP and colorectal cancer patients, *Science* **253**, 665–669.

O'BRIEN, S. J., MENOTTI-RAYMOND, M., MURPHY, W. J., NASH, W. G., WIENBERG, J. et al. (1999), The promise of comparative genomics in mammals, *Science* **286**, 458–481.

OKUI, M., IDE, T., MORITA, K., FUNAKOSHI, E., ITO, F. et al. (1999), High-level expression of the *Mnb/Dyrk1A* gene in brain and heart during rat early development, *Genomics* **62**, 165–171.

OLSCHWANG, S., SEROVA-SINILNIKOVA, O. M., LENOIR, G. M., THOMAS, G. (1998), PTEN germ-line mutations in juvenile polyposis coli, *Nature Genet.* **18**, 12–14.

ORLOWSKI, S., GARRIGOS, M. (1999), Multiple recognition of various amphiphilic molecules by the multidrug resistance P-glycoprotein: molecular mechanisms and pharmacological consequences coming from functional interactions between various drugs, *Anticancer Res.* **19**, 3109–3123.

PAPADOPOULOS, N., NICOLAIDES, N. C., WEI, Y. F., RUBEN, S. M., CARTER, K. C. et al. (1994), Mutation of a mutL homolog in hereditary colon cancer, *Science* **263**, 1625–1629.

PAPADOPOULOS, N., NICOLAIDES, N. C., LIU, B., PARSONS, R., LENGAUER, C. et al. (1995), Mutations of GTBP in genetically unstable cells, *Science* **268**, 1915–1917.

PARSONS, R., LI, G. M., LONGLEY, M. J., FANG, W. H., PAPADOPOULOS, N. et al. (1993), Hypermutability and mismatch repair deficiency in RER⁺ tumor cells, *Cell* **75**, 1227–1236.

PATIL, N., PETERSON, A., ROTHMAN, A., DE JONG, P. J., MYERS, R. M., COX, D. R. (1994), A high resolution physical map of 2.5 Mbp of the Down syndrome region on chromosome 21, *Hum. Mol. Genet.* **3**, 1811–1817.

PEARSON, W. R., LIPMAN, D. J. (1988), Improved tools for biological sequence comparison, *Proc. Natl. Acad. Sci. USA* **85**, 2444–2448.

PEDEMONTE, S., SCIALLERO, S., GISMONDI, V., STAGNARO, P., BITICCHI, R. et al. (1998), Novel germline APC variants in patients with multiple adenomas, *Genes Chromosomes, Cancer* **22**, 257–267.

PELTOMAKI, P., AALTONEN, L. A., SISTONEN, P., PYLKKANEN, L., MECKLIN, J. P. et al. (1993), Genetic mapping of a locus predisposing to human colorectal cancer, *Science* **260**, 810–812.

PETERSEN, G. M., BRENSINGER, J. D., JOHNSON, K. A., GIARDIELLO, F. M. (1999), Genetic testing and counseling for hereditary forms of colorectal cancer, *Cancer* **86**, 2540–2550.

PICOULT-NEWBERG, L., IDEKER, T. E., POHL, M. G., TAYLOR, S. L., DONALDSON, M. A. et al. (1999), Mining SNPs from EST databases, *Genome Res.* **9**, 167–174.

POIRIER, J., DELISLE, M. C., QUIRION, R., AUBERT, I., FARLOW, M. et al. (1995), Apolipoprotein E4 allele as a predictor of cholinergic deficits and treatment outcome in Alzheimer disease, *Proc. Natl. Acad. Sci. USA* **92**, 12260–12264.

POPE, F. M., NICHOLLS, A. C., MCPHEAT, J., TALMUD, P., OWEN, R. (1985), Collagen genes and proteins in osteogenesis imperfecta, *J. Med. Genet.* **22**, 466–478.

POWELL, S. M., ZILZ, N., BEAZER-BARCLAY, Y., BRYAN, T. M., HAMILTON, S. R. et al. (1992), APC mutations occur early during colorectal tumorigenesis, *Nature* **359**, 235–237.

POWELL, S. M., PETERSEN, G. M., KRUSH, A. J., BOOKER, S., JEN, J. et al. (1993), Molecular diagnosis of familial adenomatous polyposis, *N. Engl. J. Med.* **329**, 1982–1987.

RAMJEESINGH, M., HUAN, L. J., WILSCHANSKI, M., DURIE, P., LI, C. et al. (1998), Assessment of the efficacy of *in vivo* CFTR protein replacement therapy in CF mice, *Hum. Gene Ther.* **9**, 521–528.

RASMUSSEN, S. A., FRIEDMAN, J. M. (2000), NF1 gene and neurofibromatosis 1, *Am. J. Epidemiol.* **151**, 33–40.

REDDY, P. H., WILLIAMS, M., TAGLE, D. A. (1999), Recent advances in understanding the pathogenesis of Huntington's disease, *Trends Neurosci.* **22**, 248–255.

RIORDAN, J. R., ROMMENS, J. M., KEREM, B., ALON, N., ROZMAHEL, R. et al. (1989), Identification of the cystic fibrosis gene: cloning and characterization of complementary, *Science* **245**, 1066–1073.

RISCH, N., MERIKANGAS, K. (1996), The future of genetic studies of complex human diseases, *Science* **273**, 1516–1517.

ROMMENS, J. M., IANNUZZI, M. C., KEREM, B., DRUMM, M. L., MELMER, G. et al. (1989), Identification of the cystic fibrosis gene: chromosome walking and jumping, *Science* **245**, 1059–1065.

ROSS, D. T., SCHERF, U., EISEN, M. B., PEROU, C. M., REES, C. et al. (2000), Systematic variation in gene expression patterns in human cancer cell lines, *Nature Genet.* **24**, 227–235.

ROUGEULLE, C., GLATT, H., LALANDE, M. (1997), The Angelman syndrome candidate gene, UBE3A/E6-AP, is imprinted in brain, *Nature Genet.* **17**, 14–15.

SANDHU, J. S., KEATING, A., HOZUMI, N. (1997), Human gene therapy, *Crit. Rev. Biotechnol.* **17**, 307–326.

SCHATZKIN, A., LANZA, E., CORLE, D., LANCE, P., IBER, F. et al. (2000), Lack of effect of a low-fat, high-fiber diet on the recurrence of colorectal adenomas. Polyp Prevention Trial Study Group, *N. Engl. J. Med.* **342**, 1149–1155.

SCHROCK, E., DU MANIOR, M. S., VELDMAN, T., SCHOELL, B., WIENBERG, J. et al. (1996), Multicolor spectral karyotyping of human chromosomes, *Science* **273**, 494–497.

SCHROEDER, S. A., GAUGHAN, D. M., SWIFT, M. (1995), Protection against bronchial asthma by CFTR delta F508 mutation: a heterozygote advantage in cystic fibrosis, *Nature Med.* **1**, 703–705.

SCHWEIZER, J., ZYNGER, D., FRANCKE, U. (1999), *In vivo* nuclease hypersensitivity studies reveal multiple sites of parental origin-dependent differential chromatin conformation in the 150 kb SNRPN transcription unit, *Hum. Mol. Genet.* **8**, 555–566.

SCRIVER, C. R., WATERS, P. J. (1999), Monogenic traits are not simple: lessons from phenylketonuria, *Trends Genet.* **15**, 267–272.

SCULLY, R., CHEN, J., PLUG, A., XIAO, Y., WEAVER, D. et al. (1997), Association of BRCA1 with Rad51 in mitotic and meiotic cells, *Cell* **88**, 265–275.

SHENG, H., SHAO, J., KIRKLAND, S. C., ISAKSON, P., COFFEY, R. J. et al. (1997), Inhibition of human colon cancer cell growth by selective inhibition of cyclooxygenase-2, *J. Clin. Invest.* **99**, 2254–2259.

SLAGER, S. L., HUANG, J., VIELAND, V. J. (2000), Effect of allelic heterogeneity on the power of the transmission disequilibrium test, *Genet. Epidemiol.* **18**, 143–156.

SORAVIA, C., BAPAT, B., COHEN, Z. (1997), Familial adenomatous polyposis (FAP) and hereditary nonpolyposis colorectal cancer (HNPCC): a review of clinical, genetic and therapeutic aspects, *Schweiz. Med. Wochenschr.* **127**, 682–690.

SPARKS, A. B., MORIN, P. J., VOGELSTEIN, B., KINZLER, K. W. (1998), Activation of β-catenin–Tcf signalling in colon cancer by mutations in β-catenin or APC, *Science* **275**, 1787–1790.

SPIELBERG, S. P. (1996), N-acetyltransferases: pharmacogenetics and clinical consequences of polymorphic drug metabolism, *J. Pharmacokinet. Biopharm.* **24**, 509–519.

SPIELMAN, R. S., MCGINNIS, R. E., EWENS, W. J.

(1993), Transmission test for linkage disequilibrium: the insulin gene region and insulin-dependent diabetes mellitus (IDDM), *Am. J. Hum. Genet.* **52**, 506–516.

STRACHAN, T., READ, A. P. (1999), *Human Molecular Genetics*. Oxford: BIOScience.

STRAND, M., PROLLA, T. A., LISKAY, R. M., PETES, T. D. (1993), Destabilization of tracts of simple repetitive DNA in yeast by mutations affecting DNA mismatch repair, *Nature* **365**, 274–276.

STRAUSBERG, R. L., BUETOW, K. H., EMMERT-BUCK, M. R., KLAUSNER, R. D. (2000), The cancer genome anatomy project: building an annotated gene index, *Trends Genet.* **16**, 103–106.

The European Polycystic Kidney Disease Consortium (1994), The polycystic kidney disease 1 gene encodes a 14 kb transcript and lies within a duplicated region on chromosome 16, *Cell* **77**, 881–894.

THOMAS, H. J., WHITELAW, S. C., COTTRELL, S. E., MURDAY, V. A., TOMLINSON, I. P. et al. (1996), Genetic mapping of hereditary mixed polyposis syndrome to chromosome 6q, *Am. J. Hum. Genet.* **58**, 770–776.

THUN, M. J., NAMBOODIRI, M. M., CALLE, E. E., FLANDERS, W. D., HEATH, C. W., Jr. (1993), Aspirin use and risk of fatal cancer, *Cancer Res.* **53**, 1322–1327.

TILGHMAN, S. M. (1999), The sins of the fathers and mothers: genomic imprinting in mammalian development, *Cell* **96**, 185–193.

VASEN, H. F., MECKLIN, J. P., KHAN, P. M., LYNCH, H. T. (1991), The International Collaborative Group on Hereditary Non-Polyposis Colorectal Cancer (ICG-HNPCC), *Dis. Colon Rectum* **34**, 424–425.

VELCULESCU, V. E., ZHANG, L., VOGELSTEIN, B., KINZLER, K. W. (1995), Serial analysis of gene expression, *Science* **270**, 484–487.

VU, T. H., HOFFMAN, A. R. (1997), Imprinting of the Angelman syndrome gene, UBE3A, is restricted to brain, *Nature Genet.* **17**, 12–13.

WANG, Y., CORTEZ, D., YAZDI, P., NEFF, N., ELLEDGE, S. J., QIN, J. (2000), BASC, a super complex of BRCA1-associated proteins involved in the recognition and repair of aberrant DNA structures, *Genes Dev.* **14**, 927–939.

WARTHIN, A. S. (1913), Heredity with reference to carcinoma, *Arch. Intern. Med.* **12**, 546–555.

WEATHERALL, D. (2000), Academia and industry: increasingly uneasy bedfellows, *Lancet* **355**, 1574.

WHITE, R., WOODWARD, S., LEPPERT, M., O'CONNELL, P., HOFF, M. et al. (1985), A closely linked genetic marker for cystic fibrosis, *Nature* **318**, 382–384.

WIEDEMANN, H. P., STOLLER, J. K. (1996), Lung disease due to alpha 1-antitrypsin deficiency, *Curr. Opin. Pulm. Med.* **2**, 155–160.

WIJNEN, J., DE LEEUW, W., VASEN, H., VAN DER KLIFT, K. H., MOLLER, P. et al. (1999), Familial endometrial cancer in female carriers of MSH6 germline mutations, *Nature Genet.* **23**, 142–144.

WILLEY, T. F. (1999), Opening Remarks: U.S. Senate Committee on Commerce, Science and Transportation Subcommittee on Science, Technology and Space Hearing on Federal R & D – April 15, 1999, *The Journal of the Association of University Technology Managers* **IX**.

WINTER, R. M., TICKLE, C. (1993), Syndactylies and polydactylies: embryological overview and suggested classification, *Eur. J. Hum. Genet.* **1**, 96–104.

YANG, P., WENTZLAFF, K. A., KATZMANN, J. A., MARKS, R. S., ALLEN, M. S. et al. (1999), Alpha1-antitrypsin deficiency allele carriers among lung cancer patients, *Cancer Epidemiol. Biomarkers Prev.* **8**, 461–465.

4 Genomics and Human Disease

ROGER C. GREEN

St. John's, Canada

1 Introduction

With the availability of the complete human genome sequence, the relationship between genomics and human disease is about to change dramatically. Up to this point it could be argued that genomics has got more out of the relationship than has human disease. This is because the study of inherited diseases has played a key role in the mapping and identification of many human genes. This situation is about to change. It was always envisaged that the Human Genome Project (HGP) would ultimately benefit humankind by providing greater insights into mechanisms of human disease. It is becoming clear, however, that the impact of genomics may be greater than most people could have imagined about 15 years ago, when the concept of the HGP was being formulated.

The final data on the human genome will tell us a number of things:

- how many genes there are [it seems that what might have appeared to be a simple task – just counting the number of genes – has turned out to be quite difficult. Even now that the first draft of the complete human DNA sequence is available, we still do not have even an approximate gene count. A factor of 4 still separates the latest high and low estimates (LIANG et al., 2000; EWING and GREEN, 2000; APARICIO, 2000)],
- where these genes are to be found on the 24 chromosomes, and
- the DNA sequence of all the genes.

This phase of the HGP has been termed *structural genomics* and should be completed in 2003. Later, after analyzing the DNA sequence of particular genes from many unrelated individuals and populations, we will learn about the amount of variability in the DNA sequence of individual genes.

The genome data will neither tell us about the function of genes, nor about what disease processes they may influence. However, building on the foundation of structural genomics, the next phase of genomic investigation –

termed *functional genomics* – will provide us with information on the role of each gene.

We now realize that virtually all human diseases have a component affected by the unique combination of genes that we inherit from our parents. At one extreme are the classical genetic diseases that are directly caused by inherited mutations in single genes, e.g., cystic fibrosis, sickle-cell anemia, and Huntington disease. At the other end of the spectrum are such conditions as infectious disease and trauma, where environmental factors are most important. Nevertheless, even in these cases, we know that susceptibility to infection is influenced by genetic variation (HURME et al., 1998; ROGER, 1998; IKEDA and YU, 1998), and it is plausible that our chance of being involved in an accident may also be influenced by genetic factors which modulate behavior.

As many of the genes that cause the classical genetic diseases are becoming mapped and cloned, the emphasis is shifting to analyzing the genes responsible for common diseases such as heart disease, psychiatric disease, diabetes, and arthritis. It is in these areas where we will see major advances over the next decade or two. Human molecular genetics is a fairly new discipline. One of the challenges to be faced is the fact that many physicians now in practice have only minimal knowledge of this new area of medicine and its implications.

On June 26, 2000, the first "Working Draft" of the human genome was announced with great fanfare. There was much discussion in the mainstream news media about the tremendous potential – for good and evil – that this achievement represents. Even when the final sequence is available in 2003, however, this will not be the end of the saga. It will not even be the beginning of the end. It may be, however, the end of the beginning (with apologies to W. S. CHURCHILL). This chapter will review what we are likely to witness as the adventure continues to unfold.

2 Identification of Disease-Associated Genes

2.1 How Diseases Genes have been Identified in the Recent Past

Identification of disease-causing genes has been a rather tedious business. First, the approximate position of the gene must be mapped. This means identifying on which chromosomal region the gene is located (BOTSTEIN, 1999). Genetic mapping for single-gene (Mendelian) traits is usually conducted by linkage analysis. This involves the identification of DNA markers, previously placed on the genetic map, that co-segregate with the disease in one or more families.

2.1.1 The Importance of Accurate Family Information

Families with the relevant genetic disease must be ascertained. The structure of their pedigree (family tree) must be uncovered, preferably several generations back from the present. This involves talking to family members about their recollections, and also conducting research into historical archives that record information on births, marriages, and census data. The medical records of any affected, or potentially affected, individual must be thoroughly checked to confirm or rule out the diagnosis in question. Living individuals must receive a clinical evaluation to determine their status. A single error in diagnosis will often invalidate the data set for the entire family. Blood samples for DNA extraction must be collected from the appropriate family members. These may be supplemented with DNA extracted from archived pathology specimens in the case of deceased family members.

2.1.2 Determination of Genotypes

The family DNA samples are tested to determine the genotypes of a few hundred polymorphic markers that are evenly distributed along all the chromosomes. These polymorphic markers are known regions of DNA in which the base sequence varies slightly in different individuals in the population. There may be as many as a dozen different variants (called alleles) at each marker locus. The presence of these polymorphisms on every chromosome permits us to determine which copies of each chromosome a child received from its two parents.

2.1.3 Statistical Analysis

The next step is to compare the pattern of transmission of the disease in a family with the pattern of transmission of each of the several hundred marker regions. Where these patterns match is an indication of the region in which the disease gene is located. Often a "haplotype" of particular alleles of several markers will co-segregate with the disease in a family. A haplotype is simply the linear arrangement of the alleles of several adjacent markers in a particular chromosomal region.

The statistical analysis is accomplished by computer-aided determination that calculates the log of the odds ratio (LOD scores) indicating the statistical significance of the association between marker locus and disease locus (LATHROP et al., 1985). Linkage analysis is useful only where the family is quite large such that DNA samples can be obtained from several affected individuals in the family. LOD scores from several families can be combined to increase the statistical significance, but this will be misleading if different genes are responsible for the disease in different families. Because of this possibility, one large family is generally more useful than several small families.

Linkage mapping is capable, at best, of locating the disease gene to a genetic region of about one centiMorgan, which is approximately one million base pairs of DNA. However, because the mapping resolution depends on random recombination events occurring during gametogenesis, it is not unusual for the candidate region to be as large as several million base pairs.

2.1.4 Positional Cloning

After linkage mapping, the next task is to look through these millions of base pairs that have co-segregated with the known markers. The genes in this region must be identified and a specific mutation in one of the these genes must be shown to cause the disease. This process is termed *positional cloning* (Anonymous, 1995) and has been a very time consuming procedure. One of the first disease genes mapped by linkage analysis was the gene for Huntington disease. The gene was mapped to the end of chromosome 4 in 1983 (GUSELLA et al., 1983), but it then took 10 years to isolate and clone the gene (The Huntington's Disease Collaborative Research Group, 1993).

Recently, the process of cloning genes has accelerated considerably, thanks to the availability of better *physical maps* which show the order in which anonymous cloned fragments of DNA are arranged along each chromosome (HUDSON et al., 1995). The cloning process is still hampered, however, by the lack of information concerning the location of genes on the physical map.

2.2 The Availability of the Genome Sequence will Change the Way in which Disease Genes are Identified

The completion of the structural genomic component of the HGP will provide information that will greatly facilitate the identification of disease-associated genes. The HGP will provide us with the precise location and base sequence of every gene. At this stage, however, we will still not know the function of many genes. From the base sequence alone, we will be able to predict the amino acid structure of the resulting protein. By comparing the structure of a novel protein to the structure of known proteins, we can often make an educated guess as to the cellular location and functional role of the new protein. We will still be ignorant of any role the gene may play in a disease process, however.

By examining the complement of messenger RNA (mRNA) that is produced in different tissues, we will know in which range of tissues each gene is expressed. This can be done by directly examining the mRNA in each tissue by Northern blot analysis, by scanning tissue-specific cDNA libraries for specific gene sequences, or by using DNA chip technology. These approaches will facilitate the linking of genes to diseases.

2.2.1 Strategies Based on Prior Genetic Mapping: "Positional Candidate" Gene Approach

The mapping of disease genes to narrow chromosomal regions, as outlined in Sect. 2.1, will continue to provide important clues to the location of disease genes. However, we will now have a complete list of all the genes that are located within this candidate region and we will know their "normal" sequence. As we gain more insight into the normal function of these genes and in which tissues they are expressed, we will be able to select those that may plausibly be involved in the disease process we are studying. Eventually the gene from an affected person must be sequenced and compared to the reference sequence of that gene. If a difference – capable of altering the protein function – is found, it must be established that every affected person in the family carries the same mutation, and that the mutation is not found in unaffected members or in the general population. This combination of genetic mapping followed by examination of candidate genes within the region is known as the *positional candidate* method of locating disease genes.

2.2.2 Direct Analysis of Candidate Genes

The main drawback to the positional candidate method is that it still relies on some type of preliminary genetic mapping using families that have a particular disease. This mapping process is very time-consuming since it involves accessing data from a large number of family members. As we develop more information about the human genome, and as technological advances in sequence identification continue, we may be able to jump past the mapping step altogether in many cases. Within a few years we will know the DNA sequence of every gene. We will know the functional significance of most of these genes. We will have information about variations in the DNA sequence within defined populations. We will know in which tissues, and possibly under what circumstances, these genes are turned on to produce proteins.

With this additional information it will be possible to identify candidate genes which plausibly could be involved in a genetic disease. The DNA of candidate genes from affected patients will then be sequenced to look for evidence of mutations. With further improvements in DNA sequencing technology, including use of gene chip microarrays, it will be possible to scan hundreds of candidate genes in a relatively short time. Thus, from a number of candidates, one or more disease-associated genes will be identified.

2.3 Problems of Genetic Heterogeneity

Sickle-cell anemia was one of the first inherited diseases in which the molecular basis was discovered (INGRAM, 1959). Subsequently it was found that a single DNA mutation was responsible for virtually all cases, and that the detection of this mutation could be used as a way to identify carriers of the sickle-cell trait. It was, therefore, envisaged that mutations involved in other inherited diseases would follow this paradigm. Alas, this proved not to be the case. Indeed, sickle-cell disease appears exceptional in this regard. Often an inherited disease can be caused by a mutation in any of several different genes (locus heterogeneity), and usually there will be a large number of different possible mutations in each gene (allelic heterogeneity).

The extreme example of locus heterogeneity is retinitis pigmentosa, which is a progressive eye disease that leads to loss of vision. The pattern of inheritance can be dominant, recessive or sex-linked. More than 20 different chromosomal regions have been linked to this disease. Several genes have already been identified, but most remain unknown.

Locus heterogeneity often results from a defect in a biochemical pathway that involves several different genes. A mutation in any one of the genes has the potential to disrupt the pathway and produce the same disease phenotype. As an example, the most common inherited form of cancer, hereditary non-polyposis colon cancer (HNPCC), results from a defect in one of the DNA repair systems within the cell. Several different proteins are required for this specific repair mechanism to work correctly. It is known that HNPCC can result from a mutation in at least 5 of these proteins, yet the clinical features are similar in all families (LYNCH et al., 1999; PETERSEN et al., 1999; SYNGAL et al., 1999).

Locus heterogeneity poses particular problems during the linkage mapping phase of gene hunting. If data from several different families are pooled, then locus heterogeneity can make it impossible to identify any statistically significant associations with marker loci. The availability of very large kindreds is a definite asset in overcoming this problem, since it may be possible to map a gene using only one family. One large kindred is, therefore, much more useful than a collection of smaller families with the same disease.

2.4 The Importance of Founder Populations

Genetically isolated populations have proven to be a useful resource in discovering disease genes (SHEFFIELD et al., 1998). These are populations which originated from a relatively small number of founders, have been geographically or culturally isolated for a prolonged period of time, and have experienced population "bottlenecks" (PELTONEN, 2000). Such populations demonstrate increased genetic homogeneity, exhibiting a higher frequency of certain rare monogenic diseases than is present in a mixed population, while other such diseases may be entirely absent. Well characterized populations of this type are found in Finland (PELTONEN, 2000; DE LA CHAPELLE, 1993), in French Canada (HEYER et al., 1997; DE BRAEKELEER, 1991), and in Iceland (HELGASON et al., 2000). Indeed, Iceland has developed a controversial industry centered on finding disease genes within the population of the island nation (ANNAS, 2000).

Analysis of linkage disequilibrium is a particularly useful technique in mapping disease genes within genetically isolated populations. Linkage disequilibrium is the non-random association of specific marker alleles with a disease susceptibility gene. Linkage disequilibrium is typically employed in the context of association analysis where the frequency of particular haplotypes surrounding a candidate disease gene in affected individuals is compared to the frequency of those same haplotypes in a control population. In isolated populations, where affected individuals are relatively recent descendants of a common mutation-bearing founder, there will have been an insufficient number of generations to eliminate, by genetic recombination, the association between the mutant gene and the marker alleles (SERVICE et al., 1999). This technique has often been employed to fine map the location of a genetic locus first identified by linkage analysis.

2.5 Diseases Caused by Mutation in a Single Gene

Mendelian Inheritance in Man, edited by VICTOR A. MCKUSICK, is a catalog of human genes and genetic disorders. The first edition was published in 1966 and listed some 1,500 phenotypes that were each presumed to represent the manifestation of mutation in a single gene. Today the online version, OMIM (*http://www3.ncbi.nlm.nih.gov/Omim/*), is the recognized "bible" of human Mendelian disease and currently contains over 11,700 separate entries, of which over 8,400 represent different genetic loci. The majority of the entries in OMIM represent inherited diseases that are caused by mutation in a single gene, such as cystic fibrosis, Huntington disease, and muscular dystrophy. With very few exceptions, each of these diseases individually is rare among the population as a whole. However, given the number of different diseases, it is estimated that 3–5% of the population will be diagnosed with a "recognized" genetic disease. Due to the problems of heterogeneity (see Sect. 2.3) and the rarity of many Mendelian

diseases, it may yet be a considerable time before we have a picture of the full range of genetic loci and individual mutations that are responsible.

2.6 Polygenic Diseases

The success of the HGP will have a much greater impact on the elucidation of the etiology of common multifactorial diseases than it will on that of the monogenic diseases. Sorting out the genetic factors that are involved in the common complex disorders, such as schizophrenia or diabetes, remains a challenging proposition. For any multifactorial disease, we are faced with the need to identify multiple genes, each having a small effect on the phenotype. It is the combination of mutations in several genes – coupled with environmental factors – that is responsible for the onset of the disease. Indeed even the term "mutation" may not be appropriate to describe the genetic variations involved. Whereas the mutation in single-gene diseases often has a profound effect on the function of the encoded protein, it is likely that the "mutations" in the polygenic diseases will be found to result in much more subtle changes in protein function. We should perhaps instead use the term "polymorphisms" to describe the variations in gene structure. The effects of these polymorphic variants on protein function will be quantitative rather than qualitative. For example, the binding coefficient of an enzyme–substrate complex or the affinity of a growth factor for a receptor may be affected.

Offshoots of the HGP may also begin to clarify the role of variable penetrance of genes, that seems to play a part in the elusive relationship between genotype and phenotype in multifactorial diseases. It has become apparent that linkage studies have low power in mapping genes with low penetrance. A more productive approach to the identification of susceptibility genes has been the use of association studies, which will become standard procedure with the completion of the HGP.

2.6.1 Candidate Gene Approach

As our knowledge of functional genomics expands in the wake of the final determination of the human genome sequence, we will, for the first time, be aware of the function of the entire set of human genes. This will provide a tremendous boost to selecting candidate genes for the association studies which will form the mainstay of identifying the genes involved in complex diseases.

In a parallel development, the availability of a reference sequence of the entire genome will provide the basis for studying the nature of sequence variation, particularly single nucleotide polymorphisms (SNPs), in human populations. Two companies, Human Genome Sciences Inc. and Compugen, have announced a partnership to identify over 500,000 human cSNPs – which are SNPs located in protein-coding regions of genes – using data from more than 800 unrelated individuals (*http://www.cgen.com/about/press/press020300.htm*). They have estimated that there are, on average, four single nucleotide differences per individual messenger RNA of any two people. Thus the availability of many more candidate genes for association studies will be enhanced by knowledge of the cSNPs occurring within each gene. The association of individual cSNPs with the disease phenotype will provide additional evidence for the genetic basis of complex diseases. In fact, non-synonymous cSNPs, which affect the amino acid sequence of proteins, are expected to be the most frequent type of variation associated with the change in a disease phenotype.

2.6.2 Linkage Disequilibrium Approach

Programs to detect and map SNPs in the human genome are underway with the aim of establishing a comprehensive SNP map of the genome during the next two years (TAILLON-MILLER et al., 1999; A Database of Single Nucleotide Polymorphisms *http://www.ncbi.nlm.nih.gov/SNP/index.html*). Such SNP maps will facilitate a strategy involving the use of very dense maps of polymorphic markers, such that every gene would be covered or narrowly flanked by one or a few markers. Only SNPs, evenly distributed every few hundred bases throughout the whole genome, can provide the marker density required. The effect of a functional variant of a gene will then be indirectly detected through linkage disequilibrium between SNPs (not necessarily cSNPs) and the gene locus (RISCH and MERIKANGAS, 1996; COLLINS et al., 1997). SNP typing is a powerful tool for genetic analysis, and will enable us to discover the association of specific genetic loci with a wide range of disease traits. As was the case with the monogenic traits, the use of founder populations will accelerate such discoveries (RISCH, 2000).

As our knowledge of gene sequences and sequence variation in populations increases, we will pinpoint more of the genes and proteins that are important in common, complex diseases. This will in turn lead to an understanding of the etiology of disease at the molecular level. Diseases will be classified on a molecular and biological basis rather than on a collection of signs and symptoms. This will pave the way for a more rational approach to therapy.

3 Diagnostics and Genetic Testing

3.1 Molecular Diagnosis of Monogenic Disorders

The purpose of genetic testing is usually to identify carriers of genetic defects that could predispose the carrier, or the carrier's children, to an inherited disease. There are many possible reasons for conducting a genetic test. For dominant traits it may be to predict who is susceptible to a disease later in life, so that the appropriate interventions can take place to prevent or ameliorate the condition. Such interventions may take the form of clinical screening programs to detect the disease at an early stage (BRADLEY and EVERS, 1997), of prophylactic surgery, or of pharmaceutical intervention (RUSCHOFF et al., 1998; GRANN et al.,

2000). For recessive traits, carrier screening is conducted to determine, if a couple is at risk of bearing an affected child, or to determine prenatally the genotype of the fetus. For conditions in which there is no effective intervention, e.g., Huntington disease, the demand for genetic testing is comparatively low.

There are two basic approaches to conducting genetic testing:

1. Direct detection of the mutation
   The preferred method is to directly test for the actual mutation that is causing the disease. Direct testing enables us to test any member of a family and to determine unequivocally whether or not they are carrying the mutation. However, the specific mutation in each family must first be identified, and this is not a simple task.
2. Linkage testing
   If the precise mutation is not known, then the DNA tests must depend on linkage analysis which requires following the pattern of inheritance of the disease in the family. This pattern of disease inheritance is compared with the pattern of inheritance of polymorphic DNA markers that are known to be in close proximity to the gene being investigated. Once it is established which gene is responsible for the disease in a given family, linkage analysis can be used to predict whether a person at risk has in fact inherited the mutation-bearing chromosome.

Fig. 1 outlines the factors determining which of the two methods, if either, can be used for genetic testing. Both methods have their drawbacks. Linkage analysis is useful only where the family is quite large. DNA samples must be obtained from several affected individuals in the family before linkage testing is informative. On the other hand, direct mutation testing depends on prior knowledge of the specific gene and the specific mutation that is responsible for the disease in each family.

Linkage testing is falling out of favor because of the complexity of the analysis and the potential for ambiguities in the results. Yet our ability to provide direct mutation testing is

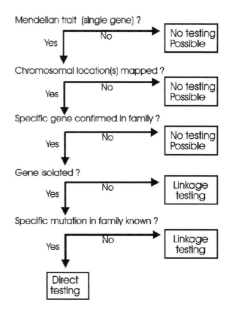

Fig. 1. Genetic testing flow diagram. Flow chart showing the factors that determine what type of genetic testing, if any, is available in a given family situation.

limited by our capacity to identify specific family mutations in a cost-effective manner. Nevertheless, once the mutation is found, it becomes relatively simple to design a test specific for that particular mutation.

How will the Human Genome Project affect this situation? First, the primary objective of the HGP – the reference sequence of the normal gene – will provide the starting point for detecting disease-causing mutation. Second, the next part of the HGP will make us aware of polymorphisms in the normal gene sequence that are inconsequential as far as causing disease is concerned, and will provide us with cSNP maps. Such documentation of "normal" variation will provide a basis for the subsequent identification of pathologic variations.

The significance of some mutations is fairly obvious: if a mutation has a profound effect on the final protein, it is usually evident that this mutation will be responsible for causing disease. Such mutations include most deletions and insertions of DNA bases, "nonsense" mutations – which introduce a "stop" codon into the mRNA thus causing truncation of the protein molecule, and "splice-site" mutations which lead to abnormal processing of the mRNA molecule and production of an aberrant protein.

More difficult to interpret are the "missense" mutations which cause the substitution of one amino acid for another. Missense mutations usually involve the change of a single DNA base to another and are, therefore, relatively frequent. However, the discovery of such a change in the DNA sequence of a patient does not necessarily mean that the mutation is responsible for the genetic disease. Many such amino acid substitutions may have no appreciable effect on the function of the protein. Such "neutral" mutations may, therefore, be regarded as naturally occurring polymorphisms (in fact single nucleotide polymorphisms, cSNPs).

3.1.1 Mutation Scanning Technology

Mutation scanning refers to the process of analyzing one or more DNA sequences or genes for the presence of any possible mutation. Mutation scanning will be the rate-limiting factor in our capacity to offer comprehensive DNA diagnostic services. Advances in this area will occur more as a result of spin-offs from the HGP, rather than as a consequence of the data collection itself. These spin-offs include the rapid advances in the development of the technology used to support the HGP. Many of these technologies can be adapted to enhance our capacity to scan for mutations once a gene has been identified as being involved in a disease process.

Over the past several years, a number of techniques have been used to detect mutations. These include single-strand conformation analysis, heteroduplex analysis, mismatch cleavage, and protein truncation testing (for reviews of these and other techniques, see COTTON, 1997; VAN OMMEN et al., 1999).

3.1.1.1 Single-Strand Conformation Analysis (SSCA)

SSCA (Fig. 2) is based on the sequence-dependent mobility of single-stranded DNA in non-denaturing polyacrylamide gels. A change in the base sequence can cause a shift in the mobility of the DNA strand. This analysis works best with DNA fragments of less than 300 bp. The DNA to be analyzed is first amplified by PCR. The two strands of the DNA product are separated. The single-stranded DNA is allowed to fold into a three-dimensional structure which is dependent on its base composition. If the single strands are then run on a non-denaturing polyacrylamide gel, the two strands will usually migrate at different rates, even though they have the same molecular mass. Any change in base composition as a result of mutation may further modify the mobility of the fragment in the gel. The detection sensitivity for point mutations may approach 80–90% when using short PCR fragments (<200 bp).

3.1.1.2 Heteroduplex Analysis

Heteroduplex analysis involves comparing the mobility of normal double-stranded DNA with "heteroduplex" DNA that contains one normal strand and a complementary strand

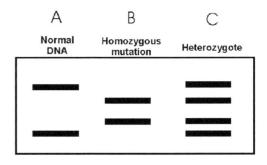

Fig. 2. Representation of SSCA gel. Each band corresponds to a single strand of DNA. Because of sequence differences, the two complementary strands in lanes A and B migrate at different rates in a non-denaturing gel. The heterozygous mutation carrier will show the band pattern in lane C.

containing the mutated sequence. The base mismatch in the heteroduplex DNA brings about a change in the conformation of the molecule which usually causes a decrease in the mobility of the heteroduplex DNA when compared with the normal (homoduplex) DNA. The effective DNA size range for heteroduplex analysis is up to 800 bases, depending on the mutation.

The DNA target is amplified by PCR (Fig. 3). The DNA double helix is separated by raising the temperature to 95 °C. During subsequent cooling, the complementary DNA strands re-anneal, but the non-complementary DNA strands also re-anneal to form heteroduplex DNA. The mismatch in the heteroduplex DNA causes it to have a different three-dimensional shape than homoduplex DNA as a result, the mobility of the heteroduplex DNA will be different (usually less) than that of the homoduplex DNA.

3.1.1.3 Enzyme Mismatch Cleavage

This approach involves preparing a PCR product for the gene under investigation from an unknown sample, hybridizing it with labeled, wild-type DNA, and exposing the resulting hybrid DNA duplex to a DNA-cleaving enzyme such as T4 Endonuclease VII (Fig. 4). The enzyme scans the double-stranded DNA, binds to single-stranded "bubbles" in the heteroduplex DNA that arise from the

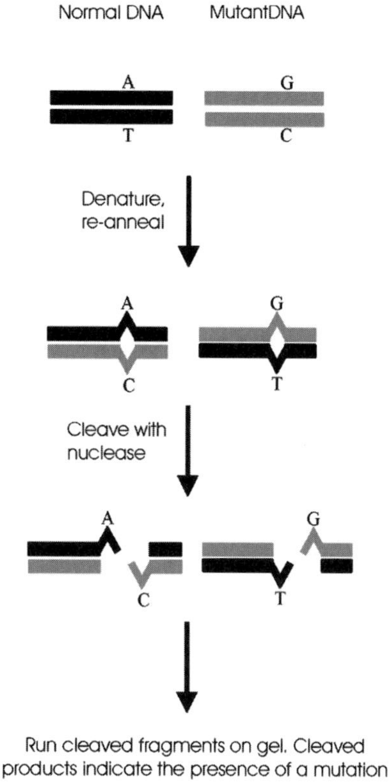

Fig. 4. Enzyme mismatch cleavage. The formation of cleavage products smaller than the original PCR product indicates the presence of a mutation. The size of the cleaved fragment gives some indication of location of the mutation.

presence of a mismatch, and cleaves on the 3′ side of the mismatched base pair. The fragments resulting from the cleavage are then separated by electrophoresis. The formation of new cleavage products indicates the presence of a mutation. The size of the fragments gives some indication of location of the mutation.

3.1.1.4 Protein Truncation Test (PTT)

The PTT is used to detect mutations that cause premature termination of translation, thus resulting in a truncated protein. Such mutations include frameshift insertions and dele-

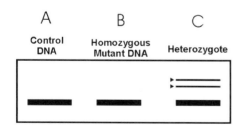

Fig. 3. Representation of heteroduplex gel. Each band corresponds to a double-stranded DNA molecule. In lanes A and B the band represents homoduplex DNA. The arrows in lane C indicate heteroduplex DNA molecules, consisting of one wild-type strand and the complementary mutant strand.

tions, nonsense, and splice-site mutations. Messenger RNA, prepared from a fresh blood sample or an immortalized cell line, is used as a template for cDNA synthesis. The cDNA is then amplified by using PCR primers that permit the subsequent *in vitro* transcription and translation of the PCR product using a rabbit reticulocyte lysate system. The products are then run on an acrylamide gel to check the size of the protein produced. An abnormally short protein indicates the presence of a truncating mutation.

The above techniques share several features, none of them positive:

(1) They are all relatively laborious and expensive (on the basis of each mutation detected);

(2) they generally require that the gene be subdivided into its different coding regions (exons) prior to analysis, and most significantly;

(3) none of them is capable of detecting all the possible mutations that may be present in a gene.

3.1.1.5 Complete Gene Sequencing

The method touted as the "gold standard" for mutation scanning is complete gene sequencing. This is somewhat of a misnomer since only the coding regions (exons) are usually sequenced, together with small flanking sequences that contain the splice sites crucial for the correct processing of the primary RNA transcript into the mature mRNA. The exons – numbering from a few up to 70 or more – usually represent only a small percentage of the total gene sequence. Complete gene sequencing is capable of detecting most sequence variants. (It will not detect large deletions or duplications spanning one or more exons, although this has often been overlooked.) Until now, complete gene sequencing has been too expensive for use as the primary mutation scanning technique. However, the combination of newly developed sequencing technologies and the genomic data obtained from the HGP may soon render most other mutation scanning methods redundant.

Contrary to the predictions made 10 or 12 years ago, the basic method of DNA sequencing has not changed at all. Instead, the sequencing of the entire genome has been accomplished by automating and scaling up the existing techniques. Single manually-poured slab gels have given way to robotic stations feeding arrays of 96-capillary gels which are capable of sequencing large numbers of DNA samples on a 24/7 basis with little operator intervention. Once the final polished genome sequence is obtained – the current prediction is sometime in 2003 – this vast amount of sequencing capability will be turned to analyzing sequence variation within populations, including the detection of disease-associated mutations. Complete gene sequencing is already commercially available for certain genes, including the BRCA1 and BRCA2 breast/ovarian cancer genes (Myriad Genetics Inc: http://www.myriad.com/gt.html).

Gel-based sequencing may be supplemented by the use of array-based methods (gene chips, see Chapter 9, this volume). Applications for the use of oligonucleotide microarrays for resequencing and mutation analysis are being developed (HACIA, 1999; TONISSON et al., 2000). They are not yet robust enough, however, for the routine automated detection of heterozygous mutations in genomic DNA. The false negative rate is still somewhat high.

3.1.2 Mutation Screening

Mutation screening is defined as the testing of DNA samples from multiple individuals for the presence of a specific single mutation. Mutation screening is much easier to implement than mutation scanning. Existing techniques include exploiting the creation or deletion of a restriction enzyme site, use of allele-specific oligonucleotides to bind selectively to the region of the mutation, and use of allele-specific PCR primers. The newer technologies that will take over include use of single nucleotide primer extension assays (SYVANEN, 1999), oligonucleotide microarrays for SNP detection (HACIA, 1999), and combinations of the two (KURG et al., 2000).

3.1.3 Ethical and Social Issues

When considering mutation screening, the *capability* of screening must not be confused with the *desirability* of screening. The objectives of the screening program must be clear, the interventions that will be required once the results of the screening are released must be carefully defined (SEAMARK and HUTCHINSON, 2000), and the people to be screened must be fully aware of the implications of screening and must give an explicit informed consent for testing. The last points usually necessitate the involvement of a trained genetic counsellor in the process (PARENS, 1996; MATHEW, 1999), since most other health care professionals may be unaware of all the potential implications. The availability of such counsellors may soon be quite inadequate to cope with the flood of genetic tests that will eventually become available.

Mutation screening can be offered to at-risk individuals or to groups of related individuals. In some cases offering screening to whole populations or sub-populations may be appropriate (KRONN et al., 1998). It may also be conducted using fetal cells obtained by chorionic villus biopsy or amniocentesis. Pre-implantation testing of the embryo has been conducted in the context of *in vitro* fertilization (DE VOS et al., 1998; KULIEV et al., 1999), but is unlikely to become widespread in the near future.

3.1.4 Confirmation of Clinical Diagnosis

So far, this has been a relatively less common use of DNA testing. As our knowledge of genomics increases, however, testing specific genes for mutations will more frequently be used to confirm a clinical diagnosis or to make a primary diagnosis. This will be especially relevant for diseases with limited locus and allelic heterogeneity – conditions involving only one or two possible genes, with a limited number of specific mutations. In many centers, DNA testing is already part of the diagnostic procedure for conditions such as sickle-cell anemia, cystic fibrosis, and hemochromatosis.

3.2 Molecular Diagnosis in Common Complex Diseases

Despite the promise of cataloging low-penetrance genetic variants that modify the risk to common diseases (see Sect. 2.6), the current list of confirmed variants is remarkable for its brevity. There have been a large number of studies attempting to find such variants, and many of them report associations. In most cases, however, the analysis of the statistical probability of association has not been rigorous, or the associations have not been replicated in subsequent studies.

A notable discovery was the association of the E4 allele of apolipoprotein E with Alzheimer's disease (CORDER et al. 1993; ROSES, 1997). This finding initiated a flood of publications concerning the role of apolipoprotein E in Alzheimer's disease, and indicates what we may expect to happen as other such associations are made. In other disease, susceptibility loci have been established, but the gene has not yet been identified. These include migraine (NYHOLT et al., 1998a, b), type II diabetes (HANIS et al., 1996; SHAW et al., 1998), and psoriasis (ENLUND et al., 1999). In the case of cardiovascular disease, the significance of polymorphisms in angiotensinogen (JEUNEMAITRE et al., 1992) and angiotensin converting enzyme (CAMBIEN et al., 1992) was recognized some time ago.

Association studies have been most successful in mapping susceptibility genes to the human leukocyte antigen (HLA) region of chromosome 6. The HLA region contains about 200 genes, many of which are involved in the immune response. Associations with genes in this region have been described for over 100 diseases (PILE, 1999; THORSBY, 1997). In some cases the HLA gene is directly involved in the etiology of the disease, e.g., ankylosing spondylitis (ALLEN et al., 1999). In other diseases, the association appears due to linkage disequilibrium (i.e., it is secondary to the occurrence of the true susceptibility gene at a nearby locus).

Thus in the present situation there is virtually no genetic testing available for predisposition to the common polygenic diseases. We are instead in a period of waiting: waiting for susceptibility genes to be discovered, waiting for the variants of these genes to be exposed, and waiting for the functional significance of these variants to be revealed (see Sect. 2.6). With the sequencing phase of the HGP swiftly coming to a close, expectations are running high that we will not have to wait too much longer. Once we have this information, the potential to screen for genetic predisposition to the common diseases is inspiring, perhaps overwhelming. For any given disease or group of diseases it will be possible to design a DNA microarray that will simultaneously test for the presence of thousands of possible cSNPs in any number of susceptibility genes. How will the health care system – not to mention the person being tested – cope with this mountain of information? How can we determine the significance of the interactions between different variants in multiple susceptibility genes? Should we even be doing this kind of susceptibility testing? The answers to these and numerous other relevant questions will reveal themselves over the course of the next decade or two. It will be an interesting time.

4 Treatment

It is not an overstatement to assert that the Human Genome Project will greatly increase our understanding of the biological and molecular basis of most human disease. This leap in our appreciation of the underlying basis of disease will provide the framework from which to develop more rationale treatment strategies. Most of the benefits will be seen in diseases not usually considered to be primarily "genetic" in origin, that is to say the common complex diseases. Disease gene identification will give rise to the development of genotype-specific therapies. Ironically, the HGP will have little immediate impact on gene therapy for the monogenic inherited diseases.

4.1 How to Find New Targets for Drug Therapy

So many targets, so little time. This could be the perception of the Human Genome Project from the viewpoint of International Pharma. In fact, sorting out the wheat from the chaff is likely to consume a considerable portion of the pharmaceutical research budget over the next few years (for a review of pharmacogenetics, see ROSES, 2000). There are two main strategies for using information from the HGP to develop novel therapeutic agents. The first is to identify genes in which variation leads to a disease phenotype. The second is to use a brute-force approach to mine data on genomics, tissue-specific gene expression, and proteomics (WANG and HEWICK, 1999).

4.1.1 Disease-Associated Genes as Drug Targets

This is a top-down approach. The major benefit of using disease-associated genes is the obvious one: the gene, or rather the protein produced from the gene, must be involved in the etiology of the disease. This does not necessarily mean, however, that the target protein will ultimately prove useful for therapeutic intervention. Nevertheless, knowledge gained about the underlying biology of the disease may lead to other proteins that may be better targets. The best targets are proteins belonging to such molecular families as enzymes or receptors for which high-throughput protocols are already available for screening potential drugs (ROSES, 2000).

4.1.2 Any Gene as a Drug Target

The second approach is essentially a bottom-up design employing a shotgun or brute-force strategy. Computer algorithms have been developed to permit "mining" the vast amount of data from the HGP together with complementary proteomic data. The approach here is to identify novel families of genes and proteins on the basis of structural or functional similarities. From these molecular families, proteins or other molecules will be selected that have the potential to be good targets for drug screening protocols. Once such targetable genes are discovered, the final step will be to determine what possible role that gene may have in human disease. This can be accomplished by examining tissue-specific expression, by examining the occurrence of variants in susceptible individuals, and by correlating the information from the top-down discovery of disease-associated genes whenever possible.

4.2 Variations in Response of Patients to Drug Therapy

It has long been recognized that different people respond differently to any given drug. Indeed, some diseases are sub-classified based on the response of a patient to a drug. The types of response include:

- favorable and effective response,
- low or no response,
- over-response (overdose),
- occurrence of adverse reactions to the drug.

A common assumption is that much of the variation is due to genetic factors, but in most cases we have little insight into the nature of such factors. Polymorphisms in genes coding for receptors, transporters, and metabolizing pathways can all modulate these effects. Known variations in drug metabolizing pathways include:

(1) primaquine sensitivity, a trait attributed to glucose-6-phosphate dehydrogenase deficiency,
(2) the cytochrome P-450 CYP2D6 polymorphism that results in poor, extensive, and ultrarapid metabolizers of at least 30 drugs, and
(3) the polymorphic (NAT2*) acetylation of aromatic amine drugs, such as isoniazid and sulfasalazine (for a review, see EVANS and RELLING, 1999).

Collection of pharmacogenomic data during clinical drug trials is becoming more common. By applying whole genome SNP mapping to clinical trial data, it may be possible to find genomic regions associated with efficacy or common adverse responses. Analysis of cSNPs will eventually lead to a personal pharmacogenetic profile that will help in predicting how an individual patient will respond to possible drug therapy. This will help narrow down the number of drugs to be considered in treating that patient. However, we are still a long way from a pharmacogenetic DNA chip that will predict the response of an individual to a wide range of commonly indicated drugs.

4.3 Gene Therapy: Promises and Problems

As soon as scientists realized that a defective gene could cause a disease, they began to speculate about the possibility of fixing the broken gene. Then about 10–15 years ago, when we actually began to identify and sequence disease-associated genes, prophecies that we would see some kind of gene therapy within the next several years were common. This was not to be. Despite choosing diseases where we had a good understanding of the underlying biology and where we could (or at least thought we could) access the critical cell or tissue, there has been no "Eureka" moment in the gene therapy community.

Gene therapy is the use of genetic material to treat human disease. The most obvious targets are the inherited monogenic diseases, in which a single gene has gone wrong. However, many other possibilities exist, including using gene transfer to modulate the immune response, to mediate cell killing (in cancer), for example), or to activate a pro-drug specifically within the target tissue. A vector, often a disabled virus, must be used to transfer the therapeutic gene into the target tissue or cell population (for reviews, see SMITH, 1999; PALU et al., 1999).

The difficulties with gene therapy are technical in nature, involving such mundane problems as getting enough gene into the target cell, getting a large enough proportion of the target cells to take up the gene, and making sure the new gene is expressed over a prolonged period of time. We have hardly begun to think about questions like how to ensure that the new gene responds to the normal regulatory signals. Since these problems are unrelated to knowing the genetic origins of the disease, the HGP will have little immediate effect on gene therapy research.

There may be some light at the end of the tunnel, however. After a decade of unsuccessful attempts to cure one form of severe combined immunodeficiency disease by using gene therapy, researchers have apparently achieved success in treating another genetic type of the same disease (BUCKLEY, 2000; CAVAZZANA-CALVO et al., 2000).

5 Conclusions

The knowledge gained from the Human Genome Project, when combined with new genomic and proteomic technologies, has the capacity to transform the practice of medicine. We will eventually be able to characterize and treat diseases on a biological and molecular basis, rather than by empiricism. The impact on human health may surpass that of antibiotics in decreasing morbidity and increasing life expectancy. It is difficult to predict the time it will take to fully implement these benefits (remembering the lesson of gene therapy), but it will certainly take longer than was the case for antibiotics. Meanwhile, it is vital to educate the existing cadre of physicians and other health care workers about the implications of the new discoveries, so that they may take the best advantage of subsequent developments.

It is not unreasonable to imagine that, at some time in the not too distant future, the standard practice will be to obtain the sequence of all of the coding DNA in every individual's genome. An individual profile of all known cSNPs and other coding variants could possibly substitute for the complete sequence. This information, carried on a microchip on our health card, will define each of us in molecular terms with respect to our disease susceptibilities and our likely response to drug therapy. As one might imagine, the social and ethical considerations of such a capability will be the subject of considerable discussion. The public as a whole will rightly be wary of a

move in this direction. Thus we have an obligation to also educate a wider audience with respect to the science and the implications.

It is to be hoped that the philosophical discussions will be able to keep pace with the speed of our scientific discoveries.

Acknowledgement
I would like to thank ELIZABETH LILES for help in preparation of the manuscript.

6 References

ALLEN, R. L. et al. (1999), The role of HLA-B27 in spondyloarthritis, *Immunogenetics* **50**, 220–227.

ANNAS, G. J. (2000), Rules for research on human genetic variation – lessons from Iceland, *N. Engl. J. Med.* **342**, 1830–1833.

Anonymous (1995), Positional Cloning Approach Expedites Gene Hunts, *Human Genome News* **6**, 1.

APARICIO, S. A. (2000), How to count ... human genes, *Nature Genet.* **25**, 129–130.

BOTSTEIN, D. (1999), Of genes and genomes, *Ann. N. Y. Acad. Sci.* **882**, 32–41.

BRADLEY, B. A., EVERS, B. M. (1997), Molecular advances in the etiology and treatment of colorectal cancer, *Surg. Oncol.* **6**, 143–156.

BUCKLEY, R. H. (2000), Gene therapy for human SCID: Dreams become reality, *Nature Med.* **6**, 623–624.

CAMBIEN, F. et al. (1992), Deletion polymorphism in the gene for angiotensin-converting enzyme is a potent risk factor for myocardial infarction, *Nature* **359**, 641–644.

CAVAZZANA-CALVO, M. et al. (2000), Gene therapy of human severe combined immunodeficiency (SCID)-X1 disease, *Science* **288**, 669–672.

COLLINS, F. S. et al. (1997), Variations on a theme: cataloging human DNA sequence variation, *Science* **278**, 1580–1581.

CORDER, E. H. et al. (1993), Gene dose of apolipoprotein E type 4 allele and the risk of Alzheimer's disease in late onset families, *Science* **261**, 921–923.

COTTON, R. G. H. (1997), *Mutation Detection*. Oxford: Oxford University Press.

DE BRAEKELEER, M. (1991), Hereditary disorders in Saguenay-Lac-St.-Jean (Quebec, Canada), *Hum. Hered.* **41**, 141–146.

DE LA CHAPELLE, A. (1993), Disease gene mapping in isolated human populations: the example of Finland, *J. Med. Genet.* **30**, 857–865.

DE VOS, A. et al. (1998), Pregnancy after preimplantation genetic diagnosis for Charcot-Marie-Tooth disease type 1A, *Mol. Hum. Reprod.* **4**, 978–984.

ENLUND, F. et al. (1999), Psoriasis susceptibility locus in chromosome region 3q21 identified in patients from southwest Sweden, *Eur. J. Hum. Genet.* **7**, 783–790.

EVANS, W. E., RELLING, M. V. (1999), Pharmacogenomics: translating functional genomics into rational therapeutics, *Science* **286**, 487–491.

EWING, B., GREEN, P. (2000), Analysis of expressed sequence tags indicates 35,000 human genes, *Nature Genet.* **25**, 232–234.

GRANN, V. R. et al. (2000), Prevention with tamoxifen or other hormones versus prophylactic surgery in BRCA1/2-positive women: a decision analysis, *Cancer J. Sci. Am.* **6**, 13–20.

GUSELLA, J. F. et al. (1983), A polymorphic DNA marker genetically linked to Huntington's disease, *Nature* **306**, 234–238.

HACIA, J. G. (1999), Resequencing and mutational analysis using oligonucleotide microarrays, *Nature Genet.* **21**, 42–47.

HANIS, C. L. et al. (1996), A genome-wide search for human non-insulin-dependent (type 2) diabetes genes reveals a major susceptibility locus on chromosome 2, *Nature Genet.* **13**, 161–166.

HELGASON, A. et al. (2000), mtDNA and the origin of the Icelanders: deciphering signals of recent population history, *Am. J. Hum. Genet.* **66**, 999–1016.

HEYER, E. et al. (1997), Seventeenth-century European origins of hereditary diseases in the Saguenay population (Quebec, Canada), *Hum. Biol.* **69**, 209–225.

HUDSON, T. J. et al. (1995), An STS-based map of the human genome, *Sience* **270**, 1945–1954.

HURME, M. et al. (1998), Gene polymorphisms of interleukins 1 and 10 in infectious and autoimmune diseases, *Ann. Med.* **30**, 469–473.

IKEDA, M., YU, D. T. (1998), The pathogenesis of HLA-B27 arthritis: role of HLA-B27 in bacterial defense, *Am. J. Med. Sci.* **316**, 257–263.

INGRAM, V. M. (1959), Abnormal human haemoglobin. III. The chemical difference between normal and sickle cell haemoglobins, *Biochim. Biophys. Acta* **36**, 402–411.

JEUNEMAITRE, X. et al. (1992), Molecular basis of human hypertension: role of angiotensinogen, *Cell* **71**, 169–180.

KRONN, D. et al. (1998), Carrier screening for cystic fibrosis, Gaucher disease, and Tay-Sachs disease in the Ashkenazi Jewish population: the first 1000 cases at New York University Medical Center, New York, NY, *Arch. Intern. Med.* **158**, 777–781.

KULIEV, A. et al. (1999), Birth of healthy children after preimplantation diagnosis of thalassemias, *J. Assist. Reprod. Genet.* **16**, 207–211.

KURG, A. et al. (2000), Arrayed primer extension: solid-phase four-color DNA resequencing and mutation detection technology, *Genet. Test.* **4**, 1–7.

LATHROP, G. M. et al. (1985), Multilocus linkage analysis in humans: detection of linkage and estimation of recombinations, *Am. J. Hum. Genet.* **37**, 482–498.

LIANG, F. et al. (2000), Gene index analysis of the human genome estimates approximately 120,000 genes, *Nature Genet.* **25**, 239–240.

LYNCH, H. T. et al. (1999), Clinical impact of molecular genetic diagnosis, genetic counseling, and management of hereditary cancer: Part II: Hereditary nonpolyposis colorectal carcinoma as a model, *Cancer* **86**, 2457–2463.

MATHEW, C. G. (1999), DNA diagnostics: goals and challenges, *Br. Med. Bull.* **55**, 325–339.

MCKUSICK, V. A. (Ed.) (1966), *Mendelian Inheritance in Man.* Baltimore, MD: Johns Hopkins University Press.

NYHOLT, D. R. et al. (1998a), Familial typical migraine: linkage to chromosome 19p13 and evidence for genetic heterogeneity, *Neurology* **50**, 1428–1432.

NYHOLT, D. R. et al. (1998b), Evidence for an X-linked genetic component in familial typical migraine, *Hum. Mol. Genet.* **7**, 459–463.

PALU, G. et al. (1999), In pursuit of new developments for gene therapy of human diseases, *J. Biotechnol.* **68**, 1–13.

PARENS, E. (1996), Glad and terrified: On the ethics of BRCA1 and 2 testing, *Cancer Invest.* **14**, 405–411.

PELTONEN, L. (2000), Positional cloning of disease genes: advantages of genetic isolates, *Hum. Hered.* **50**, 66–75.

PETERSEN, G. M. et al. (1999), Genetic testing and counseling for hereditary forms of colorectal cancer, *Cancer* **86**, 2540–2550.

PILE, K. D. (1999), Broadsheet number 51: HLA and disease associations, *Pathology* **31**, 202–212.

RISCH, N. J. (2000), Searching for genetic determinants in the new millennium, *Nature* **405**, 847–856.

RISCH, N., MERIKANGAS, K. (1996), The future of genetic studies of complex human diseases, *Science* **273**, 1516–1517.

ROGER, M. (1998), Influence of host genes on HIV-1 disease progression, *FASEB J.* **12**, 625–632.

ROSES, A. D. (1997), A model for susceptibility polymorphisms for complex diseases: apolipoprotein E and Alzheimer disease, *Neurogenetics* **1**, 3–11.

ROSES, A. D. (2000), Pharmacogenetics and the practice of medicine, *Nature* **405**, 857–865.

RUSCHOFF, J. et al. (1998), Aspirin suppresses the mutator phenotype associated with hereditary nonpolyposis colorectal cancer by genetic selection, *Proc. Natl. Acad. Sci. USA* **95**, 11301–11306.

SEAMARK, C. J., HUTCHINSON, M. (2000), Controversy in primary care. Should asymptomatic haemochromatosis be treated? *Br. Med. J.* **320**, 1314–1317.

SERVICE, S. K. et al. (1999), Linkage-disequilibrium mapping of disease genes by reconstruction of ancestral haplotypes in founder populations, *Am. J. Hum. Genet.* **64**, 1728–1738.

SHAW, J. T. et al. (1998), Novel susceptibility gene for late-onset NIDDM is localized to human chromosome 12q, *Diabetes* **47**, 1793–1796.

SHEFFIELD, V. C. et al. (1998), Use of isolated inbred human populations for identification of disease genes, *Trends Genet.* **14**, 391–396.

SMITH, A. E. (1999), Gene therapy – where are we? *Lancet* **354** (Suppl. 1), SI1–SI4.

SYNGAL, S. et al. (1999), Interpretation of genetic test results for hereditary nonpolyposis colorectal cancer: implications for clinical predisposition testing, *JAMA* **282**, 247–253.

SYVANEN, A. C. (1999), From gels to chips: "minisequencing" primer extension for analysis of point mutations and single nucleotide polymorphisms, *Hum. Mut.* **13**, 1–10.

TAILLON-MILLER, P. et al. (1999), Efficient approach to unique single-nucleotide polymorphism discovery, *Genome Res.* **9**, 499–505.

The Huntington's Disease Collaborative Research Group (1993), A novel gene containing a trinucleotide repeat that is expanded and unstable on Huntington's disease chromosomes, *Cell* **72**, 971–983.

THORSBY, E. (1997), Invited anniversary review: HLA associated diseases, *Hum. Immunol.* **53**, 1–11.

TONISSON, N. et al. (2000), Unravelling genetic data by arrayed primer extension, *Clin. Chem. Lab. Med.* **38**, 165–170.

VAN OMMEN, G. J. et al. (1999), The human genome project and the future of diagnostics, treatment, and prevention, *Lancet* **354** (Suppl. 1), SI5–10.

WANG, J. H., HEWICK, R. M. (1999), Proteomics in drug discovery, *Drug Discov. Today* **4**, 129–133.

5 Pharmaceutical Bioinformatics and Drug Discovery

CHANDRA S. RAMANATHAN

DANIEL B. DAVISON

Wallingford, CT, USA

1 Introduction

Bioinformatics is the discipline at the intersection of biology, computer science, and mathematics. It has developed over the last 20 years by accumulation of researchers from those, and other, disciplines. However, in large part, there is no good definition of exactly what a bioinformatician does. If one were to ask 10 people in the field for a definition, one would get at least 10 different answers. It is also well known that there is a critical shortage of trained bioinformatics personnel, but it is less clear what training is needed. This is, in part, because of the rapid rate of change in the field, its underlying analyses, and its data. In this chapter, we will address those aspects of bioinformatics that contribute to drug discovery and drug development in the pharmaceutical and biotechnology industries.

Bioinformatics has developed from a number of academic disciplines, and this has provided a great deal of synergism. However, training in academia does not necessarily provide the background needed for a successful bioinformatics career in industry. This is due to a variety of factors, the chief one being the scale of analyses performed. It is a truism of biology that, until very recently, progress was made "one gene, one protein, one postdoc" at a time. However, in the analysis of human and other genomes for drug targets, there is not enough time or resources to take that approach. Instead, pharmaceutical bioinformaticians must work on hundreds to millions of protein or nucleic acids sequences simultaneously. This bulk-data-processing approach is usually not encountered academically other than in genome sequencing centers. Another difference is that all of the data generated need to be stored and converted into knowledge for experimentalist colleagues, so that they do not have to learn detailed algorithmic interpretation of bioinformatics output.

Bioinformatics can be thought of as consisting of two main interdisciplinary sub-fields: the research and development work required to build the software and database infrastructure and the computation-based research devoted to understanding and solving biological questions. The emphasis of this chapter will be on the application of bioinformatics tools and databases to identify and characterize novel gene targets, which is of interest to pharmaceutical companies. The classical drug discovery paradigm (Fig. 1) depends on the knowledge of functional activity of a protein. This functional knowledge is used either to isolate a target or develop a screen against the target. A pharmaceutical bioinformatician must design a system for converting bulk data and analyses into experimentalist-accessible knowledge. While we describe the current practice of bioinformatics in drug discovery, we also discuss knowledge generation.

1.1 Databanks, Algorithms, and Databases

It is now a cliché to note that sequence information is growing exponentially – but it is still true. The primary source of information for bioinformaticians is a small set of core databanks, and databanks derived from those data. In this chapter we will draw a sharp distinction between a (possibly formatted) collection of information, a *databank*, and how it is stored, which may be in some kind of database management system (RDBMS), such as Oracle. The terms are typically used interchangeably, but it is critical to understand this

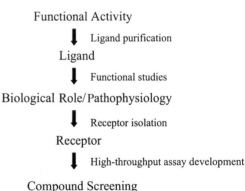

Functional Activity

↓ Ligand purification

Ligand

↓ Functional studies

Biological Role/Pathophysiology

↓ Receptor isolation

Receptor

↓ High-throughput assay development

Compound Screening

↓ Medicinal/Combinatorial chemistry

Drug Development

Fig. 1. "Classical" drug discovery.

difference. A flat file collection of data is substantially less useful than one organized in a RDBMS. For analyses where high throughput is required, such as drug discovery, flat file organization of data can be a major bottleneck in an analysis pipeline.

Most pharmaceutical companies wish to have the best possible intellectual property (IP) position, so most are interested in novel genes. Public and proprietary databanks are the chief sources of data, whether GenBank, PIR, Swiss-Prot, Incyte, Celera, CurGen, or many other sources (CAPONE et al., 1996; FANNON, 1996; BAINS, 1998). In the next section, we will discuss a procedure for novel gene discovery.

1.2 Databanks for Mining Novel Genes

The discovery of eukaryotic drug targets via bioinformatics can be considered "gene-centric" drug discovery, as opposed to the more traditional target-centric drug discovery. In our paradigm for gene-centric drug discovery, identification of a novel (or orphan) gene is the first step in this model of the drug discovery process (Fig. 2). A "protein", or "target", of interest to a company involved in drug discovery could be

(1) a novel protein related to a protein or family of proteins of interest,

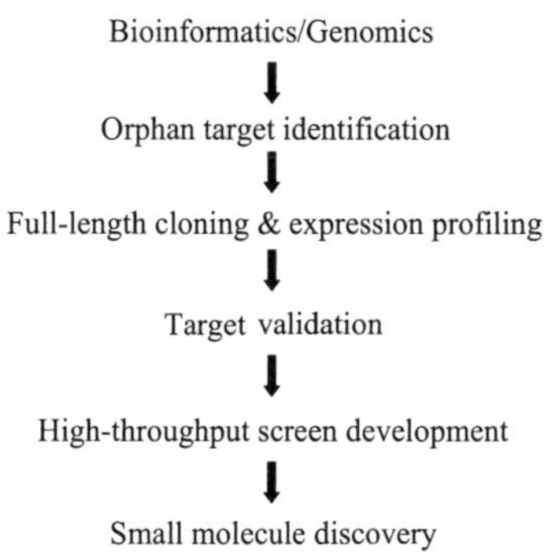

Bioinformatics/Genomics

↓

Orphan target identification

↓

Full-length cloning & expression profiling

↓

Target validation

↓

High-throughput screen development

↓

Small molecule discovery

Fig. 2. Bioinformatics/genomics-based drug discovery.

(2) a novel protein which has a desirable expression profile,
(3) a known or unknown protein which has been shown to be differentially expressed in a biological condition of unmet medical need,
(4) a mutation in the gene or genes associated with a disease,
(5) the human homolog of a model organism protein which exhibits a tractable phenotype.

Currently, there are around 500 such molecular targets (DREWS, 2000). It is expected that this number will increase by about an order of magnitude over the next decade.

The last decade saw an explosion in the generation and availability of genomic data. The Human Genome Project (*http://www.nhgri.nih.gov*) finished the draft human genome sequence on June 26, 2000. Also, many microbial genomes, and the complete genomes of bakers' yeast (*Saccharomyces cerevisiae*), fruit fly (*Drosophila melanogaster*), and nematode (*Caenorhabditis elegans*) have been published. We are also beginning to obtain other types of data, such as gene expression, single nucleotide polymorphisms and other mutations, protein–protein interactions, and regulatory networks. In addition, to be maximally effective in drug discovery, molecular biological data have to be integrated with biochemical, chemical, and clinical data. Those companies that make the most effective use of this complex mesh of data will be the most successful. The knowledge of the most interest to pharmaceutical companies is the identification of drug targets, functional characterization of genes, and the elucidation of gene–disease relationships.

When sequence databanks were first created the amount of data was small and the entries were manually created and entered as text files. As new types of data were captured or created, new data repositories were created using a variety of file formats (BAKER and BRASS, 1998). Over the past 10 years the management of biological information has truly come of age, becoming increasingly integrated into the scientific process. It is now difficult to think of an experimental strategy that does not involve some data from scientific databanks.

The flood of sequence data has led to problems with quality control and the speed with which the data can be searched and analyzed (FLEISCHMANN et al., 1995; BLUT et al., 1996; MEWES et al., 1997). The majority of biological databanks now need to be re-engineered and updated to cope with this quantity of information.

As stated by BAKER and BRASS (1998), databanks can be classified into four categories, depending on the source of the data:

(1) Primary databanks: They contain one principal kind of information (e.g., sequence data) which may be derived from many sources, such as large-scale sequencing projects, individual submission, literature, and other databanks. Examples for primary databanks include protein databanks such as the Protein Information Resource (PIR) (*http://www-nbrf.georgetown.edu/pir/*) and Swiss-Prot (*http://expasy.hcuge.ch/sprot*).

(2) Secondary databanks: They contain one principal kind of information (e.g., alignment data) which is derived solely from other databases. The data may be a straightforward subset of another database or may be derived by analysis of another database. Examples for secondary databases include motif and pattern databases such as: BLOCKS (*http://www.blocks.fhcrc.org/*) derived from PROSITE (*http://expasy.hcuge.ch/sprot/prosite.html*) and PRINTS (*http://www.biochem.ucl.ac.uk/bsm/dbbrowser/PRINTS/*) derived from OWL (*http://www.biochem.ucl.ac.uk/bsm/dbbrowser/OWL/OWL.html*). These databanks are discussed below.

(3) Knowledge databanks and true databases: These are specialist databanks containing related information from many sources, such as the literature, expert input, and other databanks. Examples of knowledge databanks include structural databases such as SCOP (*http://scop.mrc-lmb.cam.ac.uk/*) from PDB (*http://www.rcsb.org/*) and the *E. coli* biochemical pathway database, EcoCyc (*http://ecocyc.panbio.com/ecocyc/*).

(4) Integrated databank systems: It is a combination of primary or secondary databanks. The examples of integrated databanks include those corporate genomic databases developed by pharmaceutical and biotechnology companies as single-stop source for genomic information for their biologists.

The two main databanks that are widely used for novel gene discovery are the Expressed Sequence Tag (EST) sequence division and High-Throughput Genomic sequence division of GenBank. These two are the high-impact databases and they are described next.

1.3 Expressed Sequence Tag (EST) Databanks

ESTs provide the direct window onto the expressed genome (RAWLINGS and SEARLS, 1997). They are single-pass, partial sequences of 50–500 nucleotides made from cDNA libraries. The main public source of these sequences is dbEST (*http://www.ncbi.nlm.nih.gov/dbest/*) and the EST clones are available from the I.M.A.G.E. consortium ("Integrated Molecular Analysis of Genomes and their Expression" consortium; URL: *http://www-bio.llnl.gov/bbrp/image/image.html*). As of February 2001, there are 7.3 million public domain ESTs (Tab. 1). EST databanks and cDNA sequencing are now used widely as part of both academic and commercial gene discovery projects. With the availability of high-performance computers and the sharp decline in their cost, large-scale analysis and review of EST sequence data, particularly taking into account quality scores from the sequencing runs, are now possible. These data are also linked with other genomic information in databases such as EGAD (AARONSON et al., 1996).

One of the success stories of using EST data is the identification and subsequent cloning of a candidate gene for chromosome 1 familial Alzheimer's disease, STM2 (LAHAD-LEVY et al., 1995). Another is the identification of the human homolog (hMLH1) of the bacterial

Tab. 1. Number of Available ESTs in the Public Domain (as of February 2001)

Organism	Number of ESTs
Homo sapiens (human)	3,148,771
Mus musculus and *Mus domesticus* (mouse)	1,929,354
Rattus spp. (rat)	263,234
Caenorhabditis elegans (nematode)	109,215
Drosophila melanogaster (fruit fly)	116,471
Danio rerio (zebrafish)	79,237

DNA mismatch repair gene *mutL*. Missense mutations in hMLH1 has been associated with chromosome-3-linked hereditary nonpolyposis colorectal cancer (PAPADOPOULOS et al., 1994; BRONNER et al., 1994). Biotechnology companies like Human Genome Sciences (*http:// www.humangenomesciences.com*) and Incyte (*http://www.incyte.com*) have also capitalized upon this success and potential for novel gene discovery from EST databanks.

EST sequences are generated by shotgun sequencing methods. The sequencing is random and a sequence can be generated several times. This results in a huge amount of redundancy in the database. Large-scale bioinformatics and experimental comparative genomics are complex and time consuming. One challenge is to eliminate the redundancy in the EST databases. Sequence–cluster databases such as UniGene (NCBI News, 1996), EGAD (Expressed Gene Anatomy Database, AARONSON et al., 1996) and STACK (The South Africa National Bioinformatics Institute: Sequence Tag Alignment and Consensus Knowledgebase, *http://www.sanbi.ac.za/Dbases.html*; MILLER et al., 1999) address the redundancy problem by clustering sequences that are sufficiently similar that one may reasonably infer that they are derived from the same gene. Many companies, e.g., Celera (*http://www. celera.com*) and Incyte have their own clustering software. A commercial software package available for clustering is based on the d^2 algorithm (BURKE et al., 1999) and is currently marketed by DoubleTwist (*http://www. doubletwist.com/*) and Electric Genetics (*http://ambient.sanbi.ac.za*).

1.4 Human Genome Sequencing

The main public domain source for the human genome data is from the Human Genome Project (HGP, *http://www.nhgri.nih.gov*). It is an international research program designed to construct detailed genetic and physical maps of the human genome, and to determine the complete nucleotide sequence of human DNA. It will also identify the estimated 25,000–100,000 genes (DAVISON et al., 2000) within the human genome. An extension of the HGP is to compare human sequences with the genomes of other organisms used as model systems. The scientific products of the HGP will comprise a resource of detailed information about the structure, organization, and function of human DNA and the information that constitutes the basic set of inherited "instructions" for the development and functioning of a human being. The draft version of the human genome was completed in 2000, with the full sequence a year or so later.

It is estimated that 3% of the total human genome encodes proteins, although estimates of gene number vary by a factor of 5 (APARICIO, 2000). A major challenge in the analysis of genomic DNA sequence is to locate protein-coding regions, as well as identifying important functional sites. The current laboratory methods are only adequate for characterizing loci of special interest (e.g., disease genes). They are quite laborious and not suitable for analyzing and annotating megabases of sequence. Computational methods can provide an alternative that can be used to characterize and annotate these megabase long sequences, either in an automated or semi-automated way (CLAVERIE, 1997).

Several computational tools have been developed in recent years to tackle the gene prediction problem. A listing of selected gene identification resources, freely available for academic use, is given in Tab. 2. It is important to distinguish between two different goals in gene finding research. The first is to provide computational methods to aid in the annotation of the large volume of genomic data that is produced by genome sequencing efforts. The second is to provide a computational model to help elucidate the mechanisms involved in transcription, splicing, polyadenylation, and other critical pathways from genome to proteome. No single computational gene finding approach will be optimal for both goals (HAUSSLER, 1998). Most gene finding tools perform quite adequately at identifying internal exons. However, locating initial and terminal exons, as well as locating multiple genes in a long stretch of DNA is still quite problematic. One method currently used is to trust exons identified by two or more programs (e.g., see the analysis of the chromosome 21 sequence, HATTORI, et al., 1999). One such example is the comparison of some gene finding programs for 1.4 MB genomic region BRCA2 on human chromosome 13q. A combination of FGENESHM (SOLOVYEV and ASALAMOV, unpublished data; see *http://dot.imgen.bcm.tmc. edu:9331/gene-finder/Help/fgenesh.html*) and GenScan (BURGE and KARLIN, 1997) and a combination of FGENESHM and FGENES

have been shown to be superior in this gene prediction when compared to other methods (*http://genomic.sanger.ac.uk/gf/braca2.html*). However, results in gene prediction vary considerably, and these methods should only be considered a guide to reasonably reliable *internal* exons. They are not yet an accurate prediction of gene structure.

The programs listed in Tab. 2 share a number of limitations. One is that they are not trained to predict some types of non-coding regions, which can be biologically important in some cases. As a consequence, non-coding parts of RNA genes, such as XIST (BROCKDORFF et al., 1992), H19 (PFEIFER et al., 1996), IPW (WEVRICK and FRANCKE, 1997), and NTT (LIU et al., 1997) are transparent to the current gene prediction programs. These genes are known to play a role in transcription inactivation or imprinting.

A number of organizations are developing programs to cluster ESTs, then use those clusters to guide the alignment of the ESTs against genomic DNA (DoubleTwist, Inc, Oakland, CA, USA; Electric Genetics, Bellville, South Africa) or for prediction of exons and genes (Neomorphic, Inc, Berkeley, CA, USA).

Tab. 2. Internet Resources for Gene Identification

Gene Identification Method	URL
GENSCAN (BURGE and KARLIN, 1997)	*http://genomic.stanford.edu/GENSCANW.html*
MZEF (ZHANG, 1997)	*http://sciclio.cshl.org/genefinder*
Genie (KULP et al., 1996)	*http://www-hgc.lbl.gov/inf/genie.html*
FGENES (SOLOVYEV et al., 1995)	*http://dot.imgen.bcm.tmc.edu:9331/gene-finder/gf.html*
FGENES-M	*http://dot.imgen.bcm.tmc.edu:9331/gene-finder/gf.html*
FGENESH	*http://dot.imgen.bcm.tmc.edu:9331/gene-finder/gf.html*
GeneID (GUGIO et al., 1992)	*http://www.imim.es/GeneIdentification/Geneid/geneid_input.html*
GeneLang (DONG and SEARLS, 1994)	*http://cbil.humgen.upenn.edu/~sdong/genlang.html*
GRAIL-II (MILANESI et al., 1993)	*http://combio.ornl.edu/Grail-bin/EmptyGrailForm*
HMMgene (KOUGH, 1997)	*http://www.cbs.dtu.dk/services/HMMgene/*
GeneMark.HMM (BORODOVSKY and McINNICH, 1993)	*http://genemark.biology.gatech.edu/GeneMark/hmmchoice.html*
Glimmer (DELCHER et al., 1999)	*http://www.cs.jhu.edu/labs/compbio/glimmer.html*

2 Computational Strategy for Novel Gene Discovery

A simple strategy for *high-throughput* identification of novel targets is given in Fig. 3. The first step is selecting a probe, which can be a sequence or a motif. Probes can be obtained from Entrez (*http://www.ncbi.nlm.nih.gov/*) and Hidden Markov Model (HMM) profiles from the Pfam databank (*http://pfam.wustl.edu/*). This sequence or profile is searched against a sequence databank. The quality of the databank searched determines the quality of the predicted novel proteins. Sequencing errors will lead to the inference of false positives, i.e., the prediction of a novel gene when it does not actually exist. *The choice of search program used depends on what information is to be gained from the probe sequence.* The third step is the interpretation of these results and identifying the positive hits for subsequent analysis. Promising hits are searched back against the protein databanks. We call this process a "retro-search" or a "back search". The EST/genomic hits are retro-searched against a

non-redundant protein databank and a patent databank to ensure they are novel. Novels are evaluated to determine whether they are candidates for further biological evaluation. The search programs and methods for characterizing the database hits are described in detail in the next section.

3 Bioinformatics Methods for Identifying Novel Targets

Molecular sequence is not very informative in and of itself. It is only in the context of other biological sequences and their evolutionary context that sequence data become truly illuminating. Sequence similarity analysis, allowing inference of evolutionary homology, is the single most powerful tool for functional and structural inference available in bioinformatics today. Methods can be divided into two categories, sequence-based searching and pro-

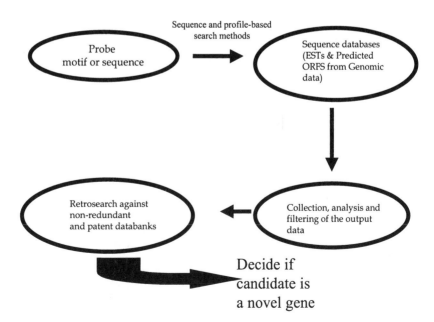

Fig. 3. A simple protocol for identifying novel genes.

file-based searching. These approaches can be used to identify novel targets in EST or genomic databanks.

3.1 Sequence-Based Search Methods

What one data one searches depends on what one data is desired from the query. Homology (evolutionary relationship) can be inferred from sequence similarity results. From the observation of similarity, one might be able to infer function.

Fundamentally, database searches are approximate string matching. In a high-level view, it is a simple operation: a sequence of interest, the query, is aligned with each of the sequences, the targets, in a databank. In practice, there are three ways to perform this action. First, the Smith–Waterman (SW) algorithm (SMITH and WATERMAN, 1981) is guaranteed to do a mathematically complete search – but only for the parameters you choose. Changing the parameters will change the mathematically optimal result. SW is a relatively slow method, although special-purpose supercomputers from Paracel (*http://www.paracel.com*), Compu-Gen (*http://www.compugen.com*), and Time-Logic (*http://www. timelogic.com*) are available that considerably shorten the search times. Two "heuristic" (approximate) search programs are the Basic Local Alignment and Search Tool (BLAST, ALTSCHUL et al., 1991) and FASTA (PEARSON and LIPMAN, 1988). To understand how these programs work, consider a comparison of two sequences using graph paper. With one sequence written across the top, and the other down the left-hand side, a dot is placed wherever the character in the top and left columns match. This "dot plot", when viewed from far away, would show some diagonal lines, from upper left to lower right. These diagonals are the longer matches between the two sequences. BLAST and FASTA are programs that very quickly find the "best" diagonals between a pair of sequences. "Best" is defined by the parameters used in a search. Using different parameters would give different "best" alignments. When iterating through a databank, these methods save a considerable amount of time. It should be carefully under-

stood, though, that BLAST and FASTA optimize different objective functions. Although their output appears similar, they were developed for different sequence searching problems. A general, but detailed explanation of these programs can be found in an excellent paper by S. ALTSCHUL (1988). BLAST is the most widely used tool for protein and nucleic acid sequence comparison. It runs one to two orders of magnitude faster than a Smith–Waterman program on the same hardware. It is important to realize that one is trading speed for sensitivity. In general, FASTA is more sensitive than BLAST in detecting distantly related protein sequences. The recent implementation of Position-Specific-Iterated BLAST (PSI-BLAST, ALTSCHUL et al., 1997) is an important contribution for the identification of weak similarities. PSI-BLAST takes as input a single protein sequence and compares it to a protein databank, using BLAST2, which permits alignments to have gaps. (The original BLAST program did not allow an alignment to have gaps.) PSI-BLAST then constructs a multiple sequence alignment. From that alignment, it builds a type of consensus sequence, called a profile, from statistically significant local alignments. The original query sequence serves as a template for the multiple sequence alignment and profile, whose lengths are identical to that of the query. That profile is compared to the protein databank, again seeking the best possible local alignments. These steps are iterated a number of times or until convergence. This methodology is similar to the "Pattern-Induced Multiple Alignment" program (PIMA, SMITH and SMITH, 1992), and appears to be about as sensitive.

3.1.1 Programs and Comparisons

Whether to use a protein or DNA sequence query depends upon the biological information desired. A list of the varieties of BLAST is presented in Tab. 3. If the sequence is a protein, or codes for a protein, then the search should almost always be performed at the protein level because proteins (with a 23-letter alphabet) allow one to detect far more distant similarities than does DNA (with a 4-letter alphabet) (STATES et al., 1991). It is also worth-

Tab. 3. BLAST Programs[a]

Program	Query	Database	Comments
blastn	DNA	DNA	tuned for very-high-scoring matches, not distant relationships
blastp	protein	protein	to find homologous proteins
blastx	DNA (translated)	protein	analysis of new DNA sequences and ESTs for finding matching and novel proteins
tblastn	protein	DNA (translated)	querying protein probes of interest against DNA database (e.g., ESTs for finding novel proteins)
tblastx	DNA (translated)	DNA (translated)	useful for gene structure and EST analysis, computationally intensive

[a] Similar programs are available for FASTA. FASTA programs are available at *ftp://ftp.virginia.edu/pub/fasta*. In blastx, tblastn and tblastx methods, DNA sequences or databases are translated to all six reading frames and the searches are done at protein level.

while to perform a protein search using chemical properties of amino acids (charge, hydrophobicity, size, etc.) also help to assess similarity rather than identity. A wide variety of amino acid weight matrices have been constructed using many different considerations (HENIKOFF and HENIKOFF, 2000). These matrices are discussed below.

In DNA-versus-DNA comparisons, the length of a statistically significant match is much longer than in a protein sequence comparison because of the shorter alphabet – matches are much more likely to occur by chance because there are fewer symbols to choose from. Typically one looks for identity. In cases of very distant similarity, it may be more useful to compare the sequences at the purine–pyrimidine level. In this method, A and G become R, C and T become Y. While finding a statistically significant match is lowered because of the size of the alphabet, any matches that are found will be biologically significant. Nucleotide sequence comparisons are best conducted, if the user is interested in information contained in the untranslated region of a gene. Recent large-scale comparison methods PIPMAKER (SCHWARTZ et al., 2000) and

MUMmer (DELCHER et al., 1999) are very useful for comparing long regions of sequence (10 kilobases or greater) against each other.

3.1.2 Databanks to Search

A list of commonly used peptide and nucleic acid databases and a brief description of their contents are given in Tabs. 4 and 5.

3.1.3 Search Parameters

There are a number of parameters available when searching. While some are algorithm-specific, there are a few general comments that can be made. Word sizes, weight matrices, and match and gap penalties and filtering are the most often available parameters. *Word size* reflects the sensitivity of a search. The higher this number, the faster a search goes, but the less sensitive it is. Conversely, the smaller this number, the slower the search, and the more sensitive it is. For most applications a word size (also called *ktup*) of 6 (for nucleic acids) or 2 (for amino acids) is as large as one would want.

Tab. 4. Some Peptide Sequence Databanks Available at NCBI

Nr	all non-redundant GenBank CDS translations + PDB + SwissProt + PIR + PRF
Month	subset of nr that is new or modified in the last 30 days
Yeast	yeast (*Saccharomyces cerevisiae*) protein sequences
E. coli	*E. coli* genomic CDS translations
Pdb	sequences derived from the 3-dimensional structure Brookhaven Protein Data Bank

Tab. 5. Nucleotide Sequence Databanks Available at NCBI

Nr	all non-redundant GenBank + EMBL + DDBJ + PDB sequences (but no EST, STS, GSS, or phase 0, 1 or 2 HTGS sequences)
Month	subset of nr that is new or modified in the last 30 days
dbest	non-redundant Database of GenBank + EMBL + DDBJ EST Divisions
dbsts	non-redundant Database of GenBank + EMBL + DDBJ STS Divisions
htgs	htgs unfinished High Throughput Genomic Sequences: phases 0, 1, and 2 (finished, phase 3 HTG sequences are in nr)
Yeast	yeast (*Saccharomyces cerevisiae*) genomic nucleotide sequences
E. coli	*E. coli* genomic nucleotide sequences
pdb	sequences derived from the 3-dimensional structure
gss	Genome Survey Sequence, includes single-pass genomic data, exon-trapped sequences, and Alu PCR sequences

There are a wide variety of weight matrices available, and a thorough discussion of them would require a separate and very long chapter of this volume. A recent review is available (HENIKOFF and HENIKOFF, 2000). The most widely used protein scoring (or weight) matrix now is BLOSUM62 (HENIKOFF and HENIKOFF, 1992). If BLOSUM matrices are not available, try PAM250 (DAYHOFF, 1972). Both of these matrices provide a long "look back time" (DOOLITTLE, 1992) and will detect very distantly related proteins acceptably well. No program can provide reliable searching below 15% similarity, however. An excellent discussion of using PAM matrices to reasonably cover the protein search space is by ALTSCHUL (1991). Most search programs provide default match and gap penalties. Gap penalties usually fall into two parts, the cost for opening a gap, and the cost for extending a gap. Manipulation of these parameters is useful when you want to discourage gaps, or when you know your sequence will have a large number, or long, gaps. Making the gap penalty larger will make gaps less likely, while lowering it will make gaps more likely. One could spend hours changing the parameters, which also gives different results. A good rule to follow is that if you are matching a short sequence against a very long one, set the gap penalties lower. This avoids "rubber band stretching" the smaller sequence along the larger. Finally, most search methods will offer some method of filtering. It is critically important that users always filter their data. The statistics of databank searching assume that unrelated sequences will look essentially random with respect to each other. How-

ever, there are certain patterns in sequences, which violates this rule. The most common exceptions are long runs of same or similar amino acids (like polyalanine, polyproline, or a hydrophobic stretch). Such regions of sequence will obtain extremely high scores. This is another example where a match with statistical significance is not necessarily biologically significant. For this reason, the default option is to include filtering in the NCBI BLAST server. The SEG (WOOTON and FEDERHEN, 1993) program is used for masking proteins, and DUST (TATUSOV and LIPMAN, unpublished data) program is used for masking DNA sequences. These programs are not guaranteed to filter all low-complexity sequences, but they generally do a very good job and visual inspection of matching segments will show where problems are. If one were looking for repetitive sequence, perhaps a microsatellite repeat, it would not be a good idea to mask.

Other useful parameters include those that change the number of scores and number of alignments presented. Typically, the defaults are adequate. However, for some query sequences, where there are a large number of significant hits in the databank, one may need to change the default number of matches from 20 to several hundred to find the more distantly related sequences.

3.1.4 Interpretation of Results

Result interpretation is the most challenging part of the search process. Scores calculated by the program, using statistical measures, are only guidelines. In cases of weak similarity, or of alignments with low statistical significance, biological knowledge and experience must be used for the interpretation. A biologically significant alignment is by no means necessarily statistically significant. For instance, short nucleotide-binding sites, such as the glucocorticord binding site, is too short to be statistically significant. However, when found in the proper context, it is very significant biologically. A statistically significant alignment is virtually always biologically significant, provided the query and library sequences have been properly filtered. The usual measure to use when interpreting results is the expectation, or E value (BRENNER et al., 1998). The E value of a match measures the expected (average or mean) number of sequences in a database that would achieve the specified score by chance. E values of 10^{-5} and smaller usually suggests meaningful similarity. Since E values are dependent on the length of the databank, they will change over time. This means that a search done today yielding a significant alignment may not yield a significant alignment next year, when the databank is larger. Specifying an arbitrarily large size for the database can compensate for this. In BLAST, the option is $-Y$. By doing so, the analyst can guarantee that today's result can be meaningfully compared to last year's. One should look at the actual alignment and check the regions of the query matching with the target. If this region happens to be of functional importance, and residues shown to have a functional role are matching, then the alignment may be biologically significant, even if it has a high E value.

The presence of sequence similarity allows the inference of homology, and the homology can help us infer whether the sequences share function. For the past 100 or more years in biology, homology has meant descent by common ancestry, so a sequence cannot be 35% homologous any more than a person can be 35% pregnant. Molecular biologists typically use "similarity" and "homology" interchangeably, but most other disciplines do not.

The inferring of function from the matched sequences needs be done cautiously. If the score is good, and the alignment matches the entire protein, then there is a very good chance that they share the same or a related function.

If only a part of the target sequence matches with the query, they might just share a domain. This further implies that the contribution is one aspect of the overall function of the query protein. This situation occurs frequently with multi-domain proteins. One should be cautious before making any functional conclusion. An EST matching with a zinc finger domain of a nuclear hormone receptor need not be a gene encoding a nuclear hormone receptor (NHR). It may be any DNA binding domain shared by many families of proteins.

There is another problem, that of transitive erroneous annotation. A protein (e.g., "baseballase" from species X) appears similar to a protein from species Y by sequence comparison. An annotation is entered in the GenBank entry for sequence Y that it is similar to "baseballase". The similarity may extend over the entire protein, but much more often it is only a domain or a part of a domain. Future searches of proteins from species Z show a minor degree of statistically significant similarity to species Y "baseballase". Therefore, the species Z protein is labeled a "baseballase" – even if it matches a different domain. This chain of incorrect inference happens very frequently and requires great caution. Transitive annotation is the single largest cause of grief in drug discovery bioinformatics. *Every annotation should be regarded with great suspicion.*

Many sequences are highly diverged during evolution and they cannot be detected by simple sequence similarity search methods. Thus, failure to find a significant match does not indicate that no homologs exist in the databank. A computational tool that goes beyond the simple pair-wise sequence similarity should also be used. This leads us to an another search method, based on profiles.

3.2 Profile-Based Methods

Analysis derived from a multiple sequence alignment can reveal gene structure or function that otherwise would not be clear from simple pair-wise sequence alignment (EDDY, 1998). Software packages are available that can take a multiple sequence alignment and build a *profile* from it. As stated by SEAN EDDY (1998), a profile incorporates position-specific information that is derived from the frequency with which a given residue (amino acid or nucleic acid base) is seen in an aligned column. Component residues of an active site, ligand binding pockets, or functional motifs tend to be well-conserved in sequence families. With a profile, which includes both conserved and less-conserved residues, a much more sensitive databank search is possible.

Much of the new software for profile searches are based on statistical models called Hidden Markov Models (HMMs) (SONNHAMMER et al., 1998). This section is a brief introduction to profile-based HMM methods, and a comprehensive review is available (DURBAN et al., 1998). Profile-based searches can be done in two ways, by using publicly available HMM profiles, or by creating a new HMM profile from one's aligned sequence data. When searching novel members of sequence families, analysts frequently focus on specific areas of interest. These could be a specific sequence or a domain of interest.

Pfam is a databank of protein domain family profiles. It is freely available at *http://www. sanger.ac.uk/Software/Pfam* and *http://www. cgr.ki.se/Pfam* (Europe) and at *http://pfam. wustl.edu* (USA). Using these publicly available HMM profiles is convenient, if the domain of interest is already present in the Pfam databank.

Pfam contains expertly curated multiple sequence alignments for each family. These multiple sequence alignments are used to create HMM profiles. These profiles are then used in turn to identify protein domains in uncharacterized sequences. Pfam contains functional annotation, literature references, and databank links for each family. Pfam profiles are built from two kinds of alignments. The "seed" alignment contains representative members of the family, and is generally hand-curated. The full alignment contains all members that can be detected and semi-automatically aligned. All sequences are taken from pfamseq, which is a non-redundant set composed of SWISS-PROT and SP-TrEMBL.

The Pfam distribution contains a number of files: Pfam-A.seed, Pfam-A.full, Pfam, Pfam-Frag, SwissPfam, Pfam-B, diff and Pfamseq. Following BATEMAN et al. (2000), Pfam-A.seed and Pfam-A.full contain the seed and full an-

notation respectively, in a marked up alignment format called Stockholm format. The Pfam file contains the library of Pfam profile HMMs. We can search any given sequence against this file to find any Pfam domain present in the query sequence. The Pfam models are iteratively defined. They start with clear homologs (in the strict sense of the word) and incorporate increasingly distant family members as a multiple sequence alignment is constructed. PfamFrag is a library of profile HMMs designed specifically to find matches to protein fragments; SwissPfam is a file containing the domain organization for each protein in the databank; Pfam-B contains the data for Pfam-B families in Stockholm format. Sequences that were not available when Pfam-A was generated are clustered and aligned automatically and are released as Pfam-B. The file *diff* contains the changes between releases to allow incremental updates of Pfam-derived data; and pfamseq contains the underlying sequence database, in FASTA format. The Pfam package contains the above mentioned files and executables are available for a number of different operating systems.

HMMER is a freely distributed implementation of profile HMM software for protein sequence analysis. It is available at *http://hmmer. wustl.edu*. There are currently 9 programs in the HMMER package. These programs can be used to search a protein database or create a new HMM profile. If the domain of interest is not present in the Pfam database, then the user must create a new HMM profile for the desired domain. A flow chart of the process of creating a new profile is presented in Fig. 4. The description of the 9 programs and an online manual for HMMER is available at the HMMER URL given above. For nucleic acid analysis, a new package called Wise2 is available at the Sanger Center, UK (*http://www. sanger.ac.uk/Software/Wise2/index.shtml*). It can compare a single protein or a profile HMM to a genomic DNA sequence, and predict a gene structure. The genomic sequence analysis algorithm is called Genewise and the corresponding one for ESTs is called ESTwise.

The previous sections presented an overview of the sequence-based and profile-based methods. The next step in an analysis is to use this information in the identification of novel

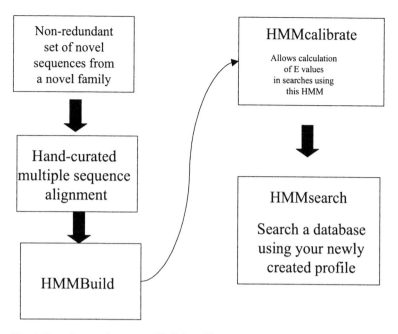

Fig. 4. Steps in creating a new HMM profile.

proteins. The next section describes the application of these methods in novel gene discovery.

3.3 Identification of a Novel Protein: An Example

To illustrate how to use these search methods, and apply the strategy for identifying novel genes described above, we now present an example using G-Protein Coupled Receptors (GPCRs). They are excellent drug targets, and approximately 50–60% of marketed drugs are GPCRs (HORN and VRIEND, 1998). With recent advances in genomics, more information on the functional role of GPCRs is available. An increasing number of mutations in GPCRs are being found to be associated with diseases. Fig. 5 gives an overview of a protocol that can be used to identify novel GPCRs. Note, however, that this method can be applied with any protein family. One must always use both sequence-based and profile-based search methods to ensure that a search is as complete. It is especially important not to miss

any weak similarity hits, or be distracted by transitive annotation errors. This type of process can be easily automated by using PERL scripts (CHRISTIANSEN and TORKINGTON, 1998). Designing a user-friendly web interface for accessing the search output data would be useful to help experimentalists browse through the results.

4 Protein Classification and Functional Assignment

So far, we have discussed various ways for finding novel proteins related to a query of interest. An another aspect of the problem is when one has a set of nucleotide sequences and would like to make a (putative) functional assignment. The set of sequences can be from any source such as a cDNA library, a subtractive hybridization study, a transcriptional profiling study, a sequence from a chromosomal region implicated in some disease, or genomic

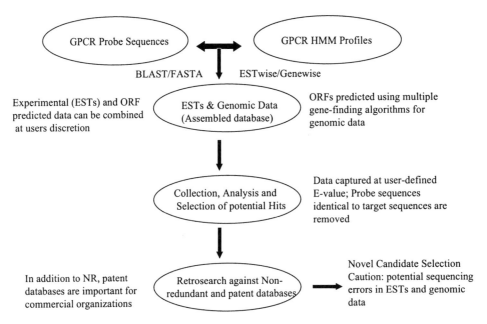

Fig. 5. Application of bioinformatics techniques for identifying novel GPCRs.

data from the Human Genome Project. In the case of genomic data, gene finding tools have to be applied first to locate presumptive exons. Those resulting exons then can be analyzed for functional assignment. As mentioned earlier, the Genewise suite of programs (*http://www. sanger.ac.uk/Software/Wise2/index.shtml*) can take a genomic sequence as input and search them against the HMM profile libraries. We will assume that the user has no knowledge about the function of the sequence.

An analysis pipeline is presented in Fig. 6. The input sequences considered here are EST sequences and genomic data. The EST sequences must first be masked to remove repetitive elements, housekeeping, immunoglobin gene, and low-quality sequence regions. Then each screened EST must be compared against existing sequence contigs and BACs. Each EST will either become a part of an existing contig or form a novel singleton cluster. Similarly, genomic sequences must be masked and the ORFs predicted from gene-finding programs compared to existing EST databases. A new EST sequence can merge two clusters, be added to a cluster, or form a new singleton cluster. Combining EST data with

predictions from genomic data is at the user's discretion. It is policy at many organizations to draw a sharp distinction between experimentally generated data and predicted data.

Next, we analyze the masked sequences using the sequence-based approaches described in Sect. 3.1. Partial matches with databank sequences have to be very carefully examined, since there is a high probability of matching to a domain in a multi-domain protein. Since ESTs and exons predicted from genomic data are most likely to be partial sequences, one can expect only partial matches to database sequences. Such sequences should also be carefully examined since the ESTs and working draft genomic data will have frequent sequencing errors.

In addition to the sequence-based approach, a profile-based approach should also be used. The simple answer to questions like "which method should I use" or "which of the databanks should I use" is, "use as many methods and databanks as possible". When using sequence databanks, make sure they are current. Sequence data are being added at a considerable rate, so a nightly update of sequence databanks is essential. A combination of non-

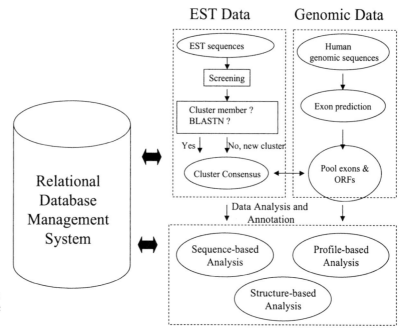

Fig. 6. Bioinformatics analysis protocol for an unknown EST/genomic sequence.

redundant databanks (protein and nucleic acids) and patent databanks is very good in mining for novel proteins.

4.1 Motif- and Profile-Based Methods

Although domain databanks have considerable overlap, each has its particular strength (HOFMANN, 1998). Domains critical for signal transduction are likely to be found in PRO-SITE profiles (BAIROCH et al., 1997) or SMART (SCHULTZ et al., 1998). Pfam excels in extracellular domains, while PROSITE patterns are good at identifying enzyme classes by their active site motif. Even where there is overlap, the redundancy assists in evaluating the reliability of one's prediction. A list of motif databases and their FTP addresses are given in Tab. 6.

The PROSITE pattern library (BAIROCH et al., 1997; BAIROCH, 1991) was one of the pioneering efforts in collecting descriptors for functionally important motifs. It is based on regular expression grammars (SEARLS, 1997) which emphasize only the most highly conserved residues in a protein family. A PRO-

Tab. 6. Freely Available Motif Databanks

Databases	FTP Site for Downloading[a]
PROSITE patterns	*ftp://ftp.expasy.ch/databases/prosite*
PROSITE profiles	*ftp://ftp.isrec.isb-sib.ch/sib-isrec/profiles*
pfam	*ftp://ftp.sanger.ac.uk/pub/databases/Pfam*
BLOCKS	*ftp://ncbi.nlm.nih.gov/repository/blocks*
PRINTS	*ftp://ftp.biochem.ucl.ac.uk/pub/prints*
ProDom	*ftp://ftp.toulouse.inra.fr/pub/prodom*

[a] Use may be restricted for commercial users.

SITE pattern does not attempt to describe a complete domain or an entire protein. It attempts to identify the most diagnostic residue combinations, such as those residues conserved in the catalytic site of an enzyme. A benefit of PROSITE is that all motifs have extensive documentation with excellent annotation. The main disadvantage of PROSITE is that its syntax is too rigid to represent highly divergent protein families. PROSITE's short patterns do not contain enough information to yield statistically significant matches in large databanks available now. Since 1995, the PROSITE pattern library has been supplemented by the PROSITE profile library to overcome these problems. PROSITE profiles rank in between a sequence-to-sequence comparison and the matching of a regular expression to a sequence. Unconserved regions in a profile get lower weight scores rather than being totally ignored. PROSITE profiles try to cover complete domains wherever possible. PRODOM (SONNHAMMER and KAHN, 1994; CORPET et al., 1999) was the first comprehensive collection of complete domains. It is constructed from SWISS-PROT in a fully automated manner. Pfam, which evolved from ProDom, is a collection of HMMs that are conceptually related to PROSITE profiles (SONNHAMMER et al., 1997; EDDY, 1998). Because of their information-rich descriptors, both PROSITE profiles and Pfam collections are able to detect even very distant instances of a motif not otherwise detectable. The documentation available with Pfam profiles is minimal, but frequently contains pointers to the corresponding PROSITE entry. Pfam models and PROSITE searches are inter-convertible and combination searches are available.

Profile collections focusing on specific functional families are also available. The SMART databank is an independent collection of 181 HMMs focusing on protein domains involved in signal transduction. BLOCKS (PIETROKOVSKI et al., 1996) and PRINTS (ATTWORD et al., 1999) are two motif databanks that represent protein or domain families by one or more short ungapped multiple alignment fragments. The PRINTS motifs are also refined iteratively, resulting in an increased sensitivity compared with BLOCKS. Both BLOCKS and PRINTS have good annotation.

In all the above databanks, the number of available domain descriptors is not an indicator of complete coverage and should not be used as an estimate of quality. Each motif and profile databank, and its associated search methods, differs in selectivity and sensitivity. For example, the relatively small pre-release library of PROSITE profiles detected 70 significant domain matches in a collection of 880 unknown yeast genes. Pfam 3.0 reported 36 matches, whereas the PROSITE pattern collection found 114.

Once a novel protein is identified, a challenging analysis is assigning that protein to a particular subfamily. A domain search can reveal whether the query protein belongs to a superfamily such as GPCRs or Nuclear Hormone Receptors. The next question is identification of other proteins closely related to the query protein. Characterization of orphan proteins to the sub-family level is of paramount importance to drug discovery scientists. One approach to this problem is to study the evolutionary relationships among the family members by phylogenetics (PIETROKOVSKI et al., 1996). An excellent introduction to molecular phylogenetics is provided by HILLIS et al. (1993). The merits and pitfalls of various phylogenetics methods are summarized in useful reviews (AVISE, 1994; SAITOU, 1996; LI, 1997; SWOFFORD et al., 1996). There are three methods commonly used in phylogenetic analysis of protein and nucleic acid sequences. They are distance, maximum likelihood, and parsimony. All begin with a carefully constructed multiple sequence alignment (MSA). While automated methods are fine for initial MSA, in practice each alignment must be checked by hand. Distance matrix methods are fastest and most often used in very large-scale (greater than about 50 sequences) phylogenetic analysis. It should be noted that distance methods "summarize" the sequences, in the sense that no matter how many positions there are in a sequence, each is reduced to a single number. It is commonly not a preferred method among specialists.

Parsimony uses position-specific information in a multiple sequence alignment. It is sensitive, with longer running time than distance methods. One can analyze up to about 30 sequences with parsimony methods. Maximum likelihood uses a different model. It takes into account every sequence, every sequence change, and a specific (user-specified) model of sequence evolution. It estimates the likelihood the data fitting a particular tree. It is the slowest of the three methods. Tab. 7 gives information about the sources of information in the web for some of the phylogenetics programs.

4.2 Structure-Based Methods

Structure is evolutionarily conserved to a greater extent than sequence. Even if two sequences do not share obvious sequence similarity, they might share aspects of the same 3D structure (called a "fold"). So, fold recognition methods have tremendous potential in characterizing unknown proteins. Although there are many genes, the number of domain sequence families is much smaller. The number of folds is almost certainly an order of magnitude less, as is the number of *architectures*. The architecture refers to the packing of sheets and helices in a structure regardless of sequential connectivity. It is commonly expected that the number of structural supersecondary motifs, which together constitute a fold, is very small. Biological complexity is achieved by using local variation, together with a combinatorial mixture at primary, secondary, tertiary, and quaternary levels. This includes combining domains to create different proteins and combining proteins to make different complexes. Genome-wide efforts to determine 3D protein structures, or at least one representative 3D structure for all protein families, are under way (SMAGLIK, 2000). Representative structures will allow modeling of related sequences and help infer the structures of all proteins. In turn, structural insights will help in functional elucidation, which in turn can help to determine the biological role of the protein under normal physiological and pathological conditions.

5 Disease–Target Gene Relationship

Over the last decade, a great deal of effort has been put into creating a physical map of the human genome – ordering genes within the genome by placing landmarks (sequence tagged sites, STSs) with which to navigate. In addition to providing an excellent framework for the complete sequencing of the human genome, the physical map has assisted in directly identifying about 100 disease-causing genes (a technique called positional cloning, *http://www.nhgri.nih.gov/Policy_and_public_affairs/Communications/Fact_sheets/positional_cloning.html*). One of the most difficult challenges ahead is to locate genes involved in diseases that have a complex (i.e., non-simple Mendelian) pattern of inheritance, such as those that contribute to diabetes, asthma, cancer, and mental illness. In all these cases, no particular gene can determine whether a person will get the disease. It is likely that more than one mutation, in more than one pathway, is required before the disease is expressed. A number of genes may each make a subtle contribution to a person's susceptibility to a disease. Genes may also affect how a person reacts to environmental factors. Unraveling these networks of events will undoubtedly be a challenge for some time to come. A list of selected databanks pertinent to genetic epidemiology is given in Tab. 8. Such databanks are invaluable because they contain information regarding

Tab 7. Selected Phylogenetic Program Resources

Program	URL
PAUP	*http://onyx.si.edu/PAUP/*
PHYLIP	*http://evolution.genetics.washington.edu/phylip.html*
MacClade	*http://phylogeny.arizona.edu/macclade/macclade.html*
TreeView	*http://taxonomy.zoology.gla.ac.uk/rod/treeview.html*

Tab. 8. Databanks Pertinent to Genetic Epidemiology

Database	URL	Description
Human Disease Susceptibility Genes		
Online Mendelian Inheritance in Man (OMIM)	*http://www.ncbi.nlm.nih.gov/*	detailed information of disease susceptibility genes and inherited phenotypes
GeneCards	*http://bioinformatics.weizmann. ac.il/cards/*	database of human genes, their products and their involvement in diseases
Cardiff Human Gene Mutation Database	*http://www.uwcm.ac.uk/uwcm/ mg/hgmd0.html*	collection of genetic mutations responsible for human inherited diseases
Statistical Genetics		
Human Genetic Analysis Resource (HGAR)	*http://darwin.mhmc.cwru.edu/*	distributes Statistical Analysis for Genetic Epidemiology (SAGE) Software
Laboratory of Statistical Genetics-Rockefeller University	*http://linkage.rockefeller.edu*	linkage programs
Genome Mapping and Sequencing		
Database of Expressed Sequence Tags (dbESTs)	*http://ncbi.nlm.nih.gov/*	information of expressed sequences derived from numerous tissues and cell types
Database of Sequence Tagged Sites (dbSTS)	*http://ncbi.nlm.nih.gov/*	sequencing and mapping data on short genomic landmark sequences
European Molecular Biology Laboratory (EMBL)	*http://www.ebi.ac.uk/*	annotated DNA sequence information
GenBank	*http://ncbi.nlm.nih.gov/*	NIH genetic sequence database containing annotated DNA sequence information
Gene Map of the Human Genome	*http://ncbi.nlm.nih.gov/*	maps expressed genes (ESTs) to specific regions of the genome
IMAGE Consortium	*http://www.bio.llnl.gov/bbrp/ image/image.html*	make a variety of DNA sequence, EST clones, mapping and gene expression data available to public
Unique Gene Sequence Collection (UniGene)	*http://ncbi.nlm.nih.gov/*	assembled DNA sequences to identify and map new human genes
Genetic Markers		
Cooperative Human Linkage Center	*http://www.chlc.org/*	contains statistically rigorous genetic maps enriched for highly variable microsatellite markers
Marshfield Medical Research Institute	*http://www.marshmed.org/ genetics*	mammalian genotyping services to qualified applicants
Whitehead Institute for Biomedical Research-MIT	*http://www.genome.wi.mit.edu/*	Human Single Nucleotide Polymorphisms (SNPs) database contains genetic maps showing genomic locations of SNPs

genes and molecular defects that contribute to human disease, methods for detection of numerous disease-susceptibility mutations and polymorphisms, and comprehensive description of disease phenotypes (ELLSWORTH and MANOLIO, 1999).

An excellent source of information regarding genes and their relevance to diseases is the Online Mendelian Inheritance in Man (OMIM; *http://www.ncbi.nlm.nih.gov/omim*) (MCKUSICK, 1998). It is an electronic version of the catalog of human genes and genetic disorders edited by VICTOR MCKUSICK. OMIM contains textual information from the literature published on most human genetic-related diseases. Since the online version of OMIM is housed at NCBI, links to MEDLINE and sequence retrieval sources like ENTREZ are provided from all references cited with each OMIM entry. A daunting task in this area is to integrate the diverse data – genomic sequence data, mutation information, disease conditions, mapping information, functional validation data, and experimental information.

6 Conclusion

Over the last decade, bioinformatics has become a rapidly evolving discipline of its own, contributing essential tools for biologists to mine vast amounts of genomic data. The current challenge is to identify genes from the just-released draft of the human genome. This is not a once-in-a-lifetime opportunity, this is a once-in-a-forever opportunity. As genes are identified, bioinformatics techniques can help determine, or give insights into, the gene structure–function relationship. With advances in genomics, protein and RNA expression analysis, and Single Nucleotide Polymorphism (SNP) data associated with genes are being added to our knowledge of the human genome. Pathway relationships, which connect the expression and action of genes, will be the next focus of attention in drug discovery and academia. Bioinformatics will play a key role in converting these data into knowledge. It is this knowledge which is of strategic value to research organizations. Already, the integration

of bioinformatics, molecular biotechnology, and epidemiological methods of assessing disease risk is rapidly expanding our ability to identify genetic influences on complex human diseases. These technological advances are likely to have a profound impact on knowledge of etiology of complex diseases and reveal novel approaches to disease treatment and prevention. This combination of data sources is the true definition of bioinformatics for the future. The era of personalized medicines and designing the "right" drug for the "right" patient rushing toward us. Rather than being a sunset for a niche discipline, bioinformatics in drug discovery will have increasing amounts and types of data to integrate to improve experimentation in pharmaceutical development.

7 References

AARONSON, J. S., ECKMAN, B., BLEVINS, R. A., BARKOWSKI, J. A., MYERSON, J. et al. (1996), Toward the development of gene index to the human genome: an assessment of the nature of high-throughput EST sequence data, *Genome Res.* **6**, 829–845.

Advancing genomic research: The UniGene collection (1996), *NCBI News*.

ALTSCHUL, S. F. (1991), Amino acid substitution matrices from an information theoretic perspective, *J. Mol. Biol.* **219**, 555–565.

ALTSCHUL, S. F. (1998), Fundamentals of database searching, *Trends Guide to Bioinformatics* (Trends Supplement), pp. 7–9. Amsterdam: Elsevier.

ALTSCHUL, S. F., MADDEN, T. L., SCHAFFER, A. A., ZHANG, J., ZHANG, Z. et al. (1997), Gapped BLAST and PSI-BLAST: a new generation of protein database search programs, *Nucleic Acids Res.* **25**, 3389–3402.

APARICIO, S. A. J. R. (2000), How to count … human genes, *Nature Genet.* **25**, 129–130.

ATTWOOD, T. K., FLOWER, D. R., LEWIS, A. P., MABEY, J. E., MORGAN, S. R. et al. (1999), PRINTS prepares for the new millennium, *Nucleic Acids Res.* **27**, 220–225.

AVISE, J. C. (1994), Molecular markers, natural history and evolution. New York: Chapman & Hall.

BAINS, W. (1996), Company strategies for using bioinformatics, *Trends Biotechnol.* **14**, 312–317.

BAIROCH, A. (1991), PROSITE: a dictionary of sites and patterns in proteins, *Nucleic Acids Res.* **19**

(Suppl.), 2241–2245.

BAIROCH, A., BUCHER, P., HOFMANN, K. (1997), The PROSITE database, its status in 1997, *Nucleic Acids Res.* **25**, 217–221.

BAKER, P. G., BRASS, A. (1998), Recent developments in biological sequence databases, *Curr. Opin. Biotechnol.* **9**, 54–58.

BATEMAN, A., BIRNEY, E., DURBIN, R., EDDY, S. R., HOWE, K. L., SONNHAMMER, E. L. (2000), The Pfam protein families database, *Nucleic Acids Res.* **28**, 263–266.

BENTON, D. (1996), Bioinformatics – principles and potential of a new multidisciplinary tool, *Trends Biotechnol.* **14**, 261–272.

BORODOVSKY, M., MCININCH, J. (1993), GeneMark: Parallel gene recognition for both DNA strands, *Computers Chem.* **17**, 123–133.

BRENNER, S. (1998), Practical database searching, *Trends Guide to Bioinformatics* (Trends Supplement), pp. 9–12. Amsterdam: Elsevier.

BRENNER, S. E., CHOTHIA, C., HUBBARD, T. J. P. (1998), Assessing sequence comparison methods with reliable structurally identified distant evolutionary relationships, *Proc. Natl. Acad. Sci. USA* **95**, 6073–6078.

BROCKDORFF, N., ASHWAORTH, A., KAY, G. F., MCCABE, V. M., NORRIS, D. P. et al. (1992), The product of the mouse *Xist* gene is a 15 kb inactive X-specific transcript containing no conserved ORF and located in the nucleus, *Cell* **71**, 515–526.

BRONNER, C. E., BAKER, S. M., MORRISON, P. T., WARREN, G., SMITH, L. G. et al. (1994), Mutation in the DNA mismatch repair gene homologue hMLH1 is associated with hereditary non polyposis colon cancer, *Nature* **368**, 258–261.

BULT, C. J., WHITE, O., OLSEN, G. J., ZHOU, L. X., FLEISCHMANN, R. D. et al. (1996), Complete sequence of the methanogenic archaeon, *Science* **273**, 1058–1073.

BURGE, C. S., KARLIN, K. (1997), Prediction of gene structures in human genomic DNA, *J. Mol. Biol.* **268**, 78–94.

BURKE, J., DAVISON, D., HIDE, W. (1999), d2_cluster: a validated method for clustering EST and full-length cDNA sequences, *Genome Res.* **9**, 1135–1142.

BURSET, M., GUIGO, R. (1996), Evaluation of gene structure prediction programs, *Genomics* **34**, 353–367.

CAPONE, M. C., GORMAN, D., CHING, E. P., ZOLTNIK, A. (1996), Identification through bioinformatics of cDNAs encoding human thymic shared AG-1/ stem cell AG-2 – a new member of the human Ly-6 family, *J. Immunol.* **157**, 969–973.

CHRISTIANSEN, T., TORKINGTON, N. (1998), *Perl Cookbook*. Sebastopol: O'Reilly and Associates.

CLAVERIE, J. M. (1997), Computational methods for the identification of genes in vertebrate genomic

sequences, *Hum. Mol. Genet.* **6**, 1735–1744.

CORPET, F., GOUZY, J., KAHN, D. (1999), Recent improvements of the ProDom database of protein domain families, *Nucleic Acids Res.* **27**, 263–267.

DAYHOFF, M. O., ECK, R. V., PARK, C. M. (1972), A model of evolutionary change in proteins. in: *Atlas of Protein Sequence and Structure* **5**, pp. 89–99. Washington, DC: National Biomedical Research Foundation.

DELCHER, A. L., HARMON, D., KASIF, S., WHITE, O., SALZBERG, S. L. (1999), Improved microbial gene identification with GLIMMER, *Nucleic Acids Res.* **27**, 4636–4641.

DELCHER, A. L., KASIF, S., FLEISCHMANN, R. D., PETERSON, J., WHITE, O., SALZBERG, S. L. (1999) Alignment of whole genomes. *Nucleic Acids Res.* **27**, 2369–2376.

DOOLITTLE, R. F. (1992), Reconstructing history with amino acid sequences, *Protein Sci.* **1**, 191–200.

DONG, S., SEARLS, D. B. (1994), Gene structure prediction by linguistic methods, *Genomics* **23**, 540–551.

DREWS, J. (2000), Drug Discovery: A Historical Perspective, *Science* **287**, 1960–1964.

DUNHAM, I., HUNT, A. R., COLLINS, J. E., BRUSKIEWICH, R., BEARE, D. M. et al. (1999), The DNA sequence of human chromosome 22, *Nature* **402**, 489–495.

DURBIN, R., EDDY, S. R., KROGH, A., MITCHISON, G. (1998), *Biological Sequence Analysis: Probabilistic Models of Proteins and Nucleic Acids*. Cambridge: Cambridge University Press.

EDDY, S. R. (1998), Multiple-alignments and sequence searches, *Trends Guide to Bioinformatics* (Trends Supplement), pp. 15–18. Amsterdam: Elsevier.

EDDY, S. R. (1998), Profile hidden Markov models, *Bioinformatics* **14**, 755–763.

ELLSWORTH, D. L., MANOLIO, T. A. (1999), The emerging importance of genetics in epidemiologic research III: Bioinformatics and statistical genetic methods, *Ann. Epidemiol.* **9**, 207–224.

FANNON, M. R. (1996), Gene expression in normal and disease states-identification of therapeutic targets, *Trends Biotechnol.* **14**, 294–298.

FENG, D. F., CHO, G., DOOLITTLE, R. F. (1997), Determining divergence times with a protein clock: update and reevaluation, *Proc. Natl. Acad. Sci. USA* **94**, 13028–13033.

FLEISCHMANN, R. D., ADAMS, M. D., WHITE, O., CLAYTON, R. A., KIRKNESS, E. F. et al. (1995), Whole-genome random sequencing and assembly of *Haemophilus influenzae* Rd., *Science* **269**, 496–512.

GELFAND, M. S., ROYTBERG, M. A. (1993), Prediction of exon–intron structure by dynamic programing approach, *BioSystems* **30**, 173–182.

GUIGO, R., KNUDSEN, S., DRAKE, N., SMITH, T. (1992),

Prediction of gene structure, *J. Mol. Biol.* **226**, 141–157.

HATTORI, M., FUJIYAMA, A., TAYLOR, T. D., WATANABE,H., YADA, T. et al. (2000), The DNA sequence of human chromosome 21, *Nature* **405**, 311–319.

HAUSSLER, D. (1998), Computational gene finding, *Trends Guide to Bioinformatics* (Trends Supplement), pp. 12–15. Amsterdam: Elsevier.

HENDERSON, J., SALZBERG, S., FASMAN, K. H. (1997), Finding genes in DNA with hidden Markov model, *J. Comput. Biol.* **4**, 127–142.

HENIKOFF, S., HENIKOFF, J. G. (1992), Amino acid substitution matrices from protein blocks, *Proc. Natl. Acad. Sci. USA* **89**, 10915–10919.

HENIKOFF, S., HENIKOFF, J. G. (2000), Amino acid substitution matrices, *Adv. Protein Chem.* **54**, 73–97.

HERSHKOVITZ, M. A., LEIPE, D. D. (1998), Phylogenetic Analysis, in: Bioinformatics: *A Practical Guide to the Analysis of Genes and Proteins* (BAXEVANIS, A. D., OUELLETTE, B. F. F., Eds.), pp. 189–230. New York: John Wiley & Sons.

HIETER, P., BASSETT, Jr., D. E., VALLE, D. (1996), The yeast genome – a common currency, *Nature Genet.* **13**, 253–255.

HILLIS, D. M., ALLARD, M. W., MIYAMOTO, M. M. (1993), Analysis of DNA sequence data: Phylogenetic inference, *Methods Enzymol.* **224**, 456–487.

HOFFMANN, K. (1998), Protein classification and functional assignment, *Trends Guide to Bioinformatics* (Trends Supplement), pp. 18–21. Amsterdam: Elsevier.

HORN, F., VRIEND, G. (1998), G protein-coupled receptors *in silico*, *J. Mol. Med.* **76**, 464–468.

HUANG, X., ADAMS, M. D., ZHOU, H., KERLAVAGE, A. R. (1997), A tool for analyzing and annotating genomic sequences, *Genomics* **46**, 37–45.

HUTCHINSON, G. B., HAYDEN, M. R. (1992), The prediction of exons through an analysis of spliceable open reading frames, *Nucleic Acids Res.* **20**, 3453–3462.

KROGH, A., MLAN, I. S., HAUSSLER, D. (1994), A hidden Markov model that find genes in *E. coli* DNA, *Nucleic Acids Res.* **22**, 4768–4778.

KROGH, A. (1997), Two methods for improving performance of an HMM and their application for gene finding, in: *5th Int. Conf. Intelligent Systems for Molecular Biology* (GAASTERLAND, T., Ed.), pp. 179–186. Menlo Park, CA: AAAI Press.

KULP, D., HAUSSLER, D., REESE, M. G., EACKMAN, F. H. (1996), A generalized hidden Markov model for the recognition of human genes in DNA, *Proc. 5th Int. Conf. Intelligent Systems for Molecular Biology*, pp. 134–142. Menlo Park, A: AAAI Press.

LAHAD-LEVY, E., WASCO, W., POORKAJ, P., ROMANO, D. M., OSHIMA, J. et al. (1995), Candidate gene for the chromosome 1 familial Alzheimer's disease locus, *Science* **269**, 973–977.

LI, W. H. (1997), *Molecular Evolution*. Sunderland, MA: Sinauer Associates.

LIU, A. Y., TORCHIA, B. S., MIGEON, B. R., SILICIANO, R. F. (1997), The human NTT gene: Identification of a novel 17-kb noncoding nuclear RNA expressed in activated $CD4^+$ T cells, *Genomics* **39**, 171–184.

MCKUSICK, V. A. (1998), *Mendelian Inheritance in Man. Catalogs of Human Genes and Genetic Disorders* 12th Edn. Baltimore, MD: Johns Hopkins University Press.

MEWES, H. W., ALBERMANN, K., BAHR, M., FRISHMAN, D., GLEISSNER, A. et al. (1997), Overview of the yeast genome, *Nature* **387** (Suppl.), 7–65.

MILANESI, L., KOLCHANOV, N., ROGOZIN, L., KEL, A., TIROV, I. (1993), *Guide to Human Genome Computing*, pp. 249–312. Cambridge: Academic Press.

MILLER, R. T, CHRISTOFFELS, A. G., GOPALAKRISHNAN, C., BURKE, J., PTITSYN, A. A. et al. (1999), A comprehensive approach to clustering of expressed human gene sequence: the sequence tag alignment and consensus knowledge base, *Genome Res.* **9**, 1143–1155.

PAPADOPOULOS, N., NICOLAIDES, N. C., WEI, Y. F., RUBEN, S. M., CARTER, K. C. et al. (1994), Mutation of a *mutL* homologue in hereditary colon cancer, *Science* **263**, 1625–1629.

PEARSON, W. R., LIPMAN, D. J. (1988), Improved tools for biological sequence comparison, *Proc. Natl. Acad. Sci. USA* **85**, 2444–2448.

PFAM. Copyright © 1996–1999 The Pfam Consortium.

PFEIFER, K., LEIGHTON, P. A., TILGHMAN, S. M. (1996), The structural H19 gene is required for transgene imprinting, *Proc. Natl. Acad. Sci. USA* **93**, 13876–13883.

PIETROKOVSKI, S., HENIKOFF, J. G., HENIKOFF, S. (1996), The Blocks database – a system for protein classification, *Nucleic Acids Res.* **24**, 197–200.

RAWLINGS, C. J., SEARLS, D. B. (1997), Computational gene discovery and human disease, *Curr. Opin. Genet. Devel.* **7**, 416–423.

SAITOU, N. (1996), Reconstruction of gene trees from sequence data, *Methods Enzymol.* **266**, 427–449.

SALZBERG, S., DELCHER, A., FASMAN, K., HENDERSON, J. (1997), A decision tree system for finding genes in DNA, *Technical Report*, Department of Computer Science, Johns Hopkins University, Baltimore, MD.

SCHULTZ, J., MILPETZ, F., BORK, P., PONTING, C. P. (1998), SMART, a simple modular architecture research tool: identification of signaling domains, *Proc. Natl. Acad. Sci. USA* **95**, 5857–5864.

SCHWARTZ, S., ZHANG, Z., FRAZER, K. A., SMIT, A. et al. (2000), PipMaker – a web server for aligning two genomic DNA sequences, *Genome Res.* **10**,

577–586.

SEARLS, D. B. (1997), Linguistic approaches to biological sequences, *Comput. Appl. Biosci.* **13**, 333–344

SINGH, G. B. (2000), Computational approaches to gene prediction, in: *Methods in Molecular Biology 132: Bioinformatics Methods and Protocols* (MISENER, S., KRAWETZ, S. A., Eds.), pp. 351–364. Totowa, NJ: Humana Press.

SMAGLIK, P. (2000), Genomics initiative to decipher 10,000 protein structures, *Nature* **407**, 549.

SMITH, R. F., SMITH, T. F. (1992), Pattern-induced multi-sequence alignment (PIMA) algorithm employing secondary structure-dependent gap penalties for use in comparative protein modeling, *Protein Eng.* **5**, 35–41.

SMITH, T. F., WATERMAN, M. S. (1981), Identification of common molecular subsequences, *J. Mol. Biol.* **147**, 195–197.

SNYDER, E. E., STORMO, G. D. (1993), Identification of coding regions in genomic DNA sequences: an application of dynamic programming and neural networks, *Nucleic Acids Res.* **21**, 607–613.

SOLOVYEV, V. V., SALAMOV, A. A., LAWRENCE, C. B. (1995), Identification of human gene structure using linear discriminant functions and dynamic programming, *Proc. 3rd Int. Conf. Intelligent Systems for Molecular Biology*, pp. 367–375. Menlo Park, A: AAAI Press.

SONNHAMMER, E. L., EDDY, S. R., BIRNEY, E., BATEMAN, A., DURBIN, R. (1998), Pfam: multiple sequence alignments and HMM-profiles of protein domains, *Nucleic Acids Res.* **26**, 320–322.

SONNHAMMER, E. L., EDDY, S. R., DURBIN, R. (1997), Pfam: a comprehensive database of protein domain families based on seed alignments, *Proteins* **28**, 405–420.

SONNHAMMER, E. L., KAHN, D. (1994), Modular arrangement of proteins as inferred from analysis of homology, *Protein Sci.* **3**, 482–492.

STATES, D. J., GISH, W., ALTSCHUL, S. (1991), Improved sensitivity of nucleic acid database searches using application-specific scoring matrices, *Methods: A Companion to Methods in Enzymology* **3**, 66–70.

SWOFFORD, D. L., OLSEN, G. J., WADDELL, P. J., HILLIS, D. M. (1996), Phylogenetics inference, in: *Molecular Systematics* (HILLIS, D. M., MORITZ, C., MABLE, B. K., Eds.), pp 407–514. Sunderland, MA: Sinauer Associates.

THOMAS, A., SKOLNICK, M. H. (1994), A probabilistic model for detecting coding regions in DNA sequences, *IMA J. Math. Appl. Med. Biol.* **11**, 149–160.

THRONTON, J. M. (1998), The future of bioinformatics, *Trends Guide to Bioinformatics* (Trends Supplement), pp. 30–31. Amsterdam: Elsevier.

U.S. Human Genome Project (1990), The first five years. FY 1991–1995. *NIH Publication* No. 90–1590.

VITERBI, A. J. (1967), Error bounds for convolutional codes and an asymptotically optimal decoding algorithm, *IEEE Trans. Informt. Theory* **13**, 260–269.

WEVRICK, R., FRANCKE, U. (1997), An imprinted mouse transcript homologous to the human imprinted in Prader–Willi syndrome (IPW) gene, *Hum. Mol. Genet.* **6**, 325–332.

WOOTTON, J. C., FEDERHEN, S. (1993), Statistics of local complexity in amino acid sequences and sequence, *Comput. Chem.* **17**, 149–163.

ZHANG, M. Q. (1997), Identification of protein coding regions in the human genome by quadratic discriminant analysis, *Proc. Natl. Acad. Sci. USA* **94**, 565–568.

6 Agri-Food and Genomics

Isobel A. P. Parkin

Stephen J. Robinson

Andrew G. Sharpe

Kevin Rozwadowski

Dwayne Hegedus

Derek J. Lydiate

Saskatoon, Canada

1 Introduction

The history of crop domestication by mankind has paralleled the development of civilization. Agricultural production was a prerequisite for establishing settled communities and marked our transition from hunter-gatherers. Today's crop varieties differ dramatically from the crops of our ancestors. Selection over approximately 10,000 years, both conscious and inadvertent, has concentrated over 90% of the world's crop production on a mere 15 plant species (WILSON et al., 1996).

Agricultural sustainability is now a major concern. With limited land available for cultivation it will be necessary to increase output while minimizing the environmental damage. Properly applied biotechnology research will make a significant contribution to increasing the sustainability of crop production.

1.1 Targets for Crop Improvement

The production of food, fiber and animal fodder throughout the world has been greatly influenced by the genetic improvement of crop plants through plant breeding, gradually increasing the yield and (unfortunately) concurrently reducing the genetic variability of the established crops. Nowadays plant breeders' targets for crop improvement still focus on the ever-increasing need for higher yield potentials, greater yield stability, and increased quality.

Yield improvement requires the accurate selection of the optimum parents from which to develop superior populations. The inheritance of yield is polygenic and the trait is strongly influenced by environmental factors, but components contributing to yield are often easy to measure and manipulate. For example, the introduction of the dwarf phenotype into wheat and rice cultivars has been largely responsible for the dramatic improvement in world food production termed the "green revolution" (LAW et al., 1978; CHANG and LI, 1980). This yield improvement was produced by the redirection of assimilates from elongating stems into the developing seeds (KHUSH, 1999).

Adaptations affecting flowering time have been used to extend growing areas, e.g., the development of day-length neutral wheat allows cultivation in an Australian climate (PUGSLEY, 1983). Factors that reduce post-harvest losses during storage increase agricultural returns (TUITE and FOSTER, 1979) and the development of varieties with increased resistance to pests and pathogens reduces the need for chemical inputs and improves profitability and sustainability. Examples of these are varieties that possess an increased resistance to biotic stresses (ROMMENS and KISHORE, 2000) such as fungal (LANDEO et al.,1995) and bacterial (CHEN et al., 1996) pathogens and insect pests (SCHULER et al., 1998). Increased tolerance to abiotic stresses such as drought (BOYER, 1996), salt (DVORAK et al., 1994; FLOWERS and YEO, 1995) heat (PAULSEN, 1994) and cold (THOMASHOW, 1994) have also contributed significantly to improving crop productivity and the land area available for cultivation.

Quality traits are also important targets for crop improvement. The removal/reduction of anti-nutritional compounds from crops to open up new markets or new applications (HARVEY and DOWNEY, 1964; STEFANSSON, 1976) increasing protein content (SUMBERG et al., 1983; LOFFLER et al., 1983) and increasing oil content (SCARTH et al., 1995a, b) in seed are all-important selectable characteristics.

1.2 An Urgent Need for further Crop Improvement

Over recent time, the replacement of traditional varieties of wheat and rice with high yielding modern varieties has increased yield dramatically without a proportional increase in the area of land under cultivation. The yield of both wheat and rice doubled over a period of 25 years from 308 million tons and 257 million tons in 1966 to 541 million tons and 520 million tons in 1990, respectively (KHUSH, 1999). The yield potential of modern varieties is only realized when they are used in conjunction with management practices that include using optimized sowing dates, adequate irrigation, fertilizer application, soil management, and integrated pest management. It has been estimated that without the application of agrochemicals, the yield of today's crops would be reduced by 50% (DeWit and Van Vioten-Doting, 1993).

However, the past decade of crop improvement has failed to reproduce the dramatic increases in crop productivity witnessed between 1966 and 1990. The strong pressures applied by plant breeders to mold crop phenotypes have reduced the variation available at loci controlling a wide range of traits. In addition, increased use of chemical inputs is undesirable both on economic grounds, as it is not prudent to increase input costs greater than can be remunerated by increased yield, and on environmental grounds with phosphates and nitrates applied to the soil leaching into the surface water (Smith et al., 2001a, b) and pesticide residues negatively effecting biodiversity.

The 20th century has witnessed extraordinary population growth and extrapolating from its present rate of growth, it is predicted that the world population will reach 9 billion by 2054 (U.N., 1999). The future ability of the world to generate the increase in agricultural productivity necessary to sustain this anticipated population size is a growing concern.

Plant breeders are attempting to introduce novel variation into crops through the use of wild relatives. This necessitates lengthy crossing programs to remove detrimental alleles dispersed throughout the genomes of these wild donor plants, a process which can be accelerated through the application of marker-assisted selection. Alternative biotechnology approaches for increasing crop yield potential have been developed which involve the introduction of cloned novel genes directly into the plant genome. The quantum leap in gene identification and gene function analysis promised by genomics research will provide new opportunities for the identification and indeed the synthesis of improved alleles for a wide variety of genes controlling crop productivity. Genomics will contribute significantly to providing the variation to fuel future crop improvement.

2 Gene Discovery and Gene Function

2.1 Use of Model Systems

Polyploidy, the doubling of chromosome number, through either genome fusion or genome duplication has played a major role in the evolution of crop plants with many notable examples, maize (Gaut et al., 2000), soybean (Shoemaker et al., 1996), wheat (Gale and Devos, 1998), cotton (Jiang et al., 1998), and canola (Parkin et al., 1995). Polyploids gain a natural advantage over their diploid relatives, the increased genome size and concomitantly the increased cell size leads to a more robust and generally larger plant species and the duplicate gene copies buffer recessive mutations but allow for rapid diversification. Polyploidy, although advantageous to the plant, is inconvenient for researchers; the large duplicated genomes exacerbate attempts at map-based gene cloning and hamper the identification of gene function through knockout technologies. In an attempt to circumvent the problem, crop researchers are turning to model genomes, plant species with small genomes, rapid generation times, and relatively limited amounts of repetitive DNA.

Arabidopsis thaliana has become established as the model dicot, with over 40 years of genetic, molecular, and biochemical research already in place, and 115.4 megabases of the estimated 125 megabase genome sequence has recently been completed (*Arabidopsis* Genome Initiative, 2000). *Arabidopsis* is a crucifer closely related to oilseed crop species such as *Brassica napus* (canola) and *Sinapis alba* (mustard), allowing the direct transfer of biological information from the model plant to agronomically important crops. Although the efficacy of using *Arabidopsis* as a model for distantly related species, including those across the monocot–dicot divide has been considered, the role of a model monocot is generally afforded to rice, itself an important crop, forming the nutritional base for the diet of almost 50% of the world's population. Rice has a genome size of approximately 3–4 times that of *Arabidopsis* and an international effort to sequence the whole genome of rice has been initiated (reviewed in SASAKI and BURR, 2000).

Comparative mapping in plants allows the identification of conserved regions of the genome from diverse species (reviewed in SCHMIDT, 2000). Such comparative analysis has identified conservation of gene content and gene order between the genomes of almost all economic grass crops, including wheat, rice, and maize (MOORE et al., 1995) and between genomes of the Solanaceae crops (TANKSLEY et al., 1992; PRINCE et al., 1993). The most extensive comparative mapping data has been compiled for the cereals (the Poaceae). 19 conserved linkage segments have been identified within the rice genome which can be rearranged to reconstruct the genomes of 11 different grass species (GALE and DEVOS, 1998). More recently, the extent of colinearity at the level of base pairs, micro-synteny, has been studied for a number of species. Comparative sequence analysis between rice, sorghum, and maize localized the regions of conservation to the exonic regions of genes and also suggested that the considerable differences in genome sizes between the species are largely the result of expansion of areas of repetitive DNA (BENNETZEN, 2000).

Understanding the pattern of colinearity between genomes of related species facilitates the transfer of technologies and information across the boundaries imposed by thousands of years of evolution. Through the alignment of model genomes, and by extension their associated genomic resources, with the more complex genomes of crop plants, it will be possible to facilitate the cloning of agronomically important genes in crops by shuttling between the model genome and the more recalcitrant crop genome. In regions displaying strong conservation of gene order, it will be possible to identify candidate genes for traits of interest. A homolog of FLC, an *Arabidopsis* flowering time gene, has been identified as a candidate for a major vernalization responsive flowering time gene in *Brassica rapa* (OSBORN et al., 1997). In any region showing colinearity it will be possible to isolate useful markers from the model plant for gene tagging and marker-assisted breeding in crop species. A notable result of colinearity studies was the identification of the *Arabidopsis* ortholog of the cereal "green revolution" allele, the *Gibberellin Insensitive (GAI)* gene, which led to higher yielding dwarf wheat cultivars (PENG et al., 1999).

2.2 Technologies for Identifying Novel Genes and their Functions

2.2.1 Genome Sequencing

Whole genome sequencing has become an established technology in recent years primarily because of developments in high-throughput sequencing technologies (SPURR et al., 1999). An established method for whole genome sequencing requires the use of ordered large insert DNA libraries that are usually constructed in BAC (WOO et al., 1994) or P1 (LIU et al., 1995) vectors. Once a minimum tiling path has been developed for the ordered library the individual clones are cleaved and the smaller fragments are sub-cloned and sequenced (LIN et al., 1999). The sequence alignment of these smaller fragments is a relatively simple task (providing there is little repetitive DNA present) and these methods have now been used to sequence prokaryotic organisms (COLE et al., 1998) and eukaryotic organisms such as *Caenorhabditis* (*C. elegans* Sequencing Consortium, 1998), *Drosophila* (ADAMS et al., 2000) and *Arabidopsis* (*Arabidopsis* Genome Initiative, 2000).

An alternative to this approach is random whole genome shotgun sequencing, which has recently been employed to sequence many prokaryotic (NELSON et al., 2000) genomes as well as the *Drosophila* (MYERS et al., 2000) genome. Although this approach is faster than the clone-by-clone method, it does require a large investment in high-throughput sequencing equipment to cope with the large volume of sequencing samples and a powerful computer capability for the manipulation and alignment of sequence data (BRODER and VENTER, 2000). Additionally, it has yet to be seen how this approach would perform when attempting to sequence polyploid organisms, such as the majority of crops species, where multiple and sometimes very closely related genomes reside in the same nucleus (PARKIN et al., 1995). The fact that the large genomes of many crops also contain considerable amounts of different repeat elements (BENNETZEN, 1998) would make any large-scale genome sequencing effort a daunting task, although techniques are now available to screen out these sequences from genomic libraries (RABINOWICZ et al., 1999)

It is likely that any genome sequencing efforts in important crop species, such as wheat, maize, and soybean, will be limited to regions of high interest and utilize instead the sequence data from the *Arabidopsis* (model dicotyledon) and rice (model monocotyledon) genome sequencing projects (*Arabidopsis* Genome Initiative, 2000; SASAKI and BURR 2000).

2.2.2 ESTs and Microarrays

The development of large sets of partial sequence data from randomly chosen cDNA clones, more commonly known as Expressed Sequence Tags (ESTs), was first carried out for humans at the beginning of the last decade (ADAMS et al., 1991) and then for a range of other species (MARRA et al., 1998). The large-scale production of ESTs was driven by the rationale that this approach would provide a rapid way to identify genes in any organism. Indeed, the speed and relative ease with which ESTs can be generated has meant that large numbers have now been generated for the majority of economically important crop species by a mixture of public and commercial organizations (PENNISI, 1998).

The general strategy for an EST project involves the construction of cDNA libraries from a variety of tissues at different stages of development, and the subsequent large-scale sequencing of clones from these libraries. However, a natural consequence of this approach is that genes that are highly expressed in the different tissues tend to be overly represented in the final set of ESTs (MARRA et al., 1998). To counter this effect normalization techniques have been developed to reduce the amount of redundancy in the libraries (BONALDO et al., 1996). Additionally, subtractive methods have been developed that allow enrichment for genes that are only expressed under certain conditions, such as drought or

pathogen attack in the case of plants (ROBERTS and PRYOR, 1995; PIH et al., 1997).

ESTs provide a wealth of potential uses for researchers. Initially, the putative function of ESTs can be identified by using algorithms to search for homology between the ESTs and genes of known function (ALTSCHUL et al., 1990). They are also an instant source of "off the shelf" genes and can be used to design PCR primers for the isolation of the genomic copy of the gene (BEHAL and OLIVER, 1997). An estimation of redundancy within the EST data set can be calculated by cluster analysis (HUANG and MADAN, 1999), and non-redundant ESTs can be then be represented on nylon membrane "macroarrays" or glass slide "microarrays" for monitoring gene expression at a global level (FREEMAN et al., 2000). Clustering analysis and sophisticated algorithms can identify nucleotides that distinguish members of multigene families of particular importance in polyploid crop species (MARTH et al., 1999). The development of new molecular markers can also be accomplished by re-sequencing different EST alleles to identify SNPs (SMITH et al., 2000) which can then be assayed by a range of different methods. Comparative mapping of ESTs homologous to target regions in model species, such as *Arabidopsis* and rice, can also provide valuable information for fine-mapping experiments (SILLITO et al., 2000). ESTs can also be employed to provide anchor points for the alignment of large insert BAC libraries (YUAN et al., 2000) and also help to annotate genomic sequence data by providing evidence of transcription (MARRA et al., 1998).

Serial Analysis of Gene Expression (SAGE) is an alternative approach to microarrays for monitoring global gene expression. SAGE involves the production of short nucleotide tags from expressed genes that are then concatenated and sequenced sequentially (VELCULESCU et al., 1995). A large amount of data can be generated very quickly and it is particularly powerful when either a large amount of genomic or EST data are available for the target organism. These methods can be exploited in plant research to identify suites of genes involved in a variety of important agronomic traits such as yield, resistance to environmental stresses, and resistance to pests and pathogens. These methods can also monitor changes in global gene expression in newly developed transgenic lines. Using only a small pre-selected subset of 2,375 genes from *Arabidopsis*, representing about 10% of the total number of expected genes, SCHENK et al., (2000) were able to show that almost one-third were regulated by either pathogen inoculation, or three different elicitors implicated in the plant's defence response to pathogens. Of these more than 20% were coordinately regulated by more than one treatment, demonstrating that an extensive degree of cross talk exists between disease resistance signalling pathways (FEYS and PARKER, 2000).

2.2.3 Bioinformatics

The rapid accumulation of DNA sequence information and functional expression analysis data necessitated the evolution of a new area of biology dedicated to the computational mining or "*in silico*" analysis of this valuable data. Bioinformatics can take many guises but it basically gives researchers the tools to store, access, analyze, and search the volumes of data being generated. Many novel genes are being uncovered through the systematic searching of available genomic sequence and their putative function is being assigned through sequence identity algorithms. For the model dicot *Arabidopsis thaliana*, the recent publication of the sequence of chromosomes 2 and 4 displayed the power of computational biology (LIN et al., 1999; MAYER et al., 1999). For the combined sequence, which covers 32% of the whole genome, 7,781 putative protein-coding regions were identified using a combination of gene prediction programs and database searches. The availability of such large stretches of contiguous sequence enabled a comprehensive study of the level of gene duplication within the model genome. A large number of genes (60% of those found on chromosome 2) were found to be repeated tandemly and large chromosomal duplications were also observed. The tandem duplication of genes could form an important reservoir for allelic diversity or perhaps represent a mechanism for ensuring synchronized expression of genes. The comparative analysis of the predicted proteins from *Arabidopsis* with previously identified proteins from various lineages, from yeast to human, identified a subset of genes specific to the plant lineage and large numbers of genes that exhibited remarkable sequence conservation with human proteins, emphasizing how function bridges evolutionary barriers. A powerful new development in bioinformatics is the assignment of putative function through not only sequence identity, but also using sequence-to-structure-to-function approaches, which exploits the conserved structural motifs found at the active sites of proteins (reviewed in SKOLNICK and FETROW, 2000).

2.2.4 Proteomics

Proteomics is the systematic identification and characterization of the proteins (the proteome) of a cell, tissue, or organelle (reviewed in PANDEY and MANN, 2000). The study of global gene expression using microarrays or SAGE is reliant on the observed phenotypic changes being the result of dynamic changes in particular mRNA populations. Although extremely powerful, these types of analyses attempt to simplify the relationship between genes and their protein products, which are in fact the active agents in the cells. A complementary approach is proteomics. Through the application of 2-D gel electrophoresis, which separates proteins based on their charge and mass, thousands of proteins can be analyzed at one time. Each protein can be identified through a combination of microsequencing using mass spectrometry and bioinformatics. This allows the identification of the primary gene products and their post-translational modifications. These new technologies are allowing researchers to identify numerous novel proteins controlling important biological processes, including plant proteins expressed in response to drought, anoxia, and other environmental changes (reviewed in ZIVY and DE VIENNE, 2000). Proteomics has also been applied to the study of mutant lines, allowing the analysis of both the direct and indirect consequences of metabolic changes caused by a single gene mutation (ZIVY and DE VIENNE, 2000).

It is unlikely that the differential display techniques, which utilize this combination of 2-D electrophoresis and mass spectrometry, will ever surpass the sensitivity of microarray technology. However, proteomics is not limited to the study of expression patterns, but also encompasses high-throughput functional biochemical analyses of the proteome. By elucidating protein–protein interactions using mass spectrometry and the yeast-two-hybrid system it is possible to gain insight into the functions of many unknown proteins. For example, the yeast two-hybrid system has been used to identify plant proteins which interact with viral proteins upon infection of a plant. This work is beginning to elucidate the mechanisms by which the plant recognizes and prevents the replication of invading viruses (LEONARD et al., 2000).

2.2.5 Functional Genomics

The application of microarrays, SAGE analysis, and high-throughput proteomics to the study of agrinomically important traits will identify a plethora of genes whose expression pattern appears to coincide with the observation of a desired phenotype. It will be necessary to quickly confirm the function of these novel genes, in some cases bioinformatics will prove invaluable in suggesting a function, for many others the function will remain a mystery. From the 25,498 predicted proteins identified from the published genomic sequence of *Arabidopsis* only in the range of 69% could be assigned a possible function based on homology to known genes and less than 10% of these had their function confirmed experimentally (*Arabidopsis* Genome Initiative, 2000). Gene knockout systems have been employed in order to exploit this wealth of information and to assist in identifying gene function. So-called "functional genomics" describes the identification of gene function through the systematic cessation of expression of each and every gene from a target genome. This strategy involves the development of saturated populations of mutagenized plant lines, each line carrying mutated copies of one or more specific genes. These populations can be used for "reverse genetics", where the function of a particular

gene is determined by characterizing the phenotype of a line carrying a mutated copy of that gene, or "forward genetics", where a population of mutant lines is screened to identify lines which are defective in a particular biological process, hence, identifying genes controlling that process.

In plants a number of methods have been employed to generate mutants (FELDMAN et al., 1994):

(1) Ionizing radiation, using X-ray or neutron bombardment of plant cells generates large deletions of chromosomal DNA or reciprocal translocations and other major chromosomal rearrangements. Such mutants tend to be deficient for a number of genes, which makes it possible to find mutants after screening relatively few treated plants, but more difficult to identify the particular gene responsible for a phenotype.

(2) Chemical mutagenesis, using alkylating agents such as ethyl methane sulfonate (EMS). Most commonly EMS causes alkylation of guanine, which allows it to pair with thymine instead of cytosine during replication thus causing single base pair changes in the DNA sequence. EMS tends to cause a large number of mutations per line which means a smaller population is necessary to reach saturation than insertional mutagenesis. The single base pair also changes can induce subtle variations in alleles and phenotypes. However, as with ionizing radiation, identification of the affected gene requires a lengthy mapping process.

(3) Insertional mutagensis, using either transposable elements or the T-DNA insertion element of the plant pathogenic soil bacterium, *Agrobacterium*. The stable integration of such elements into the genome either within the coding region of a gene or within the upstream regulatory elements can knockout the function of that gene. Since the sequence of the integrated DNA is known, it is a simple procedure to amplify or clone the genomic DNA flanking the insertion site and hence identify the affected gene.

It can be very powerful to utilize variants of the three methods, e.g., ISHITANI et al. (1997), identified hundreds of *Arabidopsis* mutants which were involved in the pathways controlling the plant's response to osmotic and cold stress by utilizing a combination of chemical mutagenesis and reporter gene expression. Populations of enhancer trap lines have been developed which allow the recovery of genes expressed at specific stages in cell development (MARTIENSSEN, 1998).

A number of populations of *Arabidopsis* mutagenized lines have been developed and made publicly available (ANDERSON and WILSON, 2000). The utility of these resources has been further enhanced by the development of associated databases containing the sequences of the DNA flanking the insertion sites (PARINOV et al., 1999). This allows the researcher the luxury of searching *in silico* for the disruption of a particular gene.

3 Methods to Introduce Novel Genes

Historically, all improvements in crop genotype have been made through traditional plant breeding approaches, where the introgression of genes controlling useful traits is hampered by the concurrent introduction of undesirable background genotype from the donor parent. The application of molecular markers, which can be used to both select for genes of interest and against the unwanted donor genotype is accelerating the work of the plant breeders.

At present the only real mechanism to introduce individual novel genes under the control of specific regulatory elements is through plant transformation. The future development of an effective gene targeting technology will be of enormous benefit in exploiting information generated by genomics programs and applying it to crop improvement. The technology will enable precision engineering of plant genomes to accomplish three general goals: gene inactivation, gene modification, and gene replacement. Effective manipulation of biochemical pathways and developmental programs to improve plant performance through precise and stable modifications will be possible.

3.1 Accelerating Traditional Breeding

The application of the information and tools derived from genomics projects to the acceleration of traditional plant breeding is key to future crop development (MAZUR et al., 1999). An important step is the development of robust genetic markers to facilitate selection during plant breeding programs. Ideally, markers need to be closely linked to the genes controlling the traits of interest, which can be either simply inherited (MUDGE et al., 1997) or under more complex control (TANKSLEY, 1993). Markers can be used to efficiently select for desirable alleles as well as simultaneously eliminating potentially deleterious donor alleles (VIERLING et al., 1996). Marker-assisted breeding has the potential to maximize the use of wild germplasm in breeding, as it reduces the length of backcrossing programs, typically from 8–10 generations down to 2–3 generations (LYDIATE et al., 1993).

The initiation of large-scale DNA sequencing in a range of important crop species has generated large amounts of EST and genomic sequence data (PENNISI, 1998). This sequence data can be mined for variation that can then be utilized for marker development. One class of highly variable sequences are microsatellites or Simple Sequence Repeats (SSRs). SSRs are sequences of mono-, di-, tri-, or tetra-nucleotide repeats that were first described in humans (HAMADA and KAKUNAGA, 1982) but have subsequently been identified in many species, including plants (POWELL et al., 1996). These sequences are generally ideal for use as genetic markers because they are multiallelic, co-dominant, easily detected, multiplexable, abundant, and robust (RAFALSKI and TINGEY, 1993). They are currently the most readily amenable markers for whole genome marker-assisted selection. However, they are relatively expensive to develop because of the need for genomic sequence data (RAFALSKI and TINGEY, 1993). Public projects to develop SSR markers in a variety of crop species have been undertaken (RODER et al., 1998; MCCOUCH et al., 1997; MOULE et al., 2000) and a number of consortiums have also been established to develop large numbers of SSR markers for important crop species (SHARPE and LYDIATE, 1998; SHAROPOVA et al., 2000). SSRs are generally found in intergenic regions of the genome, however, they can be found closely associated with genes, and they have been identified in EST collections (GROENEN et al., 1998; CARDLE et al., 2000). Indeed, it has also been postulated that they could play a major role in altering gene expression (KASHI et al., 1997). It has also been reported that DNA flanking SSRs is highly polymorphic (GRIMALDI and CROUAU-ROY, 1997; COLSON and GOLDSTEIN, 1999) and this variation could also be exploited for marker development.

The presence of large sets of ESTs for many crop species is another source of sequence variation that can be exploited for the generation of genetic markers. The variation is usually in the form of Single Nucleotide Polymorphisms (SNPs) that can be identified by analyzing sequence alignments of clustered ESTs (GU et al., 1999). SNPs found in clustered EST data that has been generated from multiple genotypes of a particular organism can represent variation between different alleles of genes. However, the sequence variation can also represent the variation between different members of multi-gene families (MARTH et al., 1999) and this variation will be particularly evident in the polyploid genomes of crop species. It will be important to distinguish between these two types of polymorphism to ensure that developed SNP markers assay the correct genomic locus. The representation of 3'-UTRs of transcripts in EST collections will be useful to distinguish between different copies of multi-gene families. These regions are not under the same selective constraints as coding sequences and are frequently just as polymorphic as random non-transcribed genomic regions (GIORDANO et al., 1999). They are often sufficiently divergent between different members of a gene family to allow locus-specific analysis. Once identified, locus-specific SNPs can be assayed by a variety of high-throughput methodologies which include gel or capilliary electrophoresis (SEE et al., 2000), mass spectrometry (TANG et al., 1999),

single base extension (HIRSCHHORN et al., 2000), pyrosequencing (AHMADIAN et al., 2000), fluorescence resonance energy transfer (Mein et al., 2000), denaturing HPLC (GIORDANO et al., 1999) and high-density oligonucleotide arrays (CHO et al., 1999).

3.2 Transgenic Modifications

The advent of plant transformation technologies in the early 1980s has enabled the introduction of novel genes into over 50 different plant species (DUNWELL, 2000). Genetic engineering has obvious benefits, evolutionary barriers do not limit the pool of potentially useful genes, a number of different desirable genes can be introduced in a single transformation event, and the concurrent introgression of unwanted alleles is eliminated. Genetic transformation of plants to express novel "transgenes" is made possible through the ability of differentiated plant cells to be grown in tissue culture to regenerate whole plants in conjunction with techniques that facilitate the transfer and integration of foreign DNA into plant cells. The gene of interest is cloned along with specific regulatory elements into an expression vector, this vector is then used to transform plant cells. For dicot species, the most commonly used technique to genetically modify plant cells is *Agrobacterium*-mediated transformation (reviewed in ZUPAN et al., 2000). *Agrobacterium* is a soil pathogen that during infection of a plant transfers a segment of its own DNA (T-DNA) into the genome of its host. The expression vector containing the gene of interest incorporates the elements of the T-DNA that induces the pathogen to integrate the T-DNA along with the novel gene into the genome of the plant.

The first transgenic plant to appear in the market place was the "Flavr savr"® tomato, which has reduced levels of the enzyme polygalacturonase, whose function is to soften the cell wall (SHEEHY et al., 1988). This modification is designed to increase the shelf life of the tomato fruits. Since that time, genetically modified plants have been developed to address a number of issues and consequently their use has expanded dramatically. It is estimated that over 70% of canola grown in Canada during

1999 was genetically modified, the majority of the modifications found in these commercial varieties provide resistance to a particular herbicide (DUNWELL, 2000). Endogenous plants can be incredibly invasive causing yield losses that are only limited by numerous applications of herbicide throughout the growing season (DE BLOCK et al., 1987). The transgenic varieties provided an alternative that reduced herbicide applications, diminishing the likely evolution of herbicide tolerant weeds, and increased farmers' yields. To further increase the yield potential in crop plants, effective and sustained control of pests and pathogens is very important. It has been estimated that the world-wide crop losses caused by pathogens amount to 12–13% of the potential crop production despite the use of pesticides (JAMES et al., 1990). To date, the development of plant varieties with enhanced insect resistance has focused on the introduction of genes expressing endotoxins from *Bacillus thuringiensis* (*Bt*) (reviewed in HILDER and BOULTER, 1999). Varieties of *Bt* corn, potato, and cotton have been commercialized and show significant protection against European corn borer, Colorado potato beetle, and bollworm infestations, respectively. There is some concern that overuse of *Bt* transgenic lines will lead to the development of *Bt* resistant insects. This is presently being managed through agronomic practices. However, in the long term it will be necessary to identify alternative genes conferring resistance to insects (SHAH et al., 1995). Several proteins have been identified which provide some protection against insect infestation, by retarding insect development when expressed at high enough levels, including protease inhibitors, α-amylases and lectins. None of these alternatives have yet shown the level of control afforded by *Bt*, so it is hoped that the new genomics technologies will allow insights into the mechanisms of insect resistance and help to identify new targets for resistance genes. Although there has been some success in the development of transgenic plant varieties with resistance to viruses, through a mechanism known as pathogen-derived resistance (PDR), there has been less progress in the development of transgenic varieties with resistance to fungal pathogens (reviewed in DEMPSEY et al., 1998). Conventional breeding has led to the

development of a number of successful fungal pathogen resistant varieties, however, the lengthy breeding process cannot react adequately to a constantly evolving pathogen population and there are presently no sources of natural resistance to a number of fungal pathogens. Two potentially promising strategies are presently being investigated (reviewed in MELCHERS and STUIVER, 2000). The first strategy attempts to slow the growth of the fungi by the expression of enzymes such as chitinase and glucanase, which can degrade the cell walls of fungal pathogens. The second strategy takes advantage of the plant's natural defense response. Plants can, when they recognize a pathogen, resist attack through the hypersensitive response (HR), where cell death at the point of infection prevents spread of the disease. HR can only be triggered, if the plant carries a resistance gene that specifically recognizes a particular pathogen avirulence (Avr) protein. Plants are now being engineered to activate the HR to multiple fungal pathogens (MELCHERS and STUIVER, 2000). In addition, some cloned plant resistance genes have shown remarkable durability in their host plants, recognizing multiple races of specific pathogens (DEMPSEY et al., 1998). Genomics is likely to facilitate the identification of further resistance genes with similar capabilities and it is possible that through their study the complexities of durable resistance can be addressed and consequently exploited in the field.

Future trangenic developments to increase yield potential may impact plant physiology. The mechanism of seed dispersal in the Brassicacae occurs by a process called fruit dehiscence or pod shatter. In spite of the agronomic value of controlling this character (typical seed losses of 8–12% have been recorded upon premature dehiscence in canola) few studies have reported on the genes that regulate this process (KADKOL et al., 1984; CHILD et al., 1998). The consequences of the seed loss are twofold. Firstly, there is a reduction in the yield potential of the crop and secondly, the residual seed germinates to contaminate crops in subsequent growing seasons (WHITEHEAD and WRIGHT, 1989). Recently, the Shatterproof genes *SHP1* and *SHP2* which are required for fruit dehiscence in *Arabidopsis thaliana* have been identified and cloned (LILJEGREN et al., 2000). These genes may play a future role in controlling pod shatter in crops such as canola.

Recent exciting developments in transgenic engineering of plants will have a direct and positive impact on human health. Over half the world's population is reliant upon rice as a source of nutrients. However, rice is deficient in certain essential nutrients, in particular vitamin A. It is estimated that a quarter of a million children in Southeast Asia go blind each year due to this nutritional deficiency (SOMMER, 1988). The development of "Golden Rice" aims to address this problem. This transgenic rice variety has been successfully engineered to express the entire vitamin A biosynthetic pathway (YE et al., 2000). Scientific advances in molecular biology and immunology have led to an improved understanding of numerous diseases and to the development of novel strategies using plants and plant viruses for expression and delivery of vaccines. In the 1990s vaccinogens (antigens capable of eliciting an immune response) for a number of devastating diseases, including hepatitis and cholera, were successfully expressed in a variety of plants (MASON et al., 1996; HEIN et al., 1996, RICHTER et al., 2000). The first human clinical trials for a transgenic plant-derived vaccinogen have been performed with promising results (TACKET et al., 1998). The development of plants as a delivery mechanism for vaccines has a number of benefits; plants are an inexpensive means of generating the required proteins, they eliminate the risk of contaminating animal pathogens, they provide a heat-stable environment which ensures safe transport, and finally they provide a mechanism for oral delivery, circumventing the potential hazards associated with injections (reviewed in WALMSLEY and ARNTZEN, 2000).

3.3 Gene Targeting

Gene targeting can result from homologous recombination between an introduced DNA molecule and the homologous genomic locus. In this case, a reciprocal exchange of genetic sequences occurs between the two DNA molecules. Alternatively, it may result from a gene conversion event, which leads to the adapta-

tion of the sequence of one strand to the sequence of another strand. Gene conversion involves local copying of genetic information from one strand to another and is not necessarily associated with cross-overs. A targeted and specific gene inactivation system will facilitate stable and heritable gene silencing with more reliability than conventional systems such as anti-sense RNA and cosuppression. The ability to inactivate target genes in an efficient manner will also obviate the need for generating large populations of insertionally inactivated plants for each crop to evaluate gene function. Secondly, targeted modification of gene sequences will make *in vivo* protein engineering a reality, in essence enabling engineering of new genetic variation in a directed and predictable fashion. Thirdly, gene targeting will enable facile replacement and exchange of genes and promoters in the genome. A novel gene could be placed adjacent to a promoter driving a desirable expression pattern in its normal chromosomal context resulting in predictable levels and patterns of gene expression. This will make conventional transgenic modifications obsolete because the current techniques result in random integration into the genome and wide variation in transgene expression levels due to position effects.

The tremendous potential for biological information and biotechnological applications from gene targeting has resulted in concerted efforts to develop the technology for various species. Effective gene targeting methods were first developed in yeast (reviewed in ROTHSTEIN, 1991) and led to an explosion in biological information by facilitating gene knockouts and subtle gene modifications. The value of this technique and resulting information is evident from large-scale application of gene targeting to create a population of yeast mutants with each open reading frame inactivated (WINZELER et al., 1999). Effective gene targeting has also been achieved in mammals enabling functional genomics though gene knockouts (THOMAS and CAPECCHI, 1987).

In contrast to the success in other systems, gene targeting in plants remains an arduous and unreliable task. In a lower plant, the moss *Physcomitrella patens*, gene targeting is very efficient with frequencies of 90% or greater being achieved (SCHAEFER and ZRYD, 1997;

GIRKE et al., 1998). Two methods currently hold promise in higher plants: *Agrobacterium tumefaciens*-mediated delivery of engineered T-DNA gene targeting substrates and particle bombardment of plant tissues with chimeric RNA/DNA hybrid molecules as gene targeting substrates.

Use of *Agrobacterium* and T-DNA for gene targeting is an obvious approach with great promise given the availability of *Agrobacterium* strains capable of infecting all major crop species, and the technical ease of engineering T-DNA molecules. However, successful application of this system to gene targeting is very limited. The general system involves engineering a T-DNA cassette with fragments of DNA homologous to the target genomic locus flanking a disruption cassette, typically a selectable marker. The T-DNA cassette is transferred to plant cells using *Agrobacterium* and the flanking homology fragments target the T-DNA to the correct genomic locus. A subsequent recombination event transfers the disruption cassette into the plant genome thereby insertionally inactivating the target gene. Using a combination of positive and negative selectable markers true gene targeting events are enriched from the background of random integration of the T-DNA throughout the genome.

However, results to date have been disappointing. Using a strategy of reactivating a defective selectable marker placed in the tobacco genome, OFFRINGA et al. (1990, 1993) demonstrated the tenability of T-DNA-mediated gene targeting. But even with a strong selectable phenotype to identify gene targeting events, the frequency of actual gene targeting was very low ($\sim 10^{-5}$). Targeted inactivation of native genomic loci has been demonstrated in *Arabidopsis* with frequencies of 1/750 transgenic lines or 2/2,580 transgenic calli being obtained (MIAO and LAM, 1995). Gene targeting frequency with T-DNA cassettes can be increased through artificially generating DNA lesions at target loci (PUCHTA et al., 1996), and some increase in gene targeting frequency has been achieved by increasing the recombination potential of plant cells (REISS et al., 2000). However, the low frequency of gene targeting currently achieveable using T-DNA limits application of this technology to genes with an easily selectable or screenable phenotype.

The second successful method of gene targeting in plants utilizes hybrid RNA/DNA molecules to catalyze specific gene conversion events. These molecules consist of two 20–30 bp complementary oligonucleotides, one RNA and the other DNA, which are homologous to the genomic locus of interest except for a single base mismatch (YOON et al., 1996). The gene targeting substrate is transferred to plant cells by particle bombardment. Some of the substrate enters the nucleus, and the homology of the RNA/DNA molecule enables it to pair with the target genomic locus. By an as yet undefined mechanism, the base change encoded by the RNA/DNA hybrid molecule is transferred to the genomic sequence thereby creating a specifically altered gene *in vivo*. The genomic base change may be engineered to create a non-sense mutation thereby shutting down functional expression of the gene of interest, or alter a specific amino acid and create a protein with altered biochemical properties. Both of these possibilities have recently been conducted in maize where activity of a selectable marker was altered and a native gene was changed to confer heritable herbicide resistance in regenerated maize plants (ZHU et al., 1999, 2000). However, only a very low frequency was obtained (10^{-5}). Reliance on biolistics to deliver the gene targeting substrate greatly limits exploitation of this technology. The low frequency of gene targeting raises the problem of screening large numbers of calli, before regenerating plants, for the successful targeting event and, therefore, severely limits general application of the technology.

Developing a system that could generate hybrid molecules of large size and effectively deliver them to the nucleus with limited tissue culture steps would greatly improve the application of this technology. At present, applications appear to be limited to modifying genes with easily selectable or screenable phenotypes thereby preventing general application in functional genomics and crop improvement programs. It is to be hoped that further research in design and delivery of gene targeting substrates and increased understanding of mechanisms of DNA recombination and repair in plants will eventually lead to a realistically useful system for targeted gene modifications in plants. Gene targeting will then be a powerful tool for applying information garnered from crop genomics programs to crop improvement.

4 Conclusion

The knowledge gained from crop genomics when it is properly applied, will provide breeders, farmers and indeed mankind with the genetic variation needed to fuel future crop improvement. The variation available through traditional means is unlikely to produce the higher yielding varieties with improved resistance to pests, disease and environmental stresses and reduced reliance on chemical fertilizers that mankind and the environment will need for survival.

Improvements in food quality, increased nutritional value and new niche products will also be facilitated through modifications to gene expression patterns and gene content guided by acquired knowledge about gene function. However, to be developed and implemented safely these potential improvements will need to be carefully regulated and tested.

5 References

ADAMS, M. D., KELLEY, J. M., GOCAYNE, J. D., DUBNICK, M., POLYMEROPOULOS, M. H. et al. (1991), Complementary DNA sequencing: expressed sequence tags and the human genome project, *Science.* **252**, 1651–1656.

ADAMS, M. D., CELNIKER, S. E., HOLT, R. A., EVANS, C. A., GOCAYNE, J. D. et al (2000), The genome sequence of *Drosophila melanogaster, Science* **287**, 2185–2195.

ALTSCHUL, S. F., GISH, W., MILLER, W., MYERS, E. W., LIPMAN, D. J. (1990), Basic local alignment search tool, *J. Mol. Biol.* **215**, 403–410.

AHMADIAN, A., GHARIZADEH, B., GUSTAFSSON, A. C., STERKY, F., NYREN, P. et al. (2000), Single-nucleotide polymorphism analysis by pyrosequencing, *Anal. Biochem.* **280**, 103–110.

ANDERSON, M., WILSON, F. (2000), Growth, maintenance and use of Arabidopsis genetic resources,

in: *Arabidopsis*: A Pratical Approach, Wilson, Z. A. (Ed.), pp 1–28. Oxford: Oxford University Press.

Arabidopsis Genome Initiative (2000), Analysis of the genome sequence of the flowering plant *Arabidopsis thaliana, Nature* **408**, 796–815.

BEHAL, R. H., OLIVER, D. J. (1997), Biochemical and molecular characterization of fumarase from plants: purification and characterization of the enzyme–cloning, sequencing, and expression of the gene, *Arch, Biochem, Biophys.* **348**, 65–74.

BENNETZEN J. L. (1998), The evolution of grass genome organisation and function, *Symp. Soc. Exp. Biol.* **51**, 123–126.

BENNETZEN, J. L. (2000), Comparative sequence analysis of plant nuclear genomes. Microcolinearity and its many exceptions, *Plant Cell.* **12**, 1021–1030.

BONALDO, M. F., LENNON, G., SOARES, M. B. (1996), Normalization and subtraction: two approaches to facilitate gene discovery, *Genome Res.* **6**, 791–806.

BOYER, J. S. (1996), Advances in drought tolerance in plants. *Adv. Agron.* **56**, 187–218.

BRODER, S., VENTER, J. C. (2000), Whole genomes: the foundation of new biology and medicine, *Curr. Opin. Biotechnol.* **11**, 581–585.

CARDLE, L., RAMSAY, L., MILBOURNE, D., MACAULAY, M., MARSHALL, D., WAUGH, R. (2000), Computational and experimental characterization of physically clustered simple sequence repeats in plants, *Genetics* **156**, 847–854.

CHANG, T. T., LI, C. C. (1980), Genetics and breeding, in: *Rice Production and Utilisation* (LUH, B. S. Ed.), pp. 87–146. Westport, CT: AVI Publishing.

CHEN, Z., YIN, S., LU, F., LI, Z. (1996), Screening and utilisation of antagonistic bacteria for rice sheath blight control, in: *Advances of Biological Control of Plant Diseases*, (TANG, W. H., COOK, R. J., ROVIRA, A. D., Eds.), pp. 1–5. China Agricultural University Press. Beijing, China.

CHILD, R. D., CHAUVAUX, N., JOHN, K., ULVSKOV, P., ONCKELEN, H. A. (1998), Ethylene biosynthesis in oilseed rape pods in relation to pod shatter. *J. Exp. Bot.* **49**, 829–838.

CHO, R. J., MINDRINOS, M., RICHARDS, D. R., SAPOLSKY, R. J., ANDERSON, M. et al. (1999) Genome-wide mapping with biallelic markers in *Arabidopsis thaliana, Nature Genet.* **23**, 203–207.

C. elegans Sequencing Consortium (1998), Genome sequence of the nematode *C. elegans*: a platform for investigating biology, *Science* **282**, 2012–2018.

COLE, S. T., BROSCH, R., PARKHILL, J., GARNIER, T., CHURCHER, C. et al. (1998), Deciphering the biology of *Mycobacterium tuberculosis* from the complete genome sequence, *Nature* **393**, 537–544.

COLSON I, GOLDSTEIN D. B. (1999), Evidence for

complex mutations at microsatellite loci in *Drosophila*, *Genetics* **152**, 617–627.

DE BLOCK, M., BOTTERMAN, J., VANDEWIELE, M., DOCKX, J., THOEN, C. et al. (1987), Engineering herbicide resistance in plants by expression of a detoxifying enzyme, *EMBO J.* **6** (9), 2513–2518.

DEMPSEY, D'M. A., SILVA, H., KLESSIG, D. F. (1998) Engineering disease and pest resistance in plants, *Trends Microbiol.* **6**, 54–61.

DEWIT, P. J. G. M., VAN VIOTEN-DOTING, L. (1993) General introduction to biotechnology in plant breeding and crop protection, in: *Developing Agricultural Biotechnology in the Netherlands* (VUIJK, D. H., DEKKERS, J. J., VAN DER PLAS, H. C., Eds.), Wageningen: Purdoc Scientific Publ.

DUNWELL, J. M. (2000), Transgenic approaches to crop improvement, *J. Exp. Botany.* **51**, 487–496.

DVORAK, J., NOAMAN, M. M., GOYAL, S. (1994), Enhancement of the salt tolerance of *Triticum turgidum* L. by the *Kna1* locus transferred from *Triticum aestivum* L. chromosome 4D by homoeologous recombination, *Theor. Appl. Genet.* **87**, 872–877.

FELDMAN, K. A., MALMBERG, R. L., DEAN, C. (1994), Mutagenesis in *Arabidopsis*, in: *Arabidopsis* (MEYEROWITZ, E. M., SOMERVILLE, C. R., Eds.), Cold Spring Harbor, NY: Laboratory Press.

FEYS, B. J., PARKER, J. E. (2000), Interplay of signaling pathways in plant disease resistance, *Trends Genet.* **16**, 449–455.

FLOWERS, T. J., YEO, A. R. (1995), Breeding for salinity resistance in crop plants, Where next? *Aust. J. Plant Physiol.* **22**, 875–884.

FREEMAN W. M., ROBERTSON D. J., VRANA K. E. (2000), Fundamentals of DNA hybridization arrays for gene expression analysis. Biotechniques **29** (5), 1042–1055.

GALE, M. D., DEVOS, K. M. (1998), Plant comparative genetics after 10 years, *Science* **282**, 656–659.

GAUT, B. S., LE THIERRY D'ENNEQUIN, M., PEEK, A. S., SAWKINS, M. C. (2000), Maize as a model for the evolution of plant nuclear genomes, *Proc. Natl. Acad. Sci. USA* **97**, 7008–7015.

GIORDANO, M., OEFNER, P. J., UNDERHILL, P. A., CAVALLI SFORZA, L. L., TOSI, R., RICHIARDI, P. M. (1999), Identification by denaturing high-performance liquid chromatography of numerous polymorphisms in a candidate region for multiple sclerosis susceptibility, *Genomics* **56**, 247–253.

GIRKE, T., SCHMIDT, H., ZAHRINGER, U., RESKI, R., HEINZ, E. (1998), Identification of a novel delta 6-acyl-group desaturase by targeted gene disruption in *Physcomitrella patens*, *Plant J.* **15**, 39–48.

GRIMALDI, M. C., CROUAU-ROY, B. (1997), Microsatellite allelic homoplasy due to variable flanking sequences, *J. Mol. Evol.* **44**, 336–340.

GROENEN, M. A., CROOIJMANS, R. P., VEENENDAAL, A., CHENG, H. H., SIWEK, M., VAN DER POEL, J. J. 9

(1998), A comprehensive microsatellite linkage map of the chicken genome, *Genomics* **49**, 265–274.

GU, Z, HILLIER, L, KWOK, P. Y. (1998), Single nucleotide polymorphism hunting in cyberspace, *Hum. Mutat.* **12**, 221–225.

HAMADA, H., KAKUNAGA, T. (1982), Potential Z-DNA forming sequences are highly dispersed in the human genome, *Nature* **298**, 396–398.

HARVEY, B. L., DOWNEY, R. K. (1964), The inheritance of erucic acid content in rapeseed (*Brassica napus*), *Can. J. Plant Sci.* **44**, 104–111.

HEIN, M. B., YEO, T. C., WANG, F., STURTEVANT, A. (1996), Expression of cholera toxin subunits in plants, *Ann. NY Acad. Sci.* **25**, 50–56.

HILDER, V. A., BOULTER, D. (1999), Genetic engineering of crop plants for insect resistance – a critical review, *Crop Protect.* **18**, 177–191.

HIRSCHHORN, J. N., SKLAR, P., LINDBLAD-TOH, K., LIM, Y. M., RUIZ-GUTIERREZ, M. et al. (2000), SBE-TAGS: An array-based method for efficient single-nucleotide polymorphism genotyping, *Proc. Natl. Acad. Sci. USA* **97**, 12164–12169.

HUANG, X., MADAN, A. (1999), CAP3: A DNA sequence assembly program, *Genome Res.* **9**, 868–877.

ISHITANI, M., XIONG, L., STEVENSON, B., ZHU, J. K. (1997), Genetic analysis of osmotic and cold stress signal transduction in *Arabidopsis*: interactions and convergence of abscisic acid-dependent and abscisic acid-independent pathways, *Plant Cell* **9**, 1935–1949.

JAMES, W. C., TENG, P. S., NUTTER, F. W. (1990), Estimated losses of crops from plant pathogens, in: *CRC Handbook of Pest Management* Vol. 1. (PIMENTAL, D., Ed.), pp. 15–50. Boca Raton, FL: CRC Press.

JIANG, CX., WRIGHT, R. J., EL-ZIK, K. M., PATERSON, A. H. (1998), Polyploid formation created unique avenues for response to selection in *Gossypium*, *Proc. Natl. Acad. Sci. USA* **95**, 4419–4424.

KADKOL, G. P., MACMILLAN, R. H., BURROW, R. P., HALLORAN, G. M. (1984), Evaluation of *Brassica* genotypes for resistance to shatter. I. Development of a laboratory test, *Euphitica* **33**, 63–73.

KASHI, Y., KING, D., SOLLER, M. (1997), Simple sequence repeats as a source of quantitative genetic variation, *Trends Genet* **13**, 74–78.

KHUSH, G. S. (1999), Green revolution, preparing for the 21st century, *Genome* **42**, 646–655.

LANDEO, J. A., GASTELO, M., PINEDO, H., FLORES, F. (1995), Breeding for horizontal resistance to late blight in potato free of R-genes, in: *Phytophthora Infestans* (DOWLEY, L. J., BANNON, E., COOKE, E., KEANE, R. L., O'SULLIVAN, T., Eds.), pp. 268–274. Dublin: Bole Press.

LAW, C. N., SNAPE, J. W., WORLAND, A. J. (1978), The genetical relationship between height and yield in wheat, *Heredity* **40**, 133–151.

LEONARD, S., PLANTE, D., WITTMANN, S., DAIG-
NEAULT, N., FORTIN, M. G., LALIBERTE, J. F. (2000),
Complex formation between potyvirus VPg
and translation eukaryotic initiation factor 4E
correlates with virus infectivity, *J. Virol.* **74**,
7730–7737.

LILJEGREN, S. J., DITTA, G. S., ESHED, Y., SAVIDGE, B.,
BOWMAN, J. L., YANOFSKY, M. F. (2000), Shatter-
proof MADS-box genes control seed dispersal in
Arabidopsis, *Nature* **404**, 766–770.

LIN, X., KAUL, S., ROUNSLEY, S., SHEA, T. P., BENITO,
M. I. et al. (1999), Sequence and analysis of chromosome 2 of the plant *Arabidopsis thaliana*, *Nature* **402**, 761–768.

LIU, Y. G., MITSUKAWA, N., OOSUMI, T., WHITTIER, R.
F. (1995), Generation of a high quality P1 library
of *Arabidopsis* suitable for chromosome walking,
Plant J. **7**, 351–358.

LOFFLER, C. M., BUSCH, R. H., WIERSMA, J. V. (1983),
Recurrent selection for grain protein percentage
in hard red spring wheat, *Crop. Sci.* **23**, 1097–1101.

LYDIATE, D., SHARPE, A., LAGERCRANTZ, U., PARKIN,
I. (1993). Mapping the *Brassica* genome, *Outlook
on Agriculture* **22**, 85–92.

MARRA, M. A., HILLIER, L., WATERSTON, R. H.
(1998), Expressed sequence tags – ESTablishing
bridges between genomes, *Trends Genet.* **14**,
4–7.

MARTH, G. T., KORF, I., YANDELL, M. D., YEH, R. T.,
GU, Z. et al. (1999), A general approach to sin-
gle-nucleotide polymorphism discovery, *Nature
Genet.* **23**, 452–456.

MARTIENSSEN, R. A. (1998), Functional genomics:
Probing plant gene function and expression with
transposons, *Proc. Natl. Acad. Sci. USA* **95**,
2021–2026.

MASON, H. S., BALL, J. M., SHI, J. J., JIANG, X., ESTES,
M. K., ARNTZEN, C. J. (1996) Expression of Nor-
walk virus capsid protein in transgenic tobacco
and potato and its oral immunogenicity in mice,
Proc. Natl. Acad. Sci. USA **93**, 5335–5340.

MAYER, K., SCHULLER, C., WAMBUTT, R., MURPHY,
G., VOLCKAERT, G. et al. (1999), Sequence and
analysis of chromosome 4 of the plant *Arabidopsis thaliana*, *Nature* **402**, 769–777.

MAZUR, B., KREBBERS, E., TINGEY, S. (1999), Gene
discovery and product development for grain
quality traits, *Science* **285**, 372–375.

MCCOUCH, S. R., CHEN, X., PANAUD, O., TEMNYKH, S.,
XU, Y. et al. (1997), Microsatellite marker development, mapping and applications in rice genetics and breeding, *Plant Mol. Biol.* **35**, 89–99.

MEIN, C. A., BARRATT, B. J., DUNN, M. G., SIEGMUND,
T., SMITH, A. N. et al. (2000), Evaluation of single
nucleotide polymorphism typing with invader on
PCR amplicons and its automation, *Genome Res.*
10, 330–343.

MELCHERS, L. S., STUIVER, M. H. (2000), Novel genes

for disease-resistance breeding, *Curr. Opin. Plant
Biol.* **3**, 147–152.

MIAO, Z. H., LAM, E. (1995), Targeted disruption of
the TGA3 locus in Arabidopsis thaliana. *Plant J.*
7, 359–365.

MOORE, G., DEVOS, K. M., WANG, Z., GALE, M. D.
(1995), Cereal genome evolution. Grasses, line up
and form a circle, *Curr. Biol.* **5**, 737–739.

MOULE, C., EDWARDS, K. J., TRICK, M. (2000) Development of *Brassica* microsatellite markers, *Plant
and Animal Genome* VIII (Abstract 175). New
York: Scherago International.

MUDGE, J., CREGAN, P. P., KENWORTHY, J. P., KEN-
WORTHY, W. J., ORF, J. H., YOUNG, N. D. (1997), Two
microsatellite markers that flank the major soybean cyst nematode resistance locus, *Crop Sci.* **37**,
1611–1615.

MYERS, E. W., SUTTON, G. G., DELCHER, A. L., DEW, I.
M., FASULO, D. P. et al. (2000), A whole-genome
assembly of *Drosophila*, *Science* **287**, 2196–2204.

NELSON, K. E., PAULSEN, I. T., HEIDELBERG, J. F.,
FRASER, C. M. (2000), Status of genome projects
for nonpathogenic bacteria and archaea, *Nature
Biotechnol.* **18**, 1049–1054.

OFFRINGA, R., DE GROOT, M. J., HAAGSMAN, H. J.,
DOES, M. P., VAN DEN ELZEN, P. J., HOOYKAAS, P. J.
(1990), Extrachromosomal homologous recombination and gene targeting in plant cells after *Agrobacterium*-mediated transformation, *EMBO J.*
9, 3077–3084.

OFFRINGA, R., FRANKE-VAN DIJK, M. E., DE GROOT,
M. J., VAN DEN ELZEN, P. J., HOOYKAAS, P. J.
(1993), Nonreciprocal homologous recombination between *Agrobacterium* transferred DNA
and a plant chromosomal locus, *Proc. Natl. Acad.
Sci. USA* **90**, 7346–7350.

OSBORN, T. C., KOLE, C., PARKIN, I. A., SHARPE, A. G.,
KUIPER, M. et al. (1997), Comparison of flowering
time genes in *Brassica rapa*, *B. napus* and *Arabidopsis thaliana*, *Genetics* **146**, 1123–1129.

PANDEY, A., MANN, M. (2000), Proteomics to study
genes and genomes, *Nature* **405**, 837–846.

PARINOV, S., SEVUGAN, M., YE, D., YANG, W.-C., KU-
MARAN, M., SUNDARESAN, V. (1999), Analysis of
flanking sequences from dissociation insertion
lines: A database for reverse genetics in *Arabidopsis*, *Plant Cell* **11**, 2263–2270.

PARKIN, I. A. P., SHARPE, A. G., KEITH, D. J., LYDIATE,
D. J. (1995), Identification of the A and C genomes of amphidiploid *Brassica napus* (oilseed
rape), *Genome* **38**, 1122–1131.

PAULSEN, G. M. (1994), High temperature responses
of crop plants, in: *Physiology and Determination
of Crop Yield* (BOOTE, K. J., BENNETT, J. M., SIN-
CLAIR, T. R., PAULSEN, G. M., Eds.), pp 365–389.
Madison, WI: American Society of Agronomy.

PENG, J., RICHARDS, D. E., HARTLEY, N. M., MURPHY,
G. P., DEVOS, K. M. et al. (1999), "Green revolu-

tion" genes encode mutant gibberellin response modulators, *Nature* **400**, 256–261.

PENNISI, E. (1998), A bonanza for plant genomics, *Science* **282**, 652–654.

PIH, K. T., JANG, H. J., KANG, S. G., PIAO, H. L., HWANG, I. (1997), Isolation of molecular markers for salt stress responses in *Arabidopsis thaliana*, *Mol. Cells* **7**, 567–571.

POWELL, W., MACHRAY, G. C., PROVAN, J. (1996), Polymorphism revealed by simple sequence repeats, *Trend Plant Sci.* **1**, 215–222.

PUCHTA, H., DUJON, B., HOHN, B. (1996), Two different but related mechanisms are used in plants for the repair of genomic double-strand breaks by homologous recombination, *Proc. Natl. Acad. Sci. USA* **93**, 5055–5060.

PUGSLEY, A. T. (1983), The impact of plant physiology on Australian wheat breeding, *Euphytica* **32**, 743–748.

PRINCE, J. P., POCHARD, E., TANKSLEY, S. D. (1993) Construction of a molecular linkage map of pepper and a comparison of synteny with tomato, *Genome* **36**, 404–417.

RABINOWICZ, P. D., SCHUTZ, K., DEDHIA, N., YORDAN, C., PARNELL, L. D. et al. (1999), Differential methylation of genes and retrotransposons facilitates shotgun sequencing of the maize genome, *Nature Genet.* **23**, 305–308.

RAFALSKI, J. A., TINGEY, S. V. (1993), Genetic diagnostics in plant breeding: RAPDs, microsatellites and machines, *Trends Genet.* **9**, 275–280.

REISS, B., SCHUBERT, I., KOPCHEN, K., WENDELER, E., SCHELL, J., PUCHTA, H. (2000), RecA stimulates sister chromatid exchange and the fidelity of double-strand break repair, but not gene targeting, in plants transformed by *Agrobacterium*, *Proc. Natl. Acad. Sci. USA* **97**, 3358–3363.

RICHTER, L. J., THANAVALA, Y., ARNTZEN, C. J., MASON, H. S. (2000), Production of hepatitis B surface antigen in transgenic plants for oral immunization, *Nature Biotechnol.* **18**, 1167–1171.

ROBERTS, J. K., PRYOR, A. (1995) Isolation of a flax (*Linum usitatissimum*) gene induced during susceptible infection by flax rust (*Melampsora lini*), *Plant J.* **8**, 1–8.

RODER, M. S., KORZUN, V., WENDEHAKE, K., PLASCHKE, J., TIXIER, M. H. et al. (1998), A microsatellite map of wheat, *Genetics* **149**, 2007–2023.

ROMMENS, C. M., KISHORE, G. M. (2000), Exploiting the full potential of disease-resistance genes for agricultural use, *Curr. Opin. Biotechnol.* **11** (2), 120–125.

ROTHSTEIN, R. (1991), Targeting, disruption, replacement, and allele rescue: integrative DNA transformation in yeast, *Methods Enzymol.* **194**, 281–301.

SASAKI, T., BURR, B. (2000), International Rice Genome Sequencing Project: the effort to complete-

ly sequence the rice genome, *Curr. Opin. Plant Biol.* **3**, 138–141.

SCARTH, R., MCVETTY, P. B. E., RIMMER, S. R. (1995a), Apollo-low linolenic summer rape, *Can. J. Plant Sci.* **75**, 203–204.

SCARTH, R., MCVETTY, P. B. E., RIMMER, S. R. (1995b), Mercury-high erucic low glucosinalate summer rape, *Can. J. Plant Sci.* **75**, 205–206.

SCHAEFER, D. G., ZRYD, J. P. (1997), Efficient gene targeting in the moss *Physcomitrella patens*, *Plant J.* **11**, 1195–1206.

SCHENK, P. M., KAZAN, K., WILSON, I., ANDERSON, J. P., RICHMOND, T. et al. (2000), Coordinated plant defense responses in Arabidopsis revealed by microarray analysis, *Proc. Natl. Acad. Sci. USA* **97**, 11655–11660.

SCHMIDT, R. (2000), Synteny: recent advances and future prospects, *Curr. Opin. Plant Biol.* **3**, 97–102.

SCHULER, T. H., POPPY, G. M., KERRY, B. R., DENHOLM, L. (1998), Insect-resistant transgenic plants, *TIBTECH* **16**, 168–175.

SEE, D., KANAZIN, V., TALBERT, H., BLAKE, T. (2000), Electrophoretic detection of single-nucleotide polymorphisms, *Biotechniques* **28**, 710–714.

SHAH, D. M., ROMMENS, C. M. T., BEACHY, R. N. (1995), Resistance to disease and insects in transgenic plants: progress and applications to agriculture, *TIBTECH* **13**, 362–368.

SHAROPOVA, N., MCMULLEN, M. D., SCHULTZ, L. M., SCHROEDER, S. G., HOUCHINS, E. C. et al. (2000), Microsatellites in maize-development and mapping, *Plant and Animal Genome* VIII (Abstract 174). New York: Scherago International.

SHARPE, A., LYDIATE, D. (1998), The development of microsatellite markers for the improved breeding of *Brassica* crops, *Plant and Animal Genome* VI (Abstract 132). New York: Scherago International.

SHEEHY, R. E., KRAMER, M., HIATT, W. R. (1988), Reduction of polygalacturonase activity in tomato fruit by antisense RNA, *Proc. Natl. Acad. Sci. USA* **85**, 8805–8809.

SHOEMAKER, R. C., POLZIN, K., LABATE, J., SPECHT, J., BRUMMER, E. C. et al. (1996), Genome duplication in soybean (*Glycine* subgenus *soja*), *Genetics* **144**, 329–338.

SILLITO, D., PARKIN, I. A., MAYERHOFER, R., LYDIATE, D. J., GOOD, A. G. (2000), *Arabidopsis thaliana*: a source of candidate disease-resistance genes for *Brassica napus*, *Genome* **43**, 452–460.

SKOLNICK, J., FETROW, J. S. (2000), From genes to protein structure and function: novel applications of computational approaches in the genomic era, *Trends Biotechnol.* **18**, 34–39.

SMITH, E., SHI, L., DRUMMOND, P., RODRIGUEZ, L., HAMILTON, R. et al. (2000), Development and characterization of expressed sequence tags for the turkey (*Meleagris gallopavo*) genome and comparative sequence analysis with other birds, *Anim. Genet.* **31**, 62–67.

SMITH, K. A., JACKSON D. R., PEPPER, T. J. (2001a), Nutrient losses by surface run-off following the application of organic manures to arable land. 1. Nitrogen, *Environ. Pollut.* **112**, 41–51.

SMITH, K. A., JACKSON D. R., PEPPER, T. J. (2001b), Nutrient losses by surface run-off following the application of organic manures to arable land. 2. Phosphorus, *Environ. Pollut.* **112**, 53–60.

SPURR, N., DARVASI, A., TERRETT, J., JAZWINSKA, L. (1999), New technologies and DNA resources for high throughput biology, *Br. Med. Bull.* **55**, 309–324.

STEFANSSON, B. (1976), Rapeseed, *Agrologist.* **5** (2), 21–23.

SUMBERG, J. E., MURPHY, R. P., LOWE, C. C. (1983), Selection for fibre and protein concentration in a diverse Alfalfa population, *Crop Sci.* **23**, 11–14.

SOMMER, A. (1988), New imperatives for an old vitamin (A), *J. Nutr.* **119** (1), 96–100.

TACKET, C. O., MASON, H. S., LOSONSKY, G., CLEMENTS, J. D., LEVINE, M. M., ARNTZEN, C. J. (1998), Immunogenicity in humans of a recombinant bacterial antigen delivered in transgenic potato, *Nature Med.* **4**, 607–609.

TANG, K., FU, D. J., JULIEN D., BRAUN, A., CANTOR, C. R., KOSTER, H. (1999), Chip-based genotyping by mass spectrometry, *Proc. Natl. Acad. Sci. USA* **96**, 10016–10020.

TANKSLEY, S. D., GANAL, M. W., PRINCE, J. P., DE VICENTE, M. C., BONIERBALE, M. W. et al. (1992), High density molecular linkage maps of tomato and potato genomes, *Genetics* **132**, 1141–1160.

TANKSLEY, S. D. (1993), Mapping polygenes, *Annu. Rev. Genet.* **27**, 205–233.

THOMAS, K. R., CAPECCHI, M. R. (1987), Site-directed mutagenesis by gene targeting in mouse embryo-derived stem cells, *Cell* **51**, 503–512.

THOMASHOW, M. F. (1994), *Aradidopsis thaliana* as a model for studying mechanisms of plant cold tolerance, in: *Arabidopsis.* (MEYEROWITZ, E. M., SOMERVILLE, C. R., Eds.). Cold Spring Harbour, NY: Laboratory Press.

TUITE, J., FOSTER, G. H. (1979), Control of storage disease. *Ann, Rev. Phytopathol.* **17**, 343–366.

U.N. (1999), The World at Six Billion. Population division, Department of economic and social affairs, United Nations Secretariat.

VELCULESCU, V. E., ZHANG, L., VOGELSTEIN, B., KINZLER, K. W. (1995), Serial analysis of gene expression, *Science* **270**, 484–487.

VIERLING, J., FAGHIHI, J., FERRIS, V. R., FERRIS, J. M. (1996), Association of RFLP markers with loci conferring broad-based resistance to the soybean

cyst nematode (*Heterodera glycines*), *Theor. Appl. Genet.* **92**, 83–86.

WALMSLEY, A. M., ARNTZEN, C. J. (2000), Plants for delivery of edible vaccines, *Curr. Opin. Biotechnol.* **11**, 126–129.

WHITEHEAD, R., WRIGHT, H. C. (1989), The incidence of weeds in winter cereals in Great Britain, in: *Brighton Crop Prot. Conf. Weeds*, Surrey. pp. 107–112.

WILSON, T. M. A., CRUZ, S. S., CHAPMAN, S. (1996), Viruses of plants in the service of man, from crop protection to biotechnology, in: *Crop Productivity and Sustainability-Shaping the Future* (CHOPRA, V. L., SINGH, R. B., VARMA, A., Eds.), pp. 245–257. *Proc. 2nd Int. Crop Science Congress*. New Delhi: Oxford and IBH Publishing.

WINZELER, E. A., SHOEMAKER, D. D., ASTROMOFF, A. LIANG, H., ANDERSON, K. et al. (1999), Functional characterization of the *S. cerevisiae* genome by gene deletion and parallel analysis, *Science* **285**, 901–906.

WOO, S. S., JIANG, J., GILL, B. S., PATERSON, A. H., WING, R. A. (1994), Construction and characterization of a bacterial artificial chromosome library of Sorgham bicolour, *Nucleic Acids Res.* **7**, 4922–4931.

YE, X., AL-BABILI, S., KLOTI, A., ZHANG, J., LUCCA, P. et al. (2000), Engineering the provitamin A (beta-carotene) biosynthetic pathway into (carotenoid-free) rice endosperm, *Science* **287**, 303–305.

YOON, K., COLE-STRAUSS, A., KMIEC, E. B. (1996), Targeted gene correction of episomal DNA in mammalian cells mediated by a chimeric RNA.DNA oligonucleotide, *Proc. Natl. Acad. Sci. USA* **93**, 2071–2076.

YUAN, Q., LIANG, F., HSIAO, J., ZISMANN, V., BENITO, M. I. et al. (2000), Anchoring of rice BAC clones to the rice genetic map in silico, *Nucleic Acids Res.* **28**, 3636–3641.

ZHU, T., METTENBURG, K., PETERSON, D. J., TAGLIANI, L., BASZCZYNSKI, C. L. (2000), Engineering herbicide-resistant maize using chimeric RNA/DNA oligonucleotides, *Nature Biotechnol.* **18**, 555–558.

ZHU, T., PETERSON, D. J., TAGLIANI, L., ST CLAIR, G., BASZCZYNSKI, C. L., BOWEN, B. (1999), Targeted manipulation of maize genes *in vivo* using chimeric RNA/DNA oligonucleotides, *Proc. Natl. Acad. Sci. USA* **96**, 8768–8773.

ZIVY, M., DE VIENNE, D. (2000), Proteomics: a link between genomics, genetics and physiology, *Plant Mol. Biol.* **44**, 575–580.

ZUPAN, J., MUTH, T. R., DRAPER, O., ZAMBRYSKI, P. (2000), The transfer of DNA from *Agrobacterium tumefaciens* into plants: a feast of fundamental insights, *Plant J.* **23**, 11–28.

DNA Technologies

7 Genomic Mapping and Positional Cloning, with Emphasis on Plant Science

APICHART VANAVICHIT

SOMVONG TRAGOONRUNG

THEERAYUT TOOJINDA

Nakorn Pathom, Thailand

1 Introduction

Most genetic traits important to agriculture or human diseases are manifested as observable phenotypes. In many instances, the complexity of the phenotype–genotype interaction and the general lack of clearly identifyable gene products renders the direct molecular cloning approach ineffective, thus additional strategies like genome mapping are required to identify the gene(s) in question. Genome mapping requires no prior knowledge of the gene function, but utilizes statistical methods to identify the most likely gene location. To completely characterize genes of interest, the initially mapped region of a gene location will have to be narrowed down to a size that is suitable for cloning and sequencing. Strategies for the gene identification within the critical region have to be applied after the sequencing of a potentially large clone or set of clones that contains these gene(s). Tremendous success of positional cloning has been shown for cloning many genes responsible for human diseases, including cystic fibrosis and muscular dystrophy as well as plant disease resistance genes (MARTIN et al., 1993; BENT et al., 1994; GRANT et al., 1995; SONG et al., 1995).

2 Genome Mapping

Genome mapping is used to identify the genetic location of mutants, or qualitative and quantitative trait loci (QTL). Linking the traits to markers using genetic and family information of a recombinant population can identify the gene location. Through mapping, we can answer how many loci are involved, where the loci are positioned in the genome, and what contribution each allele may have to the trait. To determine relationships between marker loci and the target trait, mapping requires the following:

- segregating populations (genetic stocks),
- marker data set(s), and
- a phenotypic data set.

2.1 Mapping Populations

The crucial requirement for the success of genome mapping in plant science is the choice of a suitable segregating population. Doubled haploids (DH), recombinant inbred lines (RI), backcrosses (BC), and F_2 populations are the primary types commonly used for plant genome mapping. Each of these populations has unique strengths and weaknesses. In plant science, the populations are developed from biparental crosses or backcrosses between parents that are genetically different. The resulting F_2 population is developed by selfing the F_1 individual. The BC can be developed further by crossing the F_1 with one of the two parents used in the initial cross. The DH population is usually derived by doubling chromosomes of haploid cells obtained from the F_1 generation. Various methods have been used to produce doubled haploids in plants, such as ovary culture, anther culture, microspore culture, or chromosome elimination. The RI populations are produced by random selfing or sib-mating of individuals of the F_2 or BC_1 population until they become virtually homozygous lines. Most RI populations have been developed by a single-seed descent method.

The population type and the number of progenies determine the resolution of the linkage map. The map in turn affects the precision and accuracy of the number, location, and effect of gene/QTLs (quantitative trait loci), which can be detected. Because of their high heterozygosity, F_2 and backcross populations, though easy to develop, cannot indefinitely supply resources for DNA studies and multiple replicated experiments are not possible. The advantage of F_2 populations over other population types is the large amount of genetic information per progeny, when codominant markers are used. The DH and RI populations, on the other hand, are renewable and more permanent resources, multiple-replicated experiments for QTL analysis are feasible. Because RI populations have undergone additional cycles of recombination during selfing, while F_1-derived DH populations have undergone meiosis only once, RI populations are expected to support the generation of higher resolution maps than the DH populations. The DH, RI, and backcross populations, however, provide half of the genetic information per progeny as compared to the F_2 populations.

2.2 Molecular Markers: The Key Mapping Reagents

There are two classes of markers, which can be used to detect genetic variation, phenotypic and molecular markers. Phenotypic markers are expression-dependent. Morphological markers are phenotypic markers frequently used for genetic mapping. Phenotypic markers are not an ideal type of markers for genome mapping because of several drawbacks. First, they are limited in number, distribution, and the degree to which the loci can be used to detect polymorphisms. Secondly, other genes (epistasis and pleiotropism) frequently modify the expression of phenotypic markers. Sequence polymorphisms are especially useful for the development of molecular markers, because they are generally stable, numerous and informative, and the detection is more reproducible. Desirable molecular markers for genome mapping must have additional useful properties such as

- highly abundant and evenly distributed throughout the genome,
- highly polymorphic information content (PIC),
- high multiplex ratio,
- codominant and
- neutral.

In addition, methods developed for the detection of these markers must have:

- low start-up costs,
- robustness and high reproducibility,
- a guarantee for the transfer of the detection protocol among laboratories.

Basic causes of polymorphisms in stretches of DNA are length polymorphisms due to large insertions, deletions, or rearrangements; single nucleotide polymorphisms and effects caused by sequence repeats, especially during meiosis. A large number of molecular markers is now available, each of them with different advantages and disadvantages. Among the molecular markers, five commonly used types are RFLP (restriction fragment length polymorphism), RAPD (random amplified polymorphic DNA), AFLP (amplified fragment length polymorphism), SSR (simple sequence repeat) and SSCP (single strand conformational polymorphism) markers.

RFLPs. RFLP, the first molecular markers (BOTSTEIN et al., 1980), detects polymorphisms based on differences in restriction fragment lengths. These differences are caused by mutations or insertion-deletions, which create or delete restriction endonuclease recognition sites. RFLP assays are performed by hybridizing a chemically or radioactively labeled DNA probe to a Southern blot. RFLPs are usually specific to a single clone/restriction enzyme combination and most of them are codominant, highly locus specific, and often multiallellic. However, RFLP analyses are tedious and inefficient due to the low multiplex ratio, low genotyping throughput, high labor intensity, and the requirement for large amounts of high-quality template DNA.

RAPDs. RAPD is a PCR-based technique. A single, arbitrary oligonucleotide primer (typically 10-mers) is used to amplify genomic fragments flanked by two complementary primer-binding sites in an inverted orientation (WILLIAMS et al., 1990). At low stringency, numbers of different PCR products are generated. Polymorphisms result from either base changes that alter the primer binding site, rearrangements or insertion-deletions at or between oligonucleotide primer binding sites in the genome. The primary advantages of RAPDs are the simplicity of the experimental setup, the low overhead and experiment costs as well as the high multiplex ratio. Polymorphic DNA can be isolated and cloned as probes for hybridization or sequencing (LAROCHE et al., 2000). However, the initial cycles of amplification probably involve extensive mismatch, and a rigorous standardization of the reaction conditions is required for reliable, repeatable results. Furthermore, RAPDs do not have defined locus identity, thus it can be difficult to relate RAPD loci between different experimental populations of the same species. RAPDs can be used to identify dominant mutations.

AFLP. AFLP assays are based on a combination of restriction digestion and PCR amplification. AFLPs caused by mutations or insertion-deletions in a restriction site that create or abolish restriction endonuclease recognition sites. They are visually dominant, biallelic, and high throughput. A principal drawback of AFLPs is their time-consuming assay, because the method is based on DNA sequencing protocols. The issue of locus identity needs to be established on a case-by-case basis. An additional drawback of AFLPs is that they are reported to cause map expansion and often densely cluster in centromeric regions of the chromosome in species with large genomes. Linkage map expansion is usually attributed to poor data quality. However, AFLP technology can be used for genomic analysis in any organism without the need for formal marker development (RANAMUKHAARACHCHI et al., 2000; HUYS et al., 2000; MACKILL et al., 1996). Reproducibility and high multiplex ratio make AFLP one of today's standard methods for the characterization of markers for genome mapping.

SSRs. SSRs or microsatellites are tandemly repeated mono-, di-, tri-, tetra-, penta-, and hexa-nucleotide motifs. The SSR assay is based on the PCR amplification of tandem repeats using unique flanking DNA sequences as oligonucleotide primers. The polymorphism among individuals is due to the variation in the number of repetitive units. SSRs are codominant and often multiallellic, which helps to generate the unambiguous identification of alleles. They are also highly abundant and randomly dispersed throughout most genomes. SSRs provide an excellent framework for markers with locus identity. They can be multiplexed to achieve high throughput. A drawback of the SSR technology is that the development of SSRs is labor intensive and costly. Although, SSRs are specific for the species that they were developed for, the method has now replaced traditional RFLPs for the generation of many linkage maps, largely because they are technically simple and cheap, consume minute amounts of DNA, and they can be delivered at a rapid turn-around time and high PIC (polymorphic information content) (PANAUD et al., 1996; MCCOUCH et al., 1997).

SSCP. The SSCP assay is based on changes in the conformation of single-stranded DNA of a specific sequence containing mutations or insertion-deletions under non-denaturing conditions (ORITA et al., 1989; HAYASHI, 1991, 1992; FUJITA and SILVER, 1994). The conformation of the folded DNA molecule is dependent on intra-molecular interactions. SSCP is one of the most sensitive methods for detecting changes in nucleotide sequences of an entire fragment much larger than 1,000 bp. SSCP assays are usually performed using heat-denatured DNA on non-denaturing sequencing gels. The strength of this method for genome mapping is its simplicity, multiallellicity, codominance, and locus identity (FUKUOKA et al., 1994; BODENES et al., 1996; URQUHART and WILLIAMS, 1996). However, the development of markers is labor intensive and costly. To date, SSCPs also have not been automated.

2.3 Construction of a Linkage Map

Difference in genetic information in progenitors can be visualized by using markers based on morphological, protein, or DNA data. A large number of potential DNA markers suitable for developing high-density linkage maps have been established for various organisms (KURATA et al., 1994; CAUSSE et al., 1994; HEUN et al., 1991). To construct the linkage map, polymorphic markers are scored on the random segregating populations. The distances and orders of those markers are determined based on the frequency of genetic recombinations occurring in the population. Because two linked markers tend to be inherited together from generation to generation, the distance between markers can be estimated from the observed fraction of recombinations. Map construction basically involves five steps.

First, a single-locus analysis, which is a statistical approach to identify the data quality using a single-locus genetic model. The χ^2 method is widely used to test the marker segregation according to its expected ratio of the randomly segregating population. For example, in a BC progeny, each marker locus will segregate with a 1:1 ratio for Aa and aa, a 1:1 ratio for AA and aa in DH and RIL, and 1:2:1 ratio for AA, Aa and aa in the F_2 generation, respectively. A significant departure (segregation ratio distortion) from the expected segregation ratio may be a sign of a wrong genetic model, low data quality or non-random sampling (VOGL and XU, 2000). If the segregation ratio of each marker does not deviate from the expected ratio, the analysis can proceed to the next step (LIU, 1998).

As a second step in the map construction process, a two-locus analysis, is used to test an association or non-independence among the marker alleles located on the same chromosome. Linkage is usually established by testing the independence between the two loci in segregating populations. A goodness of fit or a log likelihood ratio has been used to test for independence of two loci. Recombination fraction, lod score (base-10 log likelihood ratio) and significant P-values are used as criteria to infer whether each pair of loci belongs to same linkage group (LIU, 1998; LARSEN, 1979).

During the third step, a three-locus analysis is used to determine the ordering of the loci or the linear arrangement of markers in a linkage group. There are two methods for finding the best locus ordering among the potential orders in each linkage group, double crossing-over and two-locus recombination fraction. The order of the three loci can be determined by finding the least occurring double recombinants. Once the double recombinant classes are identified, the order of the loci can be determined. The two-locus recombination fraction approach is used to determine the locus order by comparing the likelihood of the three possible orders. The order associated with the highest likelihood values is the most likely order. The ordering of more than three loci can be determined using the maximum likelihood approach (LATHROP et al., 1986; LIU, 1998; STRINGHAM et al., 1999).

During the fourth step, a map distance is calculated. In the process of the linkage map construction, the recombination fractions are mathematically calculated from data obtained by mapping of the population (HUHN, 2000). Based on the data, mapping methods are subsequently used to convert the recombination fraction into the map distance. Various kinds of mapping functions have been proposed. Mapping functions work only for specific conditions. There is no universal mapping func-

tion. The differences among the commonly used mapping functions are due to the assumptions about distribution of crossovers on the genome, crossover interference, and length of the chromosome segment. Genes or genetic markers are organized in a linear fashion on a map, thus their relative positions on the map can be quantified in an additive fashion. If the expected number of crossovers is one in a genome segment, then the map distance between two genes or genetic markers flanking the segment is defined as 1 Morgan (M) or 100 centi-Morgans (cM). The commonly used mapping methods are Morgan's, Haldane's, and Kosambi's. Morgan's mapping method can be appropriately applied for a small genome, which very likely has a small expected number of multiple-crossovers, as compared to a large genome (MORGAN, 1928). As the size of genome increases, the expected number of multiple-crossovers becomes larger and the map distance has to be adjusted for multiple-crossovers through Haldane's mapping method by assuming that crossovers occur uniformly (randomly) along the length of the chromosome (in the absence of crossover interference) (HALDANE, 1919). Experimental evidence has been found that crossover interference exits and crossovers occur non-randomly in larger genomes. Therefore, Kosambi's mapping method was invented to take into account the crossover interference. The rationale for Kosambi's method is that the crossover interference depends on the size of a genome segment. The interference is absent when a segment is sufficiently large. The interference increases as the segment size decreases (KOSAMBI, 1944).

The final step in the mapping procedure is the linkage map construction. During the last several decades, many approaches have been developed for building a multi-locus model. The least square method was implemented in a computer package "Joinmap" for genomic mapping. The EM algorithm uses a set of procedures for obtaining a maximum likelihood estimate. The Lander–Green algorithm greatly reduces the computational complexity of obtaining multi-locus recombination fractions from traditional approaches. This method has been widely implemented in computer packages such as MAP-MAKER (LANDER et al.,

1987), GMENDEL (LIU and KNAPP, 1990) and PGRI. The joint maximum likelihood is a method to estimate recombination fractions and crossover interference by simultaneously using a multiplicative model. This algorithm has been implemented in the computer package "GLIM" which can be used to apply generalized linear regression. The simulation approach involves the comparison of multilocus likelihoods of the data using different mapping functions. This approach can be used to identify a mapping method, which fits the data well. It was implemented in the computer program LINKAGE (LATHROP and LALOUEL, 1984).

Most linkage maps are based on population sizes ranging from 100 to 200 individuals. However, the study of larger progenies can help to enhance the map resolution and the estimation of map distances (GESSLER and XU, 1999).

3 Positional Cloning

During positional cloning, information about map positions of genes is used to conduct chromosome walking. In the initial step, flanking markers which are tightly linked to the target gene must be identified. These tightly linked markers are then used as initial points for the development of the high-resolution map around the target region, using highly polymorphic content markers. When the flanking markers are narrowed down, the next step is to construct a physical map around the target region. The candidate region can subsequently be narrowed down further, sometimes to a region being covered by a single large insert clone. After the genes were characterized by sequencing, the functional analysis by complementation in transformed plants is the most important piece of evidence for the successful identification and cloning (GIBSON and SOMERVILLE, 1994).

3.1 Successful Positional Cloning

To date more than 100 inherited disease genes in humans have been isolated (*http://*

genome.nhgri.nih.gov/clone/). Significant progress in positional cloning in plants was achieved, however, due in most part to the development of high density maps and large insert libraries in major crops such as rice (KURATA et al., 1994), *Arabidopsis* (KORNNEEF, 1994), tomato (TANKSLEY et al., 1992), and barley (SHERMAN et al., 1995). Two classic examples for gene identification and location were the cloning of the *Pto* gene in tomato (MARTIN et al., 1993) and the *Xa21* gene in rice (SONG et al., 1995). These genes are responsible for resistance against bacterial pathogens. *Arabidopsis* has become a model plant for map-based cloning, due to the simplicity of the identification of mutations, comprehensive genetic and physical maps, and the ease of gene transformation. Examples for disease resistance genes from *Arabidopsis* that have been cloned include the *RPM1* against *P. syringae* (GRANT et al., 1995), the *RPS2* against a different strain of *P. syringae* (BENT et al., 1994), *RPP13* against downy mildew fungus (BITTNER-EDDY et al., 1999), *Mlo*, against the broad spectrum fungal attack in barley (BUSCHGES et al., 1997), and *I2* against fusarium wilt in tomato (SIMONS et al., 1998). Another set of disease resistance genes will soon be identified, such as *Tm2a* against *TMV* in tomato (PILLEN et al., 1996), *Asc* against alternaria stem cancer in tomato (MESBAH et al., 1999), *Pi-b* against rice blast (MONNA et al., 1997), *Pi-ta2* also against rice blast (NAKAMURA et al., 1997), and *Rar-1* against powdery mildew in barley (LAHAYE et al., 1998). DIETRICH et al. (1997) have also reported positional cloning of a negative regulator for hypersensitive response in *Arabidopsis*. Successful positional cloning has been achieved for the gene responsible for the resistance to beet cyst nematode, *Hs1^{pro-1}* (CAI et al., 1997). The *Br* gene responsible for resistance to bruchid, grain weevil that destroys mungbean seeds, was cloned by KAGA and ISHIMOTO (1998).

Cloning attempts of plant genes responsible for stress resistance to abiotic factors such as cold, drought, flooding, etc. have faced more challenges due to their complex interaction with several genetic and non-genetic factors. This is particularly true for genes related to phytohormone activity. Many *Arabidopsis* mutants with defects in signal transduction pathways were used for map-based cloning. Successful gene isolation by map-based cloning was reported for genes responsible for insensitivity to abscisic acid, *ABI1* (LEUNG et al., 1994), *ABI2* (LEUNG et al., 1997), *ABI3* (GIRAUDAT et al., 1992), and *ABI4* (FINKELSTEIN et al., 1998), elongation-regulating hormone, auxin, *AXR1 & 2* (LEYSER et al., 1993), senescence-promoting ethylene, *ETR* (CHANG et al., 1993), *EIN2* and *CTR1* (KIEBER et al., 1993), gibberellic acid, *GAI* (PENG et al., 1997) and *GA1* (SUN et al., 1992). Positional cloning has been successfully used to isolate genes involved in developments that are important to agriculture such as, *d1*, dwarfism in rice (ASHIKARI et al., 1999) and a *MADBOX* gene controlling fruit dehiscence (*jointless*) in tomato (MAO et al., 2000).

3.2 Defining the Critical Region

The identification and isolation of genes by positional cloning strategies can be conceptually divided into a series of steps. The starting point is the collection of families or germplasms with defects or special traits of interest. After the critical region is identified by flanking markers, usually spanning tens of million base pairs, more refinement is achieved by using additional genetic markers that map in the vicinity. The extent to which the map region can be determined by genetic means depends on the size of the population as well as the number and informativeness of the genetic markers in the region. In the ideal case, the defective genes were caused by cytological abnormalities, which immediately establishes the critical region containing the defective genes. Of critical importance for the phenotyping screening limits is the number of progenies that can be evaluated in the study.

Genetic mapping can be narrowed down to 1–3 cM in optimal cases of human diseases or as tight as 0 cM in plants (PILLEN et al., 1996; ASHIKARI et al., 1999). The corresponding physical size, however, may vary widely due to genome size as well as regional and sex-specific differences in recombination rates. In humans, 1 cM is typically corresponding to roughly 1–3 Mb. In plants, the physical to genetic distance per 1 cM varies with genome

size (e.g., 100 kb in *Arabidopsis*, 250 kb in rice, 1,000 kb in maize). In recombination hot spots, the physical to genetic distance may be particularly small and such regions have been frequently associated with gene richness. Choices of strategies for positional cloning, as illustrated in Fig. 1, depend on the tools available for the particular organisms.

3.3 Refining the Critical Region: Genetic Approaches

Flanking markers identified in the preliminary map normally are too far to reach the target QTL by most large-DNA insert cloning technologies. The generation of more polymorphic markers within a specific region can be achieved by genetic or physical means. Recombination events near the target gene at the resolution of 0.1 cM can help facilitating gene identification.

Positional cloning is laborious and costly in organisms for which high resolution maps are not available. An ideal method would be to directly isolate region-specific markers at high density to identify overlapping genomic clones covering the genes of interest without generating a genetic map or performing a chromosome walking procedure. Chromosome landing and pooled progeny techniques are based on the identity of their chromosomal region of interest (TANKSLEY et al., 1995). In Bulk Segregrant Analysis (BSA), pooled progenies are based on their phenotypic identity. BSA has been successfully used to isolate markers surrounding the major loci (MICHELMORE et al., 1991). BSA has the same advantage as chromosome landing, as there is no requirement for the construction of a high density map. Combined with high multiplex ratio markers such as AFLP, SSR and RAPD, BSA, and chromosome landing can lower the number of DNA samples necessary to score thousands of markers (THOMAS et al., 1995). Genetically directed representational difference analysis (GDRDA) uses phenotypic pooling, combined with a subtractive method, to specifically isolate markers from a locus of interest (LISITSYN, 1994). In many cases, this method was insufficient to locate specific markers in a specific region. RFLP subtraction has been

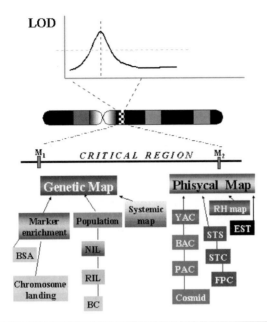

Fig. 1. Genetic and physical approaches to refine the critical region.

used to isolate large numbers of randomly located RFLPs spanning the mouse genome (ROSEMBERG et al., 1994) and *Volvox carteri* (CORRETTE-BENNETT et al., 1998). Three RFLP markers linked to *recA* at 0 cM were isolated by CORRETTE-BENNETT et al. (1998), but it is not clear if the methodology can be applied in plant or animal genomes where repetitive sequences and retrotransposons are abundant.

3.4 Refining the Critical Region: Physical Approaches

The decision to start the physical mapping process depends on the mapping tools available for a particular organism. In humans, typically a 1–3 Mb interval can be reached using YAC physical mapping. In other cases, where the physical map is not well refined, the critical region must be narrowed down to as small a distance as possible, using the wealth of polymorphic markers in the particular region. Typically, mapping the closest genetic markers is used to initiate clone isolation (e.g., YAC, BAC, or PAC). If necessary, new markers can be generated from the ends of the clone which can then be used to screen the next adjacent overlapping clones, a strategy called "chromosome walking". Additional markers such as STS and EST, which were identified to be located in the critical region, can be used to assist clone isolation. In an ideal case, such as illustrated in Fig. 2, the entire critical genomic interval between the flanking markers can be isolated in YAC or BAC clones. Fine mapping with such specific markers can significantly narrow down the critical region, and gene isolation can be done with less effort than physical mapping.

3.5 Cloning Large Genomic Inserts

Cloning large genomic fragments facilitates access to genes known by map position, physi-

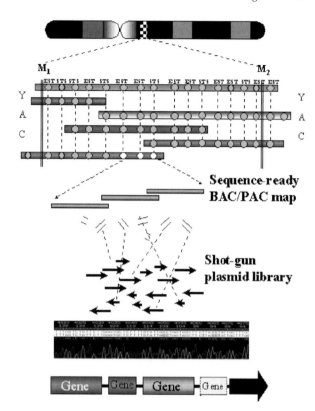

Fig. 2. Refinement of the critical region by physical map and sequencing.

cal mapping, and large-scale genome sequencing. The two most powerful cloning vectors are yeast artificial chromosome (YAC) and bacterial artificial chromosomes (BAC). The development of YAC allows cloning DNA inserts of more than 500 kb (BURKE et al., 1987). However, cloning in YACs normally generates a high degree of chimerism and rearrangements, which limits the usability of this method in physical mapping and map-based cloning (BURKE et al., 1987; BURKE, 1990; NEIL et al., 1990; GREEN et al., 1991). On the other hand, BACs, which utilize the *E. coli* single-copy fertility plasmid (F plasmid) can maintain up to 350 kb inserts, with little or no rearrangement or chimerism, respectively (SHIZUYA et al., 1992; WOO et al., 1994; JIANG et al., 1995; CAI et al., 1995). An alternative to cloning in BACs, the PAC system is an artificial chromosome vector developed based on bacteriophage P1 (PIERCE et al., 1992). Both BAC and PAC as compared to YAC, have much higher cloning efficiencies, improved fidelity, and are easier to

handle. Because of the high stability and ease of use, the BAC cloning system has emerged as the vector of choice for the construction of large insert libraries such as human (KIM et al., 1996), bovine (CAI et al., 1995), *Arabidopsis* (CHOI et al., 1995), rice (ZHANG et al., 1996), and sorghum (WOO et al., 1994). To add even more capabilities to the BAC system, the T-DNA locus and origin of replication from *Agrobacterium tumefaciens* were engineered into a binary bacterial artificial chromosome (BIBAC) vector to make it an ideal plant transformation vector with capacity of replicating in both *E. coli* and *A. tumefaciens* (HAMILTON et al., 1996). The original vectors have been modified subsequently to improve the transformation efficiency by electroporation, cloning features and selection specificity (SHIBATA and LIU, 2000). Recently, rice and wheat genomic libraries were constructed, using the improved BIBAC vector (SHIBATA and LIU, 2000). Using the improved BIBAC vector, a gene causing *FILAMENTOUS FLOWER* was isolated (SAWA et al., 1999). The entire 150 kb human genomic fragment was transferred into tobacco plants (HAMILTON et al., 1996). Recently, an additional P-lox site has been inserted into the pBACwich vector to allow site specific recombination when using particle bombardment for plant transformation (CHOI et al., 2000). The cre-lox system was successfully used to mediate the recombination between *Arabidopsis* and *Nicotiana tabacum* chromosomal regions (KOSHINSKY et al., 2000). Such improvements can now be applied to understand the function of genes with the help of positional cloning.

3.6 Radiation Hybrid Map

A radiation hybrid map (RH map) can be constructed based on radiation induced breakage between loci. The map distance between markers, estimated by the frequency of chromosome breakage, is proportional to the physical distance between the markers. The map resolution, depending on the amount of radiation, used is on average between 100–1,000 kb. Unlike physical maps, RH mapping is more random, because chromosome breakage does not depend on the frequency of restriction enzymes or a cloning bias. Nonetheless, monomorphic markers such as ESTs and STSs can be assigned to a RH map. With the high resolution achieved by RH mapping and the high density markers, RH mapping is a very powerful tool for positional cloning. Positional cloning of hyperekplexia, a human autosomal dominant neurological disorder, was done within 6 weeks after the critical region was identified and unrelated candidate genes were eliminated using RH mapping (WARRINGTON et al., 1994). RH mapping is extremely useful in any chromosomal region where recombination is suppressed. Publicly available RH mapping panels have been maintained at Genethon (*http://www.genethon.fr/genethon_en.html*), the Whitehead Institute (*http://www-genome.wi.mit.edu/*), and Stanford University (*http://shgc-www.stanford.edu/Mapping/rh*).

3.7 Identification of Genes within the Refined Critical Region

The most challenging step in positional cloning is the identification of genes in the critical region. There are strategies for the identification of genes in large genomic clones such as YAC, BAC, or PAC clones, as listed in Tab. 1. In the ideal case the trait is cosegregated with the marker, or an EST derived from high-resolution mapping. In cases where the critical region cannot be narrowed down further and no other known marker is available, small DNA fragments from a YAC or BAC clones contain-

Tab. 1. Strategies for Identification of Genes in Large Genomic Regions

Mapping Approaches
  Cosegregation of tightly linked EST to the trait of interest

Positional Syntenic Approaches
  Comparison in syntenic region of model organisms for genes or mutations

Exon Prediction Approaches
  cDNA selection
  Identification of CpG island
  Exon trapping, poly A signal trapping

Sequence-Based Approaches
  Sequence comparison to known genes or ESTs
  Gene prediction

Functional Analysis
  Transgenic knockout, transformation
  Transposon tagging

ing the gene can be used to hybridize in the orthologous region in such systems as the human/mouse/rat system, *Arabidopsis*–tomato–*Brassica*, or the rice–maize–wheat system. Detection of cross-hybridization reveals conserved sequences between these species and, therefore, genes or ESTs identified in such syntenic region are likely to have biological function. The presence of CpG islands nearby often marks the 5′ ends of genes and can be subsequently used for gene isolation. CpG islands, exon trapping, and direct cDNA selection are complementary approaches that can be used to identify exons in genomic sequences.

Gene Detection by CpG Island. Its unusual G + C% rich DNA first distinguished CpG islands from other genomic sequences (BIRD, 1987), more than 60% of human genes contain CpG islands, both in promotors and at least a part of one exon (LARSEN et al., 1992).

Exon Trapping. The presence of consensus sequences at "splice junctions" allows the isolation of adjacent exons by "exon amplification" or "exon trapping" (DUYK et al., 1990). In some cases, the entire internal exon can be captured (NEHLS et al., 1994). In a similar fashion, several genes causing human diseases have been isolated by trapping their terminal exons using the poly A tail signal (KRIZMAN and BERGET, 1993).

Direct cDNA selection. The candidate genomic clones can be used either as template to screen cDNA libraries constructed from the target tissues or subtractive hybridization. Using a method called "direct cDNA selection", the target genomic clones are fixed on an affinity matrix to capture cDNAs by homology-based hybridization. The success of the method relies on the source and quality of the cDNAs that are about to be captured. The genes can only be captured, if they are expressed in the tissue from which mRNA or cDNA libraries were isolated. Some genes showing low levels of expression or being absent in the target tissue may be difficult or impossible to isolate by direct cDNA selection.

4 Comparative Approaches to Gene Identification

Comparative mapping combines genetic information accumulated from related species. Usually, linkage maps of each species have been constructed with various kinds of molecular markers and used independently for particular genetic purposes. Comparisons among maps of related species, using low copy number sequences as a probe to hybridize with genomic DNA indicated the substantially conserved orders of the DNA sequences among their genomes. This is well documented in the grass family, which diverged ~60 million years ago (AHN and TANKSLEY, 1993; VAN DEYNZE et al., 1995; GUIMARAES et al., 1997). The mammals including human and mouse evolved from a common progenitor ~70 million years ago and are also documented to show conserved orders of DNA sequences (CARVER and STUBBS, 1997). The existence of conserved gene orders (colinearity) and contents in related species indicates that new genes are rarely created within evolutionary time frames of at least ten million years. Most new genes probably arise from gene duplication and/or gene modification of currently existing genes (BENNETZEN, 2000). Therefore, the colinearity can allow gene predictions across families and the extrapolation of mapping data from one organism to the other. Candidate genes can thus also be easily isolated and predicted from species for which well established linkage maps do not exist.

4.1 Genome Cross-Referencing in the Grass Family

During evolution, rice (430 Mb), maize (3,500 Mb), and wheat (16,000 Mb) have been separated 50–60 million years ago (MOORE et al., 1993). The only genes which are extensively conserved in the orthologous region of genomes, are those that diverged several million years ago (AVRAMORA et al., 1996). Comparative maps have been constructed between rice and maize (AHN and TANKSLEY, 1993), oat and maize (ANANIEV et al., 1997), among rice strains, as well as wheat and maize based on a common set of cDNAs (AHN et al., 1993), and rice and barley (DUNFORD et al., 1995; KILIAN et al., 1995). Being phylogenetically 60 million years apart, members of the grass family still share extensive syntenies in a number of regions. Therefore, the idea that the map position of one species can be used to identify and compare orthologous alleles across species is feasible for grasses (BENNETZEN and FREELING, 1993; PATERSON et al., 1995).

In positional cloning experiments, micro-colinearity can be extremely useful for cloning genes from species with large genomes, such as wheat and maize, using information from small genome species like rice based on their synteny. KILIAN et al. (1995) compared a 6.5 cM region in barley's chromosome 1 containing the barley rust resistance gene *Rpg1* with the 2.5 cM syntenic region in rice chromosome 6 and found that the order of RFLP markers was conserved. In the case of the *Adh1* locus, composition and arrangement of genomic DNA fragments were compared between maize YAC and sorghum BAC clones, containing the orthologous loci (SPRINGER et al., 1994; AVRA-MOVA et al., 1996). Because of a 75 kb stretch of highly repetitive elements in maize, chromosome walking to the *Adh1* gene was made possible by cross-referencing to the sorghum BAC. In a similar case, synteny was reported in the *sh2-a1* homologous regions between maize, sorghum, and rice, where the distance between the two genes was 140 kb in maize but only 19 kb in rice and sorghum (CHEN et al., 1997). These studies reveal that small rearrangements, including frequent insertions of transposons or retrotransposons can occur without significant rearrangement of the orthologous region (BENNETZEN, 2000).

Another striking demonstration of colinearity in the grass family is the gene, which causes dwarfism. The so-called "green revolution" genes were found in wheat (*Rht*, reduced height), maize (*d8*, dwarf) and rice (*d*, dwarf). The sequence comparison revealed the *Rht* and maize *d8* were orthologs of the *Arabidopsis GAI*, which is the gibberellin-insensitive mutant (PENG et al., 1999). Comparative mapping between wheat and rice using RFLP markers linked to *Rht-D1b* and between rice and

maize using RFLP markers linked to *D8* clearly showed colinearity among wheat chromosome D4, rice chromosome 3, and the maize chromosome 1. Additionally, one of the spontaneous rice dwarf mutants, *d1*, was isolated by positional cloning and was found to be an ortholog to the alpha subunit of G-protein, which is related to the *GAI* mutant found in *Arabidopsis* (ASHIKARI et al., 1999). However, the *d1* mutant was mapped to chromosome 5. Due to the fact that in rice at least 54 dwarf mutants have been identified, at least one of them may actually be located in the syntenic region of chromosome 3. Although comparative mapping is a powerful tool for gene finding in large syntenic regions, the extensive gene or segmental duplication in the reference genome may obstruct such comparison and may lead to the false assignment of a syntenic region.

4.2 Bridging Model Organisms for Positional Cloning

Genomes of model organisms have played a critical role in the positional cloning of human diseases. Humans and mice are the mammalians which are most extensively mapped. The map location of more than 3,000 genes and syntenic regions has been identified and displayed at (*http://www3.ncbi.nlm.nih.gov/Homology/*). These maps are also linked to the MGD (Mouse Genome Database) and OMIM (Online Mendelian Inheritance in Man), (*http://www3.ncbi.nlm.nih.gov/omim*), a catalog of human genes and genetic disorders. In order to integrate maps of the two different species, type I markers such as ESTs have been the most useful tool for anchoring loci during comparative mapping (DEBRY and SELDIN, 1996).

The complete genomic sequence and annotated set of genes for budding yeast (*http://www.stanford.edu/Saccharomyces*) has opened another possibility for the cross-referencing of human genes (BASSETT et al., 1996). Several human disease genes have been cloned in this way, examples include *MDR1*, the multidrug resistance gene in humans, which encodes a protein required for the pheromone factor involved in yeast mating (RAYMOND et al., 1992).

The human neurofibromatosis type 1 gene can complement the function of defective *IRA* in yeast (BALLESTER et al., 1990). A tremendous conservation of genes still exists between human and fly (MOUNKES et al., 1992).

4.3 Positional Cloning in the Genomic Era

Genome projects can dramatically simplify the long, tedious process of positional cloning. Physical mapping can also be tedious and costly. For most genome projects, BAC-end sequences or sequence tag connectors (STC) from a 10–20 X BAC library can tremendously simplify the physical map construction. "End walking" can be done *in silico*, initiated by several rounds of Blast searching the sequences of flanking markers or the next BAC end sequences against the respective BAC-end sequence database until the critical region is totally overlapped with a BAC contig. In addition, a BAC fingerprint database, where contigs are being assembled, can be used to confirm and anchor the genetic map by using well-mapped molecular information as illustrated in Fig. 3.

of DNA sequences. Such high-quality sequence will soon be available for human, *C. elegans*, and *Arabidopsis* (ROUNSLEY et al., 1998) among others. Taking the human genome project as an example, the availability of the comprehensive high resolution map and physical map that assembles EST, STC, STS, STR into large genomic contigs allows human disease genes to be assigned precisely to the critical region in a short timeframe. Genes can be predicted more accurately where genomic sequences are available using modern computational tools including GenScan, GeneMarkHMM, Xgrail, and Glimmer. All the genes identified by homology or by prediction in the critical region can become candidates. It is essential to understand how those candidate genes function. To prove that one of those candidate genes is the responsible gene, it has to be demonstrated that the mutation in the gene is genetically associated with the phenotype. At this stage the availability of single nucleotide polymorphisms (SNP) will improve the rate of mutation discovery in the candidate genes tremendously. However, the ultimate proof that the candidate gene is the correct one requires evidence that the normal phenotypic form can be complemented by the susceptible gene or vice versa.

4.4 Predicting Candidate Genes from Genomic Sequence

The most direct approach to gene identification in a genomic region involves the analysis

4.5 ESTs: Key to Gene Identification in the Critical Region

As a sequence-based marker, ESTs play a crucial role in both gene-based physical map

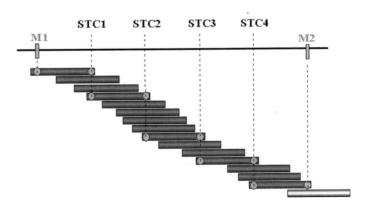

Fig. 3. *In silico* physical mapping by bridging sequence tag connectors.

Physical Map by STC

construction and candidate gene identification. In recent years, a massive scale of EST production has been conducted at the Institute for Genomic Research, TIGR (*http://www.tigr. org/*), and over 100,000 ESTs were released by TIGR and others to be maintained in the "dbEST" database (*http://www.ncbi.nlm.nih. gov/dbEST/*) at NCBI. Among the largest publicly available EST collections, more than 500,000 ESTs have been produced at Washington University in St. Louis, supported by Merck and Company (*http://genome.wustl.edu/est/ esthmpg.html*). Because the current ESTs are only 97% accurate and short, unedited, single-pass reads, they are clustered into "Tentative Consensus Sequences" at TIGR (*http://www. tigr.org/*) or "Uniquegene Cluster" at NCBI (*http://www.ncbi.nlm.nih.gov/UniGene/*).

ESTs have been also produced from organisms other than human. One of the largest collections is the 300,000 mouse ESTs funded by Howard Hughes Medical Institute (*http:// genome.wustl.edu/est/mouse_esthmpg.html*). Other model organisms are *C. elegans* (*http:// ddbj.nig.ac.jp/htmls/c-elegans/html/*), *Arabidopsis thaliana* (*http://genome-www.stanford.edu/ Arabidopsis/*), rice (*http://www.staff.or.jp/*), and *Drosophila melanogaster* (*http://fly2.berkeley. edu/*). These homologous or orthologous resources are curated at "HomoloGene" at NCBI (*http://www.ncbi. mlm.nih.gov/HomoloGene/*).

Identification of ESTs on physical and RH maps of a critical region can help to immediately identify candidate genes and simplify the positional cloning of particular genes by leaping across taxonomic boundaries due to the conserved protein sequences.

4.6 SNP: Snipping through the Sequence Variation

Genetic variation within coding sequences is very conserved and can rarely be detected by length polymorphism. Polymorphisms corresponding to differences at a single nucleotide level, which are caused either by deletion, insertion, or substitution, are biallelic in diploids, occurring frequently and uniformly in most genomes at roughly one in every 500–1,000 bp. As genomic and EST sequences are increasing at an exponential rate in the past

few years, SNP discovery in coding regions is projected for at least 100,000 markers with the aim of identifying and cataloging all human genes and creating more or less complete human SNP maps. Because SNPs residing within coding region are rare, these mutations may correspond to defects that are associated with polymorphism at the protein level, diseases, or other phenotypes. To prove that SNPs are associated with a particular disease, haplotype analysis of SNPs in the candidate genes must be conducted among the affected and unaffected individuals from the same family. In this case, the sampling size must be large enough to reveal the statistical differences between the affected and unaffected pools and show a linkage disequilibrium. Positional cloning can be more effective and less time-consuming using SNPs not only to refine the critical region, but also to confirm the position of the real candidate genes.

5 References

AHN, S., TANKSLEY, S. D. (1993), Comparative map of the rice and maize genomes, *Proc. Natl. Acad. Sci. USA* **90**, 7980–7984.

AHN, S., ANDERSON, J. A., SORRELLS, M. E., TANKSLEY, S. D. (1993), Homologous relationships of rice, wheat and maize chromosomes, *Mol. Gen. Genet.* **241**, 481–490.

ALLARD, R.W. (1956), Formulas and tables to facilitate the calculation of recombination values in heredity, *Hilgardia* **24**, 235–278.

ANANIEV, E. V., RIERA-LIZARAZU, O., RINES, H. W., PHILLIPS, R. L. (1997), Oat–maize chromosome addition lines: A new system for mapping the maize genome, *Proc. Natl. Acad. Sci. USA* **94**, 3524–3529.

ASHIKARI, M., WU, J., YANO, M., SASAKI, T., YOSHIMURA, A. (1999), Rice gibberellin-insensitive dwarf mutant gene Dwarf 1 encodes the alpha subunit of GTP-binding protein, *Proc. Natl. Acad. Sci. USA* **96**, 10284–10289.

AVRAMOVA, Z., TIKHONOV, A., SANMIGUEL, P., JIN, Y. K., LIU, C. et al. (1996), Gene identification in a complex chromosomal continuum by local genomic cross-referencing, *Plant J.* **10**, 1163–1168.

BALLESTER, R., MARCHUK, D., BOGUSKI, M., SAULINO, A., LETCHER, R. et al. (1990), The NF1 locus encodes a protein functionally related to mammalian GAP and yeast IRA proteins, *Cell* **63**, 851.

BASSETT, D. E. Jr., BOGUSKI, M. S., HIETER, P. (1996),

Yeast genes and human disease, *Nature* **379**, 589.

BODENES, C., LAIGRET, F., KREMER, A. (1996), Inheritance and molecular variation of PCR-SSCP fragments in pedunculate oak (*Quercus robur* L.), *Theor. Appl. Genet.* **993**, 348–354.

BOTSTEIN, D., WHITE, R. L., SKOLNICK, M., DAVIS, R. W. (1980), Construction of a genetic linkage map in human using restriction fragment length polymorphisms, *Am. J. Hum. Genet.* **32**, 314–331.

BENNETZEN, J. L. (2000), Comparative sequence analysis of plant nuclear genomes: microcolinearity and its many exceptions, *Plant Cell* **12**, 1021–1029.

BENNETZEN, J. L., FREELING, M. (1993), Grasses as a single genetic system: genome composition, colinearity and compatibility, *Trends Genet.* **9**, 259–261.

BENT, A. F., KUNKEL, B. N., DAHLBECK, D., BROWN, K. L., SCHINIDT, R. et al. (1994), *RPS2* of *Arabidopsis thaliana*: a leucine-rich repeat class of plant disease resistance gene, *Science* **265**, 1856–1860.

BIRD, A. P. (1987), CpG islands as gene markers in the vertebrate nucleus, *Trends Genet.* **3**, 342–347.

BITTNER-EDDY, P., CAN, C., GUNN, N., PINEL, M., TOR, M. et al. (1999), Genetic and physical mapping of the *RPP13* locus, in *Arabidopsis*, responsible for specific recognition of several *Peronospora parasitica* (downy mildew) isolates, *Mol. Plant Microb. Interact.* **12**, 792–802.

BURKE, D. T. (1990), YAC cloning: options and problems, *GATA* **7**, 94–99.

BURKE, D. T., CARLE, G. F., OLSON, M. V. (1987), Cloning of large segments of exogenous DNA into yeast using artificial-chromosome vectors, *Science* **236**, 806–812.

BUSCHGES, R., HOLLRICHER, K., PANSTRUGA, R., SIMONS, G., WOLTER, M. et al. (1997), The barley *Mlo* gene: A novel control element of plant pathogen resistance, *Cell* **88**, 695–705.

CAI, D., KLEINE, M., KIFLE, S., HARLOFF, H. J., SANDAL, N. N. et al. (1997), Positional cloning of a gene for nematode resistance in sugar beet, *Science* **275**, 832–834.

CAI, L., TAYLOR, J. F., WING, R. A., GALLAGHER, D. S., WOO, S. S., DAVIS, S. K. (1995), Construction and characterization of a bovine bacterial artificial chromosome library, *Genomics* **29**, 413–425.

CARVER, E. A., STUBBS, L. (1997), Zooming in on the human–mouse comparative map: Genome conservation re-examined on a high-resolution scale, *Genome Res.* **7**, 1123–1137.

CAUSSE, M. A., FULTON, T. M., CHO, Y. G. et al. (1994), Saturated molecular map of the rice genome based on an interspecific backcross population, *Genetics* **138**, 1251–1274.

CHANG, C., KWOK, S., BLEECKER, A., MEYEROWITZ, E. (1993), *Arabidopsis* ethylene-response gene

ETR1: Similarity of product to two-component regulators, *Science* **262**, 539–544.

CHEN, M., SANMIGUEL, P., LIU, C. N., DE OLIVEIRA, A. C., TIKHONOV, A. et al. (1997), Microcolinearity in the maize, rice, and sorghum genomes, *Proc. Natl. Acad. Sci. USA* **94**, 3431–3435.

CHOI, S., CREELMAN, R. A., MULLET, J. E., WING, R. A. (1995), Construction and characterization of bacterial artificial chromosome library of *Arabidopsis thaliana*, *Plant Mol. Biol. Rep.* **13**, 124–128.

CHOI, S., BEGUM, D., KOSHINSKY, H., OW, D. W., WING, R. A. (2000), A new approach for the identification and cloning of genes: the pBACwich system using Cre/lox site-specific recombination, *Nucleic Acids Res.* **28**, 19e.

CORRETTE-BENNETT, J., ROSENBERG, M., PRZYBYLSKA, M., ANANIEV, E., STRAUS, D. (1998), Positional cloning without a genome map: using "Targeted RFLP Subtraction" to isolate dense markers tightly linked to the regA locus of *Volvox carteri*, *Nucleic Acids Res.* **26**, 1812–1818.

DEBRY, R. W., SELDIN, M. F. (1996), Human/mouse homology relationships, *Genomics* **33**, 337–351.

DIETRICH, R. A., RICHBERG, M. H., SCHMIDT, R., DEAN, C., DANG, J. L. (1997), A novel zinc finger protein is encoded by the *Arabidopsis LSD1* gene and functions as a negative regulator of plant cell death, *Cell* **88**, 685–694.

DUNFORD, R. P., KURATA, N., LAURIE, D. A., MONEY, T. A., MINOBE, Y., MOORE, G. (1995), Conservation of fine-scale DNA marker order in the genome of rice and the Triticeae, *Nucleic Acids Res.* **115**, 133–138.

DUYK, G. M., KIM, S., MYERS, R. M., COX, D. R. (1990), Exon trapping: A genetic screen to identify candidate transcribed sequences in cloned mammalian genomic DNA, *Proc. Natl. Acad. Sci. USA* **87**, 8995–8999.

FINKELSTEIN, R. R., WANG, M. L., LYNCH, T. J., RAO, S., GOODMAN, H. M. (1998), The *Arabidopsis* abscissic acid response locus *ABI4* encodes an APETALA2 domain protein, *Plant Cell* **10**, 1043–1054.

FUJITA, K., SILVER, J. (1994), Single-strand conformational polymorphism, *PCR Methods Appl.* **4**, S137–S139.

FUKUOKA, S., INOUE, T., MIYAO, A., ZHONG, H. S., SASAKI, T., MINOBE, Y. (1994), Mapping sequence-tagged sites in rice by single strand conformation polymorphisms, *DNA Res.* **1**, 271–274.

GESSLER, D. D., XU, S. (1999), Multipoint genetic mapping of quantitative trait loci with dominant markers in outbred populations, *Genetica* **105**, 281–291.

GIBSON, S., SOMERVILLE, C. (1994), Isolating plant genes, *Trends Biotechnol.* **12**, 306–313.

GIRAUDAT, J., HAUGE, B. M., VALON, C., SMALLE, J., PACY, F., GOODMAN, H. M. (1992), Isolation of the *Arabidopsis ABI3* gene by positional cloning,

Plant Cell **4**, 1251–1261.

GRANT, M. R., GODIARD, L., STRAUBE, E., ASHFIELD, T., LEWALD, J. et al. (1995), Structure of the *Arabidopsis RPM1* gene enabling dual specificity disease resistance, *Science* **296**, 843–846.

GREEN, E. D., RIETHMAN, H. C., DUTCHIK, J. E., OLSON, M. V. (1991), Detection and characterization of chimeric yeast artificial-chromosome clones, *Genomics* **11**, 658–669.

GUIMARAES, C. T., SILLS, G. R., SOBRAL, B. W. S. (1997), Comparative mapping of Andropogoneae: *Saccharum* L. (sugarcane) and its relation to sorghum and maize, *Proc. Natl. Acad. Sci. USA* **94**, 14261–14266.

HALDANE, J. B. S. (1919), The combination of linkage values and the calculation of distances between the loci of linked factors, *J. Genet.* **8**, 299–309.

HAMILTON, C. M., FRARY, A., LEWIS, C., TANKSLEY, S. D. (1996), Stable transfer of intact high molecular weight DNA into plant chromosomes, *Proc. Natl. Acad. Sci. USA* **93**, 9975–9979.

HAYASHI, K. (1991), PCR-SSCP: A simple and sensitive method for detection of mutation in the genomic DNA, *PCR Methods Appl.* **1**, 34–38.

HAYASHI, K. (1992), PCR-SSCP: A method for detection of mutations, *GATA* **9**, 73–79.

HEUN, M., KENEEDY, A. E., ANDERSSON, J. A. et al. (1991), Construction of an RFLP map of barley (*Hordeum vulgare* L.), *Genome* **34**, 437–447.

HUHN, M. (2000), Maximum likelihood vs. minimum chi-square – a general comparison with applications to the estimation of recombination fractions in two-point linkage analysis, *Genome* **43**, 853–856.

HUYS, G., RIGOUTS, L., CHEMLAL, K., PORTAELS, F., SWINGS, J. (2000), Evaluation of amplified fragment length polymorphism analysis for inter- and intraspecific differentiation of *Mycobacterium bovis*, *M. tuberculosis*, and *M. ulcerans*, *J. Clin. Microbiol.* **38**, 3675–3680.

JIANG, J., GILL, B. S., WANG, G. L., RONALD, P. C., WARD, D. C. (1995), Metaphase and interphase fluorescence *in situ* hybridization mapping of the rice genome with bacterial artificial chromosomes, *Proc. Natl. Acad. Sci. USA* **92**, 4487–4491.

KAGA, A., ISHIMOTO, M. (1998), Genetic localization of a bruchid resistance gene and its relationship to insecticidal cyclopeptide alkaloids, the vignatic acids, in mungbean (*Vigna radiata* L. Wilczek), *Mol. Gen. Genet.* **258**, 378–384.

KIEBER, J. J., ROTHENBERG, M., ROMAN, G., FELDMANN, K. A., ECKER, J. R. (1993), *CTR1*, a negative regulator of the ethylene response pathway in *Arabidopsis*, encodes a member of the raf family of protein kinases, *Cell* **72**, 427–441.

KILIAN, A., KUDRNA, D. A., KLEINHOFS, A., YAŇO, M., KURATA, N. et al. (1995), Rice–barley synteny and its application to saturation mapping of the bar-

ley Rpg1 region, *Nucleic Acids Res.* **23**, 2729–2733.

KIM, U. J., BIRREN, B. W., SLEPAK, T., MANCINO, V., BOYSEN, C. et al. (1996), Construction and characterization of a human bacterial artificial chromosome library, *Genomics* **34**, 213–218.

KORNNEEF, M. (1994), *Arabidopsis* genetics, in: *Arabidopsis* (MEYEROWITZ, E. M., SOMERVILLE, C. R., Eds.), pp. 5.89–5.120. Cold Spring Harbor, NY: Cold Spring Harbor Press.

KOSAMBI, D. D. (1944), The estimation of map distances from recombination values, *Ann. Eugen.* **12**, 172–175.

KOSHINSKY, H. A., LEE, E., OW, D. W. (2000), Cre-lox site-specific recombination between *Arabidopsis* and tobacco chromosomes, *Plant J.* **23**, 715–722.

KRIZMAN, D. B., BERGET, S. M. (1993), Efficient selection of 3′ terminal exons from vertebrate DNA, *Nucleic Acids Res.* **21**, 5198–5202.

KURATA, N., NAGAMURA, Y., YAMAMOTO, K., HARUSHIMA, Y., SUE, N. et al. (1994), A 300 kilobase interval genetic map of rice including 880 expressed sequences, *Nature Genet.* **8**, 365–372.

LAHAYA, T., SHIRASU, K., SCHULZE-LEFERT, P. (1998), Chromosome landing at the barley *Rar1* locus, *Mol. Gen. Genet.* **260**, 92–101.

LANDER, E. S., GREEN, P., ABRAHAMSON, J., BARLOW, A., DALY, M. J. et al. (1987), MAP-MAKER: an interactive computer package for constructing primary genetic linkage maps of experimental and natural populations, *Genomics* **1**, 174–181.

LAROCHE, A., DEMEKE, T., GAUDET, D. A., PUCHALSKI, B., FRICK, M., McKENZIE, R. (2000), Development of a PCR marker for rapid identification of the Bt-10 gene for common bunt resistance in wheat, *Genome* **43**, 217–223.

LARSEN, S. O. (1979), A general program for estimation of haplotype frequencies from population diploid data, *Comput. Programs Biomed.* **10**, 48–54.

LARSEN, F., GUNDERSON, G., LOPEZ, R., PRYDZ, H. (1992), CpG islands are gene markers in human genome, *Genomics* **13**, 1095–1107.

LATHROP, G. M., LALOUEL, J. M. (1984), Easy calculation of lod scores and genetic risks on small computers, *Am. J. Hum. Genet.* **36**, 460–465.

LATHROP, G. M., LALOUEL, J. M., WHITE, R. L. (1986), Construction of human linkage maps: likelihood calculations for multilocus linkage analysis, *Genet. Epidemiol.* **3**, 39–52.

LEUNG, J., BOUVIER-DURAND, M., MORRIS, P. C., GUERRIER, D., CHEFDOR, F., GIRAUDAT, J. (1994), *Arabidopsis* ABA response gene *ABI1*: features of a calcium-modulated protein phosphatase, *Science* **264**, 1448–1451.

LEUNG, J., MERLOT, S., GIRAUDAT, J. (1997), The *Arabidopsis* ABSCISIC ACID-INSENSITIVE 2 (*ABI2*) and (*ABI1*) genes encode redundant protein phosphatases 2C involved in abscisic acid signal transduction, *Plant Cell* **9**, 759–771.

LEYSER, H. M. O., LINCOLN, C. A., TIMPTE, C., LAMMER, D., TURNER, J., ESTELLE, M. (1993), *Arabidopsis* auxin-resistance gene *AXR1* encodes a protein related to ubiquitin-activating enzyme E1, *Nature* **364**, 161–164.

LISITSYN, N. (1995), Representational difference analysis: finding the differences between genomes, *Trends Genet.* **11**, 303–307.

LIU, B. H. (1998), *Statistical Genomics: Linkage, Mapping and QTL Analysis.* New York: CRC Press.

LIU, B. H., KNAPP, S. J. (1990), GMENDEL: a program for Mendelian segregation and linkage analysis of individual or multiple progeny populations using log-likelihood ratios, *J. Heredity* **81**, 407.

MACKILL, D. J., ZHANG, Z., REDONA, E. D., COLOWIT, P. M. (1996), Level of polymorphism and genetic mapping of AFLP markers in rice, *Genome* **39**, 969–977.

MAO, L., BEGUM, D., CHUANG, H. W., BUDIMAN, M. A., SZYMKOWIAK, E. J. et al. (2000), *Jointless* is a *MADS-box* gene controlling tomato flower abscission zone development, *Nature* **406**, 910–913.

MARTIN, G. B., BROMMONSCHENKEL, S. H., CHUNWONGSE, J., FRARY, A., GANAL, M. W. et al. (1993), Map-based cloning of a protein kinase gene conferring disease resistance in tomato, *Science* **262**, 1432–1436.

MCCOUCH, S. R., CHEN, X., PANAUD, O., TEMNYKH, S., XU, Y. et al. (1997), Microsatellite marker development, mapping and applications in rice genetics and breeding, *Plant Mol. Biol.* **35**, 89–99 (Review).

MESBAH, L. A., KNEPPERS, T. J. A., TAKKEN, F. L. W., LAURENT, P., HILLE, J., NIJKAMP, H. J. J. (1999), Genetic and physical analysis of a YAC contig spanning the fungal disease resistance locus *Asc* of tomato (*Lycopersicon esculentum*), *Mol. Gen. Genet.* **261**, 50–57.

MICHELMORE, R. W., PARAN, I., KESSLI, R. V. (1991), Identification of markers linked to disease resistance genes by bulked segregant analysis: a rapid method to detect markers in specific genomic regions by using segregating populations, *Proc. Natl. Acad. Sci. USA* **88**, 9829–9832.

MONNA, L., MIYAO, A., ZHONG, H. S., YANO, M., IWAMOTO, M. et al. (1997), Saturation mapping with subclones of YACs: DNA marker production targeting the rice blast disease resistance gene, *Pi-b*, *Theor. Appl. Genet.* **94**, 170–176.

MOORE, G., DEVOS, K. M., WANG, Z., GALE, M. D. (1995), Grasses, line up and form a circle, *Curr. Biol.* **5**, 737–739.

MOORE, G., GALE, M. D., KURATA, N., FRAVELL, R. B. (1993), Molecular analysis of small grain cereal genomes: current status and prospects, *Biotechnology* **11**, 584–589.

MORGAN, T. H. (1928), *The Theory of Genes.* New Haven, CT: Yale University Press.

MOUNKES, L. C., JONES, R. S., LIANG, B. C., GELBART, W., FULLER, M. T. (1992), A *Drosophila* model for xeroderma pigmentosum and Cockayne's syndrome: haywire encodes the fly homolog of ERCC3, a human excision repair gene, *Cell* **71**, 925.

MYERS, R. M., HEDRICK ELLENSON, L., HAYASHI, K. (1998), Detection of DNA variation, in: *Genome Analysis: A Laboratory Manual* Vol. 2 *Detecting Genes* (BIRREN, B., GREEN, E. D., KLAPHOLZ, S., MYERS, R. M., ROSHAMS, J., Eds), pp. 287–384. Cold Spring Harbor, NY: Cold Spring Harbor Laboratory Press.

NAKAMURA, S., ASAKAWA, S., OHMIDO, N., FUKUI, K., SHIMIZU, N., KAWASAKI, S. (1997), Construction of an 800-kb contig in the near-centromeric region of the rice blast resistance gene *Pi-ta²* using a highly representative rice BAC library, *Mol. Gen. Genet.* **254**, 611–620.

NEHLS, M., PFEIFER, D., BOEHM, T. (1994), Exon amplification from complete libraries of genomic DNA using a novel phage vector with automatic plasmid excision facility: Application to mouse neurofibromatosis-1 locus, *Oncogene* **9**, 2169–2175.

NEIL, D. L., VILLASANTE, A., FISHER, R. B., VETRIE, D., COX, B., TYLER-SMITH, C. (1990), Structural instability of human tandemly repeated DNA sequences cloned in yeast artificial chromosome vectors, *Nucleic Acids Res.* **18**, 1421–1428.

ORITA, M., IWAHARA, H., KANAZAVA, H., HAYASHI, K., SEKIYA, T. (1989), Detection of polymorphism of human DNA by gel electrophoresis as single-strand conformation polymorphisms, *Proc. Natl. Acad. Sci. USA* **86**, 2766–2770.

PANAUD, O., CHEN, X., MCCOUCH, S. R. (1996), Development of microsatellite markers and characterization of simple sequence length polymorphism (SSLP) in rice (*Oryza sativa* L.), *Mol. Gen. Genet.* **252**, 597–607.

PATERSON, A. H., LIN, Y. R., LI, Z., SCHERTZ, K. F., DOEBLEY, J. F. et al. (1995), Convergent domestication of cereal crops by independent mutations at corresponding genetic loci, *Science* **269**, 1714–1718.

PENG, J., CAROL, P., RICHARDS, D. E., KING, K. E., COWLING, R. J. et al. (1997), *The Arabidopsis GAI* gene defines a signalling pathway that negatively regulates gibberellin responses, *Genes Dev.* **11**, 3194–3205.

PENG, J., RICHARDS, D. E., HERTLEY, N. M., MURPHY, G. P., DEVOS, K. M. et al. (1999), "Green revolution" genes encode mutant gibberellin response modulators, *Nature* **400**, 256–261.

PIERCE, J. C., SAUER, B., STERNBERG, N. (1992), A positive selection vector for cloning high molecular weight DNA by the bacteriophage P1 system: Improved cloning efficacy, *Proc. Natl. Acad. Sci. USA* **89**, 2056–2060.

PILLEN, K., GANAL, M. W., TANKSLEY, S. D. (1996), Construction of a high-resolution genetic map and YAC-contigs in the tomato *Tm-2a* region, *Theor. Appl. Genet.* **93**, 228–233.

RANAMUKHAARACHCHI, D. G., KANE, M. E., GUY, C. L., LI, Q. B. (2000), Modified AFLP technique for rapid genetic characterization in plants, *Biotechniques* **29**, 858–859, 862–866.

RAYMOND, M., GROS, P., WHITEWAY, M., THOMAS, D. Y. (1992), Functional complementation of yeast *ste6* by a mammalian multidrug resistance *mdr* gene, *Science* **256**, 232–235.

ROSEMBERG, M., PRZYBYLSKA, M., STRAUS, D. (1994), "RFLP Subtraction": A method for making libraries of polymorphic markers, *Proc. Natl. Acad. Sci. USA* **91**, 6113–6117.

ROTHENBERG, M., ROMAN, G., FELDMANN, K. A., ECKER, J. R. (1993), *CTR1*, a negative regulator of the ethylene response pathway in *Arabidopsis*, encodes a member of the raf family of protein kinases, *Cell* **72**, 427–441.

ROUNSLEY, S., LIN, X., KETCHUM, K. A. (1998), Large-scale sequencing of plant genomes, *Curr. Opin. Plant Biol.* **1**, 136–141.

SAWA, S., WATANABE, K., GOTO, K., KANAYA, E., HAYATO, M., OKADA, K. (1999), *FILAMENTOUS FLOWER*, a meristem and organ identity gene of *Arabidopsis*, encodes a protein with a zinc finger and HMG-related domains, *Genes Dev.* **13**, 1079–1088.

SHERMAN, J. K., FENWICK, A. L., NAMUTH, D. M., LAPITAN, N. L. V. (1995), A barley RFLP map, alignment of three barley maps and comparisons to Gramineae species, *Theor. Appl. Genet.* **86**, 705–712.

SHIBATA, D., LIU, Y. G. (2000), *Agrobacterium*-mediated plant transformation with large DNA fragments, *Trend Plant Sci.* **5**, 354–357.

SHIZUYA, H., BIRREN, B., KIM, U. J., MANCINO, V., SLEPAK, T. et al. (1992), Cloning and stable maintenance of 300-kilobase-pair fragments of human DNA in *Escherichia coli* using F-factor-based vector, *Proc. Natl. Acad. Sci. USA* **89**, 8794–8797.

SIMONS, G., GROENENDIJK, J., WIJBRANDI, J., REIJANS, M., GROENEN, J. et al. (1998), Dissection of the *Fusarium I2* gene cluster in tomato reveals six homologs and one active gene copy, *Plant Cell* **10**, 1055–1068.

SONG, W. Y., WANG, G. L., CHEN, L. L., KIM, H. S., PI, L. Y. et al. (1995), A receptor kinase-like protein encoded by the rice disease resistance gene, *Xa-21*, *Science* **270**, 1804–1806.

SPRINGER, P. S., EDWARDS, K. J., BENNETZEN, J. L. (1994), DNA class organization on maize *Adh1* yeast artificial chromosomes, *Proc. Natl. Acad.* *Sci. USA* **87**, 103–107.

STRINGHAM, H. M., BOEHNKE, M., LANGE, K. (1999), Point and interval estimates of marker location in radiation hybrid mapping, *Am. J. Hum. Genet.* **65**, 545–553.

SUN, T. P., GOODMAN, H. M., AUSUBEL, F. M. (1992), Cloning the *Arabidopsis GA1* locus by genomic subtraction, *Plant Cell* **4**, 119–128.

TANKSLEY, S. D., GANAL, M. W., MARTIN, G. B. (1995), Chromosome landing: a paradigm for map-based gene cloning in plants with large genomes, *Trends Genet.* **11**, 63–68.

TANKSLEY, S. D., GANAL, M. W., PRINCE, J. P., DE VICENTE, M. C., BONIERBALE, M. W. et al. (1992), High density molecular linkage maps of the tomato and potato genome, *Genetics* **132**, 1141–1160.

THOMAS, C. M., VOS, P., ZABEAU, M., JONES, D. A., NORCOTT, K. A. et al. (1995), Identification of amplified restriction length polymorphism (AFLP) markers tightly linked to the tomato *Cf-9* gene for resistance to *Cladosporium fulvum*, *Plant J.* **8**, 785–794.

URQUHART, B. G., WILLIAMS, J. L. (1996), Sequencing of a novel cDNA and mapping to bovine chromosome3 by single-strand conformation polymorphism (SSCP), *Anim. Genet.* **27**, 438.

VAN DEYNZE, A. E., NELSON, J. C., YGLESIAS, E. S., HARRINGTON, S. E., BRAGA, D. P. et al. (1995), Comparative mapping in grasses: Wheat relationships, *Mol. Gen. Genet.* **284**, 744–754.

VOGL, C., XU, S. (2000), Multipoint mapping of viability and segregation distorting loci using molecular markers, *Genetics* **155**, 1439–1447.

WARRINGTON, J. A., BENGTSSON, U. (1994), High-resolution physical mapping of human 5q31-q33 using three methods: Radiation hybrid mapping, interphase fluorescence *in situ* hybridization, and pulse-field gel electrophoresis, *Genomics* **24**, 395–398.

WILLIAMS, J. G. K., KUBELIK, A. R., LIVAK, K. J., RAFALSHI, J. A., TINGEY, S. V. (1990), DNA polymorphisms amplified by arbitrary primers are useful as genetic markers, *Nucleic Acids Res.* **18**, 6531–6535.

WOO, S. S., JIANG, J., GILL, B. S., PATERSON, A. H., WING, R. A. (1994), Construction and characterization of a bacterial artificial chromosome library of *Sorghum bicolor*, *Nucleic Acids Res.* **22**, 4922–4931.

ZHANG, H. B., CHOI, S., WOO, S. S., LI, Z., WING, R. A. (1996), Construction and characterization of two rice bacterial artificial chromosome libraries from the parents of a permanent recombinant inbred mapping population, *Mol. Breeding* **2**, 11–24.

8 Sequencing Technology

LYLE R. MIDDENDORF

PATRICK G. HUMPHREY

NARASIMHACHARI NARAYANAN

STEPHEN C. ROEMER

Lincoln, NE, USA

1 Introduction

DNA sequencing technology is a major component of the genomics discovery pipeline. The technology is rooted in the late 1960s and early 1970s when efforts to sequence RNA took place. The nucleotide sequence of 5S-ribosomal RNA from *Escherichia coli* (BROWNLEE et al., 1967), 16S- and 23S-ribosomal RNA (FELLNER and SANGER, 1968), and R17 bacteriophage RNA coding for coat protein (ADAMS et al., 1969) are some of the early examples of RNA sequencing. A few years later SANGER reported on the sequencing of bacteriophage f1 DNA by primed synthesis with DNA polymerase (SANGER et al., 1973, 1974). In the same time period GILBERT and MAXAM (1973) reported on the DNA nucleotide sequence of the *lac* operator.

This pioneering work led to the plus/minus method reported by SANGER and COULSON (1975) which determined nucleotide sequence based on two approaches: (1) a "minus" system where four separate samples of partially double-stranded DNA fragments (containing a "full length" template " − " strand and random chain extension of an oligonucleotide primer for the " + " strand) are further incubated with DNA polymerase in the presence of only three deoxyribonucleoside triphosphates such that synthesis proceeds as far as it can until the polymerase needs to incorporate the missing nucleotide of the particular sample; and (2) a "plus" system where the four separate samples of partially double-stranded DNA fragments are further incubated in the presence of only one of the four triphosphates and then subjected to exonuclease which degrades the single-stranded overhang of the " − " strand. The DNA fragments for both approaches were then subjected to gel electrophoresis for length (and thus sequence) determination.

In 1977 SANGER reported on the use of modified nucleoside triphosphates (containing dideoxyribose sugar) in combination with the natural deoxyribonucleotides to terminate chain elongation (SANGER et al., 1977). In that same year MAXAM and GILBERT (1977) disclosed a method for sequencing DNA that utilized chemical cleavage of DNA preferentially at guanines, at adenines, at cytosines and thymines equally, and at cytosines alone. These two methods accelerated manual sequencing based on electrophoretic separation of DNA fragments labeled with radioactive markers and subsequent detection via autoradiography.

The first reports of automation of DNA sequencing occurred in the mid-1980s due to novel techniques to fluorescently label DNA (SMITH et al., 1986; ANSORGE et al., 1986, 1987; PROBER et al., 1987; BRUMBAUGH et al., 1988; KAMBARA et al., 1988; MIDDENDORF et al., 1988). This automation, in conjunction with the commencement of the human genome initiative (DELISI, 1988), spurred the explosion in genomics research that is in existence today. DNA sequencing technology is now only one tool, albeit a very important and dynamic one, in the genomics toolbox along with other tools such as DNA array and lab-on-a-chip technologies as well as automated protein analysis.

This chapter illustrates the multi-disciplinary nature of DNA sequencing technology in that the chapter organization is delineated into chemistry, biology, instrumentation, and software components. It is intended to provide an exhaustive reference structure in order to allow further in-depth investigation of each of these components, and the reader is invited to take advantage of the reference list in order to capture the fuller essence of sequencing technologies.

2 Overview of Sanger Dideoxy Sequencing

DNA sequencing is the determination of the nucleotide sequence of a specific deoxyribonucleic acid (DNA) molecule. Knowing the sequence of a DNA molecule is pivotal for making predictions about its function and facilitating manipulation of the molecule. Originally, DNA was sequenced using one of two methods. MAXAM and GILBERT (1977) devised a method that chemically cleaves DNA selectively between specific bases. SANGER et al. (1977) developed an enzymatic method based on the use of chain-terminating dideoxynucleotides.

The Sanger dideoxy method is now by far the most widely used technique for sequencing DNA. Informative texts by ALPHEY (1997) and ANSORGE et al. (1997) review many variations made to this sequencing technique, but the principle remains the same. The method depends on the synthesis of a new strand of DNA starting from a specific priming site and ending with the incorporation of a chain terminating nucleotide.

Specifically, a DNA polymerase extends an oligonucleotide primer annealed to a unique location on a DNA template by incorporating deoxynucleotides (dNTPs) complementary to the template. Synthesis of the new DNA strand continues until the reaction is randomly terminated by the inclusion of a dideoxynucleotide (ddNTP). These nucleotide analogs are incapable of supporting further chain elongation since the ribose moiety of the ddNTP lacks the 3′-hydroxyl necessary for forming a phosphodiester bond with the next incoming dNTP. This results in a population of truncated sequencing fragments of varying length.

Typically, the identity of the chain-terminating nucleotide at each position is specified by running four separate base-specific reactions each of which contains a different dideoxynucleotide (ddATP, ddCTP, ddGTP, or ddTTP). The four such fragment sets are loaded in adjacent lanes of a polyacrylamide gel and separated by electrophoresis according to the fragment size (Fig. 1). Remarkably, DNA fragments differing in length by just one nucleotide can be resolved. If a radioactive label is introduced into the sequencing reaction products, then autoradiographic imaging of the DNA band pattern in the gel can be used to deduce the DNA sequence (SANGER et al., 1977; SMITH, 1989). If the reaction products are labeled with an appropriate fluorescent dye, then an automated DNA sequencing system is used for the real-time detection of DNA fragments as they move through a portion of the electrophoresis gel that is irradiated by a laser. The fluorescence emission is collected by a detector and the resultant signal produces a band or trace pattern which correlates to a DNA sequence.

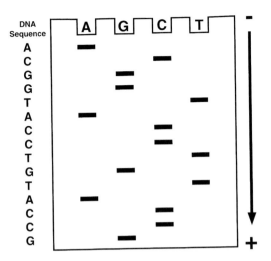

Fig. 1. DNA sequencing electrophoresis. The DNA fragments are prepared to terminate at one of four base types (A, G, C, T). A-type fragments of varying length are loaded in the "A" loading well at the top of the gel, and so forth for the G-, C-, and T-type fragments. Over time the shorter fragments in each lane migrate farther down the gel (toward the positive electrode). The DNA sequence is determined by noting the particular lane in which each succeeding band is spatially located in the vertical dimension (taken from MIDDENDORF et al., 1993).

3 Fluorescence Dye Chemistry

The original methods of DNA sequencing (MAXAM and GILBERT, 1977; SANGER et al., 1977), were implemented through the use of radioactive labels. High sensitivity and ease of labeling still make radioactive methods popular in thousands of biology laboratories around the world that practice manual radioactive DNA sequencing. However, the dangers associated with radioactivity such as health hazards and waste disposal regulations, along with the lack of automation, paved the way for the emergence of alternative non-radioactive labels (KESSLER, 1992). Most prominent among the sensitive, non-radioactive detection techniques are chemiluminescence and fluorescence. Despite excellent sensitivity, chemiluminescent methodology has not been viable for DNA sequencing due to its indirect detection limitation. Fluorescence detection (LAKOWICZ, 1999), on the other hand, employs direct detection methodology that is simple, sensitive, and easy to automate. Fluorescence methods and fluorescent dye labels have set a new standard in today's DNA sequencing community.

Several methods have been developed to sequence DNA using fluorescent labels (ANSORGE et al., 1986; SMITH et al., 1986; PROBER et al., 1987; BRUMBAUGH et al., 1988). Commercialized instruments employ one or more of the following methods for automated sequencing: four distinct dye-labeled primers with non-fluorescent terminators per DNA sample; one dye-labeled primer with non-fluorescent terminators per DNA sample; and one non-fluorescent primer with four distinct fluorescent terminators per DNA sample (see Sect. 4). This section provides a brief summary of important aspects on the advancement of chemistry of fluorescent dyes for DNA sequencing.

3.1 Fluorophore Characteristics

Fluorescence is the emission of light from electronically excited fluorophores. An electron of the fluorophore is energized into an excited orbital through the absorption of a photon where it is paired to a second electron that is in the ground-state orbital (LAKOWICZ, 1999). The excited orbital is one of several vibrational energy levels associated with one or more electronic energy states. The fluorophore is usually excited into a higher vibrational level of either the first or second electronic energy state. In a very fast process known as internal conversion the excited molecule first relaxes to the lowest vibrational level of the first electronic energy state. This is followed by a relaxation to a higher excited vibrational ground-state level with the emission of a photon. Because of the multiplicity of vibrational levels as well as electronic levels, the spectra of both absorption and emission are polychromatic and generally are mirror images of one another.

Both the absorption and emission spectra of the fluorophore depend on its chemical structure as well as the environment (solvent, pH, temperature, etc.) of the fluorophore. The spectral wavelength of fluorescence emission is generally independent of the excitation wavelength of the absorbed photons. However, because of the rapid initial non-radiative decay associated with internal conversion as well as the final decay to higher vibrational levels of the ground state, the energy of the emitted photon is less than that of the absorbed photon. This shifts the fluorescence spectra to longer wavelengths relative to the absorption spectra and is known as the Stokes' Shift (STOKES, 1852).

3.2 Commercial Dye Fluorophores

The physiological response of the human eye qualitatively defines the visible wavelength region (in nanometers or nm) of the electromagnetic spectrum. Wavelengths shorter than, but adjacent to that of the visible region, are identified as ultraviolet. Wavelengths longer than, but adjacent to that of the visible region, are identified as near infrared. The commercialized fluorescent labels that are currently in use in automated DNA sequencing are either visible dyes (450–600 nm absorption and fluorescence range) or near infrared dyes (650–860 nm absorption and fluorescence range).

The first commercialized near infrared dyes introduced for automated DNA sequencing were IRDye41 and IRDye40 (STREKOWSKI et al., 1992; NARAYANAN et al., 1995; SHEALY et al., 1995) (Fig. 2). These dyes are from the heptamethine carbocyanine dye family and nominally absorb and fluoresce near 800 nm. IRDye41 was attached to a DNA primer via a stable thiourea linkage formed by conjugating the dye to an amino linker located at the 5′ end of the primer. A phosphoramidite version (IRDye800; Fig. 3) (NARAYANAN et al., 1998) provides for direct labeling of DNA primers using an automated DNA synthesizer. For dye labeled terminator chemistry, the IRDye800 is attached to bases which are linked to a triphosphate through an acyclo bridge (Fig. 4). The incorporation of this substrate terminates DNA chain elongation in a manner similar to that obtained by using dideoxynucleotides

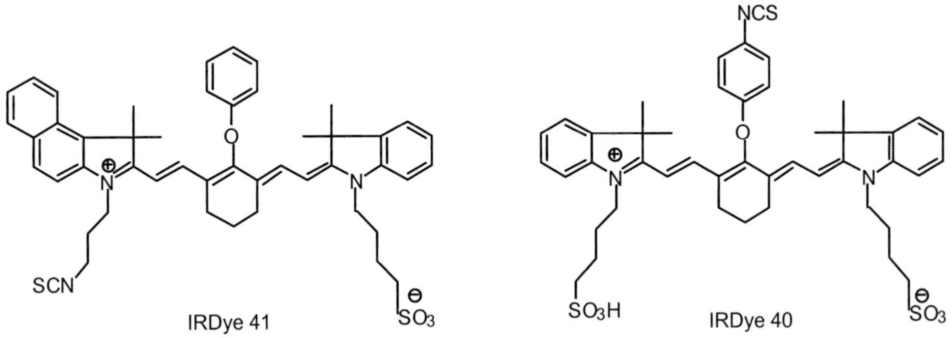

Fig. 2. Structures of IRDye41 and IRDye40. Both dyes are members of the polymethine carbocyanine dye family which is characterized by two heteroaromatic residues connected by a conjugation bridge of polyethylene units. The length of the conjugating bridge affects the absorbance and fluorescence maxima (MATSUOKA, 1990). IRDye41 and IRDye40 are heptamethine carbocyanine dyes which contain seven carbons in their conjugating bridge. The isothiocyanate (NCS) reactive functionality is used to couple the dye to a primary amine which results in a thiourea linkage.

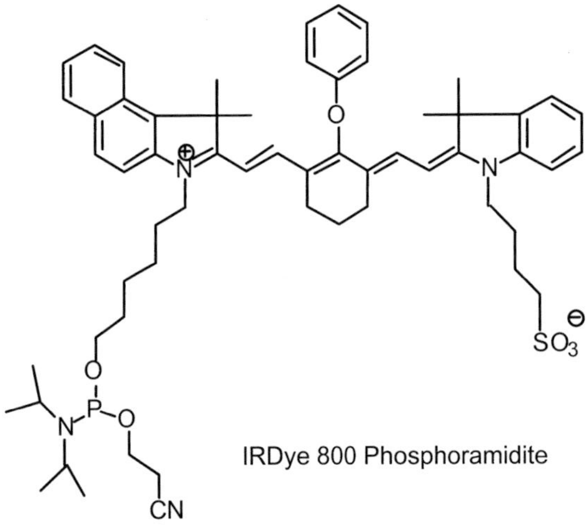

Fig. 3. Structure of IRDye800 Phosphoramidite. The amidite functionality is used to couple the dye to the 5′-OH of the 5′ terminus nucleotide of an oligonucleotide via automated DNA synthesis. See Fig. 2 legend for additional information.

(see Sect. 4). Dye properties for IRDye40, IRDye 41, and IRDye800 are listed in Tab. 1.

Commercialized near-infrared dyes that absorb and fluoresce around 650–700 nm are from the pentamethine carbocyanine dye family. They include IRDye700 (NARAYANAN et al., 1998), Cy5 (MUJUMDAR et al., 1989, 1993; ZHU et al., 1994), and Cy5.5 (TU et al., 1998). Dye properties for IRDye700, Cy5, and Cy5.5 are listed in Tab. 1 and the structures are shown in Fig. 5.

Shown in Fig. 6 are two fluorescein dye derivatives (FAM, JOE) and two rhodamine dye derivatives (TAMRA and ROX) first in use for four visible dye primer-based DNA sequencing. Fluorescein dye has also been used in single dye sequencers (ALF DNA Sequencer, Pharmacia Biotech). Shown in Fig. 7 are two rhodamine dyes (R110, R6G) which are combined with TAMRA and ROX for use in four visible color dye terminator-based DNA sequencing. Dye properties for FAM, JOE, TAMRA, ROX, R110, and R6G are listed in Tab. 1.

In order to give more even and narrower peak heights than the rhodamine dye termina-

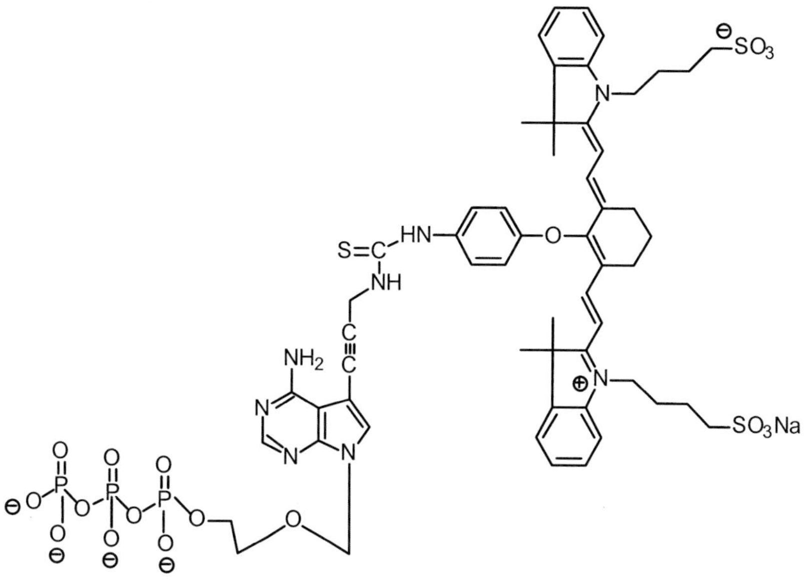

IRDye800-Acyclo-ATP

Fig. 4. Structure of IRDye800-acyclo-ATP. The dye is linked to an adenine base, which in turn is linked to a triphosphate. There is no ribose sugar. IRDye800-acyclo-CTP, IRDye800-acyclo-GTP, and IRDye800-acyclo-UTP are similarly synthesized with their respective base type. All four molecules are suitable substrates for chain elongation by DNA polymerase, but upon incorporation into the growing DNA strand, they terminate synthesis.

tors as well as reduce spectral overlap among the dyes, a family of dichlororhodamine (dRhodamine) dyes have been designed (RoSENBLUM et al., 1997). These dyes (dR110, dR6G, dTAMRA, dROX) are distinguished from R110, R6G, TAMRA, and ROX by the addition of two chlorides to the phenyl ring of the rhodamine (LEE et al., 1997). Fig. 8 shows the 4,7-dichloro-substituted R110 (dR110). Dye properties for dR110, dR6G, dTAMRA, and dROX are listed in Tab. 1.

3.3 Energy Transfer

When using four-color discrimination the spectral overlap of fluorescence emission of the four fluorophores reduces signal-to-noise and, therefore, results in less accurate and shorter read lengths. Also, because the absorp-tion spectra and molar absorptivities of the dyes are not equivalent, the use of a single excitation source for all four dyes compromises sequencing results due to widely varying fluorescence signal strength.

One method to improve the properties of dyes makes use of resonance energy transfer, an important process that occurs in the excited state of a fluorophore (FÖRSTER, 1948). Energy transfer can occur between two molecules if the emission spectrum of an absorbing fluorophore (donor) overlaps the absorption spectrum of a near-by acceptor fluorophore. The donor and acceptor molecules are coupled by a dipole–dipole interaction (LAKOWICZ, 1999). In addition to spectral overlap, the rate of energy transfer is dependent on the distance between donor and acceptor and follows an inverse relationship to the 6th power of that distance.

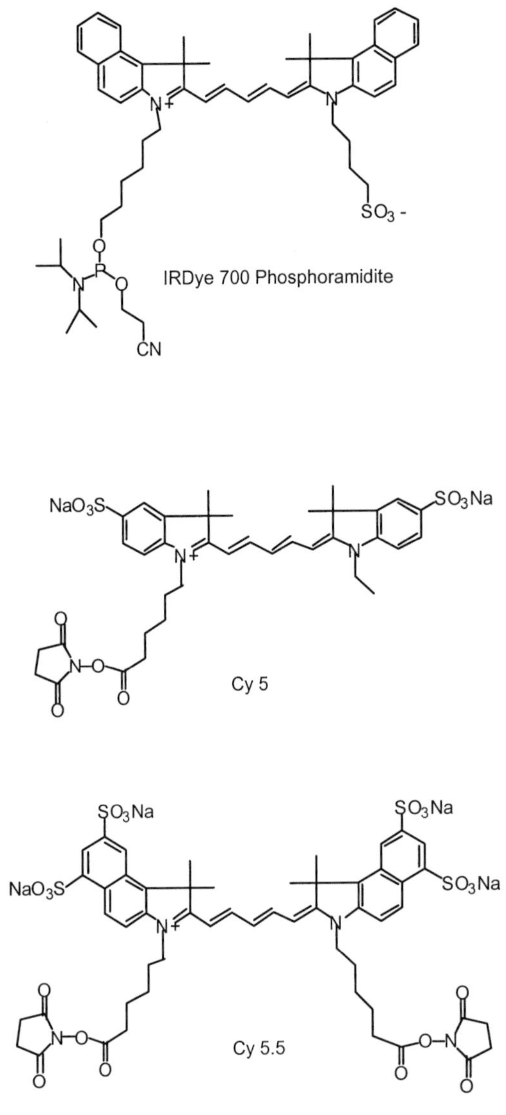

IRDye 700 Phosphoramidite

Cy 5

Cy 5.5

Fig. 5. Structures of IRD700 (phosphoramidite functionality), Cy5 (succinimidyl ester functionality), and Cy5.5 (bis-succinimidyl ester functionality). All three dyes are members of the pentamethine carbocyanine dye family which is characterized by five carbons in the conjugating bridge (see Fig. 2 legend).

An approach that involves energy transfer in labeled primer chemistry uses the oligonucleotide backbone to separate the donor and acceptor dyes (JU et al., 1995a, b, 1996;

HUNG et al., 1996a, b, 1997, 1998; METZKER et al., 1996). Another approach uses tethered donor and acceptor dyes for either labeled primers (LEE et al., 1997) or labeled terminators (ROSENBLUM et al., 1997). These tethered dyes use fluorescein as a donor dye and one of the four dRhodamine dyes (see Sect. 3.2) as an acceptor dye and are linked through 4-aminomethyl benzoic acid. The structure of a tethered fluorescein/dR110 is shown in Fig. 9.

3.4 Fluorescence Lifetime

When a fluorophore emits lights as it relaxes from an excited energy state to a ground energy state, such relaxation occurs after the molecule has spent a certain amount of time in the excited state (see Sect. 3.1). The average time spent in the excited state is known as the fluorescence lifetime of the molecule (LAKOWICZ, 1999) and it is statistically the same for all molecules having the same structure and exposed to the same environmental conditions. A common characteristic (although not necessarily assumable) is that the statistical relaxation of a fluorophore follows an exponential decay profile when examined over several excitation/relaxation cycles. For this case, the fluorescence lifetime is then specified as the exponential time constant where 63% of the relaxations occur more quickly than this lifetime average and 37% occur more slowly.

The lifetime of common visible and near infrared fluorophores ranges from 0.5–4 nanoseconds and is dependent on their chemical structure. The ability to discriminate among fluorophores is impacted by the ratio of their lifetimes as well as the number of photons available to produce the composite lifetime profile histogram (KÖLLNER and WOLFRUM, 1992; KÖLLNER, 1993).

The use of energy transfer to allow common excitation for multiple dyes has been successfully commercialized (see Sect. 3.3). As researchers examine alternative approaches for facilitating common excitation as well as increasing the number of available dye choices, the exploitation of fluorescence lifetime discrimination for DNA sequencing shows promising potential because a fluorophore's lifetime is independent of concentration and mul-

5 FAM

JOE

TAMRA

ROX

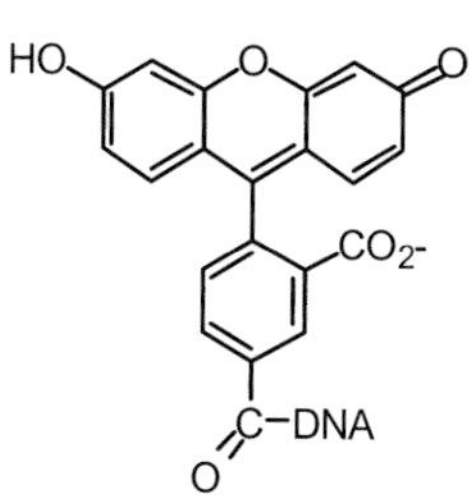

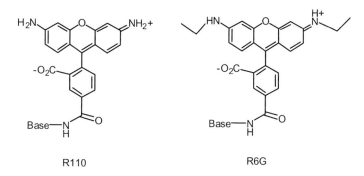

Fig. 6. Structures of dyes FAM, JOE, TAMRA, and ROX. FAM and JOE are members of the fluorescein family while TAMRA and ROX are members of the rhodamine family. All four dyes must be purified from isomers that contain alternate sites for the reactive functionality which ultimately couples the dye to DNA. Shown are the 5-isomer for FAM and the 6-isomer for JOE, TAMRA, and ROX.

R110

R6G

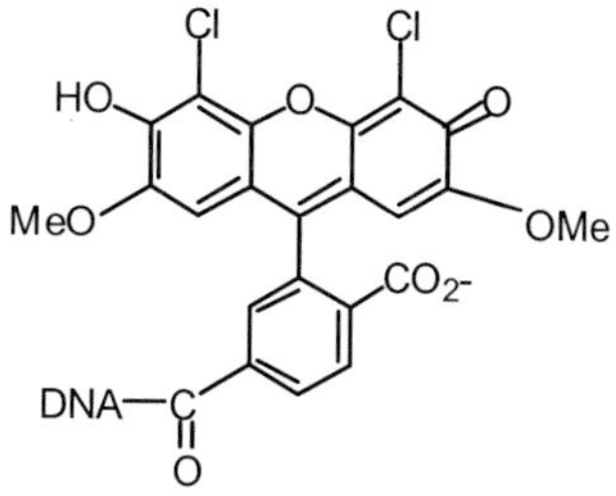

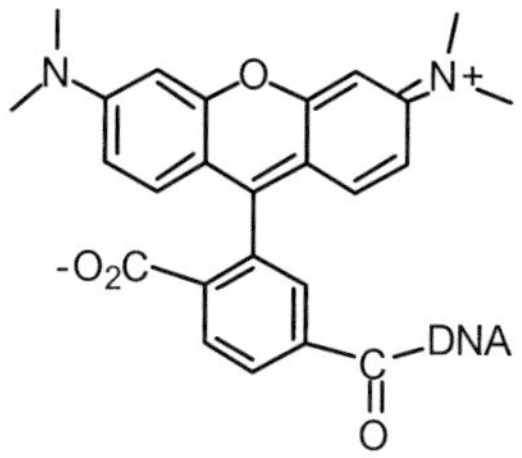

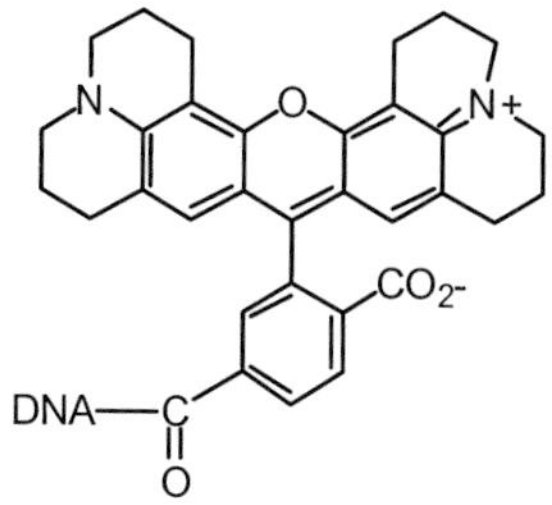

Fig. 7. Structures for R110 and R6G dyes. Both dyes are members of the rhodamine family. Shown are the 5-isomers.

tiple dyes having overlapping spectral emission can be distinguished (CHANG et al., 1993; HAN et al., 1993; SAUER et al., 1994; LEGENDRE et al., 1996; SOPER et al., 1996; MÜLLER et al., 1997; NUNNALLY et al., 1997; FLANAGAN et al., 1998). Methods for "on-the-fly" lifetime measurements of labeled DNA fragments have been described for capillary electrophoresis

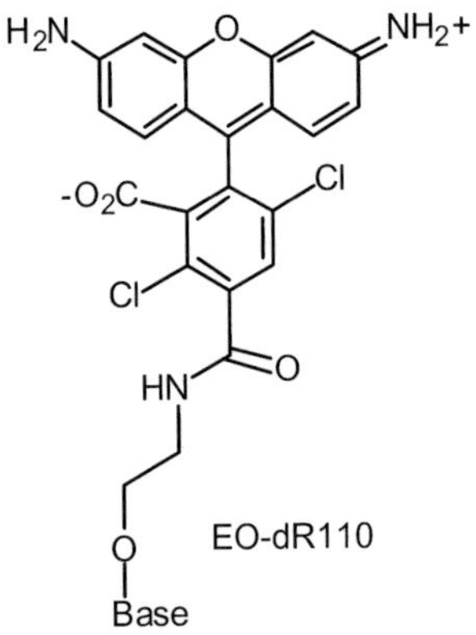

Fig. 8. Structure for dichloro-R110 dye linked to a nucleotide base. See text for the effects of adding the two chlorides to R110 (shown in Fig. 7). Similar dichloro modifications have been made to dyes TAMRA, ROX, and R6G (shown in Figs. 6 and 7).

embodiments (LI and MCGOWN, 1996; LI et al., 1997) and slab gel embodiments (LASSITER et al., 2000). Besides enabling common excitation, fluorescence lifetime discrimination also permits the use of common spectral detection optics. Both spectral and lifetime discrimina-

tion can be combined in a single design embodiment to take advantage of the strengths of each approach (PRUMMER et al., 2000; LASSITER et al., 2000).

4 Biochemistry of DNA Sequencing

The efficient completion of large DNA sequencing projects is now a reality due in great part to the development of fluorescence-based dideoxynucleotide sequencing chemistries coupled with instrumentation for real time detection of dye-labeled DNA fragments during gel electrophoresis (see Sect. 5). The commercially available automated sequencers (Sect. 5, Tabs. 2 and 3) can be divided into two groups based on the number of fluorescent dyes used in a sequencing reaction.

The first type uses the one-dye/four-lane approach in which the identity of the chain-terminating nucleotide at each position is determined by running four separate reactions each of which contains the same fluorescent

Tab. 1. Dye Absorption and Emission Properties (Aqueous Environment) for Several Commercial Dyes Available for DNA Sequencing. Absorption and Emission Maxima are Approximate and may be Dependent on Solvent, Solvent Properties (e.g., pH), and the Biomolecule to which they are Attached

Dye	Absorption Max	Emission Max	Dye Family
FAM	490–495 nm	515–520 nm	fluorescein
R110	500–505 nm	525–530 nm	rhodamine 110
dR110	NA	530–535 nm	rhodamine 110
JOE	520–525 nm	550–555 nm	dichlorodimethylfluorescein
R6G	525–530 nm	555–560 nm	rhodamine 6G
dR6G	NA	560–565 nm	rhodamine 6G
TAMRA	550–555 nm	580–585 nm	tetramethylrhodamine
dTAMRA	NA	590–595 nm	tetramethylrhodamine
ROX	580–585 nm	605–610 nm	X-rhodamine
dROX	NA	615–620 nm	X-rhodamine
Cy5	650–655 nm	665–670 nm	pentamethine carbocyanine
Cy5.5	670–675 nm	690–695 nm	pentamethine carbocyanine
IRDye700	685–690 nm	710–715 nm	pentamethine carbocyanine
IRDye40	765–770 nm	785–790 nm	heptamethine carbocyanine
IRDye41	795–800 nm	820–825 nm	heptamethine carbocyanine
IRDye800	795–800 nm	820–825 nm	heptamethine carbocyanine

NA, information not available

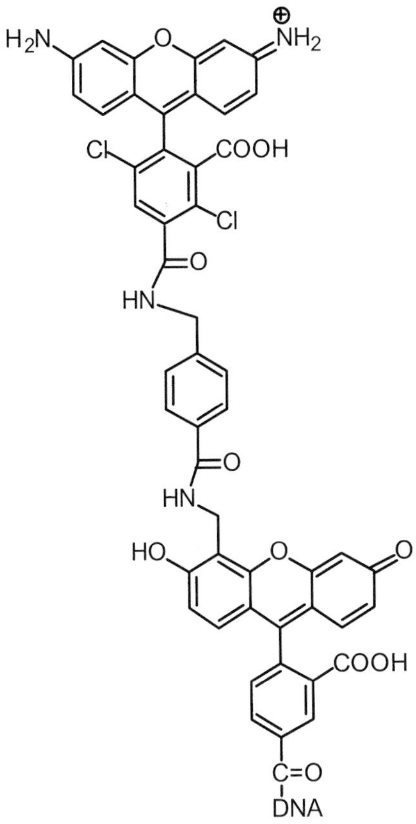

5CFB-dR110

Fig. 9. Structure of dichloro-R110 linked to 4'-aminomethylfluorescein (LEE et al., 1997; ROSENBLUM et al., 1997). This dual dye configuration permits fluorescence resonant energy transfer from fluorescein (donor) to R110 (acceptor) and is a member of a commercially available family of dyes trademarked as BigDyes™ (PE Biosystems). Other BigDyes™ are synthesized with dTAMRA, dROX, and dR6G as acceptors, all of which contain the dichloro modification.

dye but a different dideoxynucleotide (ddATP, ddTTP, ddGTP, ddCTP). The four completed sequencing reactions are loaded in separate lanes of a slab gel (Sect. 2, Fig. 1), and the automated sequencer must then be able to align the raw data from all four lanes precisely enough to determine the correct base sequence (Sect. 6.2).

The second type employs the four-dye/one-lane approach in which a single combined reaction is performed using a fluorescent label specific for each of the four dideoxynucleotides. The combined sequencing reaction can be analyzed in a single gel lane or capillary (Sect. 5.4), and the automated sequencer must first correct for the different mobilities of the four dye-labeled DNA fragment sets before calling bases (HAWKINS et al., 1992).

4.1 Sequencing Applications and Strategies

DNA sequencing is a fundamental technique in genome analysis and it has major applications which fall into two general classes:

(1) *de novo* sequencing of unknown DNA and
(2) resequencing segments of DNA for which the sequence is already known.

In both cases, the DNA to be sequenced is first cloned into a viral or plasmid vector, or is part of an amplified PCR fragment (Sect. 4.2).

The approach used to sequence unknown DNA is termed the sequencing strategy and it should provide the correct consensus sequence on both strands of the target DNA using a minimal number of sequencing reactions with minimal overlap (Sect. 4.1.1). Large-scale sequencing projects make use of one or more sequencing strategies to completely characterize the entire genome of an organism (SULSTON et al., 1992; FLEISCHMANN et al., 1995; VENTER et al., 1998; KUKANSKIS et al., 2000). On the other hand, many laboratories employ methods of resequencing to characterize the variability of smaller, known DNA segments in order to find mutations or verify recombinant DNA constructs (Sect. 4.1.2).

4.1.1 New Sequence Determination

The selection of a sequencing strategy usually depends on the size of the target DNA. For example, random shotgun sequencing is currently the method used in most large-scale DNA sequencing projects (MARTIN-GALLAR-

DO et al., 1994; VENTER et al., 1998). In shotgun sequencing, a large segment of target DNA (e.g., a medium-sized BAC clone of 100–120 kilobases) is randomly fragmented by physical shearing or enzymatic digestion to fragment sizes in the range of 1–5 kilobases. These smaller fragments are then subcloned into bacteriophage M13 or plasmid vectors (Sects. 4.2.1 and 4.2.2). The cloned inserts are sequenced from "universal" primer binding sites in the flanking vector DNA, and the resulting sequence information compiled by computer into contiguous sequence (i.e., "contigs") in order to reassemble the original large target DNA.

This method generates rapidly 95% of the desired sequence, but becomes less efficient as each subsequent random subclone is more likely to yield sequence information already obtained. Typically, each base in the target DNA sequence is read an average of four to six times during this "working draft" phase of the shotgun sequencing project. However, gaps or unresolved regions will still remain which can be filled in by directed approaches during the "finishing" phase of the sequencing project (HUNKAPILLER et al., 1991; SULSTON et al., 1992; ROACH et al., 1999; KUKANSKIS et al., 2000).

Advantages of shotgun sequencing include no requirement for prior knowledge of the insert sequence and no limitation on the size of the starting target DNA. Additionally, a high degree of parallel processing and automation can be implemented during the initial random phase, with only one or two oligonucleotide sequencing primers required.

Primer walking is a fully directed sequencing strategy. It provides an efficient way to obtain new sequence information, and is a good choice for the primary sequencing of small regions (1–3 kilobases) of genomic or cDNA clones, or as a secondary approach to achieve closure and resolve local ambiguities after an initial shotgun sequencing phase. Other approaches such as the enzymatic nested deletion method (LIU and FLEISCHMANN, 1994) or transposon insertion (MARTIN et al., 1994) have also been used for small-scale *de novo* sequencing.

The primer-directed method is initiated by sequencing the target DNA from one end using a vector-specific standard primer (VOSS et al., 1993b). A new walking primer is designed using the most distant, reliable sequence data obtained from the first sequencing reaction with the standard primer. This walking primer is then used to sequence the next unknown section of the DNA template. In theory, this primer walking process can be repeated many times to sequence extensive tracts of DNA. However, its use is generally limited to smaller projects because the successive rounds of sequence analysis, primer design, and primer synthesis are too expensive and time-consuming (ANSORGE et al., 1997).

The major benefits of primer walking are that no subcloning is required, the location and direction of each sequencing run is known, and the degree of redundancy needed to obtain final sequence is minimized. Moreover, read lengths greater than 1,000 bases have been reported (NISHIKAWA and KAMBARA, 1992; GROTHUES et al., 1993; ZIMMERMANN et al., 1994; MIDDENDORF et al., 1995; CARRILHO et al., 1996; KLEPÁRNIK et al., 1996; ROEMER et al., 1997, 1998; SALAS-SOLANO et al., 1998b; ZHOU et al., 2000) thus reducing the number of walking primers needed to finish a sequencing project.

4.1.2 Confirmatory Sequencing

The major purpose of DNA sequencing in many laboratories is to resequence small regions of interest (<1 kilobase) using cloned DNA or a PCR product as the template. Resequencing is useful for applications such as confirming plasmid constructs, screening the products of site-directed mutagenesis experiments, or comparing sequences of wild-type and mutant variants associated with genetic disease (LARDER et al., 1993; Perkin-Elmer/ABI, 1995; PLASCHKE et al., 1998). Since the target region has often been characterized, it is possible to design a primer so that the sequence of interest is within 100–150 bases of the sequencing primer. This will provide optimum resolution in the raw sequence data generated by the automated DNA sequencer, and thus the highest base calling accuracy that can be obtained (Sect. 6).

4.2 DNA Template Preparation

In the first step of a Sanger dideoxy sequencing reaction, the primer is annealed to a single-stranded DNA template (Sect. 2). DNA in this form can be purified directly from viruses such as bacteriophage M13 which have single-stranded genomes. On the other hand, double-stranded DNA such as a plasmid vector containing the target insert must first be converted to the single-stranded form either by alkali or heat denaturation prior to sequencing (CHEN and SEEBURG, 1985; ANSORGE et al., 1997).

The material presented in this section is intended to serve only as a general guide for preparing DNA templates. Specific protocols and applications can be found in several molecular biology manuals (SAMBROOK et al., 1989; AUSUBEL et al., 1992; FANNING and GIBBS, 1997; ANSORGE et al., 1997; WILSON and MARDIS, 1997a, b).

4.2.1 Single-Stranded DNA Template

Several variants of the bacteriophage M13 were constructed for the purpose of generating a DNA template for dideoxy sequencing (MESSING, 1983). The DNA to be sequenced is cloned into the double-stranded replicative form of the phage, transformed into *E. coli*, and harvested in large quantity from the culture medium in the form of phage particles containing single-stranded DNA (MESSING and BANKIER, 1989). The purified DNA is ideal for sequencing as it is single-stranded so that no complementary strand exists to compete with the sequencing primer during the annealing step. Moreover, a universal sequencing primer hybridizes to a complementary portion of the phage DNA immediately adjacent to the multiple cloning site. M13 is still used extensively for high-throughput sequencing applications (MARTIN-GALLARDO et al., 1994).

4.2.2 Double-Stranded DNA Template

Many methods have been developed to isolate and purify plasmid DNA from bacteria (SAMBROOK et al., 1989). Generally, the process involves five steps:

(1) insert foreign (target) DNA into the plasmid vector,
(2) transform a suitable bacterial strain with the recombinant plasmid,
(3) grow the bacterial culture,
(4) harvest and lyse bacteria, and
(5) purify the plasmid DNA.

For sequencing applications, double-stranded plasmid DNA containing the target sequence must be of high purity. Contaminating salt, RNA, protein, DNAses, and polysaccharides from the host bacteria can inhibit dideoxy sequencing reactions and produce low signal, high background, or spurious bands. Plasmid DNA purified through a cesium chloride gradient is suitable for sequencing provided that residual salt is removed from the DNA by ethanol precipitation. Commercial plasmid purification kits using anion-exchange resins or silica gel membrane technologies are available from Qiagen Inc. (Valencia, CA) or Promega Corp. (Madison, WI). These kits are easy to use and provide high quality DNA.

4.2.3 Vectors for Large-Insert DNA

Cloning vectors capable of replicating large DNA inserts, such as cosmids (DNA inserts 35 to 45 kilobases), P1-derived artificial chromosomes (PACs; DNA inserts from 100 to 150 kilobases), and bacterial artificial chromosomes (BACs; DNA inserts up to 300 kilobases), have been developed for use in genome mapping and large-scale DNA sequencing projects (CRAXTON, 1993; IOANNOU et al., 1994; SHIZUYA et al., 1992). These large-insert clones can be used to construct subclone libraries and then sequenced by the shotgun approach (WILSON and MARDIS, 1997b) (Sect. 4.1.1).

It is also important to sequence directly on these large DNA clones (BOYSEN et al., 1997; FAJAS et al., 1997). Sequence information from the ends of large-insert clones is used in the initial mapping phase of a sequencing project by detecting clones with overlapping sequence. Additionally, closing gaps and low quality regions in the "draft" sequence of a large-insert clone can be accomplished more efficiently by sequencing directly off the cosmid or BAC clone. This process eliminates the need to find the specific subclone sequence or to generate a new subclone library covering the gap.

4.2.4 PCR Products

The polymerase chain reaction (PCR) permits a region of DNA located between two distinct priming sites to be amplified (MULLIS and FALOONA, 1987). The product of this *in vitro* nucleic acid amplification is termed the PCR product. If equal amounts of the two primers are used, the PCR product will be a linear double-stranded DNA molecule typically less than 3 kilobases in size which can serve as template for DNA sequencing (INNIS et al., 1990; FANNING and GIBBS, 1997).

The PCR reaction mix contains significant amounts of reagents such as primers, nucleotides, enzyme, and even unwanted amplified products which must be completely removed from the PCR product before it can be successfully sequenced. Thus, the PCR product should be checked on an agarose gel to verify the presence of a single band of the expected size. Then, the PCR product is purified using a commercial PCR purification kit (e.g., Promega Corp. Wizard® DNA Clean-Up System) or by PEG precipitation (WILSON and MARDIS, 1997a). Alternatively, PCR products can be purified by agarose gel (AUSUBEL et al., 1992).

4.3 Enzymatic Reactions

4.3.1 DNA Polymerases

In the original Sanger dideoxy sequencing protocol, the Klenow fragment of *E. coli* DNA polymerase I was used for primer extension/termination reactions. The quality of the DNA sequence obtained with the Sanger method was significantly improved by the development of a modified T7 DNA polymerase (Sequenase® v2.0, United States Biochemical, Cleveland OH and Amersham Pharmacia Biotech, Piscataway NJ) which has enhanced processivity and a striking uniformity of termination patterns particularly when manganese ions are used as a cofactor (TABOR and RICHARDSON, 1987, 1989; VOSS et al., 1989). Both the Klenow fragment and modified T7 DNA polymerase catalyze the synthesis of DNA sequencing fragments in a single pass as the enzyme moves along the template DNA. However, these enzymes are also thermolabile, and thus cannot be used in cycle sequencing protocols which produce an amplification of signal by reusing repeatedly small amounts of the template DNA (CRAXTON, 1991). Modified T7 DNA polymerase is effective for sequencing difficult regions with repeats that cause premature "stops" in cycle sequencing reactions (WILSON and MARDIS, 1997a).

Cycle sequencing methods that utilize the thermostable *Thermus aquaticus* (Taq) DNA polymerase have been developed (INNIS et al., 1988; CRAXTON, 1991). The use of a thermostable DNA polymerase allows repeated rounds of high temperature DNA synthesis involving thermal denaturation of the double-stranded template DNA, primer annealing, and extension/termination of the reaction products. For each cycle, the amount of product DNA will be roughly equivalent to the amount of primed template. Thus, a significant benefit of cycle sequencing is that only small amounts of DNA template are required since the number of sequencing reaction products (i.e., "the signal") are linearly amplified during the 20–40 cycles of synthesis. For example, 20–30 ng of a small PCR product or 2–3 µg of a large BAC clone provide sufficient template DNA to complete a cycle sequencing reaction. Moreover, performing the cycle sequencing reactions at elevated temperatures minimizes sequencing artifacts due to secondary structure in the template DNA.

Until recently, the main disadvantage of cycle sequencing was the poor performance of the native Taq DNA polymerase which tends to incorporate dideoxynucleotides unevenly as compared to deoxynucleotides. As a result, sequencing patterns generated with these enzymes are not uniform (i.e., variable peak heights or band intensities) which reduces the base calling accuracy in automated DNA sequencers (Sect. 6). However, new genetically modified thermostable polymerases with a high affinity for dideoxynucleotides have been introduced (REEVE and FULLER, 1995; PARKER et al., 1996). These enzymes, Thermo Sequenase™ from Amersham Pharmacia Biotech and AmpliTaq FS™ from PE Biosystems (Foster City, CA), incorporate ddNTPs at rates similar to dNTPs resulting in uniform peak heights and, therefore, longer, more accurate sequence read lengths. Additionally, the reduced discrimination against ddNTPs that has been engineered into ThermoSequenase and AmpliTaq FS may manifest itself in the greater acceptance of fluorescent dye-labeled terminators (Sect. 2) as substrates in the enzymatic sequencing reaction (REEVE and FULLER, 1995).

4.3.2 Labeling Strategy

Automated DNA sequencing uses fluorescent dyes (Sect. 3) for the detection of electrophoretically resolved DNA fragments. There are three methods for labeling DNA sequencing reaction products:

(1) dye-labeled primer sequencing (SMITH et al., 1986; ANSORGE et al., 1986) in which the fluorescent dye is attached to the 5′ end of the oligonucleotide primer,

(2) dye-labeled terminator sequencing (PROBER et al., 1987; LEE et al., 1992) in which the fluorophores are attached to the dideoxynucleotides or a nonnucleotide terminator (ROEMER et al., 2000), and

(3) internal labeling (VOSS et al., 1993b, 1997; STEFFENS et al., 1995) in which a dye-labeled deoxynucleotide is incorporated during the synthesis of a new DNA strand.

Each labeling method has advantages and disadvantages.

Dye-labeled primer sequencing has benefited from the new DNA polymerases which do not discriminate between deoxynucleotides and dideoxynucleotides (Sect. 4.3.1). The sequencing electropherograms generated using these enzymes with dye-primers have very even peak heights which makes the base calling easy and reliable. Furthermore, signal uniformity allows heterozygote detection to be based on peak heights as well as the presence of two bases at the same position (Perkin Elmer/ABI, 1995). One disadvantage of the dye-primer method is a greater likelihood of increased background level (e.g., spurious bands) because nucleotide chains which terminate prematurely will add to the level of false terminations. Also, the four-dye/one-lane approach for automated sequencing (Sect. 4) requires four separate extension reactions and four dye-labeled primers per template.

The main advantages of dye-terminator sequencing are convenience, since only a single extension reaction is required per template, and the synthesis of a dye-labeled primer is not necessary. In fact, custom unlabeled primers with preferred hybridization sites can be used with dye terminators. Moreover, false terminations (i.e., DNA fragments terminated with a dNTP rather than a ddNTP) are not observed since these products are unlabeled. Finally, sequencing with dye terminators provides a way to read through most compressions. Presumably the large fluorophore at the 3′ end of the DNA fragment modifies or eliminates the in-gel secondary structure that causes compressions (WILSON and MARDIS, l997a). The major disadvantage of dye terminators is that the pattern of termination varies between DNA polymerases and is less uniform than for dye-labeled primers.

4.3.3 The Template–Primer–Polymerase Complex

An important factor in the relative success of a sequencing reaction is the number of template–primer–polymerase complexes formed during the course of a sequencing reaction. The formation of this complex is necessary in order to produce dye-labeled extension products. A significant number of problems associated with DNA sequencing reactions can be traced to one or more of these key elements.

For example, the ability of an oligonucleotide primer to bind to the template and interact with the DNA polymerase is a major factor in the overall signal strength of the reaction. Primers should be designed without inverted repeats or homopolymeric regions, a base composition of about 50% GC, no primer dimer formation, and one or more G or C residues at the 3′ end of the primer. These factors affect the stability of the primer–template interaction, and thus determine the number of primer–template complexes available to the DNA polymerase under a given set of conditions. For cycle sequencing with thermostable polymerases, it is important to design the primer with an annealing temperature of at least 50°C. Lower annealing temperatures tend to produce higher background and stops in cycle sequencing.

The amount of DNA template used in the dideoxy sequence reaction needs to be within an appropriate range. If the amount of template is too low, then few complexes will form and the overall signal level will be too low for automatic base calling. Additionally, higher amounts of a lower quality template (e.g., salt contaminant carried over from the DNA preparation) may be inhibitory to the DNA polymerase resulting in lower signal levels.

The most common factors which limit sequence read length and base calling accuracy in automated DNA sequencers are impure DNA template, incorrect primer or template concentrations, suboptimal primer selection and annealing, and poor removal of unincorporated dye-labeled dideoxynucleotides.

4.3.4 Simultaneous Bidirectional Sequencing

Simultaneous bidirectional sequencing (SBS), also termed "doublex" sequencing (WIEMANN et al., 1995; VOSS et al., 1997), is a sequencing method in which both strands of duplex DNA (plasmid or PCR product) are sequenced simultaneously by combining a forward and reverse primer (each labeled with a different fluorescent dye) in the same sequencing reaction. An automated DNA sequencing system with dual lasers, such as the LI-COR Model 4200 (Lincoln, NE) or the European Molecular Biology Laboratory (EMBL, Heidelberg, Germany) two-dye DNA sequencer, can be used to detect and analyze both the forward and reverse sequences of a bidirectional reaction in parallel (ROEMER et al., 1997; ANSORGE et al., 1997).

The benefits of the SBS method are threefold. First, SBS doubles the amount of sequence information from a single sequencing reaction. Second, since confirming sequence can be generated in the same reaction, it is easier to resolve ambiguities in one strand using the sequence of the complementary strand. Third, time and reagent consumption are halved by combining the forward and reverse sequencing reactions.

5 Fluorescence DNA Sequencing Instrumentation

5.1 Introduction

In principle, there are only three components of a fluorescence detection system:

(1) the excitation energy source,
(2) the fluorescent sample, and
(3) the fluorescence emission energy detector.

In practice, all of these components are sophisticated subsystems whose designs are coordinated to deliver maximum information throughput with optimized signal vs. noise discrimination (to achieve high accuracy and data quality). A brief discussion of these components is provided here to allow an overview of the parameters involved in proper instrumentation design for DNA sequencing. For a detailed description of general fluorescence-based instrumentation, a comprehensive textbook such as that authored by LAKOWICZ (1999) should be consulted. For a recent review of near-infrared fluorescence instrumentation refer to MIDDENDORF et al. (1998).

5.1.1 Excitation Energy Sources

Generally, laser-based excitation has been used for fluorescence-based DNA sequencing instrumentation, although Millipore introduced a DNA sequencer based on a white light source in 1991 (Millipore, 1991) (no longer commercialized). The most common lasers used in today's commercial DNA sequencing instrumentation are the blue/green Argon ion laser (488 nm and 514 nm excitation wavelengths) as based on early designs from the mid-1980s (SMITH et al., 1986; ANSORGE et al., 1986, 1987; PROBER et al., 1987; BRUMBAUGH et al., 1988; KAMBARA et al., 1988; MIDDENDORF et al., 1988) and far red or near infrared laser semiconductor diodes (650 nm, 680 nm, and 780 nm excitation wavelengths) (MIDDENDORF et al., 1992; YAGER et al., 1999). Additionally, the red Helium Neon laser (HeNe, 633 nm), the green frequency-doubled solid-state Neodymium:Yttrium-Aluminum-Garnet laser (Nd:YAG, 532 nm), and the green second harmonic generation laser (SHG, 532 nm, 473 nm) find their use as excitation sources for two-dimensional fluorescence scanners (PATTON, 2000) with a green laser (532 nm) also being used in a commercial capillary DNA sequencer (see Tab. 3, MegaBACE). It is necessary to combine proper geometric optics and spectral filtering to generate a highly focused excitation source with the proper wavelength necessary for compatibility with the fluorescent sample (BRUMBAUGH et al., 1988; MIDDENDORF et al., 1988, 1992, 1998; SWERDLOW et al., 1991; BRUMLEY and SMITH, 1991; HUANG et al., 1992a, b; KHETERPAL et al., 1996). The sample configuration (e.g., slab gels or capillaries) dictates additional mechanical/optical design criteria.

5.1.2 Fluorescence Samples

Dye properties such as absorption wavelength spectrum, molar absorptivity, fluorescence emission wavelength spectrum, fluorescence quantum yield, solubility, stability (e.g., temperature or light), and environmental effects (e.g., pH, quenching, temperature, solvent type) need to be all considered when designing a fluorescence system. A particular dye property critical to DNA sequencing performance is its electrophoretic mobility (TAN and YEUNG, 1997; TU et al., 1998; O'BRIEN et al., 1998a, b). For visible fluorescence (blue or green excitation), fluorophores in the fluorescein and rhodamine families are most commonly used. For far red or near infrared fluorescence the most common dyes are from the polymethine carbocyanine family (ERNST et al., 1989; MUJUMDAR et al., 1989; MATSUOKA, 1990; PATONAY and ANTOINE, 1991; MIDDENDORF et al., 1992, 1998; NARAYANAN et al., 1998). Energy transfer between acceptor and donor dyes has been successfully implemented as a strategy to manipulate compatibility between fluorophore properties and excitation sources as well as provide more even peak heights with greater color separation and, therefore, improved base calling (JU et al., 1995a, b, 1996; HUNG et al., 1996, 1998; METZKER et al., 1996; GLAZER and MATHIES, 1997; LEE et al., 1997; ROSENBLUM et al., 1997). (See Sect. 3 for a more detailed discussion of fluorescent dye chemistry.)

5.1.3 Fluorescence Detection

Three types of detectors have been used in fluorescence DNA sequencing instrumentation:

(1) photomultiplier tubes (PMTs) (SMITH et al., 1986; ANSORGE et al., 1986; PROBER et al., 1987; BRUMBAUGH et al., 1988; MIDDENDORF et al., 1988; SWERDLOW et al., 1991; HUANG et al., 1992a, b; KHETERPAL et al., 1996; BRUMLEY and LUCKEY, 1996);

(2) charge-coupled detectors (CCDs) (NORDMAN and CONNELL, 1996; UENO and YEUNG, 1994; PANG et al., 1999); and

(3) photodiode detectors (PDs), including silicon avalanche photodiodes (APDs) (MIDDENDORF et al., 1992, 1998; YAGER et al., 1999).

As in the case for the excitation subsystem, it is necessary to combine proper geometric optics, spectral filtering, and appropriate mechanical design to provide high sensitivity detection in the proper wavelength range associated with the fluorescent sample.

5.1.4 Overview of Fluorescence Instrumentation Related to DNA Sequencing

DNA samples for Sanger-based sequencing purposes (SANGER et al., 1977) (see Sects. 2 and 4) are prepared in such a way that they have three major attributes:

(1) the 5′ end of every DNA fragment within a sample begins with the same priming sequence;
(2) each DNA fragment is labeled with a fluorescent dye (or dye pair if energy transfer is used) either at or near the 5′ end or attached to the 3′ terminal dideoxynucleotide; and
(3) DNA fragments of various length, but having their 3′ terminus ending in a particular base type (A, C, T or G) are packaged into the same signal channel (e.g., they have the same type of fluorescence label with one label type for each base type or they are physically isolated from fragments terminated at the other base types such that a geometric channel can be used to distinguish base types).

The process of DNA sequencing performs three functions:

(1) maintaining the DNA samples in single-stranded form via a combination of denaturants in the gel and high temperature (45–70 °C);
(2) separation of the DNA fragments on the basis of their size with single base sizing resolution, and
(3) identification of those fragments via fluorescence optics at a "finish line" location where adequate separation among fragments has occurred.

In order to accomplish sizing, a sieving gel matrix is prepared and loaded either between two parallel glass plates (slab gels) or into a glass capillary. Slab gel matrices are generally crosslinked polyacrylamide (4–6%) whereas capillary gel matrices are non-crosslinked (such as linear polyacrylamide) (BASHKIN et al., 1996; GOETZINGER et al., 1998; HELLER, 1998; BARRON and ZUCKERMANN, 1999; SCHMALZING et al., 1999b and review references 11–15 therein). The gel thickness for slab gel sandwiches is 0.1–0.4 mm. The gel diameter for capillary gels ranges from 50–80 microns.

Both slab and capillary gels accomplish sizing (after the sample has been loaded into the gel) via a voltage gradient applied from one end of the gel to the other. The voltage gradient drives the negatively charged DNA molecules through the sieving matrix with the length of the DNA molecules determining their relative mobility. Each end of the gel is inserted into a running buffer that also contains an electrode that enables the voltage gradient across the gel. This voltage gradient may range from 30–80 volts cm^{-1} for a slab gel and from 50–250 volts cm^{-1} for a capillary gel.

For slab gels it is important to provide a method for keeping samples separated in geometric lanes. This is accomplished by one of several methods:

(1) a "comb" with multiple teeth is inserted at one end of the slab gel sandwich prior to the polymerization of the gel matrix. After the gel matrix is polymerized, the comb is removed which leaves open wells within the gel into which the DNA samples are loaded via a pipette tip;
(2) subsequent to gel polymerization, a "sharkstooth" comb is inserted into the end of the slab gel, with the sample loaded while the comb remains in place, whereby the teeth of the comb separate one sample from another;
(3) the sample is loaded into one of several wells that have been permanently fabricated into the top edge of one of the glass plates which then provides geometric isolation among samples (KOLNER et al., 1992); or
(4) the samples are first applied to a thin long membrane which is then inserted into an air gap located between the two

glass plates at one end of the sandwich (where the gel matrix has been excluded) (ERFLE et al., 1997).

For capillary gels the individual capillaries provide isolation among samples. In order to load samples into the capillaries, the loading end of the capillary is first submerged into a microwell containing the sample and the sample is then electrokinetically loaded into the capillary. After a brief period of loading, this end of the capillary is then submerged into a running buffer.

Detection at the "finish line" is accomplished by exciting the various electrophoresis channels either *en masse* (ANSORGE et al., 1987; KAMBARA et al., 1988; KAMBARA and TAKAHASHI, 1993; TAKAHASHI et al., 1994; CHEN et al., 1995; ANAZAWA et al., 1996, 1999; QUESADA et al., 1998) or one-by-one via sequential scanning or using discrete sources, with one or more laser sources (SMITH et al., 1986; PROBER et al., 1987; BRUMBAUGH et al., 1988; MIDDENDORF et al., 1988, 1992; QUESADA et al., 1991; HUANG et al., 1992a, b; KHETERPAL et al., 1996; YAGER et al., 1999). An optical microscope or individual detectors monitor any emitted fluorescence radiation from the sample (see Figs. 10–12 for one embodiment). In most cases the finish line is located towards one end of the glass enclosed gel. For capillary electrophoresis it is necessary to remove the polyimide coating of the capillary at the detection zone. One embodiment of capillary electrophoresis (see Tab. 3, Model 3700) uses a sheath flow detection scheme which monitors the sample after it exits the capillary (KAMBARA and TAKAHASHI, 1993; TAKAHASHI et al., 1994; DOVICHI, 1997).

5.2 Information Throughput

High information throughput is mandatory for addressing the accelerating demand for DNA sequencing. For the purpose of understanding the impact of all the parameters affecting throughput, one can model information throughput using the following formula (reproduced with kind permission from Kluwer Academic Publishers, Formula 1, p. 22; MIDDENDORF et al., 1998):

$$T_i = n \cdot d \cdot i/t \qquad (1)$$

where:
T_i = information $\underline{T}$hroughput
n = $\underline{n}$umber of sample channels
d = information $\underline{d}$ata per sample
i = information $\underline{i}$ndependence
t = $\underline{t}$ime per sample.

5.2.1 Sample Channels (n)

Various strategies for increasing the number of signal channels include geometric, spectral, temporal, and intensity discrimination. In almost all DNA sequencing approaches, each base type is assigned to a particular signal channel, thus requiring four signal channels per sample. However, it is theoretically necessary to have only two signal channels per DNA sample, but the use of redundant channel information reduces errors (NELSON et al., 1992, 1993; HUANG et al., 1992b).

The number of geometric channels relates to the number of lanes on a slab gel or the number of capillaries in a capillary-based DNA sequencer. As of this review, the maximum number of geometric channels currently commercially available is 96 lanes for both slab gel and capillary gel configurations (see Tabs. 2 and 3) although there are efforts to extend the number of capillaries. SCHERER et al. (1999) report on a rotary capillary array system that is designed to analyze over 1,000 sequencing separations in parallel.

Spectral discrimination using fluorescent dyes that have different wavelength properties is ubiquitous among commercial DNA sequencers and the number of dyes used ranges from two to four, although five dyes are sometimes used (ROQUE-BIEWER et al., 1998). There are two commercialized approaches to spectral discrimination:

(1) four dyes per single sample based on the early work of SMITH et al. (1986) and PROBER et al. (1987); and
(2) one dye for all four bases of each sample, but using different dyes for different samples (MIDDENDORF et al., 1998; VENTZKI et al., 1998; YAGER et al., 1999) that are loaded into the same geometric lanes.

The latter approach is based on the multiplex DNA sequencing technique developed by CHURCH and KIEFFER-HIGGINS (1988) which had its roots in the genomic sequencing approach of CHURCH and GILBERT (1984). Subsequent to CHURCH's work there were early reports of developing the technique through the use of fluorescence (YANG and YOUVAN, 1989; KAMBARA et al., 1991; WIEMANN et al., 1995; VOSS et al., 1997). Energy transfer among acceptor and donor dyes is another approach in spectral discrimination design (JU et al., 1995a, b; METZKER et al., 1996, LEE et al., 1997).

Temporal discrimination based on fluorescence lifetime has been investigated in the research community (CHANG and FORCE, 1993; HAN et al., 1993; SAUER et al., 1994; LEGENDRE et al., 1996; LI and MCGOWN, 1996; SOPER et al., 1996; LI et al., 1997; MÜLLER et al., 1997; NUNNALLY et al., 1997; FLANAGAN et al., 1998; LASSITER et al., 2000), although there yet remains to be a commercial DNA sequencer using this type of discrimination.

The use of intensity discrimination for DNA sequencing has also been limited to the research community (ANSORGE et al., 1989, 1990; HUANG et al., 1992b; PENTONEY et al., 1992; CHEN et al., 1992; STARKE et al., 1994; LI and YEUNG, 1995; WILLIAMS and SOPER, 1995; NEGRI et al., 1996).

5.2.2 Information per Channel (d)

The emphasis to increase the information per channel has been manifested in efforts to increase base read length (NISHIKAWA and KAMBARA, 1992; GROTHUES et al., 1993; ZIMMERMANN et al., 1994; MIDDENDORF et al., 1995; CARRILHO et al., 1996; KLEPÁRNIK et al., 1996; ROEMER et al., 1997, 1998; SALAS-SOLANO et al., 1998b; ZHOU et al., 2000) as well as the use of confidence values to assess the quality of each base call (EWING et al., 1998; EWING and GREEN, 1998; RICHTERICH, 1998; RICHTERICH et al., 1998). Both of these efforts are discussed in detail in Sects. 6.2 and 6.3. For a first order approximation, the base read length is related to the square root of the separation distance from the loading well to the detection location.

5.2.3 Information Independence (I)

This attribute relates to sequence alignment strategies (GAASTERLAND and SENSEN, 1996) as well as approaches to reducing systematic base calling errors that impact sequence alignment. For example, in shotgun DNA sequencing, the depth of sequence coverage impacts the amount of "draft" versus "finished" sequence due to statistical gaps between contiguous alignments (contigs) (LANDER and WATERMAN, 1988). Judicious choice of clones through tiling strategies (BATZOGLOU et al., 1999) as well as finishing strategies (ROACH et al., 1999) impact the cost/output ratio.

Another example of optimizing information dependence/independence ratios involves primer walking, where newly synthesized primers based on information from a prior DNA sequencing run are used to extend the read through a clone (VOSS et al., 1993a). Too much overlap between successive "walks" through the clone increases the overall cost per base sequenced.

The reduction of systematic errors in base calling is achieved by incorporating independent biochemical protocols such as choice of polymerase, choice of DNA strand, choice of dye chemistry (e.g., labeled primers versus labeled terminators) and choice of signal channel (e.g., four dyes/sample vs. four geometric lanes/sample) (KUKANSKIS et al., 2000).

5.2.4 Time per Sample (t)

Efforts to minimize the amount of time to obtain DNA sequence data involve three components:

(1) sample preparation,
(2) electrophoresis run times, and
(3) post-run sample information processing and analysis.

Robotics along with cycle sequencing methodology has greatly reduced the time (and cost) of sample preparation. The use of high voltage gradients (see Sect. 5.1.4) in combination with either ultrathin slab gels (BRUMLEY and SMITH, 1991; KOSTICHKA et al., 1992; CARNINCI et al., 1995; SMITH et al., 1996; YAGER et al., 1997) or capillary electrophoresis has significantly reduced run times, but at the expense of read length (LUCKEY and SMITH, 1993a; YAN et al., 1996). Efforts to increase read length for capillary DNA sequencing are under development (KLEPÁRNIK et al., 1996; CARRILHO et al., 1996; SALAS-SOLANO et al., 1998b). Extending the read length is significant in reducing the time and cost of achieving highly accurate and large contiguous regions of DNA sequence.

5.3 Instrument Design Issues

Proper design of fluorescence instrumentation for DNA sequencing involves a comprehensive analysis of both signal and noise components. The following description, although specific to the design of the LI-COR Model 4200 DNA sequencer (see Figs. 10–14), uses

principles that can be extrapolated to the design of other sequencers as well.

Fig. 10 shows the relationship among the three above-mentioned components of a fluorescence detection system (Sects. 5.1.1, 5.1.2, 5.1.3). The excitation energy source is a laser diode with associated collimating optics, spectral filtering, and focusing optics; the fluorescence sample is migrating (into the figure) through the gel that is sandwiched between a front and rear glass plate; and the fluorescence detection system consists of a collection lens, spectral filtering, focusing lens, and a cooled avalanche photodiode.

Fig. 11 shows the mechanical design required for scanning the multiple geometric channels within the gel sandwich. Other embodiments use side entry excitation for *en masse* detection (see Sect. 5.1.4) which do not require scanning or a rotating mirror to sequentially position the excitation beam across the geometric channels.

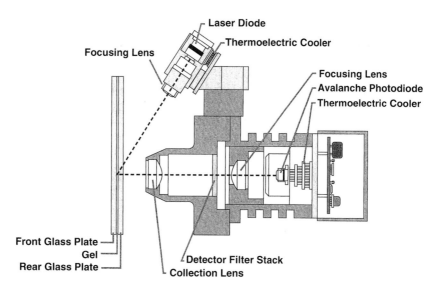

Fig. 10. Schematic of the Model 4200 single optical channel laser/microscope. The laser diode is placed at an angle such that the focused polarized excitation radiation enters the gel sandwich at Brewster's angle (56°). The detector (APD) is in an evacuated housing so that it can be cooled below the external dew point without condensation. Reproduced with kind permission from Kluwer Academic Publishers (Fig. 13, p. 33, MIDDENDORF et al., 1998).

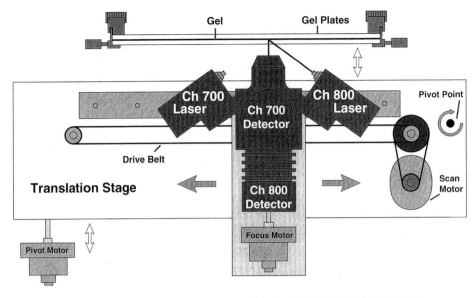

Fig. 11. A top view of the scanning platform as used in the LI-COR Model 4200 DNA sequencer. The scan motor moves the laser/detector focus motor assembly back and forth on the translation stage via the drive belt. The focus motor moves the laser/detector assembly toward or away from the glass plates. The pivot motor rotates the entire scanning assembly about the pivot point in order to align the scanning optics parallel to the gel between the glass plates. Reproduced with kind permission from Kluwer Academic Publishers (Fig. 14, p. 33, MIDDENDORF et al., 1998).

Fig. 12 shows an overview of how the three major components identified in Fig. 10 along with the scanning mechanism of Fig. 11 are integrated into a system which provides high voltage to the gel sandwich.

5.3.1 Laser Excitation and Fluorescence Emission

The relationship among fluorescence excitation, dye parameters, and fluorescence emission can be modeled by the following formula (reproduced with kind permission from Kluwer Academic Publishers, Formula 2, p. 39, MIDDENDORF et al., 1998):

$$F = \phi I_0 T (1 - 10^{-\varepsilon bc}) \approx \phi I_0 T \varepsilon b c \, [\ln 10] \quad (2)$$

where: F = fluorescence signal [watt]
ϕ = dye quantum efficiency [%]
I_0 = laser power [watt]

T = lens + filter transmission [%]
ε = decadic molar absorptivity [cm^{-1} M^{-1}]
b = path length [cm]
c = concentration [M].

Typical values of $\phi = 15\%$, $I_0 = 10$ milliwatt, $T = 50\%$, $\varepsilon = 200,000$ cm^{-1} M^{-1} (for polymethine carbocyanine dyes), and $b = 0.03$ cm predict that a concentration of 1 pM (10^{-12} molar) would generate 10 picowatts of fluorescence signal (see Fig. 13).

5.3.2 Detector Signal

The relationship between fluorescence emission and detector signal can be modeled by the following formula (reproduced with kind permission from Kluwer Academic Publishers, Formula 3, p. 40, MIDDENDORF et al., 1998):

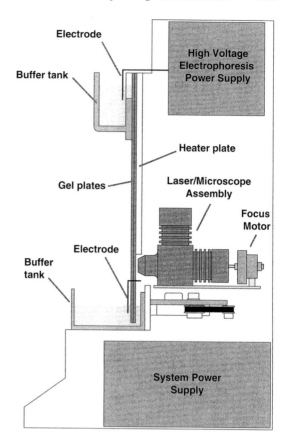

Electrode

Buffer tank

High Voltage Electrophoresis Power Supply

Heater plate

Gel plates

Laser/Microscope Assembly

Focus Motor

Electrode

Buffer tank

System Power Supply

Fig. 12. A side view of the electrophoresis apparatus and laser/microscope scanner. Two buffer tanks provide electrode contact across the gel sandwich. Platinum electrodes inserted into the buffer solutions are connected to a high voltage electrophoresis power supply. A heater plate provides regulated temperature control of the gel sandwich. The laser/microscope assembly scans in and out of the page by a belt driven scanning platform (see Fig. 11). Reproduced with kind permission from Kluwer Academic Publishers (Fig. 15, p. 34, MIDDENDORF et al., 1998).

$$S = F \cdot C \cdot T \cdot R \cdot Q \cdot G \qquad (3)$$

where: S = detector signal [amp]
F = fluorescence signal [watt]
C = collection efficiency
T = lens + filter transmission [%]
R = maximal detector responsivity [amp/watt]
Q = detector quantum efficiency [%]
G = internal detector gain.

Typical values of $C=3\%$ (numerical aperture of 0.5), $T=50\%$, $R=0.65$ amp/watt (for near infrared detection), $Q=75\%$ (for avalanche photodiode detector), and $G=250$ (for avalanche photodiode detector) predict that

a concentration of 1 pM (which generates 10 picowatts of fluorescence signal) would generate 20 picoamps of detector current (see Fig. 14).

5.3.3 System Noise

There are main sources of noise in the LI-COR Model 4200 DNA sequencer:

(1) the APD detector shot noise; and
(2) the Johnson (or thermal) noise of the electronic preamplifier.

The detector shot noise is given by the following formula:

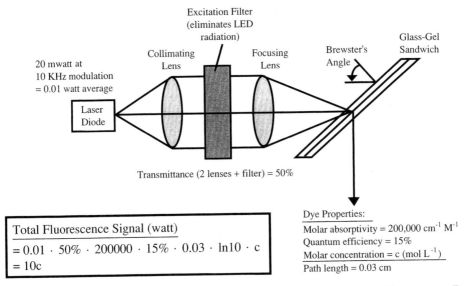

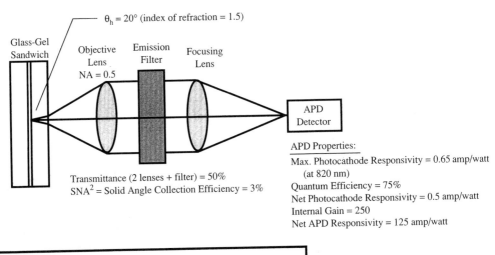

Fig. 13. Laser excitation and fluorescence emission in the LI-COR Model 4200 DNA sequencer. Reproduced with kind permission from Kluwer Academic Publishers (Fig. 22, p. 40, MIDDENDORF et al., 1998).

Fig. 14. Detection optics in the LI-COR Model 4200 DNA sequencer. Reproduced with kind permission from Kluwer Academic Publishers (Fig. 23, p. 41, MIDDENDORF et al., 1998).

$$I_s^2 = 2qB_wI_{ds} \qquad (4)$$

where: q = $1.6 \cdot 10^{-19}$ coul
B_w = bandwidth = 30 hz
I_{ds} = detector dark current =
175 pA at $-15\,°C$
then: I_s^2 = 41 fA.

The amplifier thermal noise is given by the following formula:

$$I_t^2 = 4KTB_w/R_f \qquad (5)$$

where: K = Bolzmann's constant =
$1.38 \cdot 10^{-23}$ joule/°K
T = 300 °K
B_w = bandwidth = 30 hz
R_f = 400 MΩ
then: I_t^2 = 35 fA.

The total system noise of 54 fA (10^{-15} amp) is determined from the following formula:

$$I_{sys}^2 = I_s^2 + I_t^2. \qquad (6)$$

(Formulae 4–6 are reproduced with kind permission from Kluwer Academic Publishers, Formulae 4–6, p. 42, MIDDENDORF et al., 1998.)

For a 0.4 mm thick gel and a 4.5 mm well width, the sensitivity of the LI-COR 4200 DNA sequencer is about 15 amol (MCINDOE et al., 1993). For thinner gels (0.2 cm) and narrower wells (2.25 mm), the sensitivity is about 5–10 amol (unpublished results). This compares to a sensitivity of 50–100 amol, 150–200 amol, 250–400 amol, and 250–800 amol for FAM, HEX, TAMRA, and ROX dyes, respectively in slab gel electrophoresis embodiments (NORDMAN and CONNELL, 1996).

5.4 Commercial Electrophoresis Embodiments for Fluorescence DNA Sequencing

There are currently two commercially available electrophoresis embodiments used for DNA sequencing:

(1) slab gels, and
(2) capillary gels (see Tabs. 2 and 3 for comparison of commercial products).

In addition, electrophoresis embodiments for DNA sequencing using micro-grooved channel plates are under development in both academic and industrial settings (see Sect. 5.4.3). All three embodiments use on-line detection in a "finish line" format which provides spatial information relating to the geometric channel dimension and temporal information relating to the bands within each channel. This temporal information is significantly different than that derived from spatially scanning a two-dimensional gel after stopping the electrophoresis run.

5.4.1 Slab Gels

Commercial slab gel electrophoresis systems (Tab. 2) include PE Biosystems Model 377, LI-COR Models 4200S and 4200L, Amersham Pharmacia Biotech Model ALFexpress II and SEQ4X4, Visible Genetics Models Micro-Gene Clipper and Long-Read Tower, Shimadzu Models DSQ-1000 and DSQ500, MJ Research Model BaseStation, and the Nucleo-Tech Model NucleoScan2000. There is significant variation among these systems with regard to the number of spectral channels (dyes), geometric channels (lanes), information per sample (read length, which depends on gel length), and time per read (run time).

5.4.2 Capillary Gels

Commercial capillary gel electrophoresis systems (Tab. 3) include PE Biosystems Models 310, 3100, and 3700; the Amersham Pharmacia Biotech/Molecular Dynamics MegaBACE 1000; the Beckman Coulter Model CEQ2000 and the SpectruMedix Model SCE9610. For an excellent description of capillary array electrophoresis technology please consult BASHKIN et al. (1996a), DOVICHI (1997), MARSH et al. (1997), PANG et al. (1999), BEHR et al. (1999), DOLNIK (1999), and references therein. Sample purification is important to achieve high performance in capillary DNA

Tab. 2. Comparison of Currently Commercialized Slab Gel Automated Fluorescence DNA Sequencers/Analyzers. Read Length (in Bases) is for Accuracies Ranging from 98–99%, Depending on Manufacturer. All Product Names are Trademarked under their Respective Manufacturers. Data Extracted from Internet Home Page of each Respective Manufacturer (where more Descriptive Detail can be Found) as well as BOGUSLAVSKY (2000)

Model	Company	Source	Detection	# Dyes	# Lanes	Gel Length [cm]	Read Length	Run Time
377	PE Biosystems	Ar laser	CCD	4–5	18, 36, 64, 96	36, 48 (WTR)	550, 650, 750	3, 9, 11 h
4200S	LI-COR	laser diodes	APD	2	32, 48, 64, 96	15, 31 (WTR)	400, 700	3, 6 h
4200L	LI-COR	laser diodes	APD	2	32, 48, 64, 96	15, 31, 56 (WTR)	400, 700, 1,000	3, 6, 10 h
ALFexpress	Amersham PB	HeNe laser	PD	1	40	NA	NA	NA
SEQ4X4	Amersham PB	laser diodes	PD	1	16	14	300	40 min
Clipper	Visible Genetics	laser diodes	PD	2	16	14	400	40 min
Tower	Visible Genetics	laser diodes	PD	2	16	28	800	4 h
DSQ1000	Shimadzu	Ar laser	NA	1	10 samples	NA	1,000	20 h
DSQ500	Shimadzu	Ar laser	NA	1	10 samples	NA	350	2–3 h
BaseStation	MJ Research	Ar laser	PMT	4	96	NA	500	2 h
NucleoScan	Nucleotech	solid state	NA	1	48	32	300	1 h

WTR, well-to-read distance; NA, information not available

Tab. 3. Comparison of Currently Commercialized Capillary Gel Automated Fluorescence DNA Sequencers/Analyzers. Read Length (in Bases) is for Accuracies Ranging from 98–99%, Depending on Manufacturer. All Product Names are Trademarked under their Respective Manufacturers (where more Descriptive Detail can be Found) as well as BOGUSLAVSKY (2000)

Model	Company	Source	Detection	# Dyes	# Lanes	Gel Length [cm]	Read Length	Run Time
CEQ2000	BeckmanCoulter	laser diodes	NA	4	8	NA	500	2 h
310	PE Biosystems	Ar laser	CCD	4	1	47, 61	400, 600	1, 2.8 h
3100	PE Biosystems	Ar laser	NA	4	16	NA	500, 650	1, 2.7 h
3700	PE Biosystems	Ar laser	CCD	4	96	NA	350, 500	2.3, 4 h
MegaBACE	Amersham PB	Ar laser, green laser	PMT	4	96	40 (WTR)	550, 800	2, 4 h
SCE9610	SpectuMedix	Ar laser	CCD	4–30	96	NA	NA	NA

WTR, well-to-read distance; NA, information not available

sequencing (RUIZ-MARTINEZ et al., 1998; SA-LAS-SOLANO et al., 1998a).

5.4.3 Micro-Grooved Channel Gel Electrophoresis

Instead of using capillaries for DNA separation (with associated electrokinetic loading from microwell plates), grooved channels can be etched in substrates using photolithography technology similar to that employed by the semiconductor industry (WOOLLEY et al., 1997; SCHMALZING et al., 1998, 1999a, b, 2000; SIMPSON et al., 1998; WATERS et al., 1998; LIU et al., 1999; RAMSEY, 1999; SHI et al., 1999). The January 2000 issue of *Electrophoresis* is a paper symposium on miniaturization and includes several reviews of microdevice electrophoresis, including BECKER and GÄRTNER (2000), McDONALD et al. (2000), DOLNIK et al. (2000), and CARRILHO (2000).

Loading of the sample into the grooved separation channel is significantly different from that of capillary electrophoresis. A "cross-T" interface (or variations using offsets in the junction) between a sample loading channel and the separation channel allows a sample plug to be injected into the separation channel without creating the bias towards loading only shorter fragments commonly associated with the electrokinetic loading of capillary electrophoresis. The sample can be loaded into the loading channel using electroosmotic pumping or by electrophoresis using electrodes that connect to the two ends of the loading channel.

There is also the potential for extending the lifetime of the grooved channels as compared to that of capillaries due to the ability to use high temperatures in connection with various solvents to refurbish the channels. (Using high temperatures with capillaries would damage the polyimide coating of the capillary, which is used to strengthen the capillary and reduce breakage in bending.) The micro-grooved plate embodiment is more conducive for interfacing with low volume, upstream reagent processes that require significantly less reagent quantities at reduced cost.

As of this review there were no commercialized DNA sequencers which use micro-grooved channel plates. However, a collaboration between Agilent Technologies (Palo Alto, CA) and Caliper Technologies (Mountain View, CA) has resulted in a commercial product (Agilent 2100 Bioanalyzer) that uses this technology for DNA separation of larger DNA fragments (but not for DNA sequencing). There are several academic and industrial efforts which are investigating this technology for DNA sequencing applications. Industrial efforts include reports by Agilent and Caliper (MUELLER et al., 2000), CuraGen (SIMPSON et al., 2000), and PE Biosystems (Foster City, CA) (REN et al., 1999; BACKHOUSE et al., 2000).

5.5 Non-Electrophoresis Embodiments for Fluorescence DNA Sequencing

Several programs for sequencing short fragments of DNA without using electrophoresis have been reported in the literature. The emphases of several of these programs lies in the importance of detecting single nucleotide polymorphisms (SNPs) for diagnostic applications. The efforts generally involve monitoring the extension or removal of the 3′ base, one base at a time.

One technique for the rapid sequencing of 40 kilobase or larger fragments of DNA at a rate of 100 to 1,000 bases per second has been explored by Los Alamos National Laboratory (JETT et al., 1989; DAVIS et al., 1991; HARDING and KELLER, 1992). The approach involves fluorescently labeling every base of a newly synthesized DNA strand using four types of dyes for each of the four base types, attaching the DNA strand to a solid support which is then moved into a flowing sample stream, cleaving the labeled 3′ base one by one with an exonuclease, and detecting the cleaved base using fluorescence detection.

Another technology based on the removal of bases from the 3′ end has been developed by Brenner and coworkers at Lynx Therapeutics (BRENNER, 1996a, b, 1997; BRENNER and DUBRIDGE, 1998; ALBRECHT et al., 2000; BRENNER et al., 2000). This method involves repeated cycles of ligation and cleavage of labeled probes at the 3′ terminus of target DNA. A similar method has been reported by JONES (1997).

ULMER (1997) developed a method for DNA sequencing involving exonuclease to cleave from a single DNA strand the next available single nucleotide on the strand and then detecting the cleaved nucleotide by transporting it into a fluorescence-enhancing matrix such that the natural fluorescence spectrum of the four nucleotides is exploited to determine the base sequence. In order to detect the natural fluorescence it is necessary to cool the nucleotides to cryogenic temperatures.

Several groups have investigated the method of single base extension for DNA sequencing where the extended base type is determined, one at a time, via fluorescence. MACEVICZ (1998) uses repeated cycles of ligation whereas others have used reversible terminators of polymerase extension where the terminators are labeled with a distinct, yet removable tag for each of the four base types (CANARD and SARFATI, 1994; METZKER et al., 1994).

Another ligase methodology involves the hybridization and subsequent ligation of short labeled extension oligonucleotides to a DNA template at a position adjacent to the 3′ (or 5′) end of previously hybridized oligonucleotides (BRENNAN and HEYNEKER, 1995). In this sequencing method, the labeled ligation product is formed wherein the position and type of label incorporated into the labeled ligation product provides information concerning the nucleotide residue in the DNA template with which it is base paired.

A clever version of DNA sequencing by single base extension (called pyrosequencing) that has been commercialized involves the use of pyrophosphate detection (HYMAN, 1988; NYRÉN, 1987, 1996; NYRÉN et al., 1993; RONAGHI et al., 1996, 1998a, b, 1999; AHMADIAN et al., 2000). Pyrosequencing involves the measurement of the absolute amount of natural nucleotide incorporation by detecting the amount of pyrophosphate released upon incorporation. The process utilizes a four-enzyme mixture, including DNA polymerase. The released pyrophosphate is converted to adenosine triphosphate (ATP) by ATP sulfurylase, which is then sensed by luciferase to

generate light. Apyrase is used to remove unreacted nucleotides.

A non-enzyme based technique that has shown utility for resequencing applications is sequencing by hybridization (DRMANAC et al., 1989, 1992, 1993; DRMANAC and DRMANAC, 1999). This technique involves hybridizing a library of short oligonucleotides to a DNA template and mathematically transforming the hybridization pattern into a sequence, based upon the individual sequences of the oligonucleotides from the library that actually hybridize.

5.6 Non-Fluorescence Embodiments for DNA Sequencing

Several non-fluorescence techniques for sequencing DNA have been investigated, including matrix-assisted-laser-desorption/ionization-time-of-flight (MALDI-TOF) mass spectrometry (KÖSTER et al., 1996; FU et al., 1998; KIRPEKAR et al., 1998; TARANENKO et al., 1998; GRIFFIN et al., 1999). The use of stable non-radioactive isotopes for labeling and detecting bases has been investigated for DNA sequencing with mass spectrometry (ARLINGHAUS et al., 1997, CHEN et al., 1999).

6 DNA Sequence Analysis

6.1 Introduction

The fundamental objective of data analysis is to determine in an automated fashion the DNA sequence from the fluorescence signals generated by the DNA sequence fragments associated with gel electrophoresis. The performance metrics of the data analysis software are read length, accuracy and confidence values of the resultant sequence. This is challenging due to the varying degrees of quality among multiple electrophoresis samples and runs (SERPICO and VERNAZZA, 1987).

One approach to automated sequence analysis that is dependent on minimal variation from one electrophoresis run to another requires adherence to rigid biological and electrophoresis protocols. This approach permits the implementation of an inflexible model that is relatively intolerant to data variations. However, if the data characteristics lie outside the predefined specifications, the automated analysis performance may be significantly compromised.

Another approach that requires considerably less adherence to such rigid protocols is adaptive automation. The algorithms dynamically adjust to optimally fit the data in order to be highly tolerant to the wide variability in data quality (GIDDINGS et al., 1993, 1998). In addition to evaluating local sequence parameters such as amplitude, peak time, peak width, and peak fluorescence spectra, it is important to understand the interdependence of these parameters among neighboring peaks (PARKER et al., 1996; ZAKERI et al., 1998; DAVIES et al., 1999; SONG and YEUNG, 2000). The best results from automated sequence analysis (long reads, high accuracy, and robust quality values) are obtained using a combination of prudent laboratory quality control measures and adaptive automation analysis.

During analysis the data generated by the fluorescence signal of the automated DNA sequencer is subjected to multiple processes. With each intelligent data reduction performed by the analysis software the data is further mathematically transformed such that the sequence information can be more readily and more accurately obtained. Major software deliverables of the analysis software include: lane detection and tracking, trace generation, base calling, and quality value generation.

6.2 Lane Detection and Tracking

Lane detection and tracking is the process of identifying the lane boundaries of DNA sequencing fragments throughout a complete gel image. This process is done automatically, but most analysis software packages make provision for optional visual verification and editing (retracking). Lane tracking is required for slab gel-based sequencers but is not required for capillary-based sequencers (see Sect. 5.4.2).

Lane tracking is a critical step in the sequence determination process in that its performance can directly affect the accuracy of the base calls for an entire sample or even a multi-sample gel run. Lane tracking is challenging when there is a wide range of sequence image configurations having varying degrees of quality (STARITA-GERIBALDI et al., 1993). Parameters of these various sequence image configurations include: comb sizes and types, sample loading formats, gel sizes, sequence chemistries and gel matrices. The lane tracking algorithm should also be able to effectively deal with image distortions such as non-uniform lane widths, lane drift, overlapping signals between lanes, variable background noise and signal intensities, a large signal dynamic range, and gel or image streaks/blobs.

In order for the lane tracker to effectively handle a wide range of DNA sequence loading options, a minimum set of *a priori* loading information is specified by the user for a given gel. This information includes comb type (rectangular vs. shark) and tooth size (well-to-well distance), number of samples, and sample loading format.

After initial computation of the lane boundaries, the image analysis software may, at the users discretion, display these computed boundaries graphically for manual verification or retracking. Additionally, the algorithm may perform a quality assessment of its initial lane finding performance. If the quality measure of its results is low (due to invalid sample loading information, sample loading errors, or a poor quality image), it then alerts the user while graphically displaying its suggested lane boundaries.

Under normal slab gel electrophoresis conditions the true lane tracks may contain some positional variation throughout an electrophoresis run. Such variation is different from one run to another. The adaptive algorithm responds to these (challenging) variations in an analogous manner to that of a human. Adaptive processing of stochastic images includes dynamic noise filtering, dynamic background subtraction, and pattern recognition. Artificial intelligence techniques for band feature detection and lane center determination are also utilized, including neural network schema (GOLDEN et al., 1993).

The lane detection process is initiated by the reception of the image file and the *a priori* load configuration. The image data is filtered to remove high frequency random noise and the background image "surface" is characterized according to its topology. Next, band features are detected via pattern recognition algorithms and the band center locations are determined. Using iterative optimization, the band centers are then partitioned and linked together in associated lane groups. The lane tracker is designed to dynamically make adjustments for lane drift but is restrained in making excessive adjustments. The resultant output of the lane tracker is a multiple set of lane track locations from the beginning to the end of the gel image (Fig. 15).

6.3 Trace Generation and Base Calling

After creating the lane track information (for slab gels only; not necessary for capillary gels), trace data are generated for each lane and the sequential base locations, or base calls, are determined. The objective in base calling is to extend the reading of the banding pattern as far as possible with the highest degree of accuracy, in an environment of varied degrees of gel or image quality (CHEN and HUNKAPILLER, 1992).

The accuracy of the base caller software is foremost impacted by the quality of the sequencing reactions (PARKER et al., 1996; ZAKERI et al., 1998) and the gel electrophoresis conditions (SWERDLOW et al., 1992, 1994; CARNINCI et al., 1995; DESRUISSEAUX et al., 1998).

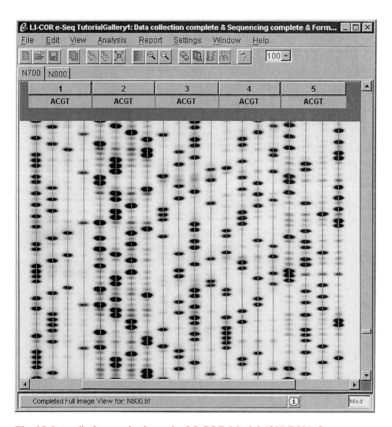

Fig. 15. Lane finder results from the LI-COR Model 4200 DNA Sequencer.

Some of the challenges to accurate base calling include: non-uniformity in band-to-band spacing, variable band spreading, non-uniform band mobility, and overlapping (poorly resolved) peaks. Additional potential sources of error include weak or variable signal strengths, ghost bands, variable band morphologies (often due to loading well distortions and/or salt gradients between the sample and the running buffer), and undesired excess signal artifacts.

For slab gels, the base caller initially transforms data within each of the two-dimensional (2D) image lane tracks into one-dimensional (1D) lane trace profiles. This step is not performed for capillary gels as the initial data is already formatted as a 1D trace. However, the 2D image data format associated with slab gels contains additional information that can be utilized to improve the base calling accuracy. For example, in reducing the dimensionality of the signal, the signal-to-noise ratio is enhanced in that summing pixel data across the lane width increases the signal linearly but the noise increases only according to the square root of the number of pixels. Additionally, pattern recognition of the full two-dimensional nature of DNA bands assists in analyzing overlapping bands. However, care must be taken in the 2D to 1D transformation process such that no signal information is lost or that distortions are not introduced when creating the 1D trace data.

A primary intra-lane image distortion that requires correction prior to dimension reduction is band tilt (due to thermal effects, non-uniform salt concentrations, or well loading errors) (KOUTNY and YEUNG, 1992). The computer algorithm dynamically calculates how much band tilt is present in each lane in order to produce undistorted lane trace profiles. The resultant lane traces are also dynamically corrected with respect to background signal levels.

A composite sequence trace is then created by overlaying each of the four associated base profiles (A, T, G, C) together (DEAR and STADEN, 1992). For purposes of displaying the composite trace, each base profile is uniquely colorized (e.g., red = T, green = A, blue = C, black = G) for visual identification. An idealized composite trace would consist of evenly spaced, non-overlapping peaks, each corresponding to the labeled fragments that terminate at a particular base in the sequence strand (EWING et al., 1998). However, the non-ideal trace requires further processing such as mobility correction and deconvolution.

Mobility correction is performed on the traces to compensate for mobility inequalities among lanes due to thermal gradients in the gel (smile) (KOUTNY and YEUNG, 1992; STARITA-GERIBALDI et al., 1993). It can also result from errors in gel preparation and loading. In order to eliminate the mobility shift effect, it is necessary for the software algorithm to dynamically examine the band signals throughout the gel. Also, in the case of loading all four base types from one sample into a single lane, it is necessary to correct for mobility shifts due to the use of different dye labels for each base type (TAN and YEUNG, 1997; TU et al., 1998; O'BRIEN et al., 1998a, b).

The next processing step is to mathematically enhance the resolution of the band signals via deconvolution (WU and MISLAN, 1992; IVES et al., 1994; RICHTERICH et al., 1998). Resolution has been shown to be the limiting factor for increasing read lengths and it degrades as a function of electrophoresis run time (NISHIKAWA and KAMBARA, 1991; SANDERS et al., 1991; ALDROUBI and GARNER, 1992; GROSSMAN et al., 1992; SLATER and DROUIN, 1992; LUCKEY et al., 1993; LUCKEY and SMITH, 1993a, b; RIBEIRO and SUTHERLAND, 1993; SLATER, 1993; MIDDENDORF et al., 1995; SLATER et al., 1995; FANG et al., 1996; NISHIKAWA and KAMBARA, 1996; YAN et al., 1996; WEISS and KIEFER, 1997; GUTTMAN et al., 1998). Mathematically, resolution is represented by the ratio of band spacing to band spreading. The result of deconvolution is to enhance the data in such a way as to increase this resolution ratio. Deconvolution techniques should not alter band positional information, but decrease band broadening, improve resolution, and decrease signal overlap as shown in Fig. 16 (ROEMER et al., 1998).

The final stage in the analysis is the identification of the individual bases via intelligent and adaptive processing of the transformed traces. These adaptive techniques characterize band signal intensities, spacing, and spreading. The performance of the LI-COR base caller on genomic DNA data is shown in Fig. 17.

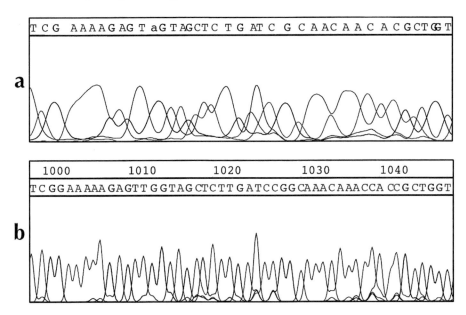

Fig. 16. Resolution enhancement by deconvolution: **a** non-deconvolved traces, **b** deconvolved trace. Both traces represent the section of the pGEM® sequence from base 1000–1040 as collected on a LI-COR Model 4200 DNA Sequencer using 66 cm gel protocols (56 cm well-to-read distance, 0.2 mm gel thickness, 100 bases per hour run speed). For the region displayed 10 errors were made by the LI-COR base caller (all deletions) when analyzing the upper, non-deconvolved trace whereas no errors resulted when analyzing the lower, deconvolved trace.

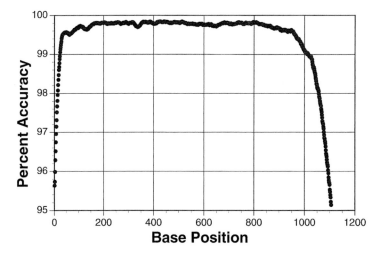

Fig. 17. LI-COR Model 4200 DNA Sequencer base caller accuracy vs. base position. The accuracy represents data from 2,200 samples of a >2.2 million base pair project where the average read length was greater than 1,000 base pairs and the average accuracy over all reads was 99.65%. (Data courtesy of Drs. P. BROTTIER, H. CRESPEAU, and P. WINCKER, GENOSCOPE National Sequencing Center, EVRY Cedex, France.)

6.4 Quality/Confidence Values

High-throughput sequencing established the need for reliable quality control measures for the sequence base calls (CHEN and HUNKA-PILLER, 1992; NICKERSON et al., 1997; EWING et al., 1998; EWING and GREEN, 1998; RIEDER et al., 1998; BUETOW et al., 1999) where an estimate of the probability of error is given for each base call. This estimate is a function of selected measurable data quality parameters and allows for automation in performing sequence assemblies, quality control, and benchmarking. Assembly software programs, such as PHRAP (EWING and GREEN, 1998), CONSED (GORDON et al., 1998), or CAP3 (HUANG and MADAN, 1999), utilize these base specific quality values to improve the accuracy of assembly by weighting the base calls according to their quality values when generating a consensus sequence (RICHTERICH, 1998).

To begin a calibration procedure, a pool of measurable sequence data characteristics or parameters that have a strong correlation to the performance of the base caller is identified. Those parameters which provide the greatest degree of discrimination between correct and incorrect base calls are then selected from the pool. Effective data quality parameters include signal-to-noise ratio, resolution, and band-to-band spacing.

Once the most effective base quality parameters have been determined, a correlation function (quality predictor) is built. The quality predictor receives as input a set of these quality parameters for each base call. The predictor's output is generated by correlating the quality parameters to a predicted accuracy (or probability of error) based on past performance under similar parameter conditions.

Proper calibration of the quality predictor requires the formation of a large database containing several millions redundant base calls along with a known consensus sequence associated with those base calls. (It is important that the redundancy is based on diverse sequence conditions.) A quality parameter set is then determined for each base call as well as

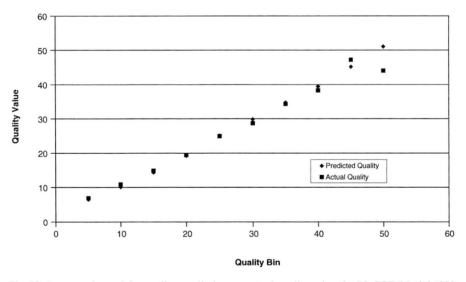

Fig. 18. A comparison of the quality prediction vs. actual quality using the LI-COR Model 4200 DNA Sequencer base caller. Predicted quality values based on 1,518,306 actual base calls were binned into 5-score bins. The actual quality values were determined by how many actual errors (based on the consensus sequence) there were for each predictive quality bin, divided by the total number of base calls in that predictive quality bin. (Data courtesy of Drs. P. BROTTIER, H. CRESPEAU, and P. WINCKER, GENOSCOPE National Sequencing Center, EVRY Cedex, France.)

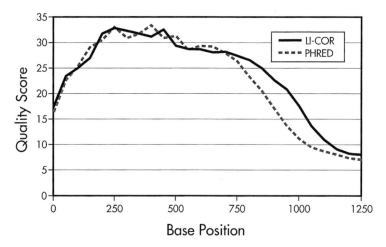

Fig. 19. Quality values vs. base position for LI-COR and PHRED base callers. These quality values reflect the actual error rates in the aligned parts of sequences in data set TitanGV2. (Data courtesy of Drs. P. BROTTIER, H. CRESPEAU, and P. WINCKER, GENOSCOPE National Sequencing Center, EVRY Cedex, France.) The graph illustrates that PHRED and LI-COR base callers perform substantially the same for the first 700 bases. After 700 bases, the Ll-COR base caller makes fewer errors than PHRED, resulting in a higher quality score.

an identification of its correctness. Using the established database of base calls (and knowledge of both correct and incorrect calls), a correlation between the accuracy and the quality parameters is statistically defined. This correlation (or quality predictor) is then stored in the form of a lookup table for further access during subsequent base calling (EWING and GREEN, 1998). As an example of the need for the large database, it can be shown that in order to make a prediction for an accuracy of 1 error out of 10,000 base calls, it is necessary to typify several tens of thousands of base calls that have a similar quality parameter set.

Results of the LI-COR Model 4200 base caller quality predictor on genomic data are shown in Figs. 18 and 19 where the quality values are defined as 10 times the logarithm (base 10) of the estimated error probability for that base call. Quality values of 20 and 30 correspond to predicted error rates of one in one hundred and one in one thousand, respectively.

7 References

ADAMS, J. M., JEPPESEN, P. G., SANGER, F., BARRELL, B. G. (1969), Nucleotide sequence from the coat protein cistron of R17 bacteriophage RNA, *Nature* **223**, 1009–1014.

AHMADIAN, A., LUNDEBERG, J., NYRÉN, P., UHLÉN, M., RONAGHI, M. (2000), Analysis of the *p53* tumor suppressor gene by pyrosequencing, *BioTechniques* **28**, 140–147.

ALBRECHT, G., BRENNER, S., DUBRIDGE, R. B., LLOYD, D. H., PALLAS, M. C. (2000), Massively parallel signature sequencing by ligation of encoded adaptors, *U.S. Patent* 6 013 445.

ALDROUBI, A., GARNER, M. M. (1992), Minimal electrophoresis time for DNA sequencing, *BioTechniques* **13**, 620–624.

ALPHEY, L. (1997), *DNA Sequencing.* New York: Springer-Verlag.

ANAZAWA, T., TAKAHASHI, S., KAMBARA, H. (1996), A capillary array gel electrophoresis system using multiple laser focusing for DNA sequencing, *Anal. Chem.* **68**, 2699–2704.

ANAZAWA, T., TAKAHASHI, S., KAMBARA, H. (1999), A capillary-array electrophoresis system using side-entry on-column laser irradiation combined with glass rod lenses, *Electrophoresis* **20**, 539–546.

ANSORGE, W., SPROAT, B. S., STEGEMANN, J., SCHWA-

GER, C. (1986), A non-radioactive automated method for DNA sequence determination, *J. Biochem. Biophys. Methods* **13**, 315–323.

ANSORGE, W., SPROAT, B., STEGEMANN, J., SCHWAGER, C., ZENKE, M. (1987), Automated DNA sequencing: Ultrasensitive detection of fluorescent bands during electrophoresis, *Nucleic Acids Res.* **15**, 4593–4602.

ANSORGE, W., VOSS, H., WIRKNER, U., SCHWAGER, C., STEGEMANN, J. et al. (1989), Automated Sanger DNA sequencing with one label in less than four lanes on gel, *J. Biochem. Biophys. Methods* **20**, 47–52.

ANSORGE, W., ZIMMERMANN, J., SCHWAGER, C., STEGEMANN, J., ERFLE, H., VOSS, H. (1990), One label, one tube, Sanger DNA sequencing in one and two lanes on a gel, *Nucleic Acids Res.* **18**, 3419–3420.

ANSORGE, W., VOSS, H., ZIMMERMANN, J. (1997), *DNA Sequencing Strategies*. New York: John Wiley & Sons.

ARLINGHAUS, H. F., KWOKA, M. N., GUO, X.-Q., JACOBSON, K. B. (1997), Multiplexed DNA sequencing and diagnostics by hybridization with enriched stable isotope labels, *Anal. Chem.* **69**, 1510–1517.

AUSUBEL, F. M., BRENT, R., KINGSTON, R. E., MOORE, D. D., SEIDMAN, J. G. et al. (1992), *Short Protocols in Molecular Biology*. New York: John Wiley & Sons.

BACKHOUSE, C., CAAMANO, M., OAKS, F., NORDMAN, E., CARRILLO, A. et al. (2000), DNA sequencing in a monolithic microchannel device, *Electrophoresis* **21**, 150–156.

BARRON, A. E., ZUCKERMANN, R. N. (1999), Review: Bioinspired polymeric materials: in-between proteins and plastics, *Curr. Opin. Chem. Biol.* **3**, 681–687.

BASHKIN, J. S., BARTOSIEWICZ, M., ROACH, D., LEONG, J., BARKER, D., JOHNSTON, R. (1996a), Implementation of a capillary array electrophoresis instrument, *J. Cap. Electrophor.* **3**, 61–68.

BASHKIN, J., MARSH, M., BARKER, D., JOHNSTON, R. (1996b), DNA sequencing by capillary electrophoresis with a hydroxyethylcellulose sieving buffer, *Appl. Theor. Electrophor.* **6**, 23–28.

BATZOGLOU, S., BERGER, B., MESIROV, J., LANDER, E. S. (1999), Sequencing a genome by walking with clone-end sequences: A mathematical analysis, *Genome Res.* **9**, 1163–1174.

BECKER, H., GÄRTNER (2000), Polymer microfabrication methods for microfluidic analytical applications, *Electrophoresis* **21**, 12–26.

BEHR, S., MÄTZIG, M., LEVIN, A., EICKHOFF, H., HELLER, C. (1999), A fully automated multicapillary electrophoresis device for DNA analysis, *Electrophoresis* **20**, 1492–1507.

BOGUSLAVSKY, J. (2000), DNA sequencers reach production scale, *Drug Discovery Devel. Mag.* January/February, 36–38.

BOYSEN, C., SIMON, M. I., HOOD, L. (1997), Fluorescence-based sequencing directly from bacterial and P1-derived artificial chromosomes, *BioTechniques* **23**, 978–982.

BRENNAN, T. M., HEYNEKER, H. L. (1995), Methods and composition for determining the sequence of nucleic acids, *U.S. Patent* 5 403 708.

BRENNER, S. (1996a), DNA sequencing by stepwise ligation and cleavage, *U.S. Patent* 5 552 278.

BRENNER, S. (1996b), DNA sequencing by stepwise ligation and cleavage, *U.S. Patent* 5 599 675.

BRENNER, S. (1997), Massively parallel sequencing of sorted polynucleotides, *U.S. Patent* 5 695 934.

BRENNER, S., DUBRIDGE, R. B. (1998), DNA sequencing by stepwise ligation and cleavage, *U.S. Patent* 5 714 330.

BRENNER, S., WILLIAMS, S. R., VERMAAS, E. H., STORCK, T., MOON, K. et al. (2000), *In vitro* cloning of complex mixtures of DNA on microbeads: physical separation of differentially expressed cDNAs, *Proc. Natl. Acad. Sci. USA* **97**, 1665–1670.

BROWNLEE, G. G., SANGER, F., BARRELL, B. G. (1967), Nucleotide sequence of 5S-ribosomal RNA from *Escherichia coli*, *Nature* **215**, 735–736.

BRUMBAUGH, J. A., MIDDENDORF, L. R., GRONE, D. L., RUTH, J. L. (1988), Continuous, on-line, DNA sequencing using multifluorescently tagged primers, *Proc. Natl. Acad. Sci. USA* **85**, 5610–5614.

BRUMLEY, JR., R. L., LUCKEY, J. A. (1996), An improved high-throughput DNA fragment analyzer employing horizontal ultrathin gel electrophoresis, *Proc. SPIE* **2680**, 349–361.

BRUMLEY, JR., R. L., SMITH, L. M. (1991), Rapid DNA sequencing by horizontal ultrathin gel electrophoresis, *Nucl. Acids Res.* **19**, 4121–4126.

BUETOW, K. H., EDMONSON, M. N., CASSIDY, A. B. (1999), Reliable identification of large numbers of candidate SNPs from public EST data, *Nature Genet.* **21**, 323–325.

CANARD, B., SARFATI, R. S. (1994), DNA polymerase fluorescent substrates with reversible 3'-tags, *Gene* **148**, 1–6.

CARNINCI, P., VOLPATTI, F., SCHNEIDER, C. (1995), A discontinuous buffer system increasing resolution and reproducibility in DNA sequencing on high voltage horizontal ultrathin-layer electrophoresis, *Electrophoresis* **16**, 1836–1845.

CARRILHO, E. (2000), DNA sequencing by capillary array electrophoresis and microfabricated array systems, *Electrophoresis* **21**, 55–65.

CARRILHO, E., RUIZ-MARTINEZ, M. C., BERKA, J., SMIRNOV, I., GOETZINGER, W. et al. (1996), Rapid DNA sequencing of more than 1,000 bases per run by capillary electrophoresis using replaceable linear polyacrylamide solutions, *Anal. Chem.* **19**, 3305–3313.

CHANG, K., FORCE, R. K. (1993), Time-resolved laser induced fluorescence study on dyes used in DNA sequencing, *Appl. Spect.* **47**, 24–29.

CHEN, W. Q., HUNKAPILLER, T. (1992), Sequence accuracy of large DNA sequencing projects, DNA Sequence, *DNA Seq.* **2**, 335–342.

CHEN, E. Y, SEEBURG, P. H. (1985), Supercoil sequencing: a fast and simple method for sequencing plasmid DNA, *DNA* **4**, 165–170.

CHEN, D., HARKE, H., DOVICHI, N. J. (1992), Two-label peak-height encoded DNA sequencing by capillary gel electrophoresis: three examples, *Nucleic Acids Res.* **20**, 4873–4880.

CHEN, D., PETERSON, M. D., BRUMLEY, JR., R. L., GIDDINGS, M. C., BUXTON, E. C. et al. (1995), Side excitation of fluorescence in ultrathin slab gel electrophoresis, *Anal. Chem.* **67**, 3405–3411.

CHEN, X., FEI, Z., SMITH, L. M., BRADBURY, E. M., MAJIDI, V. (1999), Stable isotope-assisted MALDI-TOF mass spectrometry for accurate determination of nucleotide compositions of PCR products, *Anal. Chem.* **71**, 3118–3125.

CHURCH, G. M., GILBERT, W. (1984), Genomic sequencing, *Proc. Natl. Acad. Sci. USA* **81**, 1991–1995.

CHURCH, G. M., KIEFFER-HIGGINS, S. (1988), Multiplex DNA sequencing, *Science* **240**, 185–188.

CRAXTON, M. (1991), Linear amplification sequencing, a powerful method for sequencing DNA, *Methods: a Companion to Methods in Enzymology* **3**, 20–26.

CRAXTON, M. (1993), Cosmid sequencing, in: *Methods in Molecular Biology*, Vol. 23 DNA Sequencing Protocols (GRIFFIN, H., Ed.), pp. 149–167. Totowa: Humana Press.

DAVIES, S., EIZENMAN, M., PASUPATHY, S., MULLER, W., SLATER, G. (1999), Models of local behavior of DNA electrophoresis peak parameters, *Electrophoresis* **20**, 1443–1454.

DAVIS, L. M., FAIRFIELD, F. R., HARGER, C. A., JETT, J. H., KELLER, R. A. et al. (1991), Rapid DNA sequencing based upon single molecule detection, *Genet. Anal. Tech. Appl.* **8**, 1–7.

DEAR, S., STADEN, R. (1992), A standard file format for data from DNA sequencing instruments, DNA Sequence, *DNA Seq.* **3**, 107–110.

DELISI, C. (1988), The human genome project, *Am. Sci.* **76**, 488–493.

DESRUISSEAUX, C., SLATER, G. W., DROUIN, G. (1998), The gel edge electric field gradients in denaturing polyacrylamide gel electrophoresis, *Electrophoresis* **19**, 627–634.

DOLNIK, V. (1999), Review: DNA sequencing by capillary electrophoresis, *J. Biochem. Biophys. Methods* **41**, 103–119.

DOLNIK, V., LIU, S., JOVANOVICH, S. (2000), Capillary electrophoresis on microchip, *Electrophoresis* **21**, 41–54.

DOVICHI, N. J. (1997), Review: DNA sequencing by capillary electrophoresis, *Electrophoresis* **18**, 2393–2399.

DRMANAC, R., DRMANAC, S. (1999), cDNA screening by array hybridization, *Methods Enzymol.* **303**, 165–178.

DRMANAC, R., LABAT, I., BRUKNER, I., CRKVENJAKOV, R. (1989), Sequencing of megabase plus DNA by hybridization: theory of the method, *Genomics* **4**, 114–128.

DRMANAC, R., DRMANAC, S., LABAT, I., CRKVENJAKOV, R., VICENTIC, A., GEMMELL, A. (1992), Sequencing by hybridization: towards an automated sequencing of one million M13 clones arrayed on membranes, *Electrophoresis* **13**, 566–573.

DRMANAC, R., DRMANAC, S., STREZOSKA, Z., PAUNESKU, T., LABAT, I. et al. (1993), DNA sequence determination by hybridization: a strategy for efficient large-scale sequencing, *Science* **260**, 1649–1652.

ERFLE, H., VENTZKI, R., VOSS, H., RECHMANN, S., BENES, V. et al. (1997), Simultaneous loading of 200 sample lanes for DNA sequencing on vertical and horizontal, standard and ultrathin gels, *Nucleic Acids Res.* **25**, 2229–2230.

ERNST, L. A., GUPTA, R. K., MUJUMDAR, R. B., WAGGONER, A. L. (1989), Cyanine dye labeling reagents for sulfhydryl groups, *Cytometry* **10**, 3–10.

EWING, B., GREEN, P. (1998), Base-calling of automated sequencer traces using *Phred*. II. Error probabilities, *Genome Res.* **8**, 186–194.

EWING, B., HILLIER, L., WENDL, M. C., GREEN, P. (1998), Base-calling of automated sequencer traces using *Phred*. I. Accuracy Assessment, *Genome Res.* **8**, 175–185.

FAJAS, L., STAELS, B., AUWERX, J. (1997), Cycle sequencing on large DNA templates, *BioTechniques* **23**, 1034–1036.

FANG, Y., ZHANG, J. Z., HOU, J. Y., LU, H., DOVICHI, N. J. (1996), Activation energy of the separation of DNA sequencing fragments in denaturing non-cross-linked polyacrylamide by capillary electrophoresis, *Electrophoresis* **17**, 1436–1442.

FANNING, S., GIBBS, R. A. (1997), PCR in genome analysis, in: *Genome Analysis, A Laboratory Manual* (BIRREN, B. et al., Eds.), pp. 249–299. Cold Spring Harbor, NY: Cold Spring Harbor Laboratory Press.

FELLNER, P., SANGER, F. (1968), Sequence analysis of specific areas of 16S and 23 ribosomal RNAs, *Nature* **219**, 236–238.

FLANAGAN, J. H. JR., OWENS, C. V., ROMERO, S. E., WADDELL, E., KAHN, S. H. et al. (1998), Near-infrared heavy-atom-modified fluorescent dyes for base-calling in DNA-sequencing applications using temporal discrimination, *Anal. Chem.* **70**, 2676–2684.

FLEISCHMANN, R. D., ADAMS, M. D., WHITE, O., CLAYTON, R. A., KIRKNESS, E. F. et al. (1995), Whole-genome random sequencing and assembly of *Haemophilus influenzae* Rd, *Science* **269**, 496–512.

FÖRSTER, T. (1948), Intermolecular energy migration and fluorescence, *Ann. Phys.* (Leipzig) **2**, 55–75.

FU, D. J., TANG, K., BRAUN, A., REUTER, D., DARN-HOFER-DEMAR, B. et al. (1998), Sequencing exons 5 to 8 of the *p53* gene by MALDI-TOF mass spectrometry, *Nature Biotechnol.* **16**, 381–384.

GAASTERLAND, T., SENSEN, C. W. (1996), Fully automated genome analysis that reflects user needs and preferences. A detailed introduction to the MAGPIE system architecture, *Biochimie* **78**, 302–310.

GIDDINGS, M. C., BRUMLEY, R. L., HAKER, M., SMITH, L. M. (1993), An adaptive object oriented strategy for base calling in DNA sequence analysis, *Nucleic Acids Res.* **21**, 4530–4540.

GIDDINGS, M. C., SEVERIN, J., WESTPHALL, M., WU, J,. SMITH, L. M. (1998), A software system for data analysis in automated DNA sequencing, *Genome Res.* **8**, 644–665.

GILBERT, W., MAXAM, A. (1973), The nucleotide sequence of the lac operator, *Proc. Natl. Acad. Sci. USA* **70**, 3581–3584.

GLAZER, A. N., MATHIES, R. A. (1997), Energy-transfer fluorescent reagents for DNA analyses, *Curr. Opin. Biotechnol.* **8**, 94–102.

GOETZINGER, W., KOTLER, L., CARRILHO, E., RUIZ-MARTINEZ, M. C., SALAS-SOLANO, O., KARGER, B. L. (1998), Characterization of high molecular mass linear polyacrylamide powder prepared by emulsion polymerization as a replaceable polymer matrix for DNA sequencing by capillary electrophoresis, *Electrophoresis* **19**, 242–248.

GOLDEN, J. B. III, TORGERSEN, D., TIBBETTS, C. (1993), Pattern recognition for automated DNA sequencing: I. On-line signal conditioning and feature extraction for basecalling, *Intelligent Syst. Mol. Biol.* **1**, 136–144.

GORDON, D., ABAJIAN, C., GREEN, P. (1998), Consed: a graphical tool for sequence finishing, *Genome Res.* **8**, 195–202.

GRIFFIN, T. J., HALL, J. G., PRUDENT, J. R., SMITH, L. M. (1999), Direct genetic analysis by matrix-assisted laser desorption/ionization mass spectrometry, *Proc. Natl. Acad. Sci. USA* **96**, 6301–6306.

GROSSMAN, P. D., MENCHEN, S., HERSHEY, D. (1992), Quantitative analysis of DNA sequencing electrophoresis, *Genet. Anal. Tech. Appl.* **9**, 9–16.

GROTHUES, D., VOSS, H., STEGEMANN, J., WIEMANN, S., SENSEN, C. et al. (1993), Separation of up to 1,000 bases on a modified A.L.F. DNA sequencer, *Nucleic Acids Res.* **21**, 6042–6044.

GUTTMAN, A., BENEDEK, K., KALÁSZ (1998), On the separation parameters in DNA sequencing by capillary gel electrophoresis, *Am. Lab.* (April), 63–65.

HAN, K.-T., SAUER, M., SCHULZ, A., SEEGER, S., WOLFRUM, J. (1993), Time-resolved fluorescence studies of labelled nucleosides, *Ber. Bunsenges. Phys. Chem.* **97**, 1728–1730.

HARDING, J. D., KELLER, R. A. (1992), Review: Single-molecule detection as an approach to rapid DNA sequencing, *Trends Biotechnol.* **10**, 55–57.

HAWKINS, T. L., DU, Z., HALLORAN, N. D., WILSON, R. K. (1992), Fluorescence chemistries for automated primer-directed DNA sequencing, *Electrophoresis* **13**, 552–559.

HELLER, C. (1998), Finding a universal low viscosity polymer for DNA separation (II), *Electrophoresis* **19**, 3114–3127.

HUANG, X., MADAN, A. (1999), CAP3: A DNA sequence assembly program, *Genome Res.* **9**, 868–877.

HUANG, X. C., QUESADA, M. A., MATHIES, R. A. (1992a), Capillary array electrophoresis using laser-excited confocal fluorescence detection, *Anal. Chem.* **64**, 967–972.

HUANG, X. C., QUESADA, M. A., MATHIES, R. A. (1992b), DNA sequencing using capillary array electrophoresis, *Anal. Chem.* **64**, 2149–2154.

HUNG, S. C., JU, J., MATHIES, R. A., GLAZER, A. N. (1996a), Cyanine dyes with high absorption cross section as donor chromophores in energy transfer primers, *Anal. Biochem.* **243**, 15–27.

HUNG, S. C., JU, J., MATHIES, R. A., GLAZER, A. N. (1996b), Energy transfer primers with 5- or 6-carboxyrhodamine-6G as acceptor chromophores, *Anal. Biochem.* **238**, 165–170.

HUNG, S. C., MATHIES, R. A., GLAZER, A. N. (1997), Optimization of spectroscopic and electrophoretic properties of energy transfer primers, *Anal. Biochem.* **252**, 78–88.

HUNG, S. C., MATHIES, R. A., GLAZER, A. N. (1998), Comparison of fluorescence energy transfer primers with different donor-acceptor dye combinations, *Anal. Biochem.* **255**, 32–38.

HUNKAPILLER, T., KAISER, R. J., KOOP, B. F., HOOD, L. (1991), Large-scale and automated DNA sequence determination, *Science* **254**, 59–67.

HYMAN, E. D. (1988), A new method of sequencing DNA, *Anal. Biochem.* **174**, 423–436.

INNIS, M. A., MYAMBO, K. B., GELFAND, D. H., BROW, M. D. (1988), DNA sequencing with *Thermus aquaticus* DNA polymerase and direct sequencing of polymerase chain reaction-amplified DNA, *Proc. Natl. Acad. Sci. USA* **85**, 9436–9440.

INNIS, M., GELFAND, D., SNINSKY, J., WHITE, T. (1990), *PCR Protocols: A Guide to Methods and Applications.* San Diego, CA: Academic Press.

IOANNOU, P. A., AMEMIYA, C. T., GARNES, J., KROISEL, P. M., SHIZUYA, H. et al. (1994), A new bacteriophage P1-derived vector for the propagation of large human DNA fragments, *Nature Genet.* **6**, 84–89.

IVES, J. T., GESTELAND, R. F., STOCKHAM, T. G. (1994), An automated film reader for DNA sequencing based on homomorphic deconvolution, *IEEE Trans. Biomed. Engr.* **41**, 509–518.

JETT, J. H., KELLER, R. A., MARTIN, J. C., MARRONE, B. L., MOYZIS, R. K. et al. (1989), High-speed DNA sequencing: an approach based upon fluorescence detection of single molecules, *J. Biomol. Struct. Dyn.* **7**, 301–309.

JONES, D. H. (1997), An iterative and regenerative method for DNA sequencing, *BioTechniques* **22**, 938–946.

JU, J., RUAN, C., FULLER, C. W., GLAZER, A. N., MATHIES, R. A. (1995a), Fluorescence energy transfer dye-labeled primers for DNA sequencing and analysis, *Proc. Natl. Acad. Sci. USA* **92**, 4347–4351.

JU, J., KHETERPAL, I., SCHERER, J. R., RUAN, C., FULLER, C. W. et al. (1995b), Design and synthesis of fluorescence energy transfer dye-labeled primers and their applications for DNA sequencing and analysis, *Anal. Biochem.* **231**, 131–140.

JU, J., GLAZER, A. N., MATHIES, R. A. (1996), Energy transfer primers: A new fluorescence labeling paradigm for DNA sequencing and analysis, *Nature Med.* **2**, 246–249.

KAMBARA, H., TAKAHASHI, S. (1993), Multiple-sheathflow capillary array DNA analyser, *Nature* **361**, 565–566.

KAMBARA, H., NISHIKAWA, T., KATAYAMA, Y., YAMAGUCHI, T. (1988), Optimization of parameters in a DNA sequenator using fluorescence detection, *Bio/Technology* **6**, 816–821.

KAMBARA, H., NAGAI, K., HAYASAKA, S. (1991), Real time automated simultaneous double-stranded DNA sequencing using two-color fluorophore labeling, *Bio/Technology* **9**, 648–651.

KESSLER, C. (Ed.) (1992), *Nonradioactive Labeling and Detection of Biomolecules.* Berlin: Springer-Verlag.

KHETERPAL, I., SCHERER, J. R., CLARK, S. M., RADHARKRISHNAN, A., JU, J. et al. (1996), DNA sequencing using a four-color confocal fluorescence capillary array scanner, *Electrophoresis* **17**, 1852–1859.

KIRPEKAR, F., NORDHOFF, E., LARSEN, L. K., KRISTIANSEN, K., ROEPSTORFF, P., HILLENKAMP, F. (1998), DNA sequence analysis by MALDI mass spectrometry, *Nucl. Acids Res.* **26**, 2554–2559.

KLEPÁRNIK, K., FORET, F., BERKA, J., GOETZINGER, W., MILLER, A. W., KARGER, B. L. (1996), The use of elevated column temperature to extend DNA sequencing read lengths in capillary electrophoresis with replaceable polymer matrices, *Electrophoresis* **17**, 1860–1866.

KÖLLNER, M. (1993), How to find the sensitivity limit for DNA sequencing based on laser-induced fluorescence, *Appl. Optics* **32**, 806–820.

KÖLLNER, M., WOLFRUM, J. (1992), How many photons are necessary for fluorescence-lifetime measurements? *Chem. Phys. Lett.* **200**, 199–204.

KOLNER, D. E., MEAD, D. A., SMITH, L. M. (1992), Ultrathin DNA sequencing gels using microtrough vertical electrophoresis plates, *BioTechniques* **13**, 338–339.

KÖSTER, H., TANG, K., FU, D.-J., BRAUN, A., VAN DEN BOOM, D. et al. (1996), A strategy for rapid and efficient DNA sequencing by mass spectrometry, *Nature Biotechnol.* **14**, 1123–1128.

KOSTICHKA, A. J., MARCHBANKS, M. L., BRUMLEY, R. L. JR., DROSSMAN, H., SMITH, L. M. (1992), High speed automated DNA sequencing in ultrathin slab gels, *BioTechnology* **10**, 78–81.

KOUTNY, L. B., YEUNG, E. S. (1992), Automated image analysis for distortion compensation in sequencing gel electrophoresis, *Appl. Spect.* **46**, 136–141.

KUKANSKIS, K. A., SIDDIQUEE, Z., SHOHET, R. V., GARNER, H. R. (2000), Mix of sequencing technologies for sequence closure: an example, *BioTechniques* **28**, 630–632, 634.

LAKOWICZ, J. (1999), Instrumentation for fluorescence spectroscopy, in: *Principles of Fluorescence Spectroscopy* 2nd Edn. (LAKOWICZ, J., Ed.), pp. 25–61. New York: Kluwer Academic/Plenum Publishers.

LANDER, E. S., WATERMAN, M. S. (1988), Genomic mapping by fingerprinting random clones: a mathematical analysis, *Genomics* **2**, 231–239.

LARDER, B. A., KOHLI, A., KELLAM, P., KEMP, S. D., KRONICK, M., HENFREY, R. D. (1993), Quantitative detection of HIV-1 drug resistance mutations by automated DNA sequencing, *Nature* **365**, 671–673.

LASSITER, S. J., STRYJEWSKI, W., LEGENDRE, B. L. JR., ERDMANN, R., WAHL, M. et al. (2000), Time-resolved fluorescence imaging of slab gels for lifetime base-calling in DNA sequencing applications, *Anal. Chem.* **72**, 5373–5382.

LEE, L. G., CONNELL, C. R., WOO, S. L., CHENG, R. D., McARDLE, B. F. et al. (1992), DNA sequencing with dye-labeled terminators and T7 DNA polymerase: effect of dyes and dNTPs on incorporation of dye-terminators and probability analysis of termination fragments, *Nucl. Acids Res.* **20**, 2471–2483.

LEE, L. G., SPURGEON, S. L., HEINER, C. R., BENSEN, S. C., ROSENBLUM, B. B. et al. (1997), New energy transfer dyes for DNA sequencing, *Nucleic Acids Res.* **25**, 2816–2822.

LEGENDRE, B. L., WILLIAMS, D. C., SOPER, S. A., ERDMANN, R., ORTMANN, U., ENDERLEIN, J. (1996), An all solid-state near-infrared time-correlated single photon counting instrument for dynamic life-time measurements in DNA sequencing applications, *Rev. Sci. Inst.* **67**, 3984–3989.

LI, L.-C., MCGOWN, L. B. (1996), On-the-fly frequency domain fluorescence lifetime detection in capillary electrophoresis, *Anal. Chem.* **68**, 2737–2743.

LI, L.-C., HE, H., NUNNALLY, B. K., MCGOWN, L. B. (1997), On-the-fly fluorescence lifetime detection of labeled DNA primers, *J. Chromatogr.* **695**, 85–92.

LI, Q., YEUNG, E. S. (1995), Simple two-color base-calling schemes for DNA sequencing based on standard four-label Sanger chemistry, *Appl. Spect.* **49**, 1528–1533.

LIU, L. I., FLEISCHMANN, R. D. (1994), Construction of exonuclease III generated nested deletion sets for rapid DNA sequencing, in: *Automated DNA Sequencing and Analysis* (ADAMS, M. D. et al., Eds.), pp. 65–70. London: Academic Press.

LIU, S., SHI, Y., JA, W. W., MATHIES, R. A. (1999), Optimization of high-speed DNA sequencing on microfabricated capillary electrophoresis channels, *Anal. Chem.* **71**, 566–573.

LUCKEY, J. A., SMITH, L. M. (1993a), Optimization of electric field strength for DNA sequencing in capillary gel electrophoresis, *Anal. Chem.* **65**, 2841–2850.

LUCKEY, J. A., SMITH, L. M. (1993b), A model for the mobility of single-stranded DNA in capillary gel electrophoresis, *Electrophoresis* **14**, 492–501.

LUCKEY, J. A., NORRIS, T. B., SMITH, L. M. (1993), Analysis of resolution in DNA sequencing by capillary gel electrophoresis, *J. Phys. Chem.* **97**, 3067–3075.

MACEVICZ, S. C. (1998), DNA sequencing by parallel oligonucleotide extension, *U.S. Patent* 5 750 341.

MARSH, M., TU, O., DOLNIK, V., ROACH, D., SOLOMON, N. et al. (1997), High-throughput DNA sequencing on a capillary array electrophoresis system, *J. Cap. Electrophor.* **4**, 83–89.

MARTIN, C. H., MAYEDA, C. A., DAVIS, C. A., STRATHMANN, M. P., PALAZZOLO, M. J. (1994), Transposon-facilitated sequencing: an effective set of procedures to sequence DNA fragments smaller than 4 kb, in: *Automated DNA Sequencing and Analysis* (ADAMS, M. D. et al., Eds.), pp. 60–64. London: Academic Press.

MARTIN-GALLARDO, A., LAMERDIN, J., CARRANO, A. (1994), Shotgun sequencing, in: *Automated DNA Sequencing and Analysis* (ADAMS, M. D. et al., Eds.), pp. 37–41. London: Academic Press.

MATSUOKA, M. (Ed.) (1990), *Infrared Absorbing Dyes*, pp. 19–33. New York: Plenum Press.

MAXAM, A. M., GILBERT, W. (1977), A new method for sequencing DNA, *Proc. Natl. Acad. Sci. USA* **74**, 560–564.

MCDONALD, J. C., DUFFY, D. C., ANDERSON, J. R., CHIU, D. T., WU, H. et al. (2000), Fabrication of microfluidic systems in poly(dimethylsiloxane), *Electrophoresis* **21**, 27–40.

MCINDOE, R. A., HOOD, L., BAUMGARTNER, R. E. (1993), An analysis of the dynamic range and linearity of an infrared DNA sequencer, *Electrophoresis* **17**, 652–658.

MESSING, J. (1983), New M13 vectors for cloning, *Methods Enzymol.* **101**, 20–78.

MESSING, J., BANKIER, A. T. (1989), The use of single-stranded DNA phage in DNA Sequencing, in: *Nucleic Acids Sequencing* (HOWE, C. J., WARD, E. S., Eds.), pp. 1–36. Oxford: IRL Press.

METZKER, M. L., RAGHAVACHARI, R., RICHARDS, S., JACUTIN, S. E., CIVITELLO, A. et al. (1994), Termination of DNA synthesis by novel 3'-modified deoxyribonucleoside 5'-triphosphates, *Nucleic Acids Res.* **22**, 4259–4267.

METZKER, M. L., LU, J., GIBBS, R. A. (1996), Electrophoretically uniform fluorescent dyes for automated DNA sequencing, *Science* **271**, 1420–1422.

MIDDENDORF, L. R., BRUMBAUGH, J. A., GRONE, D. L., MORGAN, C. A., RUTH, J. L. (1988), Large scale DNA sequencing, *Am. Biotechnol. Laboratory* **6**, 14–22.

MIDDENDORF, L. R., BRUCE, J. C., BRUCE, R. C., ECKLES, R. D., GRONE, D. L. et al. (1992), Continuous, on-line DNA sequencing using a versatile infrared laser scanner/electrophoresis apparatus, *Electrophoresis* **13**, 487–494.

MIDDENDORF, L. R., BRUCE, J. C., BRUCE, R. C., ECKLES, R. D., ROEMER, S. C., SLONIKER, G. D. (1993), A versatile infrared laser scanner/electrophoresis apparatus, *Proc. SPIE* **1885**, 423–434.

MIDDENDORF, L., GARTSIDE, B., HUMPHREY, P., ROEMER, S., SORENSEN, D. et al. (1995), Enhanced throughput for infrared automated DNA sequencing, *Proc. SPIE* **2386**, 66–78.

MIDDENDORF, L., AMEN, J., BRUCE, R., DRANEY, D., DEGRAFF, D. et al. (1998), Near-infrared fluorescence instrumentation for DNA analysis, in: *Near-Infrared Dyes for High Technology Applications* (DAEHNE, S. et al., Eds.), pp. 21–54. Dordrecht: Kluwer Academic Publishers.

Millipore Corporation (1991), *BaseStation™ Automated DNA Sequencer*, Lit. No. PM016, Bedford, MA.

MUELLER, O., HAHNENBERGER, K., DITTMANN, M., YEE, H., DUBROW, R. et al. (2000), A microfluidic system for high-speed reproducible DNA sizing and quantitation, *Electrophoresis* **21**, 128–134.

MUJUMDAR, R. B., ERNST, L. A., MUJUMDAR, S. R., WAGGONER, A. L. (1989), Cyanine dye labeling reagents containing isothiocyanate groups, *Cytometry* **10**, 11–19.

MUJUMDAR, R. B., ERNST, L. A., MUJUMDAR, S. R., LEWIS, C. J., WAGGONER, A. L. (1993), Cyanine dye labeling reagents: sulfoindocyanine succinimidyl esters, *Bioconj. Chem.* **4**, 105–111.

MÜLLER, R., HERTEN, D., LIEBERWIRTH, U., NEU-
MANN, M., SAUER, M. et al. (1997), Time-resolved
DNA identification in capillary gel electrophoresis with semiconductor lasers, *Proc. SPIE* **2980**,
116–126.

MULLIS, K. B., FALOONA, F. A. (1987), Specific synthesis of DNA *in vitro* via a polymerase-catalyzed
chain reaction, in: *Methods in Enzymology* (WU,
R., Ed.), pp. 335–350. London: Academic Press.

NARAYANAN, N., LITTLE, G., RAGHAVACHARI, R., PA-
TONAY, G. (1995), New near infrared dyes for applications in bioanalytical methods, *Proc. SPIE*
2388, 6–15.

NARAYANAN, N., LITTLE, G., RAGHAVACHARI, R.,
GIBSON, J., LUGADE, A. et al. (1998), New NIR
dyes: Synthesis, spectral properties and applica-
tions in DNA analyses, in: *Near-Infrared Dyes for
High Technology Applications* (DAEHNE, S. et al.,
Eds.), pp. 141–158. Dordrecht: Kluwer Academic
Publishers.

NEGRI, R., COSTANZO, G., SALADINO, R., DIMAURO,
E. (1996), One-step, one-lane chemical DNA sequencing by *N*-Methylformamide in the presence
of metal ions, *BioTechniques* **21**, 910–917.

NELSON, M., VANETTEN, J. L., GRABHERR, R. (1992),
DNA sequencing of four bases using three lanes,
Nucleic Acids Res. **20**, 1345–1348.

NELSON, M., ZHANG, Y., STEFFENS, D. L., GRABHERR,
R., VANETTEN, J. L. (1993), Sequencing two DNA
templates in five channels by digital compression,
Proc. Natl. Acad. Sci. USA **90**, 1647–1651.

NICKERSON, D. A., TOBE, V. O., TAYLOR, S. L. (1997),
PolyPhred: automating the detection and genotyping of single nucleotide substitutions using
fluorescence-based resequencing, *Nucleic Acids
Res.* **25**, 2745–2751.

NISHIKAWA, T., KAMBARA, H. (1991), Analysis of
limiting factors of DNA band separation by a
DNA sequencer using fluorescence detection,
Electrophoresis **12**, 623–631.

NISHIKAWA, T., KAMBARA, H. (1992), High resolu-
tion-separation of DNA bands by electrophoresis
with a long gel in a fluorescence-detection DNA
sequencer, *Electrophoresis* **13**, 495–499.

NISHIKAWA, T., KAMBARA, H. (1996), Characteristics
of single-stranded DNA separation by capillary
gel electrophoresis, *Electrophoresis* **17**, 1476–
1484.

NORDMAN, E., CONNELL, C. (1996), New optical design for automated DNA sequencer, *Proc. SPIE*
2680, 290–293.

NUNNALLY, B. K., HE, H., LI, L.-C., TUCKER, S. A.,
McGOWN, L. B. (1997), Characterization of visible
dyes for four-decay fluorescence detection in
DNA sequencing, *Anal. Chem.* **69**, 2392–2397.

NYRÉN, P. (1987), Enzymatic method for continous
monitoring of DNA polymerase activity, *Anal.
Biochem.* **167**, 235–238.

NYRÉN, P. J. (1996), Apyrase immobilized on para-
magnetic beads used to improve detection limits
in bioluminometric ATP monitoring, *J. Biolumin.
Chemilumin.* **9**, 29–34.

NYRÉN, P., PETTERSSON, B., UHLÉN, M. (1993), Solid
phase DNA minisequencing by an enzymatic lu-
minometric inorganic pyrophosphate detection
assay, *Anal. Biochem.* **208**, 171–175.

O'BRIEN, K. M., IRONSIDE, M. A., ATHANASIOU, M.
C., BASIT, M. A., EVANS, G. A., GARNER, H. R.
(1998a), Correcting data shifts in gel files created
by Model 377 DNA sequencers, *BioTechniques*
24, 1002–1003.

O'BRIEN, K. M., SCHAGEMAN, J. J., MAJOR, T. H.,
EVANS, G. A., GARNER, H. R. (1998b), Improving
read lengths by recomputing the matrices of
Model 377 DNA sequencers, *BioTechniques* **24**,
1014–1016.

PANG, H., PAVSKI, V., YEUNG, E. S. (1999), DNA sequencing using 96-capillary array electrophoresis,
J. Biochem. Biophys. Methods **42**, 121–132.

PARKER, L. T., ZAKERI, H., DENG, Q., SPURGEON, S.,
KWOK, P. Y., NICKERSON, D. A. (1996), AmpliTaq®
DNA polymerase, FS dye-terminator sequencing:
analysis of peak height patterns, *BioTechniques*
21, 694–699.

PATONAY, G., ANTOINE, M. D. (1991), Near-infrared
fluorogenic labels: New approach to an old problem, *Anal. Chem.* **63**, 321A–327A.

PATTON, W. F. (2000), Making blind robots see: the
synergy between fluorescent dyes and imaging
devices in automated proteomics, *BioTechniques*
28, 944–957.

PENTONEY, JR., S. L., KONRAD, K. D., KAYE, W.
(1992), A single-fluor approach to DNA sequence determination using high performance
capillary electrophoresis, *Electrophoresis* **13**, 467–
474.

Perkin-Elmer/ABI (1995), *Comparative PCR Sequencing Manual.* Foster City, CA.

PLASCHKE, J., VOSS, H., HAHN, M., ANSORGE, W.,
SCHACKERT, H. K. (1998), Doublex sequencing in
molecular diagnosis of hereditary diseases, *Bio-
Techniques* **24**, 838–841.

PROBER, J. M., TRAINOR, G. L., DAM, R. J., HOBBS, F.
W., ROBERTSON, C. W. et al. (1987), A system for
rapid DNA sequencing with fluorescent chain-
terminating dideoxynucleotides, *Science* **238**, 336–
341.

PRUMMER, M., HÜBNER, C. G., SICK, B., HECHT, B.,
RENN, A., WILD, U. P. (2000), Single-molecule
identification by spectrally and time-resolved
fluorescence detection, *Anal. Chem.* **72**, 443–447.

QUESADA, M. A., RYE, H. S., GINGRICH, J. C., GLA-
ZER, A. N., MATHIES, R. A. (1991), High-sensitivity DNA detection with a laser-excited confocal
fluorescence gel scanner, *BioTechniques* **10**, 616–
625.

QUESADA, M., DHADWAL, H., FISK, D., STUDIER, F. W. (1998), Multi-capillary optical waveguides for DNA sequencing, *Electrophoresis* **19**, 1415–1427.

RAMSEY, J. M. (1999), The burgeoning power of the shrinking laboratory, *Nature Biotechnol.* **17**, 1061–1062.

REEVE, M. A., FULLER, C. W. (1995), A novel thermostable polymerase for DNA sequencing, *Nature* **376**, 796–797.

REN, H., KARGER, A. E., OAKS, F., MENCHEN, S., SLATER, G., DROUIN, G. (1999), Separating DNA sequencing fragments without a sieving matrix, *Electrophoresis* **20**, 2501–2509.

RIBEIRO, E. A., SUTHERLAND, J. C. (1993), Resolving power: A quantitative measure of electrophoretic resolution, *Anal. Biochem.* **210**, 378–388.

RICHTERICH, P. (1998), Estimation of errors in "raw" DNA sequences: A validation study, *Genome Res.* **8**, 251–259.

RICHTERICH, P., HUMPHREY, P., AMEN, J. (1998), Optimization of LI-COR trace files for processing by PHRED and PHRAP, *Microb. Comp. Genomics* **3**, C-91 (Abstract C-10 only).

RIEDER, M. J., TAYLOR, S. L., TOBE, V. O., NICKERSON, D. A. (1998), Automating the identification of DNA variations using quality-based fluorescence re-sequencing: analysis of the human mitochondrial genome, *Nucl. Acids Res.* **26**, 967–973.

ROACH, J. C., SIEGEL, A. F., VAN DEN ENGH, G., TRASK, B., HOOD, L. (1999), Gaps in the Human Genome Project, *Nature* **401**, 843–845.

ROEMER, S. C., BRUMBAUGH, K. A., BOVEIA, V., GARDNER, J. (1997), Simultaneous bidirectional cycle sequencing, *Microb. Comp. Genomics* **2**, 206 (Abstract A-33 only).

ROEMER, S., BOVEIA, V., HUMPHREY, P., AMEN, J., OSTERMAN, H. (1998), Improvements in long read automated sequencing, *Microb. Comp. Genomics* **3**, C-67 (Abstract A-59 only).

ROEMER, S., BOVEIA, V., BUZBY, P., DIMEO, J., DRANEY, D. et al. (2000), New dye-labeled acycloterminators for automated infrared DNA sequencing, *Adv. Genome Biol. Technol.* **I** (Abstract only).

RONAGHI, M., KARAMUHAMED, S., PETTERSSON, B., UHLÉN, M., NYRÉN, P. (1996), Real-time DNA sequencing using detection of pyrophosphate release, *Anal. Biochem.* **242**, 84–89.

RONAGHI, M., UHLÉN, M., NYRÉN, P. (1998a), A sequencing method based on real-time pyrophosphate, *Science* **281**, 363–365.

RONAGHI, M., PETTERSSON, B., UHLÉN, M., NYRÉN, P. (1998b), PCR-introduced loop structure as primer in DNA sequencing, *BioTechniques* **25**, 876–884.

RONAGHI, M., NYGREN, M., LUNDEBERG, J., NYRÉN, P. (1999), Analyses of secondary structures in DNA by pyrosequencing, *Anal. Biochem.* **267**, 65–71.

ROQUE-BIEWER, M., SHARAF, M., TAYLOR, W., LAB-

RENZ, J., MENCHEN, S., TYNAN, K. (1998), Expanding the capability of sequencing and fragment analysis by the introduction of a 5th dye on a multiple capillary electrophoresis instrument, *Microb. Comp. Genomics* **3**, C-92 (Abstract C-15 only).

ROSENBLUM, B. B., LEE, L. G., SPURGEON, S. L., KHAN, S. H., MENCHEN, S. M. et al. (1997), New dye-labeled terminators for improved DNA sequencing patterns, *Nucleic Acids Res.* **25**, 4500–4504.

RUIZ-MARTINEZ, M. C., SALAS-SOLANO, O., CARRILHO, E., KOTLER, L., KARGER, B. L. (1998), A sample purification method for rugged and high-performance DNA sequencing by capillary electrophoresis using replaceable polymer solutions. A. Development of the cleanup protocol, *Anal. Chem.* **70**, 1516–1527.

SALAS-SOLANO, O., RUIZ-MARTINEZ, M. C., CARRILHO, E., KOTLER, L., KARGER, B. L. (1998a), A sample purification method for rugged and high-performance DNA sequencing by capillary electrophoresis using replaceable polymer solutions. B. Quantitative determination of the role of sample matrix components on sequencing analysis, *Anal. Chem.* **70**, 1528–1535.

SALAS-SOLANO, O., CARRILHO, E., KOTLAR, L., MILLER, A. W., GOETZINGER, W. et al. (1998b), Routine DNA sequencing of 1,000 bases in less than one hour by capillary electrophoresis with replaceable linear polyacrylamide solutions, *Anal. Chem.* **70**, 3996–4003.

SAMBROOK, J., FRITSCH, E. F., MANIATIS, T. (1989), *Molecular Cloning* 2nd Edn. Cold Spring Harbor, NY: Cold Spring Harbor Laboratory Press.

SANDERS, J. Z., PETTERSON, A. A., HUGHES, P. J., CONNELL, C. R., RAFF, M. et al. (1991), Imaging as a tool for improving length and accuracy of sequence analysis in automated fluorescence-based DNA sequencing, *Electrophoresis* **12**, 3–11.

SANGER, F., DONELSON, J. E., COULSON, A. R., KOSSEL, H., FISCHER, D. (1973), Use of DNA polymerase I primed by a synthetic oligonucleotide to determine a nucleotide sequence in phage f1 DNA, *Proc. Natl. Acad. Sci. USA* **70**, 1209–1213.

SANGER, F., DONELSON, J. E., COULSON, A. R., KOSSEL, H., FISCHER, D. (1974), Determination of a nucleotide sequence in bacteriophage f1 DNA by primed synthesis with DNA polymerase, *J. Mol. Biol.* **90**, 315–333.

SANGER, F., COULSON, A. R. (1975), A rapid method for determining sequences in DNA by primed synthesis with DNA polymerase, *J. Mol. Biol.* **94**, 441–448.

SANGER, F., NICKLEN, S., COULSON, A. R. (1977), DNA sequencing with chain-terminating inhibitors, *Proc. Natl. Acad. Sci. USA* **74**, 5463–5467.

SAUER, M., HAN, K.-T., EBERT, V., MULLER, R.,

SCHULZ, A. et al. (1994), Design of multiplex dyes for the detection of different biomolecules, *Proc. SPIE* **2137**, 762–774.

SCHERER, J. R., KHETERPAL, I., RADHAKRISHNAN, A., JA, W. W., MATHIES, R. A. (1999), Ultra-high throughput rotary capillary array electrophoresis scanner for fluorescent DNA sequencing and analysis, *Electrophoresis* **20**, 1508–1517.

SCHMALZING, D., ADOURIAN, A., KOUTNY, L., ZIAUGRA, L., MATSUDAIRA, P., EHRLICH, D. (1998), DNA sequencing on microfabricated electrophoretic devices, *Anal. Chem.* **70**, 2303–2310.

SCHMALZING, D., TSAO, N., KOUTNY, L., CHISHOLM, D., SRIVASTAVA, A. et al. (1999a), Toward real-world sequencing by microdevice electrophoresis, *Genome Res.* **9**, 853–858.

SCHMALZING, D., KOUTNY, L., SALAS-SOLANO, O., ADOURIAN, A., MATSUDAIRA, P., EHRLICH, D. (1999b), Review: Recent developments in DNA sequencing by capillary and microdevice electrophoresis, *Electrophoresis* **20**, 3066–3077.

SCHMALZING, D., BELENKY, A., NOVOTNY, M. A., KOUTNY, L., SALAS-SOLANO, O. et al. (2000), Microchip electrophoresis: a method for high-speed SNP detection, *Nucleic Acids Res.* **28**, e43 (electronic article).

SERPICO, S. B., VERNAZZA, G. (1987), Problems and prospects in image processing of two-dimensional gel electrophoresis, *Opt. Engr.* **26**, 661–668.

SHEALY, D. B., LIPOWSKA, M., LIPOWSKI, J., NARAYANAN, N, SUTTER, S. et al. (1995), Synthesis, chromatographic separation, and characterization of near-infrared labeled DNA oligomers for use in DNA sequencing, *Anal. Chem.* **67**, 247–251.

SHI, Y., SIMPSON, P. C., SCHERER, J. R., WEXLER, D., SKIBOLA, C. et al. (1999), Radial capillary array electrophoresis microplate and scanner for high-performance nucleic acid analysis, *Anal. Chem.* **71**, 5354–5361.

SHIZUYA, H., BIRREN, B., KIM, U. J., MANCINO, V., SLEPAK, T. et al. (1992), Cloning and stable maintenance of 300-kilobase-pair fragments of human DNA in *Escherichia coli* using an F-factor-based vector, *Proc. Natl. Acad. Sci. USA* **89**, 8794–8797.

SIMPSON, P. C., ROACH, D., WOOLLEY, A. T., THORSEN, T., JOHNSTON, R. et al. (1998), High-throughput genetic analysis using microfabricated 96-sample capillary array electrophoresis microplates, *Proc. Natl. Acad. Sci. USA* **95**, 2256–2261.

SIMPSON, J. W., RUIZ-MARTINEZ, M. C., MULHERN, G. T., BERKA, J., LATIMER, D. R. et al. (2000), A transmission imaging spectrograph and microfabricated channel system for DNA analysis, *Electrophoresis* **21**, 135–149.

SLATER, G. W. (1993), Theory of band broadening for DNA gel electrophoresis and sequencing, *Electrophoresis* **14**, 1–7.

SLATER, G. W., DROUIN, G. (1992), Why can we not

sequence thousands of DNA bases on a polyacrylamide gel? *Electrophoresis* **13**, 574–582.

SLATER, G. W., MAYER, P., GROSSMAN, P. D. (1995), Diffusion, joule heating, and band broadening in capillary gel electrophoresis of DNA, *Electrophoresis* **16**, 75–83.

SMITH, L. M. (1989), DNA sequence analysis: past, present, and future, *Int. Biotech. Lab.* **7**, 8–19.

SMITH, L. M., SANDERS, J. Z., KAISER, R. J., HUGHES, P., DODD, C. et al. (1986), Fluorescence detection in automated DNA sequence analysis, *Nature* **321**, 674–679.

SMITH, L. M., BRUMLEY, R. L. JR., BUXTON, E. C., GIDDINGS, M., MARCHBANKS, M., TONG, X. (1996), High-speed automated DNA sequencing in ultrathin slab gels, *Methods Enzymol.* **271**, 219–237.

SONG, J. M., YEUNG, E. S. (2000), Alternative base-calling algorithm for DNA sequencing based on four-label multicolor detection, *Electrophoresis* **21**, 807–815.

SOPER, S. A., DAVIDSON, Y. Y., FLANAGAN, J. H., LEGENDRE, JR., B. L., OWENS, C. et al. (1996), Micro-DNA sequence analysis using capillary electrophoresis and near-IR fluorescence detection, *Proc. SPIE* **2680**, 235–246.

STARITA-GERIBALDI, M., HOURI, A., SUDAKA, P. (1993), Lane distortions in gel electrophoresis patterns, *Electrophoresis* **14**, 773–781.

STARKE, H. R., YAN, J. Y., ZHANG, J. Z., MÜHLEGGER, K., EFFGEN, K., DOVICHI, N. J. (1994), Internal fluorescence labeling with fluorescent deoxynucleotides in two-label peak-height encoded DNA sequencing by capillary electrophoresis, *Nucleic Acids Res.* **22**, 3997–4001.

STEFFENS, D. L., JANG, G. Y., SUTTER, S. L., BRUMBAUGH, J. A., MIDDENDORF, L. R. et al. (1995), An infrared fluorescent dATP for labeling DNA, *Genome Res.* **5**, 393–399.

STOKES, G. G. (1852), On the change of refrangibility of light, *Phil. Trans. R. Soc. London* **142**, 463–562.

STREKOWSKI, L., LIPOWSKA, M., PATONAY, G. (1992), Facile derivatizations of heptamethine cyanine dyes, *Synth. Comm.* **22**, 2593–2598.

SULSTON, J., DU, Z., THOMAS, K., WILSON, R., HILLIER, L. et al. (1992), The *C. elegans* genome sequencing project: a beginning, *Nature* **356**, 37–41.

SWERDLOW, H., ZHANG, J. Z., CHEN, D. Y., HARKE, H. R., GREY, R. et al. (1991), Three DNA sequencing methods using capillary gel electrophoresis and laser-induced fluorescence, *Anal. Chem.* **63**, 2835–2841.

SWERDLOW, H., DEW-JAGER, K. E., BRADY, K., GREY, R., DOVICHI, N. J., GESTELAND, R. (1992), Stability of capillary gels for automated sequencing of DNA, *Electrophoresis* **13**, 475–483.

SWERDLOW, H., DEW-JAGER, K., GESTELAND, R. F. (1994), Reloading and stability of polyacrylamide slab gels for automated DNA sequencing, *Bio-*

Techniques **16**, 684–685.

TABOR, S., RICHARDSON, C. C. (1987), DNA sequence analysis with a modified bacteriophage T7 DNA polymerase, *Proc. Natl. Acad. Sci. USA* **84**, 4767–4772.

TABOR, S., RICHARDSON, C. C. (1989), Effect of manganese ions on the incorporation of dideoxynucleotides by bacteriophage T7 DNA polymerase and *E. coli* DNA polymerase I, *Proc. Natl. Acad. Sci. USA* **86**, 4076–4080.

TAKAHASHI, S., MURAKAMI, K., ANAZAWA, T., KAMBARA, H. (1994), Multiple sheath-flow gel capillary-array electrophoresis for multicolor fluorescent DNA detection, *Anal. Chem.* **66**, 1021–1026.

TAN, H., YEUNG, E. S. (1997), Characterization of dye-induced mobility shifts affecting DNA sequencing in poly(ethylene oxide) sieving matrix, *Electrophoresis* **18**, 2893–2900.

TARANENKO, N. I., ALLMAN, S. L., GOLOVLEV, V. V., TARANENKO, N. V., ISOLA, N. R., CHEN, C. H. (1998), Sequencing DNA using mass spectrometry for ladder detection, *Nucleic Acids Res.* **26**, 2488–2490.

TU, O., KNOTT, T., MARSH, M., BECHTOL, K., HARRIS, D. et al. (1998), The influence of fluorescent dye structure on the electrophoretic mobility of end-labeled DNA, *Nucleic Acids Res.* **26**, 2797–2802.

UENO, K., YEUNG, E. S. (1994), Simultaneous monitoring of DNA fragments separated by electrophoresis in a multiplexed array of 100 capillaries, *Anal. Chem.* **66**, 1424–1431.

ULMER, K. M. (1997), Methods and apparatus for DNA sequencing, *U.S. Patent* 5 674 743.

VENTER, J. C., ADAMS, M. D., SUTTON, G. G., KERLAVAGE, A. R., SMITH, H. O., HUNKAPILLER, M. (1998), Shotgun sequencing of the human genome, *Science* **280**, 1540–1542.

VENTZKI, R., STEGEMANN, J., BENES, V., RECHMANN, S., ANSORGE, W. (1998), Simultaneous loading of 200 sample lanes for DNA sequencing on vertical and horizontal, standard and ultrathin gels, *Microb. Comp. Genomics* **3**, C-57 (Abstract A-27 only).

VOSS, H., SCHWAGER, C., KRISTENSEN, T., DUTHIE, S., OLSSON, A. et al. (1989), One-step reaction protocol for automated DNA sequencing with T7 DNA polymerase results in uniform labeling, *Methods Mol. Cell. Biol.* **1**, 155–159.

VOSS, H., WIEMANN, S., GROTHUES, D., SENSEN, C., ZIMMERRNANN, J. et al. (1993a), Automated low-redundancy large-scale DNA sequencing by primer walking, *BioTechniques* **15**, 714–721.

VOSS, H., WIRKNER, U., SCHWAGER, C., ZIMMERMANN, J., STEGEMANN, J. et al. (1993b), Automated DNA sequencing system resolving 1,000 bases with fluorescein-15-dATP as internal label, *Methods Mol. Cell. Biol.* **3**, 153–155.

VOSS, H., NENTWICH, U., DUTHIE, S., WIEMANN, S., BENES, V. et al. (1997), Automated cycle sequencing with Taquenase™: Protocols for internal labeling, dye primer and "doublex" simultaneous sequencing, *BioTechniques* **23**, 312–318.

WATERS, L. C., JACOBSON, S. C., KROUTCHININA, N., KHANDURINA, J., FOOTE, R. S., RAMSEY, J. M. (1998), Multiple sample PCR amplification and electrophoretic analysis on a microchip, *Anal. Chem.* **70**, 5172–5176.

WEISS, G. H., KIEFER, J. E. (1997), Some properties of a measure of resolution in gel electrophoresis and capillary zone electrophoresis, *Electrophoresis* **18**, 2008–2011.

WIEMANN, S., STEGEMANN, J., GROTHUES, D., BOSCH, A., ESTIVILL, X. et al. (1995), Simultaneous on-line DNA sequencing on both strands with two fluorescent dyes, *Anal. Biochem.* **224**, 117–121.

WILLIAMS, D. C., SOPER, S. A. (1995), Ultrasensitive near-IR fluorescence detection for capillary gel electrophoresis and DNA sequencing applications, *Anal. Chem.* **67**, 3427–3432.

WILSON, R. K., MARDIS, E. R. (1997a), Fluorescence-based DNA sequencing, in: *Genome Analysis, A Laboratory Manual* (BIRREN, B. et al., Eds.), pp. 301–395. Cold Spring Harbor, NY: Cold Spring Harbor Laboratory Press.

WILSON, R. K., MARDIS, E. R. (1997b), Shotgun sequencing, in: *Genome Analysis, A Laboratory Manual* (BIRREN, B. et al., Eds.), pp. 397–454. Cold Spring Harbor, NY: Cold Spring Harbor Laboratory Press.

WOOLLEY, A. T., SENSABAUGH, G. F., MATHIES, R. A. (1997), High-speed DNA genotyping using microfabricated capillary array electrophoresis chips, *Anal. Chem.* **69**, 2181–2186.

WU, A., MISLAN, D. (1992), Automated DNA sequencing: an image processing approach, *Appl. Theor. Electrophor.* **3**, 223–228.

YAGER, T. D., DUNN, J. M., STEVENS, S. K. (1997), High-speed DNA sequencing in ultrathin slab gels, *Curr. Opin. Biotechnol.* **8**, 107–113.

YAGER, T. D., BARON, L., BATRA, R., BOUEVITCH, A., CHAN, D. et al. (1999), High performance DNA sequencing, and detection of mutations and polymorphisms, on the Clipper sequencer, *Electrophoresis* **20**, 1280–1300.

YAN, J. Y., BEST, N., ZHANG, J. Z., REN, H. J., JIANG, R. et al. (1996), The limiting mobility of DNA sequencing fragments for both cross-linked and noncross-linked polymers in capillary electrophoresis: DNA sequencing at 1,200 V cm^{-1}, *Electrophoresis* **17**, 1037–1045.

YANG, M. M., YOUVAN, D. C. (1989), A prospectus for multispectral-multiplex DNA sequencing, *Bio/Technology* **7**, 576–580.

ZAKERI, H., AMPARO, G., CHEN, S. M., SPURGEON, S., KWOK, P. Y. (1998), Peak height pattern in dichloro-rhodamine and energy transfer dye terminator sequencing, *BioTechniques* **25**, 406–410, 412–414.

ZHOU, H., MILLER, A. W., SOSIC, Z., BUCHHOLZ, B., BARRON, A. E. et al. (2000), DNA sequencing up to 1,300 bases in two hours by capillary electrophoresis with mixed replaceable linear polyacrylamide solutions, *Anal. Chem.* **72**, 1045–1052.

ZHU, Z., CHAO, J., YU, H., WAGGONER, A. S. (1994), Directly labeled DNA probes using fluorescent nucleotides with different length linkers, *Nucleic Acids Res.* **22**, 3418–3422.

ZIMMERMANN, J., WIEMANN, S., VOSS, H., SCHWAGER, C., ANSORGE, W. (1994), Improved fluorescent cycle sequencing protocol allows reading nearly 1,000 bases, *BioTechniques* **17**, 302–308.

9 A DNA Microarrays Fabrication Strategy for Research Laboratories

DANIEL C. TESSIER

DAVID Y. THOMAS

ROLAND BROUSSEAU

Montreal, Canada

1 Introduction

DNA microarrays (also known as DNA chips or gene chips) are powerful new tools for the study of gene expression and genetic variation. With the availability of increasing numbers of completely sequenced genomes, it is now possible to make DNA microarrays in which all the genes of an organism are represented, and to simultaneously assess the expression of all these genes. This technology is leading biomedical research into a new era of "discovery research" that will complement the "hypothesis-driven research" that has marked the phenomenal success of molecular biology in the past four decades (RAMSAY, 1998).

Pioneering early studies on the chemistry of nucleic acids (NYGAARD and HALL, 1964) showed that mRNA could be measured by hybridization to total bacteriophage DNA fixed on nitrocellulose filters by Coulombic forces. The characteristics of hybridization of RNA–DNA and DNA–DNA were established by extensive experimentation both in liquid and on solid supports (typically nitrocellulose membranes). The advent of restriction enzymes enabled the electrophoretic separation of DNA fragments and localization of mRNA to specific regions – Southern blots (SOUTHERN, 1975). The concept of DNA chips arose from the coincidence of a number of technologies and concepts. These include: methods for rapid oligonucleotide synthesis and genome sequencing; the understanding of the enzymology of DNA synthesis and the availability of such enzymes, robotic methods of arraying, the refinement of DNA chemistry for creating fluorescent hybridization probes and completely sequenced genomes. Thus the development of these technologies combined with the evolution of nucleic acid hybridization methods led to the realization that individual genes arranged separately and in order on a solid substrate could be used to monitor all the genes of an organism.

There are essentially two technologies for making DNA chips. One employs direct spatially ordered synthesis of oligonucleotides on a solid support (silica) and the other uses either post-synthesis arraying of oligonucleotides or DNA on solid supports. There are variations in the details of both technologies, but methods based on post-synthesis arraying definitely represent the most accessible technology. While initial experiments have used arrayed cDNAs and genomic DNA clones on nitrocellulose or nylon supports and detection using ^{32}P or ^{33}P labeled probes, recent improvements and accessibility of robotic technology have now made glass microscope slides the increasingly popular substrate. It is worth emphasizing at this point an unfortunate confusion that has crept into recent microarray literature for those that are familiar with these conventional hybridization studies. In the original hybridization literature, the term "probe" was used for the species of nucleic acid that carried the detection system (usually radioactively labeled) and usually present in the solution phase. In the microarray literature, "probe" often, but not invariably, refers to the DNA fixed on the solid support and it is not usually labeled. Conventional terminology will be used throughout this chapter.

The main applications of DNA chips and microarrays are in gene expression profiling (SCHENA et al., 1995; SCHENA and SHALON, 1996; LOCKHART et al., 1996; DERISI et al., 1996; HELLER et al., 1997), mutational analysis (CHEE et al., 1996; CRONIN et al., 1996), detection of single nucleotide polymorphisms (SNPs) (WANG et al., 1998), pharmacogenomics (SCHENA and DAVIS, 1999), validation of drug targets, identification of tagged biological strains and monitoring of microbial flora in soil and wastewater. Its potential does not need to be emphasized, but this technology is not an end in itself, it is first and foremost a quantitative high-throughput method of screening using genetic material as a target.

There are many hurdles in the construction of microarrays and for their use by researchers. The costs of microarrays are high, the technology for their fabrication is precise, and the parameters for their use and analysis are demanding. However, there are now a number of cost-effective strategies that can be used by the large core facilities at academic centers to make high-quality microarrays available to researchers at reasonable cost.

The fabrication of microarrays requires a source of available cloned cDNAs of an organism. These are typically available from central clone repositories or commercial sources. The plasmid DNA can be arrayed directly or the cDNA insert can be amplified by PCR using common oligonucleotide primers adjacent in the plasmid DNA. In this way, although the initial cost of acquiring the clones may be high, the cost of oligonucleotides is low. An alternative strategy is to amplify coding regions from genomic DNA using PCR and specific oligonucleotides. The advantage of this strategy is that for any sequenced genome, all identified coding or potentially coding sequences can be arrayed. This also avoids a problem inherent in EST-based strategies where some transcripts maybe extremely rare. All genes are present equally in genomic DNA. In another version of this strategy oligonucleotide sequences of 70 nucleotides based on coding sequences can be arrayed. The immediate disadvantage of this very flexible strategy is the high cost of standard oligonucleotide synthesis.

The implementation of microarray technology also requires the installation of high-throughput, high-precision instrumentation in order to carry out the various laboratory tasks involved in the production of the microarrays as well as a comprehensive bioinformatics platform to support the design, clone tracking, data collection and analysis aspects of microarrays. In this chapter, we do not address the analysis of microarray data, but instead we describe the practical implementation of a microarray facility using as an example the fabrication of *Candida albicans* microarrays.

2 The Database

Candida albicans is an opportunistic human fungal pathogen causing fatal infections in immuno-compromised individuals. The application of microarray technology is a powerful new tool in the study of the mechanism of pathogenesis, and the *Candida albicans* genome has been sequenced by a shotgun method by RON DAVIS and coworkers to a 7.5-fold redundancy (*http://www-sequence.stanford.edu/group/candida*, see acknowledgement in References). We transferred approximately 20 mil-

lion bases of the available sequence information for the organism into a database called MAGPIE (Multipurpose Automated Genome Project Investigation Environment). MAGPIE was developed by TERRY GAASTERLAND (Rockefeller University) and CHRISTOPH SENSEN (University of Calgary) for the analysis of microbial genomes and is maintained by the Canadian Bioinformatics Resource (*http://www.cbr.nrc.ca*) and hosted by the Institute for Marine Biosciences of the National Research Council of Canada (Halifax, Nova Scotia, Canada). MAGPIE is capable of identifying open reading frames (ORFs). To identify ORFs from genomic DNA, MAGPIE uses the position of any of the three termination codons or STOP codons (TAA, TAG and TGA) and then reads the sequence backwards until it reaches an in-frame "ATG" codon. The potential ORF identified is then automatically searched against GenBank (e.g., BLAST) to search for DNA and/or protein homologies. MAGPIE then continues to read in the 5′ direction until it finds the best homology for the longest ORF.

One of the challenges of chip fabrication is data handling. For this, we have developed an integrated bioinformatics platform and a set of tools to query MAGPIE, design primers, assess amplicon quality, keep track of the position of PCR products on the chips, and link with data analysis (Fig. 1). The first step was to identify all ORFs greater than 250 base pairs (>80 amino acids) and then to identify short unique flanking sequences (20 nucleotides) for each ORF to create amplicons. The primer identification tool takes into account the length of the primer, its location around the initiator and termination codons and melting temperature (T_m) while considering potential secondary binding sites, hairpin formation, and primer dimerization energies. Another important parameter was to limit the length of the amplicons to <1,500 base pairs as the efficiency of *Taq* polymerase drastically diminishes above this size. In the case of ORFs >1,500 base pairs, the primer closest to the START codon was redesigned to be within 1.5 kb of the STOP codon thus favoring the 3′ end of the gene. The rationale being that cDNAs tend to be over-represented in 3′ sequences as a consequence of using Oligo(dT) priming at the 3′ end of

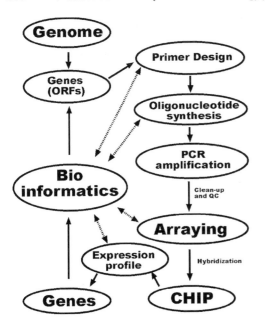

Fig. 1. Flowchart demonstrating the pivotal role of an integrated bioinformatics database in tracking sequence information from a genome all the way to microarrays and linking expression profiling data back to individual genes.

mRNAs during the probe amplification process.

Primers closest to the START codon within each ORF were classified as Forward primers and primers closest to the STOP codon were classified as Reverse primers. The production database was designed to assign batches of 96 Forward primers directly to 96-well plates for DNA synthesis. The Reverse primers corresponding to the same 96 ORFs were assigned to a second 96-well plate such that the Forward/Reverse primer pairs would superimpose well for well for every ORF. The rationale for this design was to facilitate the automation of all the subsequent liquid handling steps.

In addition to ORF identification and PCR primer design, the microarray production database automatically redesigns primers in the case of PCR failures or weak bands. It systematically assigns and tracks the origin of oligonucleotides and/or PCR products from 96-well

plates to 384-well plates all the way to their position on the microarrays and generates a key of the genes based on the spotting method used by the arrayer (number of pins, array size, number of sample, etc.). One essential component is the tracking of samples by bar-coding plates and slides. The large number of samples make this procedure absolutely essential and also simplifies the quality control procedures.

3 High-Throughput DNA Synthesis

Accessibility to high-quality oligonucleotides at a reasonable cost is an important consideration for a microarray fabrication facility. A preferred approach is to use synthetic oligonucleotides to amplify ORFs directly from genomic DNA. This latter approach, however, necessitates the synthesis of thousands of oligonucleotides per genome, creating a demand for a large number of oligonucleotides.

There are also situations where one may prefer to use oligonucleotides directly immobilized on the microarray in preference to amplicons. This situation arises, for instance, when one is developing diagnostic microarrays for a large number of pathogens. It is then much easier, faster, and safer to design oligonucleotides from available genomic sequences than to obtain the various organisms, prepare their genomic DNA under appropriate biosafety conditions and prepare the relevant amplicons.

3.1 Scale and Cost of Synthesis

The current technology of oligonucleotide synthesis, based on automated synthesizers using solid-phase phosphoramidite chemistry (CARUTHERS et al., 1987) offers a reasonable solution for the preparation of large numbers of oligonucleotides at a reasonable cost. The scale of synthesis is not a problem, inasmuch as current oligonucleotide synthesis technology generally prepares far too much material than is normally required for amplified DNA microarrays. As an example, a 5 nmole scale of synthesis, small by modern standards, produces enough material for 200 PCR reactions, given the usual quantity of 25 pmoles per reaction. Even a smaller scale could be used in order to reduce costs further, were it not for the fact that the delivery of very small volumes of reagents and the exclusion of oxygen and moisture become increasingly difficult at synthesis scales below 5 nmoles.

A 5 nmole synthesis scale is also adequate for many applications using immobilized oligonucleotide microarrays, unless large production runs are envisaged. Spotting concentration for amino-linked oligonucleotides is in the range of 10–25 nmoles per mL (SurModics/Motorola), so that a 5 nmoles synthesis would provide 200–300 µL of sample, enough to print several hundred slides.

Current reagent costs for oligonucleotides produced on modern high-throughput DNA synthesizers lie in the range of US\$ 0.10–0.15 per base. This translates to a cost per oligonucleotide of between US\$ 2.00–3.00 (for 20 bases). The oligonucleotide cost for microarrays covering a full bacterial genome (~3,000 ORFs, ~6,000 oligonucleotides) would thus fall between US\$ 15,000 and US\$ 21,000 estimating a 15% failure rate in the PCRs.

3.2 Operational Constraints

The operation of a high-throughput DNA synthesizer capable of producing 100–300 oligonucleotides per day involves the handling of significant quantities of flammable, corrosive, highly reactive, and carcinogenic materials. These chemicals are hazardous and should be handled by a trained chemist. Several of the reagents also suffer from limited stability once they are installed on the synthesizer; a chemist will pay careful attention to their age and condition in order to maintain the highest quality while simultaneously keeping costs under control.

An important advantage lies with those synthesizers which accept the use of universal synthesis support (controlled pore glass, CPG). Conventional DNA synthesizers use membranes or CPG columns bearing one of the four bases A, C, G, or T. This is fine for low numbers of syntheses, but would be intolerable in a 96-well plate format. The time saved as well as the avoidance of possible errors fully justifies the use of the universal supports, where the first base is added by the synthesizer program as opposed to being present on the solid phase synthetic support.

One commercially available high-throughput oligonucleotide synthesizer is the LCDR/MerMade (BioAutomation Corporation, *http://www.bioautomation.com*, a list of current high-throughput DNA synthesizers on the market is available in Tab. 1). The MerMade is a Liquid Chemical Dispensing Robot that was adapted to perform all the operations of DNA synthesis using classical phosphoramidite chemistry (CARUTHERS et al., 1987) in a fully automated fashion. The MerMade is equipped with a motorized XY table within a closed argon chamber providing the inert atmosphere necessary for synthesis. The XY table can hold up to two filter plates (96-well format) in which a universal control pore glass (CPG) support is loaded to allow solid-phase synthesis of the oligonucleotides. The instrument is equipped with computer-controlled valves to deliver the reagents from the bottles to the injection heads. The instrument can be set to operate at scales ranging from 5–50 nmoles with minimal adjustments. The setup time for synthesis is routinely about 1 h and the MerMade will synthesize 192 oligonucleotides (two 96-well plates of 20mers) in about 15 h without operator intervention. For a 5 nmole scale synthesis, the yield of fully deprotected oligonucleotides is ~4 nmoles or 25 µg of 20mer products. Oligonucleotide products up to 45 bases were routinely obtained without significant drop in yield or quality.

Our microarray facility has operated the LCDR/MerMade synthesizer with good success since March 2000. Using this machine, one operator consistently produces 288 oligonucleotides per day with a success rate above 80%, as defined by PCR amplification. The machine has shown no major flaws and provides a good level of user serviceability, convenience, and performance.

Tab. 1. List of Suppliers

Manufacturers of High-Throughput DNA Synthesis Instrumentation

Applied Biosystems   *http://www.appliedbiosystems.com*
BioAutomation Corporation   *http://www.bioautomation.com*
Gene Machines   *http://www.genemachines.com*
Polygen   *http://www.polygen.com*

Manufacturers of Liquid Handlers

Beckman/Coulter   *http://www.beckmancoulter.com*
Packard BioChip Technologies *http://www.packardinstrument.com*
Tomtec   *http://www.tomtec.com*
Zymark   *http://www.zymark.com*

Manufacturers of Arraying Robots

BioRobotics   *http://www.biorobotics.com*
Cartesian Technologies   *http://www.cartesiantech.com*
Engineering Services   *http://www.esit.com*
Gene Machines   *http://www.genemachines.com*
Genetic Microsystems   *http://www.geneticmicro.com*
Genetix   *http://www.genetix.co.uk*
Genomic Solutions   *http://www.genomicsolutions.com*
Intelligent Automation Systems *http://www.ias.com*
Nanogen   *http://www.nanogen.com*
Packard BioChip Technologies *http://www.packardinstrument.com*
Virtek   *http://www.virtekvision.com*

Suppliers of Glass Slide Supports for Microarrays

Amersham Pharmacia Biotech   *http://www.apbiotech.com/microarray*
Corning   *http://www.corning.com/cmt*
SurModics   *http://www.surmodics.com*
NoAb Diagnostics   *http://www.noabdiagnostics.com*
TeleChem International   *http://www.arrayit.com*

Automated Slide Handling Systems

Amersham Pharmacia Biotech   *http://www.apbiotech.com/microarray*
Genomics Perkin Elmer   *http://www.genomicsolutions.com*
Ventana   *http://www.ventanadiscovery.com*

Microarray Scanners and Data Analysis Packages

Affymetrix   *http://www.affymetrix.com*
Alpha Innotech   *http://www.alphainnotech.com*
Applied Precision   *http://www.appliedprecision.com*
Axon Instruments   *http://www.axon.com*
BioDiscovery   *http://www.biodiscovery.com*
Genetic Microsystems   *http://www.geneticmicro.com*
Genomics Solutions   *http://www.genomicsolutions.com*
GSI Lumonics Inc.   *http://www.gsilumonics.com*
Hitachi Genetic Systems   *http://www.miraibio.com*
Imaging Research   *http://www.imaging.brocku.ca*
Incyte Genomics   *http://www.incyte.com*
Informax   *http://www.informaxinc.com*
Media Cybernetics   *http://www.mediacy.com*
Molecular Dynamics   *http://www.mdyn.com*
NetGenics   *http://www.netgenics.com*
Research Genetics   *http://www.resgen.com*
Rosetta Inpharmatics   *http://www.rii.com*
Silicon Genetics   *http://www.sigenetics.com*

Tab. 1. Continued

Spectral Genomics	*http://www.spectralgenomics.com*
Spotfire	*http://www.spotfire.com*
Stanford University	*http://rana.lbl.gov*
The Institute for Genomic Research	*http://www.tigr.org/tdb/microarray*
Virtek	*http://www.virtekvision.com*
Vysis	*http://www.vysis.com*

Protocols for Microarrays	
Stanford University	*http://cmgm.stanford.edu/pbrown*
TeleChem International	*http://arrayit.com*
The Institute for Genomic Research	*http://www.tigr.org/tdb/microarray*
University of Toronto	*http://www.oci.utoronto.ca/services/microarray*

3.3 Quality Control Issues

Quality control of oligonucleotides has always presented a difficult problem. Polyacrylamide gel electrophoresis (PAGE) is inexpensive and adequate for small production numbers, but far too labor-intensive when hundreds of oligonucleotides are produced daily. High-pressure liquid chromatography (HPLC) techniques, although automatable, suffer from a throughput problem. A facility producing 288 oligos per day would need to run the QC procedure under 5 min per oligo, which is difficult to achieve with ion exchange columns. Ion exchange HPLC columns remain the only columns that give sufficient resolution to distinguish consistently the full-length oligo from its $n-1$ failure sequence, but they need to be run with salt gradients and re-equilibrated after each use, hence their comparatively low throughput. The throughput problem also affects capillary electrophoresis (CE) methods, although equilibration is more rapid in CE and an autosampling machine would possibly fulfill the requirement. Mass spectroscopy methods (MALDI-TOF) suffer from the drawback that presence of an ion of the correct molecular weight tells little about the abundance of contaminants, some of which may present widely different ionization efficacies. These last three methods (HPLC, CE, and mass spectrometry) also require expensive equipment.

A cost-effective solution at the present time lies in performing PAGE on randomly selected oligos from each synthesis plate in addition to performing agarose gel electrophoresis on each of the amplicons produced from the oligos. A failed PCR is considered always to have arisen from problems with the oligonucleotides, either at the design or synthesis stages, although we do not have evidence for this for every case. Thus to simplify and speed the operation new design and resynthesis of a different pair of oligos is then performed until the required amplicon is obtained.

4 Amplicon Generation

96-well plates containing Forward and Reverse PCR primers were used to set up PCR reactions using Biomek2000/Biomek F/X workstations (Beckman/Coulter, a list of other manufacturers of liquid handlers is provided in Tab. 1). Each *Candida albicans* ORF was amplified from genomic DNA and the amplicons purified on 96-well ArrayIt SuperFilter plates (TeleChem) to eliminate unincorporated triphosphates and primers. This purification step, although not absolutely necessary, significantly improves the subsequent binding of the amplified DNAs onto the glass slides. Following purification, the amplified products were analyzed by agarose gel electrophoresis as a quality control step that also indirectly assesses the quality of the oligonucleotides.

We have developed BandCheck, a unique bioinformatics tool to assess the quality of PCR products following agarose gel electrophoresis. It creates a virtual band pattern predicted from the database which gets superimposed on a digitized TIFF image of the actual gel. The tool allows the annotation of the gel to assess the quality of 96 PCR amplifications in less than 5 min (Fig. 2). The tool compensates for anomalous migrations as it uses molecular weight standards as a reference. PCR amplifications were scored and an overall success rate of ~85% was obtained for *Candida* genomic PCRs. New oligonucleotide primers were automatically redesigned and resynthesized to cover the 15% PCR failures or weak bands.

Each genomic DNA amplicon was quantified by O.D.$_{260\,nm}$ and yielded an average of 4 μg of DNA per 100 μL PCR reaction. Amplified DNAs were lyophilized in 96-well V-bottom plates and resuspended in spotting buffer at a concentration such that >80% of all the products were between 0.1–0.2 μg μL^{-1}. Reconstituted product from four 96-well plates were transferred to 384-well V-bottom plates for spotting. With such yields, a single PCR amplification would be sufficient to print a few thousand slides.

96-well or 384-well source plates and on the slides. The pins allow the delivery of sub-nanoliter volumes of solution at densities nearing 2,000 spots per cm^2 (~25,000 spots/25 × 75 mm microscope slides). Spotting buffer and slide surface chemistry are key factors to the successful printing and later hybridization of microarrays. We have found that a DMSO-based spotting buffer contributes to better denaturation of the DNA thus increasing the number of single-stranded molecules available for hybridization (HEDGE et al., 2000). Moreover, DNA solutions stored in DMSO buffer are less prone to evaporation due to intrinsic hygroscopic properties of this solvent. Although there are a few vendors of glass substrates for microarrays (see Tab. 1), we opted for the CMT-GAPS slides (Corning) for our DNA microarrays. These gamma-aminopropylsilane coated glass slides, in concert with a DMSO spotting buffer have been the most consistent in our hands giving more uniform spot morphology, better intensity signals and less background.

5 Microarraying

A number of microarraying robots are available on the market. One instrument was developed as a collaboration between engineers at the University of Toronto and biologists at the Ontario Cancer Institute and is available as the SDDC-2 or ChipWriter (Engineering Services Inc./Virtek; *http://www.esit.com*, *http://www.virtekvision.com*, a list of other manufacturers of arraying robots is available in Tab. 1). The SDDC-2 is equipped with a print head capable of holding up to 48 pins (Tele-Chem quill pins) and the surface platen allows for the simultaneous printing of 75 slides. Integrated inside the spotting chamber are a circulating and sonicating waterbath, and a vacuum station to clean the pins between samplings. The chamber is under a slight positive pressure of HEPA-filtered air to minimize dust and particulates. It is also temperature and humidity controlled to optimize spot morphology and minimize sample evaporation in the

6 Probing and Scanning Microarrays

A typical microarray experiment is designed to compare a test condition and an experimental condition. After RNA is extracted from biological samples, cDNAs are synthesized *in vitro* using reverse transcriptase to incorporate a specific fluorescently labeled nucleotide analog (MUJUMDAR et al., 1993) corresponding to the test condition (e.g., Cy3) and another for the experimental condition (e.g., Cy5). The use of multi-color fluorescent labels allows the simultaneous analysis of two or more biological samples or states in a single experiment. Following the reverse transcriptase reaction, the fluorescently labeled cDNA is purified to eliminate unincorporated fluorescent dye and allowed to hybridize with the surface of the microarray in a volume rarely exceeding 100 μL. Coverslips are used to prevent dehydration of the probe and the entire setup is placed in a humid chamber for the duration of

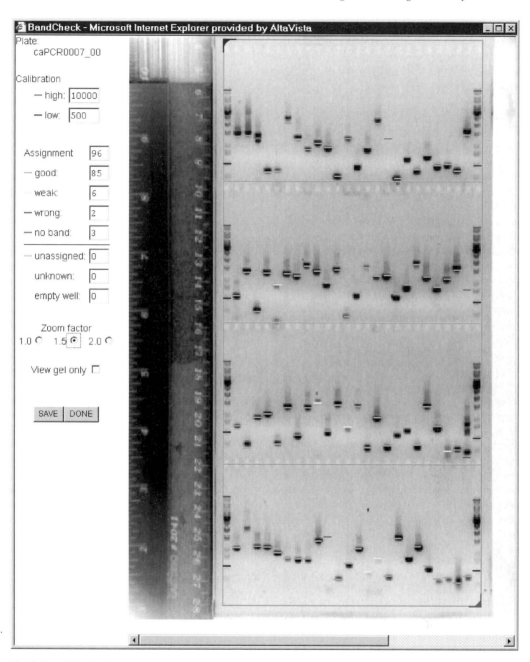

Fig. 2. BandCheck, the gel annotation tool of our microarray production database was developed to assess the quality of PCR products following agarose gel electrophoresis. A TIFF image of the gel is uploaded into the database, the expected band pattern is superimposed onto the gel, and markers are positioned over designated bands in the marker lanes for calibration. Then a simple click of the mouse on the virtual objects superimposed on the PCR bands on the gel defines these products as either good, weak, wrong, or absent. These decisions are accounted for in the database and tracked all the way to the spotting on the microarrays or redirect a failed PCR for resynthesis.

the hybridization. Once the hybridization is complete, unreacted probe is washed from the surface of the glass slide and the hybridized probe molecules are visualized by fluorescence scanning. For detailed protocols of probe preparation and clean-up, prehybridization, hybridization, and washes refer to NELSON and DENNY (1999) and HEDGE et al. (2000).

CCD cameras and confocal scanning devices from various instrumentation companies are currently being used for microarray scanning (see list of suppliers in Tab. 1). Light emitted by the excited fluorophores at the surface of the slide is converted to an electrical signal by a photomultiplier tube and captured by a detector. Confocal-type scanning devices are capable of scanning at a 5 μm resolution such as the GSI Lumonics: the ScanArray5000 (*http:// www.gsilumonics.com*, a list of other manufacturers of scanning instruments is available in Tab. 1). All these instruments allow the quantitation of the fluorescent emissions from the different spots on the microarray. Each spot in the microarray corresponding to a different gene or coding region, the quantitation software automatically compares and identifies the induced and repressed genes. Linking of the data obtained with the MAGPIE database is probably the most crucial aspect of the bioinformatics platform that we have developed. Data collection, normalization, and analysis are topics of their own and extend beyond the scope of this chapter. Complete discussions and review articles are available elsewhere (EISEN et al., 1998; EISEN and BROWN, 1999; HEDGE et al., 2000).

7 Conclusion

Microarrays are already having a large impact upon the way that research is planned and performed. Wider impact of this technology is limited by the cost of the presently available microarrays and the relatively few species for which arrays are available. We show here that it is a feasible and cost-effective solution for even small laboratories to set up to produce their own high-quality microarrays for the genome of any species. The keys to this flexibility are the ability to synthesize large numbers of high-quality oligonucleotides at modest cost and to have an integrated informatics platform to track samples and follow them through the quality control steps.

Acknowledgements

The authors would like to acknowledge the contributions of FRANÇOIS BENOIT, BILL CROSBY, PAUL GORDON, HERVÉ HOGUES, MARCO VAN HET HOOG, TRACEY RIGBY, and ANITA YIP. This paper was published as NRCC publication number 42991.

8 References

Sequence data for *Candida albicans* was obtained from the Stanford DNA Sequencing and Technology Center website at *http://www-sequence.stanford.edu/group/candida*. Sequencing of *Candida albicans* was accomplished with the support of the NIDR and the Burroughs Wellcome Fund.

CARUTHERS, M. H., BARONE, A. D., BEAUCAGE, S. L., DODDS, D. R., FISHER, E. F., MCBRIDE, L. J. et al. (1987), Chemical synthesis of deoxyoligonucleotides by the phosphoramidite method, *Methods Enzymol.* **154**, 287–313.

CHEE, M., YANG, R., HUBBELL, E., BERNO, A., HUANG, X. C. et al. (1996), Accessing genetic information with high-density DNA arrays, *Science* **274**, 610–614.

CRONIN, M. T., FUCINI, R. V., KIM, S. M., MASINO, R. S., WESPI, R. M., MIYADA, C. G. (1996), Cystic fibrosis mutation detection by hybridization to light-generated DNA probe arrays, *Hum. Mutat.* **7**, 244– 255.

DERISI, J., PENLAND, L., BROWN, P. O., BITTNER, M. L., MELTZER, P. S. et al. (1996), Use of a cDNA microarray to analyse gene expression patterns in human cancer, *Nature Genet.* **14**, 457–460.

EISEN, M. B., SPELLMAN, P. T., BROWN, P. O., BOSTEIN, D. (1998), Cluster analysis and display of genome-wide expression patterns. *Proc. Natl. Acad. Sci. USA* **95**, 14863–14868.

EISEN, M. B., BROWN, P. O. (1999), DNA arrays for analysis of gene expression, *Methods Enzymol.* **303**, 179–205.

HEDGE, P., QI, R., ABERNATHY, K., GAY, C., DHARAP, S. et al. (2000), A concise guide to cDNA microarray analysis, *Biotechniques* **29**, 548–562.

HELLER, R. A., SCHENA, M., CHAI, A., SHALON, D., BEDILION, T. et al. (1997), Discovery and analysis of inflammatory disease-related genes using cDNA microarrays, *Proc. Natl. Acad. Sci. USA* **94**, 2150–2155.

LOCKHART, D. J., DONG, H., BYNRE, M. C., FOLLETTIE, M. T., GALLO, M. V. et al. (1996), Expression monitoring by hybridization to high-density oligonucleotide arrays, *Nature Biotechnol.* **14**, 1675–1680.

MUJUMDAR, R. B., ERNST, L. A., MUJUMDAR, S. R., LEWIS, C. J., WAGGONER, A. S. (1993), Cyanine dye labeling reagents: sulfoindocyanine succinimidyl esters, *Bioconj. Chem.* **4**, 105–111.

NELSON, S. F., DENNY, C. T. (1999), Representational differences analysis and microarray hybridization for efficient cloning and screening of differentially expressed genes, in: *DNA Microarrays: A Practical Approach* (SCHENA, M., Ed.), pp. 43–58. New York: Oxford University Press.

NYGAARD, A. P., HALL, B. D. (1964), Formation and properties of RNA–DNA complexes, *J. Mol. Biol.* **9**, 125–142.

RAMSAY, G. (1998), DNA chips: State-of-the-art, *Nature Biotechnol.* **16**, 40–53.

SCHENA, M., SHALON, D., DAVIS, R. W., BROWN, P. O. (1995), Quantitative monitoring of gene expression patterns with a complementary DNA microarray, *Science* **270**, 467–470.

SCHENA, M., SHALON, D. (1996), Parallel human genome analysis: microarray-based expression monitoring of 1000 genes, *Proc. Natl. Acad. Sci. USA* **93**, 10614–10619.

SCHENA, M., DAVIS, R. W. (1999), Genes, genomes and chips, in: *DNA Microarrays: A Practical Approach* (SCHENA, M., Ed.), pp. 1–16. New York: Oxford University Press.

SOUTHERN, E. M. (1975), Detection of specific sequences among DNA fragments separated by gel electrophoresis, *J. Mol. Biol.* **98**, 503–517.

WANG, D. G., FAN, J. B., SIAO, C. J., BERNO, A., YOUNG, P. et al. (1998), Large-scale identification, mapping, and genotyping of single-nucleotide polymorphisms in the human genome, *Science* **280**, 1077–1082.

Protein Technologies

10 Two-Dimensional Gel Electrophoresis and Mass Spectrometry for Proteomic Studies: State-of-the-Art

DANIEL FIGEYS

Toronto, Canada

1 Introduction

With the completion of many sequencing projects (for examples, see *http://www.tigr. org/tdb/*) and the near future completion of the human genome sequencing projects, attention is shifting towards the post-genomic era, i.e., the utilization of the gathered genomic information to understand cellular mechanisms. Currently we do not have enough tools to bridge the gap between the large amount of information and the understanding of the function of the encoded material and ultimately the delivery of new potential drug targets.

By studying a wide variety of organisms that have been sequenced, we could obtain a glimpse of the challenges ahead of us in the post-genomic era. For example, the completion of the yeast (*Saccharomyces cerevisiae*) genome sequencing was achieved 3 years ago (GOFFEAU et al., 1996). However, a large amount of open reading frames (ORFs) (~ 30% in yeast) still remain the functions of which have eluded genomic-based approaches. The structure and regulation of gene expression in other species, e.g. human, are often more complex than yeast, making the understanding of the function of the genetic material even more difficult to decipher. It is clear that in the post-genomic era, whole system-based analyses will be necessary to understand the functions of the genetic material. Furthermore, it is indisputable that downstream information in the chain of information transfer will have to be gathered in a coherent manner.

Tools are already available for the post-genomic era. For example, quantitative monitoring of the expression level for complete sets or subsets of RNA can be gathered using the DNA/RNA array technology (BLANCHARD, 1998; DESPREZ et al., 1998; MARSHALL and HODGSON, 1998; RUAN et al., 1998; SERVICE, 1998) or from the serial analysis of gene expression (SAGE) (VELCULESCU et al., 1995; MADDEN et al., 1997; LAL et al., 1999; MATSU-MURA et al., 1999; NEILSON et al., 2000). Through differential experimentation, the changes in the expression level can then be used to assess the function of a gene. However, for complex organisms, information gathered at the RNA level often fails to represent the changes occurring at the protein level. This is due to the numerous regulation mechanisms in place during protein expression and post-expression. Reports from different laboratories (ANDERSON and SEILHAMER, 1997; GYGI and AEBERSOLD, 1999; GYGI et al., 1999) have demonstrated poor correlations between the expression levels of RNA and proteins. A report from another group (FUTCHER et al., 1999) has demonstrated some correlation by correcting the expression level of RNA. However, the protein level would be difficult to predict from their results. Nonetheless, these three studies clearly show that the information gathered at the RNA level would be difficult to use to accurately predict the level of translation of proteins and their post-translational processing.

A recent large-scale study of the yeast genome has also reinforced the need for a comprehensive analysis of gene expression (ROSS-MACDONALD et al., 1999). This study, performed by transposon tagging and gene disruption, found 31 meiotic genes detected at the protein level by in-frame *lacZ* fusion and assay for β-gal activity. Only 17 of the 31 meiotic genes were previously reported to be induced by at least 2-fold during sporulation, a result of DNA microarray studies representing all annotated ORFs in yeast screened for changes in gene expression. For the remainder, the microarray analysis failed to find any significant induction during meiosis. Therefore, the relationship between the expression level of RNA and the expression level of proteins is not only a complex matter, but it can also be misleading to rely only on the RNA expression pattern to predict cellular functions.

Currently, universal tools are not available to understand the biological function of genes. It is clear that it will be essential in the post-genomic era to utilize in a comprehensive manner a combination of bioanalytical tools for the study of the different biomolecules involved in cellular functions. Proteins are one of the fundamental biomolecules involved in cellular functions, and the term proteomics has been introduced recently (WILKINS et al., 1996) to describe the large-scale study of the proteins related to a genome. Currently, the proteomic field is in its early phase and, therefore, it is labor-intensive, requiring intense efforts on automation and the development of new technologies. Although proteomics is still not straightforward, it is increasingly generating attention as illustrated by the number of publications that over the years contained the word proteome (Fig. 1).

2 Protein Isolation and Separation

Although the field of proteomics is relatively new some of its components have been available for over 25 years. 2D gel electrophoresis has been in use for the separation of proteins for over 25 years. Mass spectrometry has been recently added to 2D gel electrophoresis creating the first-generation proteomic platform. Within these years, standard procedures were developed for the isolation and purification of proteins. These methods were the basis of the first generation of proteomic platforms, and the current state of protein isolation and separation is reviewed below.

2.1 Protein Isolation

The first and probably the most important experimental step in a proteomic study is the extraction of proteome from its medium. Great care has to be taken during the extraction of cells from their environment and during cell lysis to reduce the influence of the sample extraction protocol on the observed state of the proteome.

The diversity of samples that can be analyzed by 2D gel electrophoresis precludes the utilization of a universal protein extraction method. Fortunately, sets of well-characterized methods covering the majority of the needs in protein extraction from cells are available. The harvesting of soluble cytosolic proteins is simply performed by lysing the cells and collecting the supernatant. Different approaches are available to lyse cells (see *http://expasy.cbr. nrc.ca/ch2d/protocols/*). They can be as gentle as adding a surfactant or subjecting to an osmotic shock, or they can be more energetic such as ultrasonification, bead beater, or French press (LILLEHOJ and MALIK, 1989; DUNN and CORBETT, 1996; RABILLOUD, 1996; SHAW, 1998; HERBERT, 1999; RABILLOUD,

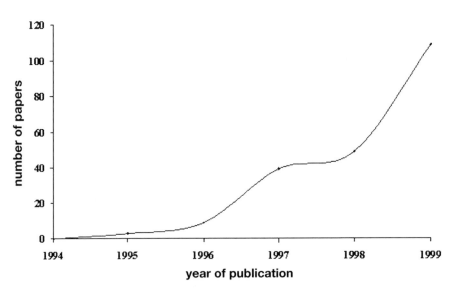

Fig. 1. Number of publications on proteomics over the recent years. The PubMed database was searched by years for the papers that contain the term proteome.

1999b). The technique used to lyse the cell is usually selected with regard to simplicity and compatibility with 2D gel electrophoresis (i.e., minimum salt and ionic surfactant contents).

Even though the approaches for protein extraction and solubilization have been intensively studied, sub-pools of the proteomes, such as hydrophobic proteins, are more difficult to extract with conventional methods. In the last few years, intense efforts have been placed in the development of new chemicals and protocols that improve the recovery of hydrophobic proteins. In particular, the work of Rabilloud (RABILLOUD, 1996; ADESSI et al., 1997; BLISNICK et al., 1998; CHEVALLET et al., 1998; RABILLOUD, 1998; RABILLOUD, 1999; RABILLOUD et al., 1999; SANTONI et al., 1999) has greatly improved the recovery of hydrophobic proteins.

Comprehensive reviews of the extraction and solubilization of proteins from biological samples for 2D gel electrophoresis (DUNN and CORBETT, 1996; RABILLOUD, 1996; SHAW, 1998; HERBERT, 1999; RABILLOUD, 1999b; RAMAGLI, 1999) have been published. However, the results of proteomic studies reported in recent years indicate that the cellular diversity within a biological sample and the protein localization within cells greatly impact the conclusions of a proteomic study. A limited number of publications has glanced at these issues, indicating that the cellular diversity and protein localization need to be taken into account. The current challenges related to the level of cellular mass complexity are presented below.

2.1.1 Homogenous Sample

Cell cultures are probably the first and easiest entry point in the field of proteomics. Large collections of cell cultures, such as the American Type Culture Collection (*www.atcc.org*), are readily available, making it easier for starting groups to establish a proteomics experiment. Access to large quantities of protein extract from cell cultures is typically straightforward. Cell cultures are generally homogeneous in nature and provide a more controllable proteome. Moreover, the culture growth and the lysis conditions are easily controllable, which significantly minimizes the batch to

batch fluctuation and artifacts in the proteome. Reproducible 2D gel patterns from cells cultures treated under the same conditions are routinely observed over time.

2.1.2 Heterogeneous Sample

It becomes more complicated when dealing with tissue samples or primary cultures, mainly because of the poor quality controls during the extraction of the samples. This can induce drastic changes in the resulting proteome. For example, human tissue used in proteomics studies is often collected from patients in hospitals with little control on the preservation of the tissues, resulting in serious artifacts in the proteome due to stress-induced responses.

Tissue samples are heterogeneous in nature (i.e., composed of different cell types, volume, and cell ratio composition), which can cause large variations in the proteome from within the same tissues and for tissues of different sources. Often, valid conclusions can only be reached after a large set of tissues from different sources is studied by proteomics. In some cases, such as human tissues, this is further complicated by the fact that samples are not readily available. When appropriate controls are in place and ample tissue supplies are available, valid results can be derived from large-scale proteomic studies.

For example, OSTERGAARD et al. (1997) have clearly illustrated that proteomic studies can be performed on a complex tissue sample. In this case, a differential proteomic study of bladder cancers (OSTERGAARD et al., 1997; CELIS et al., 1999) has provided invaluable information. In their approach, they blindly analyzed hundreds of well preserved bladder cancer samples by proteomics and were able to find different protein expression profiles related to the different stages of the disease by 2D gel electrophoresis. They then identified the differently expressed proteins by mass spectrometry, and established protein markers. Afterwards, antibodies against some of the differentially expressed proteins along with a few keratinocyte markers were used to stain different cryostat sections of biopsies obtained from bladder cystectomies revealing the different progression stages of the cancer (Fig. 2).

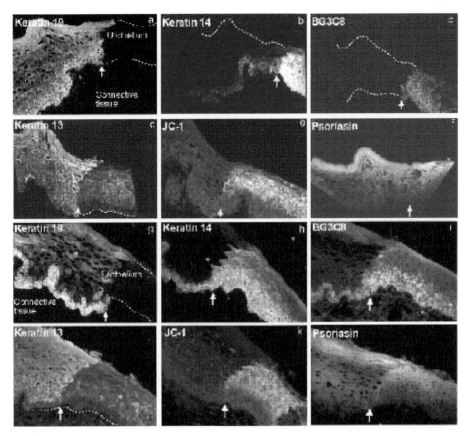

Fig. 2a–l. Identification of type 1 and 3 metaplastic lesions in cystectomy 884-1. Serial cryostat sections were reacted with antibodies against proteins differentially expressed by the normal urothelium and the SCCs (immunowalking). (**a–f**) and (**g–l**) correspond to serial sections of two different areas. White arrows indicate reference points for comparison (reproduced from CELIS et al., 1999, with permission from *Cancer Research*).

2D gel analysis of proteins isolated from complex tissue samples has been performed for over 20 years. Looking back on these efforts, it is clear that the success rate in delivering meaningful information from complex samples is relatively low, even after a comprehensive population study has been performed. In reality the proteome obtained from a primary culture or from tissue samples is composed of a multitude of proteomes, each corresponding to an individual sub-population of cells. The resulting scrambled proteome is often not decipherable, even with large population studies.

2.1.3 Sub-Population of Cells

It is becoming apparent that the separation of the sub-population of cells from complex biological samples is required to define the information provided by the different proteomes. In recent years, classical cell sorting with appropriate markers, laser capture microdissection (BANKS et al., 1999; SIRIVATANAUKSORN et al., 1999), and immunomagnetic techniques using immobilized antibodies specific to certain cell types have been investigated to separate cellular populations for proteomic studies (CLARKE et al., 1994; GOMM et al.,

1995). For example, PAGE et al. (1999) utilized an immunomagnetic technique to separate normal human luminal and myoepithelial breast cells from a set of 10 mammoplasties. This was achieved by double antibody magnetic affinity cell sorting and pull down of the magnetic bead using magnets. This approach resulted in 10 sets of purified luminal and myoepithelial cells ($5 \cdot 10^6$ to $2 \cdot 10^7$ cells per fraction), that were processed by proteomics. They identified 51 differentially expressed proteins, a first step towards new markers and drug targets. Without the fractionation of the cells, they would have obtained scrambled proteomic information and would have failed to identify the differentially expressed proteins.

2.1.4 Subcellular Components

The reduction of the proteome complexity can also be implemented at the subcellular level. The localization of the proteins is a well-known phenomenon in cell biology. The localization of the proteins is often temporally modulated. For example, the study of the growth factor receptor signaling in T cells has shown that the SH3 domains (domain that bind to proline-rich sites) of Grb2 (an adaptor protein) binds to SOS (son-of-sevenless) (SA-MELSON, 1999). The Grb2-SOS complex gets localized to the plasma membrane through the binding of the Grb2 SH2 domain (phosphotyrosine binding domain) to phosphorylated residues of LAT. The enrichment of subcellular components, such as the plasma membrane and the organelles, reduces the complexity of the proteome and provides the concentration of localized, low abundance proteins.

ENGERING et al. (1997) studied the molecular composition of the organelles in murine macrophage cell lines involved in the major histocompatibility complex (MHC) class II restricted antigen presenting and antigen uptake. MHC class II molecules are involved in presenting antigens that reside in circulation and have been taken up. In particular, ENGE-RING et al. proceeded with the infection of the murine macrophage cells with *Mycobacterium bovis* (BCG), which once taken up by phagocytosis stays alive in a phagosome. They separated the subcellular organelles involved in MHC antigen presentation in murine macrophage by free flow electrophoresis and organelle electrophoresis in a density gradient. They were then able to establish different 2D gel profiles for the separated endosomes/lysomes from the phagosomes. The results indicated an increase in the expression of three polypeptides present in the phagosomal organelles. Furthermore, it is clear that without the separation of the organelle proteomes it would have been difficult, if not impossible, to observe differences in expression patterns and to reach valid conclusions.

2.2 Protein Separation by 2D Gel Electrophoresis

Once the protein extract has been isolated a separation technique is required to separate the individual proteins. Two-dimensional gel electrophoresis (2DE) has been the method of choice for the large-scale purification of proteins in proteomic studies. The 2DE method has demonstrated the potential to separate several thousand proteins (GÖRG et al., 1988; KLOSE and KOBALZ, 1995) in a single experiment, partly addressing the complexity issues related to cell lysates. It is now believed that protein separation can be done by as little as 0.1 pI unit and 1 KDa in molecular weight.

2.2.1 Principles of 2D Gel Electrophoretic Separation

The full potential of 2D gel electrophoresis was realized with the introduction in the 1980s of the immobilized pH gradient (IPG strip) (RIGHETTI et al., 1983; RIGHETTI and GIANAZ-ZA, 1987; GÖRG, 1993; RIGHETTI and BOSSI, 1997a; RIGHETTI and BOSSI, 1997b; FICHMANN, 1999; GIANAZZA, 1999; MATSUI et al., 1999b; SANCHEZ et al., 1999). This was achieved by the co-polymerization with acrylamide of a set of monomers that carried ampholyte functionality. Therefore, by changing the concentration of the different monomers along the strip, pH gradients were covalently immobilized and stabilized into the gel. IPG strips with various pH gradients and integrated instruments to perform isoelectric focusing are now commercially available.

Typically, the protein extract, collected as described in the previous section, is used to re-swell the IPG strip prior to focusing (SANCHEZ et al., 1999). This ensures that the protein mixture is uniformly distributed across the whole IPG and limits precipitation due to the excess of proteins in specific areas of the gel. Once the strip is re-swelled, an electric field is gradually applied across the IPG (RIGHETTI and GIANAZZA, 1987). The proteins that are positively charged (i.e., in the area of the strip with pH below their pI) move towards the cathode and encounter an increasing pH until reaching their pI, at which point they will be neutral. The proteins that are negatively charged (i.e., in the area of the strip with pH above their pI) will move towards the anode and encounter a decreasing pH until reaching their pI. The end result is that every protein is concentrated and constantly focused at their respective pI.

The second dimension for the two-dimensional separation is formed by pouring and polymerizing an acrylamide solution between two glass plates spaced using a 1–1.5 mm spacer. Typically, the gel is cast the day prior to the experiment or alternatively can be stored in the refrigerator with an appropriate buffer for a few weeks. After proper equilibration, consisting of in-gel reduction and alkylation of the proteins, the first dimension IPG strip is applied to the second dimension, a large SDS polyacrylamide gel. An electric field is then applied across the gel and the proteins migrate into the second dimension where they are separated according to their molecular weight. Again, systems are commercially available to run the second dimension, and pre-cast, larger gels were recently made available (e.g., the ExcelGel 2-D Homogeneous 12.5 from Amersham Pharmacia Biotech).

2.2.2 Detection of Proteins Separated by 2D Gel Electrophoresis

A wide selection of protein detection schemes is available for the visualization of proteins separated by 2DE (RABILLOUD,

2000). Approaches have been developed to label proteins prior to the separation, to label the proteins after the first dimension (JACKSON et al., 1988; URWIN and JACKSON, 1991), and to label the proteins after the second dimension. The pre- and post-labeling of proteins are the most commonly employed techniques.

2.2.2.1 Pre-Labeling of Proteins

The labeling of proteins prior to the first dimension is routinely done by *in vivo* labeling of the proteins with radioisotopes (O'FARRELL, 1975). This is typically achieved by the incorporation of radioactive metabolites, such as ^{35}S-methionine, in a culture. The radiolabeled samples are separated by 2DE. The 2D gel pattern is then visualized by exposing a film or using a phosphor imager screen. Obviously, the radioactive *in vivo* labeling approach is not suitable for the study of humans and animals.

Fluorescence labeling using electrical neutral dyes covalently coupled to the proteins prior to the 2DE has also been reported (URWIN and JACKSON, 1993). Although the neutral dye does not perturb the 2D gel pattern, it provides limited sensitivity because of poor absorption and quantum yields. More fluorescent dyes were reported to provide nanogram (ng) levels of sensitivity for proteins separated by 2DE (UNLU et al., 1997). All the fluorescent dye approaches require the utilization of a fluorescence detection system to determine the position of the protein spots on the 2D gel.

A technique called multiphoton detection (MPD) has been developed for the exquisite sensitive detection of proteins on a 2D gel. The biological sample of interest is labeled by the *in vivo* incorporation of minute amounts of ^{125}I or ^{131}I. A dedicated system for the acquisition of 2D gel images based on MPD provides the imaging of the 2DE pattern down to attomol (amol) of protein present on the gel. Fig. 3 illustrates the 2D gel images obtained for 150 ng of protein lysate from *E. coli* previously labeled by the incorporation of ^{125}I. Furthermore, the insert shows the detection of sub-amol levels of proteins present in the gel.

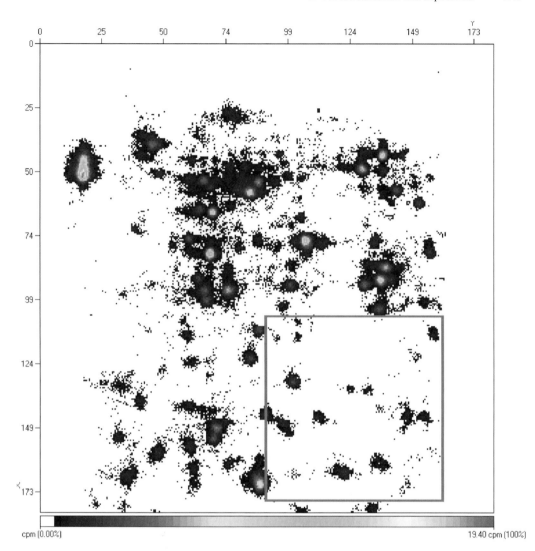

Fig. 3. 2D gel electropherogram profile obtained for 100 ng per gel of *E. coli* protein extract. The *E. coli* cells had been previously labeled using ^{125}I. Approximately 0.1 to 0.01 microCi of ^{125}I per gel were loaded per gel. A limit of detection at about 0.1 attomol per spot was reported (reproduced with permission from BioTraces Inc., *www.biotraces.org*, Dr. A. K. Drukier and Dr. H. Langen).

2.2.2.2 Post-Labeling of Proteins

The labeling of proteins after the 2D gel electrophoretic separation is the most popular approach to visualize 2DE patterns. Typically, the visualization of the proteins present on the 2D gel is obtained using either coomassie blue (about 50 ng limit of detection, LOD) (Smith, 1994; Matsui et al., 1999) or silver (about 5 ng LOD) staining (Blum et al., 1987; Rabilloud, 1990; Rabilloud et al., 1994; Rabilloud, 1999). Other staining techniques based on reverse-stain have also been popularized (Matsui et al., 1999). Although tedious, silver staining is easy to perform and is the most sensitive technique for direct visualization of proteins in gel. Fig. 4 illustrates a silver stained 2D-gel electropherogram obtained for approximately 150 μg of total protein from a *Sulfolobus solfataricus* cell lysate. Each spot on the gel can contain one or more proteins and can be manually or automatically excised from the gel for further analysis.

More sensitive and wider dynamic ranges of labeling techniques have also been described based on fluorescent labels (STEINBERG et al., 1996a, b; STEINBERG et al., 1997). All of the more sensitive techniques do not create any visible spots on the gel and require instrumentation to visualize and extract the proteins in the gels.

Clearly, the combination of both of these purely perpendicular dimensions creates the most powerful separation technique, which has, to date, provided for the separation of up to 10,000 proteins in one experiment (GÖRG et al., 1988; KLOSE and KOBALZ, 1995). Over the years the 2DE has evolved from somewhat of an "art" to a generally practicable but tedious technique that can be applied by non-specialized laboratories.

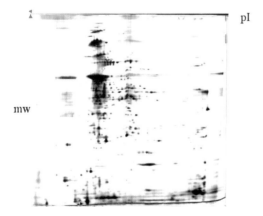

Fig. 4. 2D gel electropherogram profile obtained for 150 mg of *Sulfolobus solfataricus* protein extract. The protein extract was first loaded on a IPG strip and focussed and transferred to the second dimension. The 2D gel electropherogram was silver stained.

2.3 New Development

The current literature on proteomics clearly indicates a significant barrier for the analysis of gel-separated proteins below the low fmol level. This is an unresolved restriction for the analysis of low-abundance proteins. Different explanations were put forward for this alarming problem, such as the poor enzyme kinetics during the digestion of low abundance proteins, lack of sensitivity from the mass spectrometers, lack of material for the low abundance proteins in 2D gel and the shadowing of low-abundance proteins on 2D gel by high abundance proteins. Novel approaches have been developed to circumvent some of these issues.

Low-abundance proteins are often shadowed on 2D gel by the higher abundance proteins that tend to produce large spots on the gel. Increasing the size of the gel would allow better resolution of the different spots. The zoom 2D gel (the zoom here means IPG with narrow pI) was designed to minimize the effects of proteins shadowing on the visualization of low abundance proteins (URQUHART et al., 1997; WASINGER et al., 1997). In this technique, a series of 2D gels with different, short pI ranges is run for every sample of interest. Therefore, instead of running one IPG which covers pI 3–10 on 18 cm, 4–5 18 cm IPGs, each covering roughly 1.5 pI, units are run. The result is a series of large 2D gels each covering a specific pI range of a sample. This approach increases the resolution of proteins resulting in less shadowing of proteins. Furthermore, although the loading capability of individual IPG stays the same by selecting a small pI range there is an increased amount of proteins in this pI range.

2.4 Preparation of Isolated Proteins for Analysis by Mass Spectrometry

Recent developments in mass spectrometry has made the rapid identification and partial amino acid sequencing of minute amounts of proteins possible. Techniques were developed for the enzymatic fragmentation of 2D gel isolated proteins, either directly in the gel or after electroblotting them onto a membrane (AEBERSOLD, 1993; HESS et al., 1993; PATTERSON and AEBERSOLD, 1995; VAN OOSTVEEN et al., 1997). The gel fragmentation method, often called in-gel digestion method (ROSENFELD et al., 1992; SHEVCHENKO et al., 1996b) became the most popular approach mainly due to its simplicity. Generally, this method consists of individually excising the spots of interest from a 2D gel, either manually using a sharp razor blade or using a robotic system (Fig. 5). The individual pieces of gel are then subjected to a series of rinses by subsequent shrinking and swelling using acetonitrile and an aqueous ammonium bicarbonate buffer. Then, an enzyme solution, typically trypsin, is added to the gel and the digestion is allowed to proceed at the appropriate conditions for 3 h to overnight. The resulting peptides are extracted from the gel pieces through a series of extractions by shrinking the gel with acetonitrile. Further purification of the peptides is performed off-line or on-line depending on which mass spectrometry technique is selected.

Matrix-assisted laser desorption ionization (MALDI) time of flight (TOF) and nanoelectrospray-tandem mass spectrometry (MS/MS) require the purification of the peptide mixtures. This is currently performed using commercially available pipette tips with immobilized C18 material for mid fmol and higher levels of proteins; however, they perform poorly for lower level proteins.

3 Mass Spectrometry Tool for the Analysis of Proteins

Until recently, protein sequencing by Edman degradation was the method of choice for the identification of proteins (EDMAN and BEGG, 1967; EDMAN, 1970; HAN et al., 1977; HEINRIKSON, 1978; AEBERSOLD et al., 1986, 1987; TSUGITA, 1987; AEBERSOLD, 1990; GOOLEY et al., 1997; GRANT et al., 1997). However, Edman degradation is not easily amenable to the analysis of minute amounts of proteins and requires up to a day to sequence a peptide, and often up to a month to fully sequence a protein.

Recent developments in mass spectrometry have dramatically improved the performances for the analysis of proteins and peptides. Mass

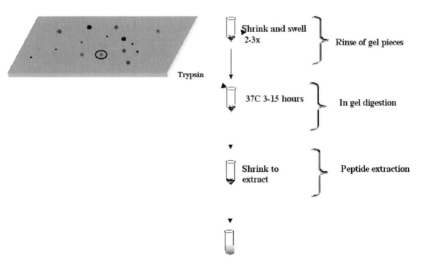

Fig. 5. Preparation for mass spectrometric analysis of proteins separated by 2D gel electrophoresis. The spot containing the protein is excised, rinsed by subsequent shrinking and swelling, soaked in an enzyme containing solution, and digested overnight. The generated peptides are then extracted by shrinking.

spectrometry requires the transfer of the analyte of interest into the gas phase, which is very unfavorable for proteins and peptides. The introduction of gentle ionization techniques, such as matrix-assisted laser desorption ionization (MALDI) and electrospray ionization (ESI) permits the efficient transfer of proteins and peptides into the gas phase and then into the mass analyzer.

Depending on the type of mass analyzer used, information on the peptide mass and fragmentation pattern related to the peptide amino acid sequences can be generated. MALDI and ESI based mass spectrometers have become the instruments of choice for the large-scale identification of proteins in proteomic studies. The complementarity of the two techniques is clearly illustrated in the industrialized application of proteomics.

3.1 MALDI-TOF-MS

Matrix-assisted laser desorption ionization (MALDI) allows the transfer of peptides from a solid state to the gas phase, while the time of flight mass spectrometer (TOF) rapidly separates peptides according to their m/z ratio. This combination of ionizer/mass analyzer is used in proteomic studies for the sensitive and rapid screening of protein digests.

Typically, proteins isolated from 2D gel electrophoresis are individually digested, and spotted on a MALDI plate for co-crystallization with a saturated matrix solution (Fig. 6). The plate, which commonly contains up to 96 digest spots, is inserted into the vacuum chamber of the mass spectrometer. A selected spot is then illuminated using a focussed pulse laser beam (1–10 ns pulse) of wavelength tuned to the absorbance wavelength of the matrix. An ionized plume of material is ejected from the plate surface, due to the rapid expansion from the solid to gas phase, and brings along the

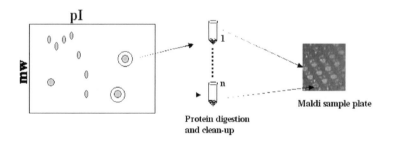

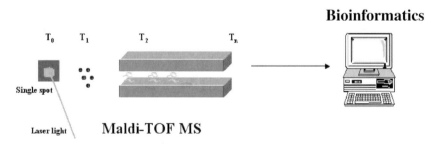

Fig. 6. MALDI-TOF analysis of protein digests for identification by peptide mass fingerprinting. The protein digests of interest are mixed with matrices and spotted on a MALDI plate. Once dried, the plate is inserted in the vacuum chamber of the MALDI-TOF MS. The spot of interest is then briefly illuminated with a laser beam which a wavelength matching the absorbance of the matrix. A plume of sample is ejected from the plate and transferred into the time of flight tube for measurement.

peptides into the gas phase. This is rapidly followed by the biasing of the MALDI plate to +20 kV to +30 kV, with respect to a grounded orifice, and positively charged peptides accelerate towards the orifice of the flight tube. All the peptides are subjected to the same electric field for the same distance (plate–orifice distance), and reach the orifice with a velocity proportional to $(z/m)^{1/2}$. Once in the field-free region of the flight tube, the peptides separate according to their m/z. The different peptides traverse the flight tube according to their m/z ratio, and hit the detector at different time intervals. The mass analyzer, triggered by the laser pulse, records the intensity observed at the detector versus the time of flight that, once the instrument is calibrated, can be readily transformed into a m/z ratio.

Generally, one laser shot ablates only a small fraction of the sample. Therefore, multiple laser shot/mass analysis cycles can be carried out on a sample generating exquisite sensitivity (fmol or less). Good quality spectra can be acquired within a few minutes for each spot on the MALDI plate. Although the technique may sound complicated, it can be fully automated with only a basic knowledge of the instrument principles.

The addition of a reflectron in the TOF reduces band broadening caused by the distribution in kinetic energy of the peptides, resulting in an improvement in resolution and mass accuracy. It is normal to obtain peptide masses accuracy down to 10–50 ppm and obtain a 5,000–10,000 resolution.

MALDI is typically forgiving of impurities compared to other ionization techniques. The protein digests are mixed to the matrix and less than 1 μL is applied onto the MALDI plates. Therefore, less than 10% of a typical tryptic digests volume is applied to the plate, making MALDI the method of choice to screen proteolytic digests.

3.1.1 Bioinformatics for Protein Identification by MALDI-TOF

Protein identification by mass spectrometry relies on the generation of information unique to a protein and the availability of correlating information in the form of DNA or protein sequences in genomic databases.

It was realized, early on, that the accurate measurement of a protein mass was insufficient for identification by protein database searches. However, it became apparent that the accurate masses of peptides, generated by proteolytic digestion of a protein, were often sufficient to identify unambiguously the protein in databases. MALDI-TOF MS became the method of choice to rapidly generate mass fingerprints of proteolytic digests.

Different softwares (HILLENKAMP et al., 1991; COTTRELL, 1994; JAMES et al., 1994; BEAVIS and CHAIT, 1996; COTTRELL and SUTTON, 1996; PAPPIN, 1997; WILKINS et al., 1997; WISE et al., 1997) were developed to search protein/DNA databases using peptide mass fingerprints. Typically, they return the proteins matched in order of the number of peptides that are matched to the measured masses. Some also provide confidence levels on the matches, and allow the identification of the components of a simple protein mixture and the identification of post-translational modifications. Most of this software is freely available for academic application using web page accesses, and can be found at *http://www.expasy.ch/tools/*.

The level of accuracy required for the identification of protein based on a peptide mass fingerprint depends mainly on the number of peptides measured, the number of contaminant masses, the number of entries in the database, and the type of database. To illustrate the potential of peptide mass fingerprinting, a simulation of the peptide mass fingerprinting approach was performed using the IF4E human protein. The protein was *in silico* trypsinized, and random mass fluctuations were added to a few peptides that were then used to search the OWL database (312,942 entries) using a peptide mass fingerprinting software. Tab. 1 illustrates the success of protein identification using various numbers of peptide masses and mass tolerance. The mass tolerance required to identify the protein based on 3 peptides is out of range for current MALDI-TOF MS. However, the software performed well for the identification of the protein based on 4 and 5 peptides. The same simulation was performed again with an addition of a different number of unrelated masses (lower entries in Tab. 1). The

Tab. 1. Results of the searches of the OWL database (312,942 entries) with different peptide masses from the protein IF4E_human and different mass tolerances using Matrix Science (Mascot) peptide mass fingerprinting software. Random noise within the mass tolerance was added to each mass. The set of masses were searched alone and also with the addition of 4 masses from another protein

	250	50	25	10
3 peptides	–	–	–	identified
3 peptides + 4 others	–	–	–	identified
4 peptides	identified	identified	identified	identified
4 peptides + 4 others	–	–	–	identified
5 peptides	identified	identified	identified	identified
5 peptides + 4 others	identified	identified	identified	identified

software failed to identify the protein when 4 related and 4 unrelated masses were used as input to the search algorithm.

MALDI-TOF MS can also be used for the generation of MS/MS spectra using what is called the post-source decay technique (KAUFMANN, 1995; KAUFMANN et al., 1996; PATTERSON et al., 1996). However, this technique significantly affects the sensitivity of the system and often generates MS/MS spectra that are difficult to interpret or to use for database searching.

Clearly, MALDI-TOF, in combination with protein database searches by peptide mass fingerprinting, is a valuable sensitive and high-throughput screening approach for proteomics. However, as the number of entries in databases increases, and for complex protein mixtures the probability of obtaining an unambiguous identification of a protein based on peptide mass fingerprinting is reduced and the PSD generation of MS/MS spectra does not provide adequate data.

3.2 ESI-MS/MS

The increasing amount and variety of information available in protein/DNA databases augment the likelihood that a protein of interest is represented in a database. However, it also increases the information required to discern the protein of interest from all the other entries in the databases. MALDI-TOF peptide

mass fingerprinting, although rapid and sensitive, is often insufficient to identify a protein from larger genomes such as human. Furthermore, it completely fails when there are no representative entries in databases and when a complex mixture of proteins is placed on the plate.

Electrospray ionization (ESI) (FENN et al., 1990) allows the transfer of analytes at atmospheric pressure from a liquid phase to gas phase. Typically, electrospray is achieved by applying an electric field between the tip of a small tube and the entrance of a mass spectrometer. Because of the electric field, the charged liquid at the end of the tip forms a cone called *Taylor cone* that minimizes the charge/surface ratio. Droplets are liberated from the end of the cone and travel towards the mass spectrometer entrance. Solvent evaporates from the droplets, which break again into smaller droplets until the solvent has disappeared, leaving charged analytes in the gas phase. Moreover, while the droplets are shrinking, the pH in the drop decreases and facilitates the protonation of the analytes. Therefore, it is common to obtain multiply charged analytes by ESI when dealing with trypsinized proteins.

ESI based MS have two advantages over MALDI-TOF. First, they can be easily coupled to different sample separation and sample introduction techniques. The second advantage is the increase in the quality of the MS/MS spectra generated from multiply charged analytes.

3.2.1 Generation of MS/MS Spectra

Tandem mass spectrometry can provide enhanced information on the individual peptide contained in a proteolytic digest, facilitating the identification of the proteins and offering the possibility of *de novo* sequencing when no representative entry is in a database.

Unlike MALDI-TOF, a wide selection of electrospray ionization tandem mass spectrometers that are based on different mass separation principles is available. The triple quadrupole mass spectrometer was initially used for proteomic studies. However, it was rapidly replaced by ion trap mass spectrometers (LCQ from Finnigan Mat and Esquire from Bruker) which provide better sensitivity and are the most inexpensive systems for automated tandem mass spectrometry. The recently introduced hybrid mass spectrometers (Qtof from Micromass and the Qstar from PE-Sciex) provide even more sensitivity and enhanced resolutions at a reasonable cost. The Fourier transform mass spectrometer is most often used in academic laboratories and provides even higher resolutions and better sensitivity; however, its high price-tag and its inability to select and isolate a peptide prior to fragmentation limits its applicability. Currently, most proteomic companies utilize ion trap mass spectrometers and QTOF or Qstar mass spectrometers for their tandem mass spectrometry platform.

Although the variety of tandem mass spectrometers is increasing, the basis of tandem mass spectrometry by collision-induced dissociation remains the same (Fig. 7). Typically in a first pass, the introduced peptides are separated according to m/z by the mass spectrometers (MS spectrum). A list of peptides with signals above a pre-established threshold is created. In a second pass, a mass window centered on a selected peptide is isolated by the mass spectrometer and the kinetic energy of the selected peptide is increased. The collision of the peptide with small gas molecules (collision induced dissociation, CID) transfers sufficient energy to break the peptide bond(s) generating charged and neutral fragments. Then in the third pass, the generated charged frag-

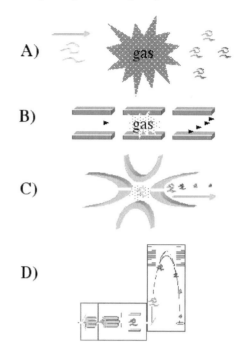

Fig. 7A–D. Collision-induced dissociation (CID) of peptides and its application in different types of electrospray based mass spectrometer. (**A**) The process of fragmentation of peptides by increasing their kinetic energy and colliding them with small gas molecules generating peptide fragments is illustrated, (**B**) the application of CID in a triple quadrupole mass spectrometer is illustrated for the isolation of a peptide, fragmentation and separation of the resulting fragments, (**C**) CID in the ion trap mass spectrometer in which the peptide of interest is trapped, fragmented and the generated fragment successively scanned out of the trap and to the detector, (**D**) CID fragmentation in a hybrid-TOF mass spectrometer in which a selected peptide is isolated, fragmented in a collision cell and the generated fragments separated in an orthogonal time of flight analyzer.

ments are separated by the mass spectrometer according to their m/z creating the MS/MS spectrum. Because the fragmentation occurs at the peptide bonds, a ladder of fragments is generated and the mass difference between two fragments of the same type in the ladder corresponds to an amino acid. The sequence of the peptide can then be reconstituted by a fragment-walk along the ladder.

The generation of MS/MS spectra is relatively slow (up to a few seconds per spectrum) and is performed in a serial manner allowing only one peptide MS/MS spectrum to be acquired at a time. Hence, the coupling of separation techniques, or low flow infusion techniques and tandem mass spectrometry via the ESI interface, allows efficient serial analysis of peptides from peptide mixtures to be performed.

3.2.2 Techniques to Introduce Peptide Mixtures to ESI-MS/MS

Different sample introduction techniques have been developed for the analysis of protein digests by ESI-MS/MS. These techniques differ in sensitivity, handling of contamination, quality of the MS/MS spectra, and throughput. Nanoelectrospray and HPLC have been accepted as the method of choice for the introduction of protein digests to ESI-MS/MS. Capillary electrophoresis (CZE) and solid phase extraction capillary electrophoresis (SPE-CZE), although not as readily applied, provide the most exquisite sensitivity.

3.2.2.1 Continuous Infusion ESI-MS/MS

The term "nanoelectrospray" generally means the continuous low flow infusion of peptides into an ESI-MS/MS (SHEVCHENKO et al., 1996a, 1997; WILM and MANN, 1996). Historically, this technique was developed on mass spectrometers that, at that time, required intense user intervention for the analysis of peptides (SHEVCHENKO et al., 1997). These instruments did not provide automated MS/MS and required the user to interpret the MS spectrum, select the peptide, switch the mass spectrometer to collision-induced dissociation, and accumulate the spectra at the appropriate energies. Even for the most experienced user, this can easily take a few minutes to perform. The nanoelectrospray approach reduces the flow rate and extends the analytical window, allowing the required user interventions for the analysis of peptide mixture. Because of the reduced flow rate, the analytes are more effi-

ciently transferred to the mass spectrometer, which significantly improves the sensitivity for the analysis of peptides and proteins.

In this technique, the peptide mixture of interest is inserted into a glass needle 1 mm in diameter that has a closed tapered end (Fig. 8). The tip is installed in front of the mass spectrometer, and a slight pressure is applied. Then the end of the tip is opened by touching it against the front plate of the MS. The peptide mixture is then continuously delivered at the end of the needle by a gentle gas pressure. The outside of the needle had been previously coated with a layer of conductive material such as gold to which a high voltage is applied. The peptide solution that reaches the tip of the nanoelectrospray needle is electrosprayed into the mass spectrometer at a low flow rate (few nL min^{-1}). Roughly 30 min to 1.5 h of analysis can be performed on 1 μL of sample.

Nanoelectrospray ionization is now a robust enough technique to be routinely utilized for the identification of proteins. Its main advantage is to provide long analytical windows for all the analytes, allowing for prolonged integration of spectra. Its main disadvantage is that while one analyte is analyzed, the other

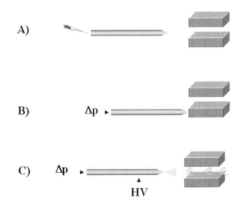

Fig. 8A–C. Procedure for the preparation of a nanoelectrospray needle and analysis by mass spectrometry. (**A**) The protein digest is pipetted in the nanoelectrospray needle. (**B**) The tip of the needle is opened by gently colliding with a plate. (**C**) The nanoelectrospray process is generated by applying a slight pressure in the rear of the sample and by applying an electric field between the tip of the needle and the entrance of a MS.

analytes are being wasted. Furthermore, it still requires skilled personnel and is very sensitive to sample contamination. Moreover, the technique has eluded automation, which reduces its widespread usage.

3.2.2.2 Separation Coupled to ESI-MS/MS

Separation techniques offer the advantage of concentrating the analytes into shorter and separated analytical windows. Instead of presenting all the analytes at the same time to the MS, separation techniques consecutively introduce analytes to the MS, which minimizes the loss of analytes. Each analyte is presented to the MS during a narrow analytical window (typically 1–30 s), during which all MS measurements on this particular analyte have to be performed. Obviously, the short analytical window precludes user intervention and requires rapid and automated mass spectrometers.

3.2.2.2.1 High Performance Liquid Chromatography–MS/MS

HPLC is a technique that concentrates the analytes and, in reverse phase mode, separates them according to their hydrophobicity. Initially, successful identification of mid pmol and higher of gel separated proteins was achieved by using conventional microbore HPLC coupled to ESI-MS. Furthermore, the full automation of the process was possible using an autosampler to sequentially introduce protein digests onto the HPLC column, a mass spectrometer to analyze the peptides separated by the HPLC, and a database search algorithm to identify the protein. DUCRET et al. (1998) have developed such a system consisting of a Microm Bioresources HPLC system equipped with a 0.5 mm C18 column, an Alcott autosampler, and a conventional ESI interface coupled to a triple quadrupole mass spectrometer (TSQ 7000^{tm} from Finnigan Mat). Once the analysis of one sample was terminated, the generated CID spectra were automatically searched against a protein database. Simultaneously, the analysis of another sample proceeded until all the samples present

on the autosampler were automatically analyzed. Using this system, they have automatically identified 90 protein-containing spots obtained from a two-dimensional separation of yeast cell lysate.

In recent years, significant improvements have been brought to the analysis of peptides by HPLC MS. In particular, the introduction of the microelectrospray interface compatible with the low flow provided by μ-HPLC columns has significantly improved the usefulness of this technique (CHERVET et al., 1996; YATES et al., 1996). Since the detection sensitivity in ESI-MS is essentially dependent on concentration, the higher sample concentration afforded by the capillary columns directly translates into a higher sensitivity of detection.

μ-HPLC columns are generally made by pressure packing a few cm of reverse-phase material into a silica capillary tube (50–150 μm i.d.) or directly into a microelectrospray needle (Fig. 9). μ-HPLC columns and packed microelectrospray needles are commercially available. These columns are then on-line with a flow-splitted HPLC system that can deliver gradients at 50–200 nL min^{-1} flow rates and coupled to the microelectrospray interface. In a typical experiment, a few μL of the protein digest is pressure-loaded onto the μ-HPLC column. Then, the HPLC system delivers a solvent gradient of increasing hydrophobicity to the column. The peptides are eluted in order of hydrophobicity, transferred to the microelectrospray interface and then to the MS. The eluting analytes successively trigger the MS to select one of the analytes and generates MS/MS spectra. The MS/MS spectra generated for the analytes are then used to search protein databases.

This was illustrated by FIGEYS et al. (1998) who coupled microelectrospray columns to a TSQ 7000, and performed the automated identification of proteins by tandem mass spectrometer. Fig. 10 shows the analysis of 390 fmol of CBS_yeast (cystathionine beta-synthase) obtained from a single silver-stained 2D gel of yeast cell lysate. The analysis was completed in about 35 min and 19 different peptide MS/MS spectra were matched to CBS_yeast. They reported that approximately 100 fmol of protein present on the gel can be analyzed with this system.

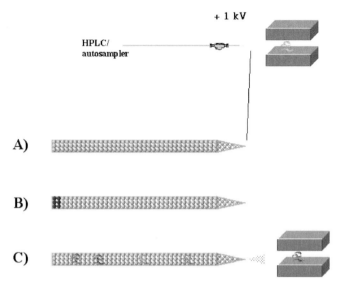

Fig. 9A–C. Diagram of a nano-HPLC column coupled to an electrospray ionization mass spectrometer. The column is either placed prior to a union followed by a transfer nanoelectrospray needle or fabricated directly into a nanoelectrospray needle (**A**). In a typical experiment the protein digest is pressure loaded on the column and accumulates at the head of the column (**B**). After appropriate rinsing the peptides are sequentially eluted from the column, at a flow rate of approximately 100–200 nL min^{-1}, by a solvent gradient generated by an HPLC system, and the peptides are transferred to the MS by microelectrospray (**C**).

It is important to keep in mind that not all the HPLC systems are efficient to drive μ-HPLC. HPLC systems can be divided into two types: low-pressure systems that mix the solvents ahead of a single pump, and high-pressure systems that mix the solvents in a mixing chamber placed after the pumps. For both systems the important parameters for μ-HPLC-MS/MS are the delay time in the gradient, the pre-pressurization performance of the pump, and the performance at lower pressure.

The low-pressure mixing systems offer low void volume after the pump. However, the ratio of solvents is achieved by mechanically proportioning the solvents using the vacuum action of the pump refill stroke. This approach performs poorly for low and high solvent ratios.

HPLC systems that perform the high-pressure mixing of the solvent in a mixing chamber placed after the pumps are more commonly employed. The pumps on an HPLC system using high-pressure mixing are often driven at 5 μL min^{-1}, and flow-splitting occurs after the

mixer, providing the nL min^{-1} required for μ-HPLC. Some of these systems have a large void volume from the pumps to the mixer and to the column. 100 μL of void volume would represent a 20 min delay in the gradient. Furthermore, depending on the type of material used in the μ-HPLC column, the back-pressure can be from a few 100 PSI to 1,000 PSI. The check-valves used in HPLC can fail to properly operate at low pressures and seriously change the gradient delivered.

3.2.2.2.2 Capillary Zone Electrophoresis–MS/MS

Capillary zone electrophoresis (CZE) coupled to mass spectrometry (SMITH et al., 1989; SMITH et al., 1991; FIGEYS et al., 1996a; VALASKOVIC et al., 1996; FIGEYS et al., 1998) has also been used for protein identification. Open tube CZE is an attractive technique because it is flow-compatible with a microelectrospray interface, it provides higher resolution of pep-

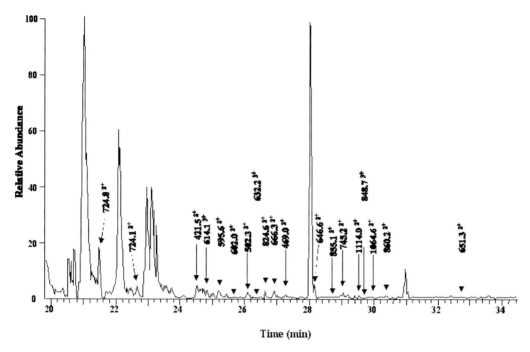

Fig. 10. Analysis by micro-HPLC-MS/MS of CBS_yeast (cystathionine beta-synthase) obtained from a single spot on a silver stained 2D gel. The HPLC column was 50 μm I.D. and 12 cm long, and the analysis was performed at a flow rate of 150 nL min^{-1}. Peaks identified as derived from CBS_yeast by database searching using the respective CID spectra are labeled with the m/z ratio and charge-state of the observed ions (reproduced from FIGEYS et al., 1998, with permission of *Electrophoresis*).

tides, and it is inexpensive. Furthermore, it separates analytes according to their m/z ratio offering a different mode of separation from reverse-phase *μ*-HPLC. The analysis of mid-amol to low fmol levels of standard protein digests have been reported using a simple CZE-MS/MS set-up (FIGEYS et al., 1996a; VA-LASKOVIC et al., 1996).

In a typical experiment, a capillary tubing is filled with a separation buffer, and one end of the tube is inserted into the microelectrospray interface. A sample reservoir containing a protein digest is installed at the other end of the capillary tubing, and an electric field is applied from the injection end to the microelectrospray interface. The applied electric field generates a bulk flow of liquid towards the microelectrospray interface (electroosmotic pumping) and an analyte dependent flow (electrophoretic force). The analytes present in the samples are driven into the capillary tubing by the electric field. The sample reservoir is then replaced by a buffer reservoir, and an electric field is applied across the capillary tubing. A bulk flow of liquid moves towards the microelectrospray due to electroosmotic pumping, while the analytes separate by electrophoresis. The separated analytes are then successively introduced into the mass spectrometer, and tandem mass spectra are acquired.

Typically, CZE techniques require a small volume (nL) injection and a high concentration (mid fmol μL^{-1} to pmol μL^{-1}) of sample to keep the efficiency of the separation. However, enzymatic digest of protein samples typically produces a dilute mixture of peptides in a large volume (few μL). Therefore, although the mass limit of detection is impressive, the sample concentration required by CZE is difficult to achieve for low abundance proteins.

Finally, CZE can be greatly affected by the presence of salts and other matrix components present in samples, making it difficult to utilize CZE as a routine technique for the analysis of protein digests by mass spectrometry.

3.2.2.2.3 Solid-Phase Extraction Capillary Zone Electrophoresis–MS/MS

Hyphenated systems based on capillary electrophoresis have been developed and perfected to efficiently concentrate digests of gel-separated proteins on-line with CZE separation and mass spectrometry (TOMLINSON et al., 1994, 1995a–c, 1996, 1997, 1998, 1999; FIGEYS et al., 1996a, b, 1997; FIGEYS and AEBERSOLD, 1997). In this technique, a small cartridge made of C18 impregnated membrane or reverse phase resin is inserted online with a CZE capillary and is used for the extraction of peptides from a protein digest (Fig. 11). Tens of µL can be easily concentrated on the system, and the cartridge can be effectively rinsed. The peptides are then eluted from the cartridge and

transferred to the separation capillary in a volume of approximately 100 nL. At the same time an electrokinetic stacking effect is simultaneously applied to further concentrate the eluted sample. Concentration factors of up to a thousand can be obtained. A significant reduction in the chemical background is also achieved due to the possibility of extensively rinsing the system before elution and the utilization of buffer compatible with low background mass spectrometry. Once in the CZE capillary, the peptides are electrophoretically separated and successively introduced into the mass spectrometer. The mass spectrometer is triggered to generate CID on the eluting peptides. The SPE-CE-MS/MS has reached a degree of maturity where it can be routinely applied for proteome analyses (see Fig. 12 for example).

3.2.3 Bioinformatic for the Identification of Proteins Based on MS/MS

Regardless of the separation technique utilized to present the peptides to the mass spec-

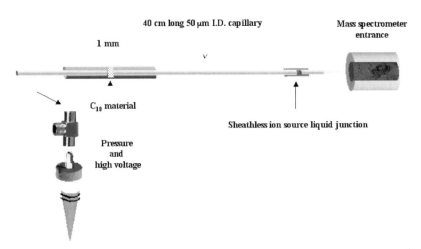

Fig. 11. Diagram of the SPE-CZE-MS/MS system. A small 1 mm long C18 cartridge for sample concentration was coupled on-line with a CZE separation system and a microESI-MS/MS system as previously described. The peptides are extracted from a large volume onto the reverse phase cartridge. After appropriate rinsing they are eluted into the CZE separation capillary in a few nL volume. The peptides are then separated and introduced into the MS by applying a high electric field across the capillary tube (reproduced from FIGEYS et al., 1998, with permission of *Electrophoresis*).

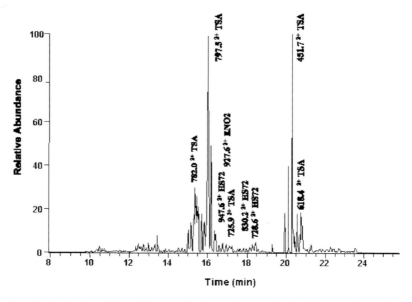

Fig. 12. Analysis by SPE-CZE-MS/MS of a single spot obtained from a silver stained 2D gel. Peaks identified as derived from TSA_yeast (thiol-specific antioxidant protein), HS72_yeast (heat shock protein ssa2), and ENO2_yeast (2-phosphogly-cerate dehydratase) by database searching using the respective CID spectra are labeled with the m/z ratio, charge-state of the observed ions and the name of the protein from which they came (reproduced from FIGEYS et al., 1998, with permission of *Electrophoresis*).

trometer, the identification of proteins based on MS/MS spectra remains the same. Typically, the protein of interest is digested with an enzyme, generally trypsin, producing a peptide mixture. MS/MS spectra for some or all of the peptides present in the mixture are generated. Two different approaches have been developed to correlate the peptide amino acid sequence information contained in the MS/MS spectra with the known sequence in protein/ DNA databases.

In the first approach, often called "sequence-tag" (Fig. 13A), the mass spacing between a few fragments in a MS/MS spectrum is determined and produces a short section (tag) of the peptide sequence. The protein provenance of the analyzed peptide is then established by searching protein databases for isobaric peptides, which contain the sequence tag and the residual masses prior and following the tag. This process is then repeated for the other MS/MS spectra until a consensus protein

is identified. Different algorithms have been developed to perform database searching using sequence tags (MANN and WILM, 1994; PATTERSON et al., 1996; WILKINS et al., 1996). Although they can rapidly search large databases using commonly used desktop computers, they still require a manual reading of the MS/MS spectra to generate the sequence tags. The labor requirement involved in this approach is a major limitation for its applicability in large-scale proteomic projects. Attempts to automate the generation of sequence tags are underway to solve this problem.

The second approach consists of correlating the information contained in uninterpreted MS/MS spectra and information contained in protein/DNA databases (Fig. 13B). An algorithm called "Sequest" (ENG et al., 1994) has been developed to automatically perform this task. In this approach, a list of the top 500 proteins which contain an isobaric peptide to the mass of the analyzed peptide is established

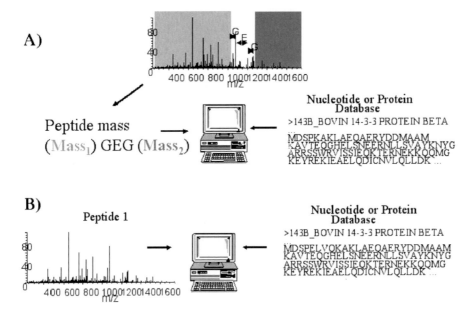

Fig. 13A, B. Two types of protein/DNA database searches using MS/MS spectra for protein identifications. (**A**) Partially interpreted spectrum: A sequence tag is generated from the spectrum, the tag in combination with the residual masses before and after the tag are used to search databases (Example ProteinProspector and SequenceTag from EMBL). (**B**) Uninterpreted spectrum: The uninterpreted spectrum (threshold and filtered) is used to search protein/DNA databases for protein identification (example: Sequest and Mascot).

from DNA/protein databases. Then, predicted spectra are generated for all the 500 isobaric peptides. The predicted spectra are rapidly correlated to the measured spectra by multiplication in the frequency domain using a fast-Fourier transform. The calculated correlation parameters indicate the quality of the match between the predicted and measured MS/MS spectra. A high cross-correlation indicates a good match with the measured spectrum. Furthermore, the difference between the cross-correlation values differentiates the top best match from the second best match. The Sequest software is fully automated and requires very little, if any, intervention from the user once the MS/MS have been generated. However, the software is computing intensive, and can rapidly paralyze the best desktop and single chip Unix systems from the results of a single experiment. Although it does not require user intervention, the Sequest software requires tremendous bioinformatics resources

for large-scale proteomic projects. Expensive multi-parallel processors and large clusters of cheaper Unix systems have been shown to improve the throughput for the commercially available version of Sequest. It turns out that the slow part of the software is to find the top 500 isobaric peptides out of unordered databases. An improved version of the software that predigests and orders the databases significantly improved the search speed and can be run on a reasonably priced computer cluster.

Other algorithms that also automatically search uninterpreted MS/MS spectra against protein and DNA databases are available and can be accessed over the Web. Mascot by Matrix Sciences (*www.matrixscience.com*), and ProteinProspector from UCSF (*http://prospector.ucsf.edu/*) are examples of such web-based MS/MS search engines. These systems typically use statistical parameters to evaluate the quality of the match between the measured MS/MS spectra and the databases. In terms of speed, the web-access softwares often outrun the Sequest search software. Furthermore, the intranet versions of these programs are significantly faster than Sequest.

Depending on the algorithms used, a single confident match between a peptide MS/MS and a protein sequence entry can be enough to identify a protein or a family of proteins. However, when the peptide identified is not unique to a protein, a higher sequence coverage is required. Every subsequent peptide MS/MS that is matched to the same protein further increases the confidence level of the identification.

In large-scale proteomic studies, the throughput of analysis is a critical factor. Therefore, once enough MS/MS spectra have been generated to unambiguously identify a protein, generating MS/MS spectra on the residual peptides can be a waste of time. However, in some cases, as when dealing with EST or protein mixtures, it is important to increase the number of MS/MS generated.

3.2.4 Identification of Genes Based on *de novo* Sequencing of Proteins

Although a tremendous amount of genomic material is available, it poorly represents all the genome pools. Furthermore, robust algorithms are not available to do similarity searches based on MS/MS spectra. Therefore, in some cases, it is essential to do *de novo* sequencing of a protein. The typical objective of *de novo* sequencing is to generate enough sequence information to be able to make a primer for sequencing of the gene related to the protein or identify it through database searches.

Edman chemical degradation can be used for *de novo* sequencing of protein by sequential analysis of the amino acids at the N-terminus of a protein (EDMAN and BEGG, 1967; EDMAN, 1970; HAN et al., 1977; HEINRIKSON, 1978; AEBERSOLD et al., 1987; AEBERSOLD, 1990; GOOLEY et al., 1997; GRANT et al., 1997). When the protein is N-terminus modified or when a higher sequence coverage is required, a more labor-intensive scheme is applied. The protein of interest is digested using an enzyme. The resulting peptides are separated by HPLC with fraction collection, and the purified peptides

are then sequenced by Edman degradation. The peptide sequences are then pieced together to reconstitute the original protein sequence.

Edman degradation is a reliable technique for obtaining amino acid sequences of pmol or higher levels of protein. The assurance that sequencing is performed on a single protein instead of a mixture of proteins is important for *de novo* sequencing. A strong point of Edman degradation analysis is that the presence of a protein mixture is readily detected. Although, it takes about 1–1.5 h per amino acid cycle, the interpretation of the results is fairly straightforward when sufficient amounts of protein are available.

In the second approach, mass spectrometry is used to generate good quality MS/MS spectra of the peptides contained in a protein digest. The sequence of the peptides can then be manually extracted from the MS/MS spectra. Although, MS/MS spectra contain sufficient information for protein identification by database searching, they often do not contain enough information to generate a sufficient stretch of a peptide sequence to design a good primer for DNA sequencing. Typically, the enzyme chosen for proteolytic digestion of proteins is trypsin, which will cleave at the carboxyl terminus of basic residues (lysine and arginine) in the protein. In reality, part of the peptide sequence is often missing from the spectrum due to the dependence of the peptide bond fragmentation on the amino acid sequences.

Although the generation of MS/MS spectra is rapid, the interpretation of the spectra can be relatively complicated when dealing with low level samples and not all MS/MS provides sufficient information. Some programs, such as Lutefisk (TAYLOR and JOHNSON, 1997; *www.hairyfatguy.com/Lutefisk/*), are available to facilitate the *de novo* sequencing of proteins by mass spectrometry, but it is still essentially a manual technique. Currently, *de novo* sequencing is not applicable for large-scale analysis. Only a few proteins can be *de novo* sequenced per week.

The partial protein sequence obtained either by Edman degradation or mass spectrometry is typically used to identify the gene. Generally, a single primer is sufficient to sequence the gene related to the proteins. Furthermore, single primers are easy to deduce from Edman or MS data. However, it is often the case that a mixture of primers is prepared due to the codon degeneration. Sometimes two primers in opposite directions are required to unambiguously identify a gene. Extracting enough reliable information from Edman is feasible, and requires the enzymatic digestion of the protein and the separation of the resulting peptide prior to Edman degradation. Although it is easier to obtain enough protein sequences for the generation of two primers by MS, it is more difficult to ensure that the two sequences are from the same protein. It often requires a series of experiments with different proteolytic enzymes to obtain high sequence coverage of the protein.

4 Conclusions

The field of proteometrics is relatively new. It is clear that 2D gel electrophoresis, mass spectrometry, and their ancillary techniques are the foundation of today's proteomics platforms. Proteomics is an evolving field that will encompass all high-throughput gathering of information related to proteins. For example, nascent efforts in high-throughput discovery of protein functions and high-throughput elucidation of protein structures represent the next phase of proteomics. These novel directions bring into play other technologies, such as NMR, X-ray crystallography, and yeast two hybrid, to name a few.

5 References

ADESSI, C., MIEGE, C., ALBRIEUX, C., RABILLOUD, T. (1997), Two-dimensional electrophoresis of membrane proteins: a current challenge for immobilized pH gradients, *Electrophoresis* **18**, 127–135.

AEBERSOLD, R. (1990), Solid-phase protein sequence analysis, *Nature* **343**, 291–292.

AEBERSOLD, R. (1993), Mass spectrometry of proteins and peptides in biotechnology, *Curr. Opin. Biotechnol.* **4**, 412–419.

AEBERSOLD, R., TEPLOW, D., HOOD, L., KENT, S. (1986), Electroblotting onto activated glass. High efficiency preparation of proteins from analytical sodium dodecyl sulfate-polyacrylamide gels for direct sequence analysis, *J. Biol. Chem.* **261**, 4229–4238.

AEBERSOLD, R. H., LEAVITT, J., SAAVEDRA, R. A., HOOD, L. E., KENT, S. B. (1987), Internal amino acid sequence analysis of proteins separated by one- or two-dimensional gel electrophoresis after *in situ* protease digestion on nitrocellulose, *Proc. Natl. Acad. Sci. USA* **84**, 6970–6974.

ANDERSON, L., SEILHAMER, J. (1997), A comparison of selected mRNA and protein abundances in human liver, *Electrophoresis* **18**, 533–537.

BANKS, R., DUNN, M., FORBES, M., STANLEY, A., PAPPIN, D. et al. (1999), The potential use of laser capture microdissection to selectively obtain distinct populations of cells for proteomic analysis – preliminary findings, *Electrophoresis* **20**, 689–700.

BEAVIS, R. C., CHAIT, B. T. (1996), Matrix-assisted laser desorption ionization mass-spectrometry of proteins, *Methods Enzymol.* **270**, 519–551.

BLANCHARD, A. (1998), Synthetic DNA arrays, *Gen. Eng.* (NY) **20**, 111–123.

BLISNICK, T., MORALES-BETOULLE, M. E., VUILLARD, L., RABILLOUD, T., BRAUN BRETON, C. (1998), Non-detergent sulphobetaines enhance the recovery of membrane and/or cytoskeleton-associated proteins and active proteases from erythrocytes infected by *Plasmodium falciparum*, *Eur. J. Biochem.* **252**, 537–541.

BLUM, H., BEIER, H., GROSS, H. J. (1987), Improved silver staining of plant proteins, RNA and DNA in polyacrylamide gels, *Electrophoresis* **8**, 93–99.

CELIS, J. E., CELIS, P., OSTERGAARD, M., BASSE, B., LAURIDSEN, J. B. et al. (1999), Proteomics and immunohistochemistry define some of the steps involved in the squamous differentiation of the bladder transitional epithelium: A novel strategy for identifying metaplastic lesions, *Cancer Res.* **59**, 3003–3009.

CHERVET, J. P., URSEM, M., SALZMANN, J. B. (1996), Instrumental requirements for nanoscale liquid chromatography, *Anal. Chem.* **68**, 1507–1512.

CHEVALLET, M., SANTONI, V., POINAS, A., ROUQUIE, D., FUCHS, A. et al. (1998), New zwitterionic detergents improve the analysis of membrane proteins by two-dimensional electrophoresis, *Electrophoresis* **19**, 1901–1909.

CLARKE, C., TITLEY, J., DAVIES, S., O'HARE, M. (1994), An immunomagnetic separation method using superparamagnetic (MACS) beads for large-scale purification of human mammary luminal and myoepithelial cells, *Epithelial Cell Biol.* **3**, 38–46.

COTTRELL, J. S. (1994), Protein identification by peptide mass fingerprinting, *Peptide Res.* **7**, 115–124.

COTTRELL, J. S., SUTTON, C. W. (1996), The identification of electrophoretically separated proteins by peptide mass fingerprinting, *Methods Mol. Biol.* **61**, 67–82.

DESPREZ, T., AMSELEM, J., CABOCHE, M., HOFTE, H. (1998), Differential gene expression in *Arabidopsis* monitored using cDNA arrays, *Plant J.* **14**, 643–652.

DUCRET, A., VAN OOSTVEEN, I., ENG, J. K., YATES III, J. R., AEBERSOLD, R. (1998), High throughput protein characterization by automated reverse-phase chromatography/electrospray tandem mass spectrometry, *Protein Sci.* **7**, 706–709.

DUNN, M. J.,CORBETT, J. M. (1996), Two-dimensional polyacrylamide gel electrophoresis, in: *High Resolution Separation and Analysis of Biological Macromolecules* Part B, *Applications* (KARGER, B. L., HANCOOK, W. S., Eds.), pp. 177–203. San Diego, CA: Academic Press.

EDMAN, P. (1970), Sequence determination, *Mol. Biol. Biochem. Biophys.* **8**, 211–255.

EDMAN, P., BEGG, G. (1967), A protein sequenator, *Eur. J. Biochem.* **1**, 80–91.

ENG, J., MCCORMACK, A. L., YATES III, J. R. (1994), An approach to correlate tandem mass spectral data of peptides with amino acid sequences in a protein database, *J. Am. Soc. Mass Spectrom.* **5**, 976–989.

ENGERING, A., LEFKOVITS, I., PIETERS, J. (1997), Analysis of subcellular organelles involved in major histocompatibility complex (MHC) class II-restricted antigen presentation by electrophoresis, *Electrophoresis* **18**, 2523–2530.

FENN, J. B., MANN, M., MENG, C. K. (1990), Electrospray ionization-principles and practice, *Mass Spectrom. Rev.* **9**, 37.

FICHMANN, J. (1999), Advantages of immobilized pH gradients, *Methods Mol. Biol.* **112**, 173–174.

FIGEYS, D., AEBERSOLD, R. (1997), High sensitivity identification of proteins by electrospray ionization tandem mass spectrometry: initial comparison between an ion trap mass spectrometer and a triple quadrupole mass spectrometer, *Electrophoresis* **18**, 360–368.

FIGEYS, D., CORTHALS, G. L., GALLIS, B., GOODLETT, D. R., DUCRET, A. et al. (1999), Data dependent modulation of solid phase extraction capillary electrophoresis for the analysis of complex peptide and phosphopeptide mixtures by tandem mass spectrometry: Application to endothelial nitric oxide synthase, *Anal. Chem.* **13**, 2279–2287.

FIGEYS, D., DUCRET, A., AEBERSOLD, R. (1997), Identification of proteins by capillary electrophoresis tandem mass spectrometry – Evaluation of an on-line solid-phase extraction device, *J. Chromatogr. A* **763**, 295–306.

FIGEYS, D., DUCRET, A., OOSTVEEN, I. V., AEBERSOLD, R. (1996a), Protein identification by capillary zone electrophoresis-microelectrospray ionization-tandem mass spectrometry at the subfemtomol level, *Anal. Chem.* **68**, 1822–1828.

FIGEYS, D., DUCRET, A., YATES, J. R. I., AEBERSOLD, R. (1996b), Protein identification by solid phase microextraction-capillary zone electrophoresis-microelectrospray-tandem mass spectrometry, *Nature Biotechnol.* **14**, 1579–1583.

FIGEYS, D., GYGI, S. P., ZHANG, Y., WATTS, J., GU, M., AEBERSOLD, R. (1998), Electrophoresis combined with novel mass spectrometry techniques: Powerful tools for the analysis of proteins and proteomes, *Electrophoresis* **19**, 1811–1818.

FIGEYS, D., ZHANG, Y., AEBERSOLD, R. (1998), Optimization of solid phase microextraction-capillary zone electrophoresis-mass spectrometry for high sensitivity protein identification, *Electrophoresis* **19**, 2338–2347.

FUTCHER, B., LATTER, G. I., MONARDO, P., MCLAUGHLIN, C.S., GARRELS, J. I. (1999), A sampling of the yeast proteome, *Mol. Cell. Biol.* **19**, 7357–7368.

GIANAZZA, E. (1999), Casting immobilized pH gradients (IPGs), *Methods Mol. Biol.* **112**, 175–188.

GOFFEAU, A., BARRELL, B. G., BUSSEY, H., DAVIS, R. W., DUJON, B. et al. (1996), Life with 6000 genes, *Science* **274**, 546, 563–567.

GOMM, J. J., BROWNE, P., COOPE, R., LIU, Q., BULUWELA, L., COOMBES, R. (1995), Isolation of pure populations of epithelial and myoepithelial cells from the normal human mammary gland using immunomagnetic separation with Dynabeads, *Anal. Biochem.* **226**, 91–99.

GOOLEY, A. A., OU, K., RUSSELL, J., WILKINS, M. R., SANCHEZ, J. C. et al. (1997), A role for Edman degradation in proteome studies, *Electrophoresis* **18**, 1068–1072.

GÖRG, A. (1993), Two-dimensional electrophoresis with immobilized pH gradients: current state, *Biochem. Soc. Trans.* **21**, 130–132.

GÖRG, A., POSTEL, W., GUNTHER, S. (1988), The current state of two-dimensional electrophoresis with immobilized pH gradients, *Electrophoresis* **9**, 531–546.

GRANT, G. A., CRANKSHAW, M. W., GORKA, J. (1997), Edman sequencing as tool for characterization of synthetic peptides, *Methods Enzymol.* **289**, 395–419.

GYGI, S., AEBERSOLD, R. (1999), Absolute quantitation of 2-D protein spots, *Methods Mol. Biol.* **112**, 417–421.

GYGI, S., ROCHON, Y., FRANZA, B., AEBERSOLD, R. (1999), Correlation between protein and mRNA abundance in yeast, *Mol. Cell. Biol.* **19**, 1720–1730.

HAN, K. K., TETAERT, D., DEBUIRE, B., DAUTREVAUX, M., BISERTE, G. (1977), Sequential Edman degradation, *Biochimie* **59**, 557–576.

HEINRIKSON, R. L. (1978), Application of automated sequence analysis to the understanding of protein structure and function, *Ann. Clin. Lab. Sci.* **8**, 295–301.

HERBERT, B. (1999), Advances in protein solubilisation for two-dimensional electrophoresis, *Electrophoresis* **20**, 660–663.

HESS, D., COVEY, T. C., WINZ, R., BROWNSEY, R.W., AEBERSOLD, R. (1993), Analytical and micropreparative peptide mapping by high performance liquid chromatography/electrospray mass spectrometry of proteins purified by gel electrophoresis, *Protein Sci.* **2**, 1342–1351.

HILLENKAMP, F., KARAS, M., BEAVIS, R. C., CHAIT, B. T. (1991), Matrix-assisted laser desorption/ionization mass spectrometry of biopolymers, *Anal. Chem.* **63**, 1193A–1203A.

JACKSON, P., URWIN, V. E., MACKAY, C. D. (1988), Rapid imaging, using a cooled charge-coupled device, of fluorescent two-dimensional polyacrylamide gels produced by labelling proteins in the first-dimensional isoelectric focusing gel with the fluorophore 2-methoxy-2,4-diphenyl-3(2H)furanone, *Electrophoresis* **9**, 330–339.

JAMES, P., QUADRONI, M., CARAFOLI, E., GONNET, G. (1994), Protein identification in DNA databases by peptide mass fingerprinting, *Protein Sci.* **3**, 1341–1350.

KAUFMANN, R. (1995), Matrix-assisted laser desorption ionization (MALDI) mass spectrometry: a novel analytical tool in molecular biology and biotechnology, *J. Biotechnol.* **41**, 155–175.

KAUFMANN, R., CHAURAND, P., KIRSCH, D., SPENGLER, B. (1996), Post-source decay and delayed extraction in matrix-assisted laser desorption/ionization-reflectron time-of-flight mass spectrometry. Are there trade-offs?, *Rapid Commun. Mass Spectrom.* **10**, 1199–1208.

KLOSE, J., KOBALZ, U. (1995), Two-dimensional electrophoresis of proteins: an updated protocol and implications for a functional analysis of the genome, *Electrophoresis* **16**, 1034–1059.

LAL, A., LASH, A. E., ALTSCHUL, S. F., VELCULESCU, V., ZHANG, L. et al. (1999), A public database for gene expression in human cancers, *Cancer Res.* **59**, 5403–5407.

LILLEHOJ, E. P., MALIK, V. (1989), Protein purification, *Adv. Biochem. Eng. Biotechnol.* **49**, 19–71.

MADDEN, S. L., GALELLA, E. A., ZHU, J. S., BERTELSEN, A. H., BEAUDRY, G. A. (1997), SAGE transcript profiles for p53-dependent growth regulation, *Oncogene* **15**, 1079–1085.

MANN, M., WILM, M. (1994), Error-tolerant identification of peptides in sequence databases by peptide sequence tags, *Anal. Chem.* **66**, 4390–4399.

MARSHALL, A., HODGSON, J. (1998), DNA chips: An array of possibilities, *Nature Biotechnol.* **16**, 27–31.

MATSUI, N. M., SMITH-BECKERMAN, D. M., EPSTEIN, L. B. (1999a), Staining of preparative 2-D gels.

Coomassie blue and imidazole-zinc negative staining, *Methods Mol. Biol.* **112**, 307–311.

MATSUI, N. M., SMITH-BECKERMAN, D. M., FICHMANN, J., EPSTEIN, L. B. (1999b), Running preparative carrier ampholyte and immobilized pH gradient IEF gels for 2D, *Methods Mol. Biol.* **112**, 211–219.

MATSUMURA, H., NIRASAWA, S., TERAUCHI, R. (1999), Technical advance: transcript profiling in rice (*Oryza sativa* L.), *Plant J.* **20**, 719–726.

NEILSON, L., ANDALIBI, A., KANG, D., COUTIFARIS, C., STRAUSS III, J. F. et al. (2000), Molecular phenotype of the human oocyte by PCR-SAGE, *Genomics* **63**, 13–24.

O'FARRELL, P. H. (1975), High resolution 2D gel electrophoresis of proteins, *J. Biol. Chem.* **250**, 4007–4021.

OSTERGAARD, M., RASMUSSEN, H. H., NIELSEN, H. V., VORUM, H., ORNTOFT, T. F. et al. (1997), Proteome profiling of bladder squamous cell carcinomas: identification of markers that define their degree of differentiation, *Cancer Res.* **57**, 4111–4117.

PAGE, M. J., AMESS, B., TOWNSEND, R. R., PAREKH, R., HERATH, A. et al. (1999), Proteomic definition of normal human luminal and myoepithelial breast, *Proc. Natl. Acad. Sci. USA* **96**, 12589–12594.

PAPPIN, D. J. (1997), Peptide mass fingerprinting using MALDI-TOF mass spectrometry, *Methods Mol. Biol.* **64**, 165–173.

PATTERSON, S. D., AEBERSOLD, R. (1995), Mass spectrometric approaches for the identification of gel-separated proteins, *Electrophoresis* **16**, 1791–1814.

PATTERSON, S. D., THOMAS, D., BRADSHAW, R. A. (1996), Application of combined mass spectrometry and partial amino acid sequence to the identification of gel-separated proteins, *Electrophoresis* **17**, 877–891.

RABILLOUD, T. (1990), Mechanisms of protein silver staining in polyacrylamide gels: a 10-year synthesis, *Electrophoresis* **11**, 785–794.

RABILLOUD, T. (1996), Solubilization of proteins for electrophoretic analyses, *Electrophoresis* **17**, 813–829.

RABILLOUD, T. (1998), Use of thiourea to increase the solubility of membrane proteins in two-dimensional electrophoresis, *Electrophoresis* **19**, 758–760.

RABILLOUD, T. (1999a), Silver staining of 2-D electrophoresis gels, *Methods Mol. Biol.* **112**, 297–305.

RABILLOUD, T. (1999b), Solubilization of proteins in 2-D electrophoresis. An outline, *Methods Mol. Biol.* **112**, 9–19.

RABILLOUD, T. (2000), Detecting proteins. Separated by 2-D gel electrophoresis, *Anal. Chem.* **72**, 48A–55A.

RABILLOUD, T., BLISNICK, T., HELLER, M., LUCHE, S., AEBERSOLD, R. et al. (1999), Analysis of membrane proteins by two-dimensional electrophoresis: comparison of the proteins extracted from normal or *Plasmodium falciparum*-infected erythrocyte ghosts, *Electrophoresis* **20**, 3603–3610.

RABILLOUD, T., VUILLARD, L., GILLY, C., LAWRENCE, J. J. (1994), Silver-staining of proteins in polyacrylamide gels: a general overview, *Cell. Mol. Biol.* **40**, 57–75.

RAMAGLI, L. S. (1999), Quantifying protein in 2-D PAGE solubilization buffers, *Methods Mol. Biol.* **112**, 99–103.

RIGHETTI, P. G., BOSSI, A. (1997a), Isoelectric focusing in immobilized pH gradients: an update, *J. Chromatogr. B Biomed. Sci. App.* **699**, 77–89.

RIGHETTI, P. G., BOSSI, A. (1997b), Isoelectric focusing in immobilized pH gradients: recent analytical and preparative developments, *Anal. Biochem.* **247**, 1–10.

RIGHETTI, P. G., GIANAZZA, E. (1987), Isoelectric focusing in immobilized pH gradients: theory and newer methodology, *Methods Biochem. Anal.* **32**, 215–278.

RIGHETTI, P. G., GIANAZZA, E., BJELLQVIST, B. (1983), Modern aspects of isoelectric focusing: two-dimensional maps and immobilized pH gradients, *J. Biochem. Biophys. Methods* **8**, 89–108.

ROSENFELD, J., CAPDEVIELLE, J., GUILLEMOT, J. C., FERRARA, P. (1992), In-gel digestion of proteins for internal sequence analysis after one- or two-dimensional gel electrophoresis, *Anal. Biochem.* **203**, 173–179.

ROSS-MACDONALD, P., COELHO, P. S., ROEMER, T., AGARWAL, S., KUMAR, A. et al. (1999), Large-scale analysis of the yeast genome by transposon tagging and gene disruption, *Nature* **402**, 413–418.

RUAN, Y., GILMORE, J., CONNER, T. (1998), Towards *Arabidopsis* genome analysis: monitoring expression profiles of 1400 genes using cDNA microarrays, *Plant J.* **15**, 821–833.

SAMELSON, L. E. (1999), Adaptor proteins and T-cell antigen receptor signaling, *Progr. Biophys. Mol. Biol.* **71**, 393–403.

SANCHEZ, J. C., HOCHSTRASSER, D., RABILLOUD, T. (1999), In-gel sample rehydration of immobilized pH gradient, *Methods Mol. Biol.* **112**, 221–225.

SANTONI, V., RABILLOUD, T., DOUMAS, P., ROUQUIE, D., MANSION, M. et al. (1999), Towards the recovery of hydrophobic proteins on two-dimensional gels, *Electrophoresis* **20**, 705–711.

SERVICE, R. F. (1998), Microchip arrays put DNA on the spot, *Science* **282**, 396–401.

SHAW, C. E. (1998), Solubilization and assay of cellular and tissue protein, *Methods Mol. Biol.* **105**, 287–293.

SHEVCHENKO, A., CHERNUSHEVICH, I., ENS, W., STANDING, K. G., THOMSON, B. et al. (1997), Rapid *de novo* peptide sequencing by a combination of nanoelectrospray, isotopic labeling and a quadrupole/time-of-flight mass spectrometer, *Rapid Comm. Mass Spectrom.* **11**, 1015–1024.

SHEVCHENKO, A., JENSEN, O. N., PODTELEJNIKOV, A. V., SAGLIOCCO, F., WILM, M. et al. (1996a), Linking genome and proteome by mass spectrometry: Large-scale identification of yeast proteins from two dimensional gels, *Proc. Natl. Acad. Sci. USA* **93**, 14440–14445.

SHEVCHENKO, A., WILM, M., VORM, O., MANN, M. (1996b), Mass spectrometric sequencing of proteins silver-stained polyacrylamide gels, *Anal. Chem.* **68**, 850–858.

SIRIVATANAUKSORN, Y., DRURY, R., CRNOGORAC-JURCEVIC, T., SIRIVATANAUKSORN, V., LEMOINE, N. (1999), Laser-assisted microdissection: applications in molecular pathology, *J. Pathol.* **189**, 150–154.

SMITH, B. J. (1994), Quantification of proteins on polyacrylamide gels (nonradioactive), *Methods Mol. Biol.* **32**, 107–111.

SMITH, R. D., LOO, J. A., BARINAGA, C. J., EDMONDS, C. G., UDSETH, H. R. (1989), Capillary zone electrophoresis and isotachophoresis-mass spectrometry of polypeptides and proteins based upon an electrospray ionization interface, *J. Chromatogr.* **480**, 211–232.

SMITH, R. D., UDSETH, H. R., BARINAGA, C. J., EDMONDS, C. G. (1991), Instrumentation for high-performance capillary electrophoresis-mass spectrometry, *J. Chromatogr.* **559**, 197–208.

STEINBERG, T. H., HAUGLAND, R. P., SINGER, V. L. (1996a), Applications of SYPRO orange and SYPRO red protein gel stains, *Anal. Biochem.* **239**, 238–245.

STEINBERG, T. H., JONES, L. J., HAUGLAND, R. P., SINGER, V. L. (1996b), SYPRO orange and SYPRO red protein gel stains: one-step fluorescent staining of denaturing gels for detection of nanogram levels of protein, *Anal. Biochem.* **239**, 223–237.

STEINBERG, T. H., WHITE, H. M., SINGER, V. L. (1997), Optimal filter combinations for photographing sypro orange or sypro red dye-stained gels, *Anal. Biochem.* **248**, 168–172.

TAYLOR, J. A., JOHNSON, R. S. (1997), Sequence database searches via *de novo* peptide sequencing by tandem mass spectrometry, *Rapid Comm. Mass Spec.* **11**, 1067–1075.

TOMLINSON, A. J., BENSON, L. M., BRADDOCK, W. D., ODA, R. P. (1994), On-line preconcentration capillary electrophoresis-mass spectrometry (PC-CE-MS), *J. High Resol. Chromatogr.* **17**, 729–731.

TOMLINSON, A. J., BENSON, L. M., GUZMAN, N. A., NAYLOR, S. (1996), Preconcentration and microreaction technology on-line with capillary electrophoresis, *J. Chromatogr. A* **744**, 3–15.

TOMLINSON, A. J., BENSON, L. M., JAMESON, S., JOHNSON, D. H., NAYLOR, S. (1997), Utility of membrane preconcentration capillary electrophoresis mass spectrometry in overcoming limited sample loading for analysis of biologically derived drug metabolites, peptides, and proteins, *J. Am. Soc. Mass Spectrom.* **8**, 15–24.

TOMLINSON, A. J., BENSON, L. M., ODA, R. P., BRAD-DOCK, W. D., RIGGS, B. L. et al. (1995a), Novel modifications and clinical applications of preconcentration-capillary electrophoresis-mass spectrometry, *J. Cap. Elec.* **2**, 97–104.

TOMLINSON, A. J., BRADDOCK, W. D., BENSON, L. M., ODA, R. P., NAYLOR, S. (1995b), Preliminary investigations of preconcentration-capillary electrophoresis-mass spectrometry, *J. Chromatogr. B, Biomed. Appl.* **669**, 67–73.

TOMLINSON, A. J., GUZMAN, A., NAYLOR, S. (1995c), Enhancement of concentration limits of detection in CE and CE-MS: A review of on-line sample extraction, cleanup, analyte preconcentration, and microreactor technology, *J. Cap. Elec.* **2**, 247–266.

TSUGITA, A. (1987), Developments in protein microsequencing, *Adv. Biophys.* **23**, 81–113.

UNLU, M., MORGAN, M., MINDEN, J. (1997), Difference gel electrophoresis: a single gel method for detecting changes in protein extracts, *Electrophoresis* **18**, 2071–2077.

URQUHART, B. L., ATSALOS, T. E., ROACH, D., BASSEAL, D. J., BJELLQVIST, B. et al. (1997), "Proteomic contigs" of *Mycobacterium tuberculosis* and *Mycobacterium bovis* (BCG) using novel immobilised pH gradients, *Electrophoresis* **18**, 1384–1392.

URWIN, V., JACKSON, P. (1991), A multiple high-resolution mini two-dimensional polyacrylamide gel electrophoresis system: imaging two-dimensional gels using a cooled charge-coupled device after staining with silver or labeling with fluorophore, *Anal. Biochem.* **195**, 30–37.

URWIN, V. E., JACKSON, P. (1993), Two-dimensional polyacrylamide gel electrophoresis of proteins labeled with the fluorophore monobromobimane prior to first-dimensional isoelectric focusing: imaging of the fluorescent protein spot patterns using a cooled charge-coupled device, *Anal. Bio-*

chem. **209**, 57–62.

VALASKOVIC, G. A., KELLEHER, N. L., MCLAFFERTY, F. W. (1996), Attomole protein characterization by capillary electrophoresis mass spectrometry, *Science* **273**, 1199–1202.

VAN OOSTVEEN, I., DUCRET, A., AEBERSOLD, R. (1997), Colloidal silver staining of electroblotted proteins for high sensitivity peptide mapping by liquid chromatography-electrospray ionization tandem, *Anal. Biochem.* **247**, 310–318.

VELCULESCU, V. E., ZHANG, L., VOGELSTEIN, B., KINZLER, K. W. (1995), Serial analysis of gene expression, *Science* **270**, 484–487.

WASINGER, V. C., BJELLQVIST, B., HUMPHERY-SMITH, I. (1997), Proteomic "contigs" of *Ochrobactrum anthropi*, application of extensive pH gradients, *Electrophoresis* **18**, 1373–1383.

WILKINS, M. R., LINDSKOG, I., GASTEIGER, E., BAIROCH, A., SANCHEZ, J. C. et al. (1997), Detailed peptide characterization using PEPTIDEMASS – a World-Wide-Web-accessible tool, *Electrophoresis* **18**, 403–408.

WILKINS, M. R., OU, K., APPEL, R. D., SANCHEZ, J. C., YAN, J. X. et al. (1996), Rapid protein identification using N-terminal "sequence tag" and amino acid analysis, *Biochem. Biophys. Res. Commun.* **221**, 609–613.

WILKINS, M. R., PASQUALI, C., APPEL, R. D., OU, K. et al. (1996), From proteins to proteomes: Large-scale protein identification by two-dimensional electrophoresis and amino acid analysis, *BioTechnology* **14**, 61–65.

WILM, M., MANN, M. (1996), Analytical properties of the nanoelectrospray ion source, *Anal. Chem.* **68**, 1–8.

WISE, M. J., LITTLEJOHN, T. G., HUMPHERY-SMITH, I. (1997), Peptide-mass fingerprinting and the ideal covering set for protein characterisation, *Electrophoresis* **18**, 1399–1409.

YATES, J. R., MCCORMACK, A. L., LINK, A. J., SCHIELTZ, D., ENG, J., HAYS, L. (1996), Future prospects for the analysis of complex biological systems using micro-column liquid chromatography electrospray tandem mass spectrome, *Analyst* **121**, R65–R76.

11 Proteome Analysis by Capillary Electrophoresis

NORMAN J. DOVICHI

SHEN HU

DAVID MICHELS

Seattle, WA, USA

ZHERU ZHANG

Syracuse, NY, USA

SERGEY N. KRYLOV

Toronto, Canada

1 Introduction

The human genome contains $3.12 \cdot 10^9$ bases. At the moment, the number of genes in the genome is not clear, with estimates ranging from ~30,000 to ~150,000 (PENNISI, 2000). The translation product of the genome forms the proteome, which is the complement of proteins expressed in an organism, tissue, or cell (WASINGER et al., 1995). Unlike the genome, which is essentially static and identical in each cell of an organism, the proteome varies from cell to cell and changes in response to the environment and during development and disease.

Two-dimensional electrophoresis, followed by mass spectrometric analysis of isolated proteins, is the workhorse tool for proteomics research (HALL et al., 1993). In most experiments, proteins are extracted from the lysate generated from several million cells. The protein extract is first separated by isoelectric focusing, which is performed in either strip or tube format. The isoelectric focusing gel is then placed at the top of an SDS/PAGE gel and proteins are separated based on size. The proteins are detected by a staining protocol that creates a two-dimensional set of spots. The location of each spot is determined by the protein's mass and isoelectric point. The intensity of each spot is related to the amount of the protein present in the cell extract. Those proteins with interesting expression patterns are cut from the gel, digested by a protease, and identified by mass spectrometry and database searching, which is the subject of Chapter 10 in this book.

Proteomics research is at a similar stage as was DNA sequencing a decade ago. Just as DNA sequencing used cumbersome slab-gels that required manual manipulation, proteome analysis by two-dimensional electrophoresis is labor intensive and very difficult to automate. Just as genomics research was revolutionized by the development of highly automated sequencing instrumentation, proteomic research requires the development of highly automated analytical instrumentation, if it is to advance rapidly.

Proteomic research differs from genomic research in one important way. DNA research benefits greatly from PCR technology, which allows amplification of specific genomic regions, so that a few copies of interesting mRNA or genomic DNA can be amplified to levels that are easily handled. Protein research does not have an analogous tool. As a result, protein research requires the use of extremely high-sensitivity analytical tools to monitor proteins expressed at low levels.

2 Capillary Electrophoresis

Capillary array electrophoresis has proven to be a powerful tool for DNA analysis (ZHANG et al., 1999). Automated instruments have been commercialized and used to sequence the vast majority of the human genome. We believe that capillary electrophoresis will play a similar role in proteome analysis.

2.1 Instrument

Capillaries are typically 50 μm inner diameter, 150 μm outer diameter, and 35 cm long. The outside of the capillary is coated with a thin layer of polymer, usually polyamide, which gives the capillaries great strength and flexibility. The total volume of the capillary is typically 1 μL and the sample volume is typically a few nanoliters. Instrumentation for capillary electrophoresis is very simple (Fig. 1). It consists of an injector to introduce the sample to the capillary, a detector, and a high-voltage power supply to drive the separation. A computer controls injection and automatically re-

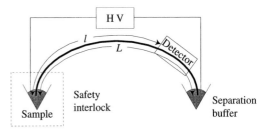

Fig. 1. Capillary electrophoreses instrument. The length from the injector to the detector is denoted by *l*, the total length of the capillary is given by *L*.

cords data. Because of the high voltage employed for separation, the operator must be protected with a safety interlock system.

2.2 Injection

Analyte is usually held in disposable microcentrifuge tubes. Injection is performed by dipping the capillary and a high-voltage electrode into the sample. In electrokinetic injection, an electric field is applied to the sample for a few seconds, drawing the protein into the capillary. Electrokinetic injection is biased; the amount of sample injected is proportional to the sample's velocity during electrophoresis (HUANG et al., 1988). More of the faster moving analytes are injected than are the slower moving components.

Alternatively, sample can be injected hydrodynamically by applying pressure to the sample or vacuum to the distal end of the capillary. Hydrodynamic injection is unbiased; the injected amount is independent of the analyte's physical properties. No matter which injection method is used, the sample volume should be less than ~0.1% of the capillary volume to preserve the separation efficiency. Conventionally, 1 nL of sample is injected onto the capillary. Fortunately, stacking methods have been developed for capillary electrophoresis, which allow larger sample volumes to be injected without overloading the capillary (BURGI and CHIEN, 1996; BURGI, 1993).

2.3 Electroosmosis

Electroosmosis is particularly important when separations are performed in uncoated capillaries. Residual silanol groups on the capillary surface take a negative charge at neutral or basic pH. These silanol groups are fixed and do not move upon application of an electric field to the capillary. Electrical neutrality requires that cationic counterions be present in a diffuse cloud very near the capillary wall. These cations migrate in an electric field toward the negative electrode. The cations draw solvent with them, transporting bulk fluid to the negative electrode. This bulk flow is called electroosmosis. Unlike pressure-driven flow in

chromatography, electroosmotic flow velocity is uniform across the capillary lumen, which minimizes peak broadening so that the separation efficiency is much higher than for liquid chromatography.

During analysis, electrophoresis and electroosmosis occur simultaneously. The overall mobility of an ion is the sum of the electrophoretic and electroosmotic mobilities. Electroosmosis is usually stronger than electrophoresis, and all components of a sample will migrate through the capillary to the detector, with cations migrating first, neutrals second, and anions last.

Electroosmosis can be reduced to negligible levels by coating the capillary. Neutral coating can be chemically bound to the wall by use of either silane or Grinard reactions (GELFI et al., 1998). Dynamic coating relies on the physical adsorption, usually of a polymer to the capillary wall; the dynamic coatings are not as robust as the chemical coatings, but may be regenerated as needed by successive treatment with the coating reagent.

2.4 Separation

The specific separation methods for proteins are discussed in Sect. 3. There, general information on capillary electrophoresis separations is presented. The capillary is filled with a separation buffer, the nature of which determines the separation mechanism. The buffer is usually a few mM in concentration. Higher ionic strength buffers suffer from excessive Joule heating, and lower ionic strength buffers do not provide adequate buffering capacity.

Once the sample has been injected into the capillary, the sample vial is replaced with a buffer-filled vial, and an electric field is applied to separate the components within the sample. The narrow diameter of the capillary minimizes the temperature rise associated with the electric field. As a result, very high potentials can be applied across the capillary without degradation of the separation due to heating. Potentials up to 30,000 V and electric fields exceeding $300 \, V \, cm^{-1}$ are routinely employed. The use of high fields results in rapid and efficient separations.

Unlike slab-gels, where proteins are visualized after a fixed run time, capillary electrophoresis is a finish-line technique, where proteins are detected after traversing a fixed distance. The time necessary for a protein to migrate through the capillary is given by

$$t = \frac{l\,L}{(\mu_{\mathrm{eof}} + \mu_{\mathrm{ep}})\,V}$$

where L is the length of the capillary (Fig. 1), l is the distance from the injection end of the capillary to the detector, μ_{ep} is the electrophoretic mobility of the analyte, μ_{eof} is the electroosmotic mobility of the buffer, and V is the applied potential (JORGENSON and LUKAS, 1981). The separation time is inversely proportional to the applied potential. For post-column detectors, $l = L$ and the separation time is proportional to the length of the capillary squared. Separation time reduces dramatically as the capillary length is shortened. Typical protein separations are completed in less than 20 min.

2.5 Detection

UV absorbance, laser-induced fluorescence, and mass spectrometry are used for protein detection. UV absorbance measurements use the capillary itself as the detection cuvette. Unfortunately, the relatively short optical pathlength across the capillary tends to produce poor sensitivity in absorbance measurements, which must be performed with relatively high concentration samples.

Mass spectrometry is usually better suited to the analysis of peptides rather than proteins; sensitivity and mass resolution fall for the higher molecular weight species. Electrospray ionization is a common interface between capillary electrophoresis and mass spectrometry (JENSEN, 1999). Unfortunately, electrospray is incompatible with capillary separations based on a surfactant. The surfactant disrupts electrospray ionization and the surfactant ions tend to contaminate the detector.

Fluorescence detection from native proteins requires the use of expensive and temperamental lasers, which are not available for commercial instruments. Instead, proteins are usually labeled with a fluorescent reagent. We find that classic fluorescent reagents, such as fluorescein isothiocyanate, are not convenient for protein labeling by capillary electrophoresis. Unreacted reagent and fluorescent impurities generate a strong background signal that can saturate the fluorescence detector. Instead, we prefer the use of fluorogenic reagents (PINTO et al., 1997; LEE et al, 1998). These reagents are non-fluorescent themselves, but produce a highly fluorescent product upon reaction with a primary amine, such as the e-amine of lysine residues. Because the reagents are non-fluorescent until they react with the amine, the background signal from unreacted reagent does not interfere in the measurement.

There is a subtle point concerning labeling chemistry. Lysine is a common amino acid, and there usually are several lysine residues in each protein. The labeling reaction rarely goes to completion and a complex mixture of reaction products is formed. A single protein with n primary amines can produce $2^n - 1$ possible fluorescent products (ZHAO et al., 1992). For example, ovalbumin has 20 lysine residues and a blocked N-terminus. There can be 1,048,575 different products generated upon reaction with a labeling reagent. These products have different mobilities and can result in a broad and complex electrophoresis pattern. A fair amount of effort has been spent in developing reaction protocols and buffer systems that minimize the effects of multiple labeling. As we show below, the use of an appropriate separation buffer eliminates the effects of multiple labeling in SDS/gel and free solution electrophoresis. However, capillary isoelectric focusing remains incompatible with fluorescence labeling technology.

3 Capillary Electrophoresis for Protein Analysis

There are three types of electrophoresis used for protein analysis. Two are similar to classic protein electrophoresis techniques: isoelectric focusing and SDS/gel electrophoresis. The third is unique to capillary electrophoresis and is based on migration in a simple buffer.

3.1 Capillary Isoelectric Focusing

STELLAN HJERTEN performed the first work on capillary isoelectric focusing (HJERTEN and ZHU, 1985; HJERTEN et al., 1987). The sample is mixed with ampholytes and used to fill the capillary. A high electric field is applied, focusing the proteins at their isoelectric point and rapidly generating a high-resolution electropherogram (RODRIGUEZ-DIAZ et al., 1997). Typically, the column is coated to minimize electroosmosis and reduce protein adsorption so that a stationary isoelectric focusing profile is formed in the capillary. Once the separation is complete, the proteins must be detected. While the proteins can be visualized with an imaging detector (WU et al., 1998), it is much more common to mobilize the proteins so that they flow through a detector at one end of the capillary. The isoelectric focusing pattern is recorded as the proteins pass through the detector.

Protein mobilization can be performed in several ways. If an uncoated capillary is used for the separation, residual electroosmosis will force the proteins to migrate past the detector. However, coated capillaries are frequently used to improve the resolution of the separation and these capillaries have negligible electroosmosis. In this case, several methods can be used to drive the contents of the capillary through the detector.

Electrokinetic mobilization of proteins can be performed by addition of a neutral salt such as sodium chloride to either the cathode or anode buffer to drive proteins from the capillary to the detector (YAO and REGNIER, 1993). This salt migrates into the capillary under the influence of an electric field, driving the focused proteins to the opposite end of the capillary.

Alternatively, one end of the capillary can be pressurized to drive the contents through the detector. The pressure must be very low to avoid peak broadening due to formation of a parabolic flow profile. The pressure can be introduced most simply by lowering the detector end of the capillary with respect to the injection end to form a siphon, which draws the contents of the capillary through the detector.

Fig. 2 presents a capillary isoelectric focusing run on a set of protein standards (pI = 5.3, 6.4, 7.4, 8.4, and 10.4). The separation was carried out using capillaries coated with polyethylene glycol, which forms a dynamic coating and reduces electroosmosis, but does not eliminate it. The cathodic reservoir contains a high pH electrolyte (20 mM NaOH) while the anodic reservoir contains a low pH electrolyte (10 mM H_3PO_4). There was sufficient residual electroosmosis to drive the focused proteins through the detector without the use of an external perturbation to the system. Detection

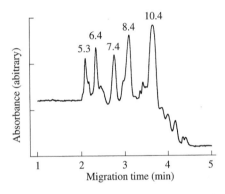

Fig. 2. Isoelectric focusing separation of a set of protein standards. Detection is by ultraviolet absorbance.

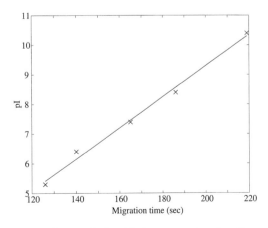

Fig. 3. Linear relationship between migration time and pI for capillary isoelectric focusing of protein standards. Data taken from Fig. 2.

was by UV absorbance at 254 nm. Migration time increased linearly with pI, $r > 0.996$ (Fig. 3).

UV absorbance or mass spectrometry is used as detection techniques for isoelectric focusing. Mass spectrometry is particularly important, because it can provide extremely high accuracy molecular mass determination for proteins. The resulting data resemble two-dimensional gels with much higher mass resolution than can be produced by SDS/PAGE (JENSEN et al., 1999). Unfortunately, Fourier transform mass spectrometers and electrospray ionization are required for analysis of the high molecular weight proteins. This technology is expensive and not routinely available. Fluorescence detection is not useful for labeled proteins; multiple labeling produces a complex mixture of reaction products that results in a complex electropherogram (RICHARDS et al., 1999).

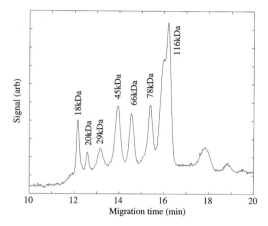

Fig. 4. SDS/PEO separation of proteins based on their size. Proteins were labeled with the fluorigenic reagent FQ and detected with laser-induced fluorescence.

3.2 SDS/Polymer Electrophoresis

SDS/PAGE is the commonly used slab electrophoresis method for the separation of proteins based on their size. Unfortunately, the crosslinked polyacrylamide used in slab-gels is not appropriate for capillary-based separations. The crosslinked material must be discarded after each run, which requires replacement of the capillary. Instead, non-crosslinked polymers are usually employed for size-based separation of proteins. These materials have relatively low viscosity and may be pumped into the capillary after each run in an automated system.

Several different polymers have been used to separate proteins in capillary electrophoresis (HU et al., 2000). Polyacrylamide is not a particularly useful polymer. It has low shelf-life and must be produced on demand. Unfortunately, the free redial polymerization is not highly reproducible, which is undesirable when comparing separations of samples with standards.

Instead, we have found that polyethylene oxide (PEO) can be used for size-based protein separations. Fig. 4 presents an SDS/PEO separation of seven size standards that had been fluorescently labeled with FQ. Only one

attomole of protein was required for this assay, which was complete in 20 min.

3.3 Free Solution Electrophoresis

Capillary-free solution electrophoresis does not have an analog in slab electrophoresis. A capillary is filled with a millimolar buffer, and proteins are separated based on their mobility in the buffer. In general, mobility is related to the size-to-charge ratio of the protein, where highly charged and small proteins have the highest mobility.

To decrease protein interaction with the capillary wall, a modest amount of anionic surfactant is often added to the separation buffer. This buffer tends to ion-pair with cationic amino acid residues that would otherwise interact with silanol groups on the capillary wall. This interaction would lead to adsorption of the protein to the capillary, which leads to band broadening and decrease in resolution. More importantly, the surfactant reduces the effects of multiple labeling to negligible levels (PINTO et al., 1997).

The separation efficiency seems to maximize when the concentration of the surfactant is below its critical micelle concentration. Fig. 5 presents a typical submicellar separation of a

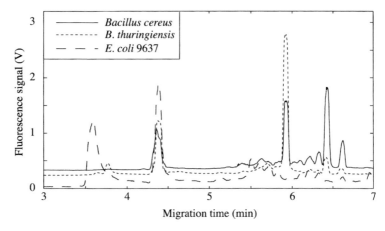

Fig. 5. Submicellar separation of fluorescently labelled protein extract from three bacteria species. Proteins were labeled with the fluorigenic agent FQ and detected with laser-induced fluorescence.

complex mixture of proteins extracted from a set of bacteria.

The use of a surfactant in the separation buffer prohibits the use of mass spectrometry for detection. The surfactant tends to contaminate the interface of the mass spectrometer. Instead, laser-induced fluorescence is a powerful detection technique, particularly when only a minute amount of protein is available.

4 Single-Cell Analysis

Capillary electrophoresis with laser-induced fluorescence detection is an extraordinarily sensitive analytical technique. We have obtained subpicomolar concentration detection limits for proteins; only a few zeptomoles of protein are used for the analysis. This extraordinary sensitivity provides opportunities to perform proteome analysis of single somatic cells (ZHANG et al., 2000). These cells are typically ~ 10 µm in diameter and 0.5 pL in volume. Assuming that the cell is 10% protein by weight and that the average molecular weight of the protein is ~ 50 kDa, then a single cell contains about 1 femtomole of protein. A typical cell may express ~ 10,000 proteins; the average protein would be present as 60,000 co-

pies or 100 zeptomoles. Of course, proteins are not expressed uniformly. Structural proteins would constitute most of the mass of proteins. The vast majority of proteins would be expressed at relatively low copy number. Nevertheless, capillary electrophoresis with laser-induced fluorescence should detect a large number of proteins from a single cell.

Fig. 6 presents the SDS/PEO analysis of the proteins in a single HT29 colon adenocarcinoma cell. In this experiment, the cell was injected into a 20 µm capillary that contained polyethylene oxide as the separation medium. The cell was lysed by contact with SDS within the separation medium. The freed proteins were fluorescently labeled with FQ, separated by SDS/PEO, and detected by laser-induced fluorescence in a sheath-flow cuvette detector. Roughly 25 components are resolved in this electropherogram, with similar resolution to that produced by conventional SDS/PAGE analysis of the cell extract from several million cells.

Relatively mild denaturation conditions were used for lysis, labeling, and separation. As a result, not all protein complexes were dissociated, and some of the very high molecular weight species may be from protein aggregates.

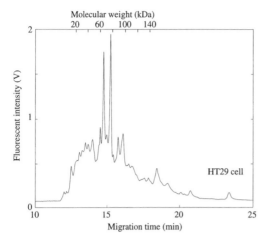

Fig. 6. SDS/PEO separation of the protein from a single HT29 human cancer cell. A single cell was injected into the separation capillary, lysed during contact with the SDS-containing buffer, labeled with the fluorigenic reagent FQ, separated by SDS/PEO electrophoresis, and detected by laser-induced fluorescence. The molecular weight scale was determined from the standard proteins in Fig. 4.

5 Two-Dimensional Separations

Capillary electrophoresis versions of isoelectric focusing, SDS/PAGE, and free solution electrophoresis have been demonstrated for protein analysis (JENSEN et al., 1999). While these techniques produce quite impressive separation power, they are one-dimensional analyses, and their ability to separate complex mixtures is limited. JORGENSON demonstrated the first two-dimensional analysis by combining capillary size exclusion chromatography with free solution electrophoresis for analysis of peptides (BUSHEY and JORGENSON, 1990; LARMANN et al., 1993; MOORE and JORGENSON, 1995)).

We are modifying JORGENSON's technology for two-dimensional capillary electrophoresis of proteins. In this method, a protein mixture is injected into the first capillary. As analyte begin to migrate from the first capillary, successive fractions are injected into a second column, where co-migrating components are separated. The electric current to the first capillary is turned off during the second column separations. This procedure is repeated to build a two-dimensional electropherogram as a raster image from successive separations of the first capillary's fractions.

As a proof-of-principle, we recently developed a two-dimensional capillary electrophoresis system. In this test instrument, both capillaries used the same free-solution electrophoresis buffer. The experiment was simply designed to test the interface between the two capillaries rather than to separate a complex mixture. Fig. 7 presents the data generated in this prototype experiment. A set of proteins was fluorescently labeled and injected into the first capillary. Electric current was applied to the first capillary for four seconds. The aliquot that migrated from the first capillary during that period was injected into a second capillary, whose entrance abutted the exit of the first capillary. The electric field across the first capillary was switched off. The fraction that was injected into the second capillary was separated by applying electric field across that capillary. The process was repeated until every component in the first capillary had been passed through the second capillary.

The data fall on a diagonal in this representation, because the two separation mechanisms were identical. Our next goal will be to replace the first dimension separation with capillary gel electrophoresis, which should produce an orthogonal separation mechanism to free solution electrophoresis. By use of the orthogonal separation mechanisms, we should be able to resolve complex mixtures into their constituents.

This analysis used fluorescently labeled proteins and laser-induced fluorescence detection. The detector produces signal-to-noise ratios greater than 10^4, which means that both major and minor components can be quantified in the same electropherogram. Furthermore, extremely small amounts of protein are required for the analysis; attomoles of protein were used to generate the data of Fig. 7.

In the near future, we hope to use an SDS/PEO column for the first capillary separation and free solution electrophoresis for the second capillary. We have resolved about 30

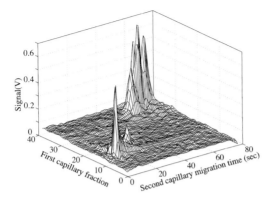

Fig. 7. Two-dimensional capillary electrophoresis separation of proteins. In this proof-of-principle experiment, submicellar electrophoresis was used for both dimensions.

components using each technology as one-dimensional electrophoresis. If the two separations are based on different mechanisms, then the resulting two-dimensional electropherogram will resolve complex mixtures into nearly a thousand components.

6 Conclusions

Capillary electrophoresis provides rapid and sensitive separation of proteins, often producing resolution similar to that produced by conventional gel-based electrophoresis methods. Unlike conventional methods, capillary electrophoresis is easily automated for analysis of large numbers of samples without requiring operator attention.

Two-dimensional separations are being developed by coupling two capillaries, where the second capillary separates fractions that migrate from the first capillary. This separation method will produce separations that are similar to IEF-SDS/PAGE in a fully automated system. Furthermore, when combined with laser-induced fluorescence detection, the separation will provide orders of magnitude higher sensitivity and dynamic range than current IEF-SDS/PAGE analysis.

When coupled with laser-induced fluorescence detection, capillary electrophoresis is a very powerful analytical tool for the study of minute samples, including single human cancer cells. The cell-to-cell variation in protein expression will be of interest in cancer prognosis, in developmental biology, and in gene expression studies.

Acknowledgement
This work was funded by a grant from the National Institutes of Health (USA).

7 References

BURGI, D. S. (1993), Large-volume stacking of anions in capillary electrophoresis using an electro-osmotic flow modifier as a pump, *Anal. Chem.* **65**, 3726–3729.

BURGI, D. S., CHIEN, R. L. (1996), Application and limits of sample stacking in capillary electrophoresis, *Methods Mol. Biol.* **52**, 211–226.

GELFI, C., CURCIO, M., RIGHETTI, P. G., SEBASTIANO, R., CITTERIO, A. et al. (1998), Surface modification based on Si-O and Si-C sublayers and a series of N-substituted acrylamide top-layers for capillary electrophresis, *Electrophoresis* **19**, 1677–1682.

HALL, S. C., SMITH, D. M., MASIARZ, F. R., SOO, V. W., TRAN, H. M. et al. (1993), Mass spectrometric and Edman sequencing of lipocortin I isolated by two-dimensional SDS/PAGE of human melanoma lysates, *Proc. Natl. Acad. Sci. USA* **90**, 1927–1931.

HUANG, X., GORDON, M. J., ZARE, R. N. (1988), Effect of electrolyte and sample concentration on the relationship between sensitivity and resolution in capillary zone electrophoresis using conductivity detection, *Anal. Chem.* **60**, 375–377.

PENNISI, E. (2000), Human genome Project. And the gene number is …? *Science* **288**, 1146–1147.

WASINGER, V. C., CORDWELL, S. J., CERPA-POLJAK, A., YAN, J. X., GOOLEY, A. A. et al. (1995), Progress with gene-product mapping of the Mollicutes: *Mycoplasma genitalium*, *Electrophoresis* **16**, 1090–1094.

ZHANG, J. Z., VOSS, K. O., SHAW, D. F., ROOS, K. P., LEWIS, D. F. et al. (1999), A multiple-capillary electrophoresis system for small-scale DNA sequencing and analysis, *Nucleic Acids Res.* **27**, e36.

Bioinformatics

12 Using the Molecular Biology Data

Evgeni M. Zdobnov

Rodrigo Lopez

Rolf Apweiler

Thure Etzold

EMBL-EBI, Cambridge, UK

1 Introduction

The goal of this chapter is to introduce the most important of the different molecular biology databases available to researchers. Recent advances in technology have resulted in an information explosion. This has risen the challenge of providing an integrated access method to these data, capable of querying cross-referenced, but highly heterogeneous databases. The SRS (ETZOLD et al., 1996) system has emerged to deal with these problems and has became a powerful tool in modern biotechnology research.

The chapter is organized in two main sections. The first section gives an overview of the following groups of databases:

- bibliographic,
- taxonomic,
- nucleic acid,
- genomic,
- protein and specialized protein databases,
- protein families, domains and functional sites,
- proteomics initiatives,
- enzyme/metabolic pathways.

The second section describes the SRS approach to integrate access to these databases and some recent developments at the EBI SRS server (*http://srs.ebi.ac.uk*).

2 Databases

2.1 Bibliographic Databases

Services that abstract the scientific literature began to make their data available in electronic form in the early 1960s. The most commonly known and the only publicly available is MEDLINE through PUBMED (*http://www.ncbi.nlm.nih.gov/PubMed/*), which covers mainly the medical literature. Other commercial bibliographic database products include: EMBASE (*http://www.embase.com*) for biomedical and pharmacological abstracts; AGRICOLA (*http://www.nalusda.gov/-*

general_info/agricola/agricola.html) for the agricultural field; BIOSIS (*http://www.biosis.org*), the inheritor of the old Biological Abstracts, for a broad biological field; the Zoological Record for the zoological literature and CAB International (*http://www.cabi.org*) for abstracts in the fields of agriculture and parasitic diseases. The reader should be aware that none of the abstracting services has a complete coverage.

2.2 Taxonomy Databases

Taxonomic databases are rather controversial since the soundness of the taxonomic classifications done by one taxonomist will be directly questioned by another! Various efforts are under way to create a taxonomic resource (e.g., "The Tree of Life" project (*http://phylogeny.arizona.edu/tree/life.html*), "Species 2000" (*http://www.sp2000.org*), International Organization for Plant Information (*http://iopi.csu.edu.au/iopi/*), Integrated Taxonomic Information System (*http://www.itis.usda.gov/itis/*), etc.). The most generally useful taxonomic database is that maintained by the NCBI (*http://www.ncbi.nlm.nih.gov/Taxonomy/*). This hierarchical taxonomy is used by the Nucleotide Sequence Databases, SWISS-PROT and TrEMBL, and is curated by an informal group of experts. Another important source of biodiversity knowledge includes the Expert Center for Taxonomic Identification (ETI, *http://www.eti.uva.nl*).

2.3 Nucleotide Sequence Databases

The International Nucleotide Sequence Database Collaboration is a joint effort of the nucleotide sequence databases EMBL-EBI (European Bioinformatics Institute, *http://www.ebi.ac.uk*), DDBJ (DNA Data Bank of Japan, *http://www.ddbj.nig.ac.jp*), and GenBank (National Center for Biotechnology Information, *http://www.ncbi.nlm.nih.gov*). In Europe, the vast majority of the nucleotide sequence data produced is collected, organized and distributed by the EMBL Nucleotide Sequence Database (*http://www.ebi.ac.uk/embl/*,

STOESSER, et al., 1999) located at the EBI, Cambridge, UK, an outstation of the European Molecular Biology Laboratory (EMBL) based in Heidelberg, Germany. The nucleotide sequence databases are data repositories, accepting nucleic acid sequence data from the community and making it freely available. The databases strive for completeness, with the aim of recording and making available every publicly known nucleic acid sequence. These data are heterogeneous, vary with respect to the source of the material (e.g., genomic versus cDNA), the intended quality (e.g., finished versus single pass sequences), the extent of sequence annotation and the intended completeness of the sequence relative to its biological target (e.g., complete versus partial coverage of a gene or a genome). EMBL, GenBank and DDBJ automatically update each other every 24 hours with new or updated sequences. The result is that they contain the same information (although there are backlogs at every site at any one time due to transfer speeds and time zone differences), but stored in different formats. Each entry in a database must have a unique identifier that is a string of letters and numbers unique to that record. This unique identifier, known as the accession number, can be quoted in the scientific literature, as it will never change. As the accession number must always remain the same, another code is used to indicate the number of changes that a particular sequence has undergone. This code is known as the sequence version and is composed of the accession number followed by a period and a number indicating which version is at hand. You should, therefore, always take care to quote both the unique identifier and the version number, when referring to records in a nucleotide sequence database.

The archival nature of the three mayor nucleotide sequence databanks means that the overall quality of the sequence is the responsibility of the authors and/or submitters. The databanks do not police or check for integrity of these data, but they make sure it is conformant with the high-quality standards agreed between the database collaborators. For these reasons these databanks are redundant and may contain sequences of low quality. For example, unrelated efforts may result in two independent submissions of the same sequence and submitters have been know to forget to check their sequences for vector contamination.

Since their conception in the 1980s, the nucleic acid sequence databases have experienced constant exponential growth. There is a tremendous increase of sequence data due to technological advances (such as sequencing machines), the use of new biochemical methods (such as PCR technology) as well as the implementation of projects to sequence complete genomes. These advances have brought along an enormous flood of sequence data. At the time of writing the EMBL Nucleotide Sequence Database has more than 10 billion nucleotides in more than 10 million individual entries. In effect, these archives currently experience a doubling of their size every year. Today, electronic bulk submissions from the mayor sequencing centers overshadows all other input, and it is not uncommon to add to the archives more than 7,000 new entries, on average, per day. You can find some statistics of the data at *http://www3.ebi.ac.uk/ Services/DBStats/*.

Sequence-cluster databases such as UniGene (*http://www.ncbi.nlm.nih.gov/UniGene*, SCHULER et al., 1996a) and STACK (Sequence Tag Alignment and Consensus Knowledgebase, *http://www.sanbi.ac.za/Dbases.html*, MILLER, et al., 1999) address the redundancy problem by coalescing sequences that are similar to the degree that one may reasonably infer that they are derived from the same gene.

Several specialized sequence databases are also available. Some of these deal with particular classes of sequence, e.g., the Ribosomal Database Project (RDP, *http://rdp.life.uiuc. edu/index.html*, MAIDAK, et al., 1999), the HIV Sequence Database (*http://hiv-web.lanl.gov/*, KUIKEN, 1999), and IMGT, the database (*http://imgt.cnusc.fr:8104/textes/info.html*, LEFRANC et al., 1999); others are focussing on particular features, such as TRANSFAC for transcription factors and transcription factor binding sites (*http://transfac.gbf.de/TRANSFAC/ index.html*, WINGENDER et al., 2000), EPD (Eukaryotic Promoter Database, *ftp://ftp.ebi. ac.uk/pub/databases/epd*, PERIER et al., 1999) for promoters, and REBASE (*http://rebase. neb.com/rebase*, ROBERTS and MACELIS, 2000) for restriction enzymes and restriction enzyme sites. (*http://megasun.bch.umontreal.ca/gobase/ gobase.html*, KORAB-LASKOWSKA et al., 1998) is a specialized database of organelle genomes.

2.4 Genomic Databases

For organisms of major interest to geneticists, there is a long history of conventionally published catalogs of genes or mutations. In the past few years, most of these have been made available in an electronic form and a variety of new databases have been developed.

There are several databases for *Escherichia coli*. CGSC, the *E. coli* Genetic Stock Center, (*http://cgsc.biology.yale.edu/top.html*, BERLYN and LETOVSKY, 1992) maintains a database of *E. coli* genetic information, including genotypes and reference information for the strains in the CGSC collection, gene names, properties, and linkage map, gene product information, and information on specific mutations. The *E. coli* Database collection (ECDC, *http:// susi.bio.uni-giessen.de/ecdc/ecdc.html*, KROGER and WAHL, 1998) in Giessen, Germany, maintains curated gene-based sequence records for *E. coli*. EcoCyc (*http://ecocyc.Pangea-Systems.com/ecocyc/ecocyc.html*, KARP et al., 2000), the "Encyclopedia of *E. coli* Genes and Metabolism" is a database of *E. coli* genes and metabolic pathways.

The MIPS yeast database (*http://www.mips. biochem.mpg.de/proj/yeast/*, MEWES et al., 2000) is an important resource for information on the yeast genome and its products. The *Saccharomyces* Genome Database (*http:// genome-www.stanford.edu/Saccharomyces*, CHERVITZ et al., 1999) is another major yeast database.

The *Arabidopsis* Information Resource (TAIR) provides genomic and literature data about *Arabidopsis thaliana* (*http://www. arabidopsis.org*, RHEE et al., 1999), while MaizeDB is the database for genetic data on maize (*http://www.agron.missouri.edu*). For other plants Demeter's genomes (*http://ars-genome.cornell.edu*) provides access to many different genome databases (mostly in ACEDB format), including *Chlamydomonas*, cotton, alfalfa, wheat, barley, rye, rice, millet, sorghum and species of Solanaceae and trees.

MENDEL is a plant-wide database for plant genes (*http://www.mendel.ac.uk*, REARDON, 1999).

ACeDB is the database for genetic and molecular data concerning *Caenorhabditis elegans*. The database management system written for ACeDB by R Durbin and J Thierry-Mieg has proved very popular and has been used in many other species-specific databases. ACEDB (spelled with a capital "E") is now the name of this database management system, resulting in some confusion relative to the *C. elegans* database. The entire database can be downloaded from the Sanger Centre (*http://www.sanger.ac.uk/Projects/C_elegans/*).

Two of the best-curated genetic databases are FlyBase (*http://flybase.bio.indiana.edu*, The FlyBase Consortium, 1999) the database for *Drosophila melanogaster* and the Mouse Genome Database (MGD, *http://www. informatics.jax.org*, BLAKE et al., 1999). ZFIN, a database for another important model organism, the zebrafish *Brachydanio rerio*, has been implemented recently (*http://zfish.uoregon. edu/ZFIN/*, WESTERFIELD et al., 1999).

There are also genetic databases available for several animals of economic importance to humans. These include pig (PIGBASE), bovine (BovGBASE), sheep (SheepBASE) and chicken (ChickBASE). All these databases are available via the Roslin Institute server (*http://www.ri.bbsrc.ac.uk/bioinformatics/ databases.html*).

Two major databases for human genes and genomics are in existence. MCKUSICK's Mendelian Inheritance in Man (MIM) is a catalog of human genes and genetic disorders and is available in an online form (OMIM, *http:// www3.ncbi.nlm.nih.gov/Omim/*, HAMOSH et al., 2000) from the NCBI. The Genome Database (GDB, *http://www.gdb.org*, LETOVSKY et al., 1998) is the major human genome database including both molecular and mapping data. Both OMIM and GDB include information on genetic variation in humans but there is also the Sequence Variation Database project at the EBI (*http://www.ebi.ac.uk/mutations/index. html*, LEHVASLAIHO et al., 2000), with links to the many sequence variation databases at the EBI; and to the SRS (Sequence Retrieval System) interface to many human mutation databases. The GeneCards resource at the

Weizmann Institute (*http://bioinfo.weizmann. ac.il/cards/*, REBHAN et al., 1998) integrates information about human genes from a variety of databases, including GDB, OMIM, SWISS-PROT and the nucleotide sequence databases. GENATLAS (*http://web.citi2.fr/GENATLAS/*, FREZAL, 1998) also provides a database of human genes, with links to diseases and maps.

A relatively new database has been created by EnsEMBL (*http://www.ensembl.org*, BUTLER, 2000), a joint project between EMBL-EBI and the Sanger Centre that strives to develop a software system, which produces and maintains automatic annotation on eukaryotic genomes. Human data are available now; worm and mouse will be added soon.

A parasite genome database (*http://www.ebi. ac.uk/parasites/parasite-genome.html*) is supported by the World Health Organisation (WHO) at the EBI, covering the five "targets" of its Tropical Diseases Research program: *Leishmania*, Trypanosomes, *Schistosoma* and Filarioidea. Databases for some vectors of parasitic diseases are also available, such as AnoDB (*http://konops.imbb.forth.gr/AnoDB/*) for *Anopheles* and AaeDB (*http://klab.agsci. colostate.edu*) for *Aedes aegypti*.

2.5 Protein Sequence Databases

The protein sequence databases are the most comprehensive source of information on proteins. It is necessary to distinguish between universal databases covering proteins from all species and specialized data collections storing information about specific families or groups of proteins, or about the proteins of a specific organism. Two categories of universal protein sequence databases can be discerned: simple archives of sequence data; and annotated databases where additional information has been added to the sequence record.

The oldest protein sequence database PIR (BARKER et al., 1999) was established in 1984 by the National Biomedical Research Foundation (NBRF) as a successor of the original NBRF Protein Sequence Database, developed over a 20 year period by the late MARGARET O. DAYHOFF and published as the "Atlas of Protein Sequence and Structure" (DAYHOFF, 1965; DAYHOFF and ORCUTT, 1979). Since 1988

the database (*http://www-nbrf.georgetown.edu*) has been maintained by PIR-International, a collaboration between the NBRF, the Munich Information Center for Protein Sequences (MIPS), and the Japan International Protein Information Database (JIPID).

The PIR release 66 (September 30, 2000) contained 195,891 entries. The database is partitioned into four sections, PIR1 (20,471 entries), PIR2 (174,756 entries), PIR3 (262 entries) and PIR4 (402 entries). Entries in PIR1 are fully classified by superfamily assignment, fully annotated and fully merged with respect to other entries in PIR1. The annotation content as well as the level of redundancy reduction varies in PIR2 entries. Many entries in PIR2 are merged, classified, and annotated. Entries in PIR3 are not classified, merged or annotated. PIR3 serves as a temporary buffer for new entries. PIR4 was created to include sequences identified as not naturally occurring or expressed, such as known pseudogenes, unexpressed ORFs, synthetic sequences, and non-naturally occurring fusion, crossover or frameshift mutations.

SWISS-PROT (BAIROCH and APWEILER, 2000) is an annotated universal protein sequence database established in 1986 and maintained collaboratively by the Swiss Institute of Bioinformatics (SIB) (*http://www.expasy.ch*) and the EMBL Outstation – The European Bioinformatics Institute (EBI) (*http://www. ebi.ac.uk/swissprot/*). It strives to provide a high level of annotation, a minimal level of redundancy, a high level of integration with other biomolecular databases as well as extensive external documentation. Each entry in SWISS-PROT gets thoroughly analyzed and annotated by biologists ensuring a high standard of annotation and maintaining the quality of the database (APWEILER et al., 1997). SWISS-PROT contains data that originate from a wide variety of organisms; release 39 (May 2000) contained around 85,000 annotated sequence entries from more than 6,000 different species.

Maintaining the high quality of SWISS-PROT requires, for each entry, a time-consuming process that involves the extensive use of sequence analysis tools along with detailed curation steps by expert annotators. It is the rate-limiting step in the production of the database.

A supplement to SWISS-PROT was created in 1996, since it is vital to make new sequences available as quickly as possible without relaxing the high editorial standards of SWISS-PROT. This supplement, TrEMBL (Translation of EMBL nucleotide sequence database), which can be classified as a computer-annotated sequence repository, consists of entries derived from the translation of all coding sequences (CDS) in the EMBL nucleotide sequence database, except for those already included in SWISS-PROT. TrEMBL is split into two main sections, SP-TrEMBL and REM-TrEMBL. SP-TrEMBL (SWISS-PROT TrEMBL) contains the entries, which should be eventually incorporated into SWISS-PROT. REM-TrEMBL (REMaining TrEMBL) contains the entries (about 55,000 in release 14) that will not get included in SWISS-PROT. It is organized in 6 subsections:

(1) *Immunoglobulins and T cell receptors:* Most REM-TrEMBL entries are immunoglobulins and T cell receptors. The integration of further immunoglobulins and T cell receptors into SWISS-PROT has been stopped, since SWISS-PROT does not want to add all known somatic recombined variations of these proteins to the database. At the moment there are more than 20,000 immunoglobulins and T cell receptors in REM-TrEMBL. SWISS-PROT plans to create a specialized database dealing with these sequences as a further supplement to SWISS-PROT, but will keep only a representative cross-section of these proteins in SWISS-PROT.

(2) *Synthetic sequences:* Another category of data which will not be included in SWISS-PROT are synthetic sequences.

(3) *Small fragments:* A subsection with protein fragments with less than 8 amino acids.

(4) *Patent application sequences:* Coding sequences captured from patent applications. Apart for a small number of entries, which have already been integrated in SWISS-PROT, most of these sequences contain either erroneous data or concern artificially generated sequences outside the scope of SWISS-PROT.

(5) *CDS not coding for real proteins:* This subsection consists of CDS translations which are most probably not coding for real proteins.

(6) *Truncated proteins:* The last subsection consists of truncated proteins, which are the results of differential splicing and fusion proteins.

TrEMBL follows the SWISS-PROT format and conventions as closely as possible. The production of TrEMBL starts with the translation of coding sequences (CDS) in the EMBL nucleotide sequence database. At this stage all annotation you can find in a TrEMBL entry comes from the corresponding EMBL entry. The first post-processing step is the reduction of redundancy (O'DONOVAN et al., 1999). One of SWISS-PROT's leading concepts from the very beginning was to minimize the redundancy of the database by merging separate entries corresponding to different literature reports. If conflicts exist between various sequencing reports, they are indicated in the feature table of the corresponding entry. This stringent requirement of minimal redundancy applies equally to SWISS-PROT + TrEMBL. The second post-processing step is the automated enhancement of the TrEMBL annotation to bring TrEMBL entries closer to SWISS-PROT standard (FLEISCHMANN et al., 1999). The method uses a Rule-based system to find SWISS-PROT entries belonging to the same protein family as the TrEMBL entry, extracts the annotation shared by all SWISS-PROT entries, assigns this common annotation to the TrEMBL entry, and flags this annotation as annotated by similarity. Currently around 20% of the TrEMBL entries get additional annotation in the automated way.

Searches in protein sequence databases have become a standard research tool in the life sciences. To produce valuable results, the source databases should be comprehensive, non-redundant, well annotated and up-to-date. The database SPTR (SWALL) was created to overcome these limitations. SPTR (SWALL) provides a comprehensive, non-redundant and up-to-date protein sequence database with a high information content. The components are

(1) the weekly updated SWISS-PROT work release. It contains the last SWISS-PROT release as well as the new or updated entries.

(2) the weekly updated SP-TrEMBL work release. REM-TrEMBL is not included in SWALL, since REM-TrEMBL contains the entries that will not be included into SWISS-PROT, e.g., synthetic sequences and pseudogenes.

(3) TrEMBLnew, the weekly updated new data to be incorporated into TrEMBL at release time.

To enable sequence comparisons against a database containing all known isoforms of proteins originating from genes undergoing alternative splicing files are provided with additional records from SWISS-PROT and TrEMBL, one for each splice isoform of each protein.

2.6 Specialized Protein Sequence Databases

The CluStr (Clusters of SWISS-PROT and TrEMBL proteins, *http://www.ebi.ac.uk/clustr*, KRIVENTSEVA et al., in press) database offers an automatic classification of SWISS-PROT and TrEMBL proteins into groups of related proteins. The clustering is based on analysis of all pairwise comparisons between protein sequences. Analysis has been carried out for different levels of protein similarity, yielding a hierarchical organization of clusters. CluStr can be used for

- prediction of functions of individual proteins or protein sets,
- automatic annotation of newly sequenced proteins,
- removal of redundancy from protein databases,
- searching for new protein families,
- proteome analysis, and
- provision of data for phylogenetic analysis.

The MEROPS database (RAWLINGS and BARRETT, 2000) provides a catalog and structure-based classification of peptidases (i.e., all pro-

teolytic enzymes). An index of the peptidases by name or synonym gives access to a set of files termed PepCards, each of which provides information on a single peptidase. Each card file contains information on classification and nomenclature, and hypertext links to the relevant entries in other databases. The peptidases are classified into families on the basis of statistically significant similarities between the protein sequences in the part termed the "peptidase unit" that is most directly responsible for activity. Families that are thought to have common evolutionary origins and are known or expected to have similar tertiary folds are grouped into clans. The MEROPS database (*http://www.merops.co.uk*) provides sets of files called FamCards and ClanCards describing the individual families and clans. Each FamCard document provides links to other databases for sequence motifs and secondary and tertiary structures, and shows the distribution of the family across the major taxonomic kingdoms.

There exists a collaboration for the collection of G-protein coupled receptors data (GPCRDB, *http://www.gpcr.org/7tm/*, HORN et al., 1998). G-protein coupled receptors (GPCRs) form a large superfamily of proteins that transduce signals across the cell membrane. At the extracellular side they interact with a ligand (e.g., adrenalin), and at the cytosolic side they activate a G protein. The data include alignments, cDNAs, evolutionary trees, mutant data and 3D models. The main aim of the effort is to build a generic molecular class specific database capable of dealing with highly heterogeneous experimental data. It is a good example for a specialized database adding value by offering an analytical view on data, which a universal sequence database is unable to provide.

YPD (HODGES et al., 1999) is a database for the proteins of *S. cerevisiae*. Based on the detailed curation of the scientific literature for the yeast *Saccharomyces cerevisiae*, YPD (*http://www.proteome.com/databases/*) contains more than 50,000 annotation lines derived from the review of 8,500 research publications. The information concerning each of the more than 6,000 yeast proteins is structured around a one-page format, the Yeast Protein Report, with additional information provided as pop-up windows. Protein classification schemas are defining the cellular role, function and pathway of each protein. YPD provides the user with a succinct summary of the function of the protein and its place in the biology of the cell. The first transcript profiling data has been integrated into the YPD Protein Reports, providing the framework for the presentation of genome-wide functional data. Altogether YPD is a very useful data collection for all yeast researchers and especially for those working on the yeast proteome.

2.7 Protein Signature Databases & InterPro

Very often the sequence of an unknown protein is too distantly related to any protein of known structure to detect its resemblance by overall sequence alignment, but it can be identified by the occurrence of sequence signatures.

There are a few databases available, which use different methodology and a varying degree of biological information on the characterized protein families, domains and sites. The oldest of these databases, PROSITE (*http://www.expasy.ch/prosite/*, HOFMANN et al., 1999), includes extensive documentation on many protein families, as defined by sequence domains or motifs. Other databases in which proteins are grouped, using various algorithms, by sequence similarity include PRINTS (*http://www.bioinf.man.ac.uk/bsm/dbbrowser/ PRINTS/PRINTS.html*, ATTWOOD, 2000), Pfam (*http://www.sanger.ac.uk/Pfam/*, BATEMAN et al., 2000), BLOCKS (*http://www. blocks.fhcrc.org/*, HENIKOFF et al., 1999) and SMART (*http://SMART.embl-heidelberg.de*, SCHULTZ et al., 2000).

These secondary protein sequence databases have become vital tools for identifying distant relationships in novel sequences and hence for inferring protein function. Diagnostically, the most commonly used secondary protein databases (PROSITE, PRINTS and PFAM) have different areas of optimum application owing to the different strengths and weaknesses of their underlying analysis methods (regular expressions, profiles, fingerprints and Hidden Markov Models). For example,

regular expressions are likely to be unreliable in the identification of members of highly divergent super-families; fingerprints perform relatively poorly in the diagnosis of very short motifs; and profiles and HMMs are less likely to give specific sub-family diagnoses. While all of the resources share a common interest in protein sequence classification, some focus on divergent domains (e.g., Pfam), some focus on functional sites (e.g., PROSITE), and others focus on families, specializing in hierarchical definitions from super-family down to sub-family levels in order to pin-point specific functions (e.g., PRINTS).

Sequence cluster databases like ProDom (*http://www.toulouse.inra.fr/prodom.html*, CORPET et al., 2000) are also commonly used in sequence analysis, e.g., to facilitate domain identification. Unlike pattern databases, the clustered resources are derived automatically from sequence databases, using different clustering algorithms. This allows them to be relatively comprehensive, because they do not depend on manual crafting and validation of family discriminators; but the biological relevance of clusters can be ambiguous and may just be artifacts of particular thresholds.

Given these complexities, analysis strategies should endeavor to combine a range of secondary protein databases, as none alone is sufficient. Unfortunately, these secondary databases do not share the same formats and nomenclature, which makes the use of all of them in an automated way difficult. In response to this InterPro – Integrated Resource of Protein Families, Domains and Functional Sites (The InterPro Consortium, in press) – has emerged as a new integrated documentation resource for the PROSITE, PRINTS, and Pfam database projects, coordinated at the EBI. InterPro (*http://www.ebi.ac.uk/interpro/*) allows users access to a wider, complementary range of site and domain recognition methods in a single package.

Release 1.2 of InterPro (June 2000) was built from Pfam 5.2 (2,128 domains), PRINTS 26.1 (1,310 fingerprints), PROSITE 16 (1,370 families), and ProDom 2000.1 (540 domains). It contained 3,052 entries, representing families, domains, repeats and PTMs encoded by 5,589 different regular expressions, profiles, fingerprints and HMMs. Provided data on

InterPro matches in known protein sequences in the SWISS-PROT and TrEMBL (BAIROCH and APWEILER, 2000) databases got named InterProMatches.

To facilitate in-house maintenance, InterPro is managed within a relational database system. For users, however, the core InterPro entries are released as XML formatted ASCII (text) file (*ftp://ftp.ebi.ac.uk/pub/databases/ interpro*).

In release 3 (March 2001) the SMART resource will also be included in InterPro. Ultimately, InterPro will include many other protein family databases to give a more comprehensive view of the resources available.

A primary application of InterPro's family, domain and functional site definitions will be in the computational functional classification of newly determined sequences that lack biochemical characterization. For instance, the EBI will use InterPro for enhancing the automated annotation of TrEMBL. InterPro is also a very useful resource for comparative analysis of whole genome (RUBIN et al., 2000) and has already been used for the proteome analysis of a number of completely sequenced organisms (*http://www.ebi.ac.uk/proteome/*, APWEILER et al., in press).

Another major use of InterPro will be in identifying those families and domains for which the existing discriminators are not optimal and could hence be usefully supplemented with an alternative pattern (e.g., where a regular expression identifies large numbers of false matches it could be useful to develop an HMM, or where a Pfam entry covers a vast super-family it could be beneficial to develop discrete family fingerprints, and so on). Alternatively, InterPro is likely to highlight key areas where none of the databases has yet made a contribution and hence where the development of some sort of signature might be useful.

2.8 Proteomics

Since the genome sequencing is proceeding at an increasingly rapid rate this leads to an equally rapid increase in predicted protein sequences entering the protein sequence databases. The term proteome is used to describe the protein equivalent of the genome, e.g., the complete set of the genome proteins. Most of these predicted protein sequences are without a documented functional role. The challenge is to provide statistical and comparative analysis, structural and other information for these sequences as an essential step towards the integrated analysis of organisms at the gene, transcript, protein and functional levels.

There are a number of existing databases that address some aspects of genome comparisons. The Kyoto Encyclopedia of Genes and Genomes (KEGG) is a knowledge base for systematic analysis of gene functions, linking genomic information with higher order functional information (*http://www.genome.ad.jp/ kegg/*, KANEHISA and GOTO, 2000). The WIT Project attempts to produce metabolic reconstructions for sequenced (or partially sequenced) genomes. A metabolic reconstruction is described as a model of the metabolism of the organism derived from sequence, biochemical, and phenotypic data (*http://wit.mcs.anl.gov/ WIT2/*, OVERBEEK et al., 2000). KEGG and WIT mainly address regulation and metabolic pathways although the KEGG scheme is being extended to include a number of non-metabolism-related functions. Clusters of Orthologous Groups of proteins (COGs) is a phylogenetic classification of proteins encoded in complete genomes (*http://www.ncbi.nlm.nih. gov/COG*, TATUSOV et al., 2000). COGs group together related proteins with similar but sometimes non-identical functions.

The Proteome Analysis Initiative has the more general aim of integrating information from a variety of sources that will together facilitate the classification of the proteins in complete proteome sets. The proteome sets are built from the SWISS-PROT and TrEMBL protein sequence databases that provide reliable, well-annotated data as the basis for the analysis. Proteome analysis data is available for all the completely sequenced organisms present in SWISS-PROT and TrEMBL, spanning archaea, bacteria and eukaryotes. In the proteome analysis effort the InterPro (*http:// www.ebi.ac.uk/interpro/*) and CluSTr (*http:// www.ebi.ac.uk/clustr/*) resources have been used. Structural information includes amino acid composition for each of the proteomes, and links are provided to HSSP, the Homology derived Secondary Structure of Proteins (*http://www.sander.ebi.ac.uk/hssp*, DODGE et al., 1998), and PDB, the Protein Data Bank (*http://oca.ebi.ac.uk/*, SUSSMAN et al., 1998), for individual proteins from each of the proteomes. A functional classification using Gene Ontology (GO) (*http://www.geneontology.org*, ASHBURNER et al., 2000) is also available. The Proteome Analysis Initiative provides a broad view of the proteome data classified according to signatures describing particular sequence motifs or sequence similarities and at the same time affords the option of examining various specific details like structure or functional classification. The Proteome Analysis Database currently contains statistical and analytical data for the proteins from 36 complete genomes and preliminary data for the human genome (*http://www.ebi.ac.uk/proteome/*, APWEILER et al., in press).

The SIB and the EBI are currently involved in a major effort to annotate, describe and distribute highly curated information about human protein sequences. It is known as the Human Proteomics Initiative (HPI, *http://www.ebi.ac.uk/swissprot/hpi/hpi.html*, APWEILER and BAIROCH, 1999).

2.9 Other Databases

The ENZYME database (*http://www.expasy.ch/enzyme/*, BAIROCH, 2000) is an annotated extension of the Enzyme Commission's publication, linked to SWISS-PROT. There are also databases of enzyme properties – BRENDA (*http://www.brenda.uni-koeln.de/brenda/*), Ligand Chemical Database for Enzyme Reactions (LIGAND *http://www.genome.ad.jp/dbget/ligand.html*, GOTO et al., 2000), and the Database of Enzymes and Metabolic Pathways (EMP). BRENDA, LIGAND and EMP are searchable via SRS at the EBI (*http://srs.ebi.ac.uk*). LIGAND is linked to the metabolic pathways in KEGG (*http://www.genome.ad.jp/kegg/kegg.html*, KANEHISA and GOTO, 2000).

Databases of two-dimensional gel electrophoresis data are available from Expasy (*http://www.expasy.ch/ch2d/*, HOOGLAND et al., 2000) and the Danish Center for Human Genome Research (*http://biobase.dk/cgi-bin/celis/*).

There are so many specialized databases, that it is not possible to mention all of them. Under the URL *http://www.expasy.ch/alinks.html* you will find a comprehensive www document that lists the databases mentioned in this document and many other information sources for molecular biologists.

3 Heterogeneity of the Data

One of the major challenges facing molecular biologists today is working with the information contained not within only one database but many, and cross-referencing this information and provide results in ways which permit to broaden the scope of a query and gain more in-depth knowledge.

Recent advances in data management such as RDBMS and OODBMS allow to implement highly sophisticated data schemas with efficient and flexible data structures and constraints on data integrity. In cases like metabolic pathways the complexity of the data forces to explore new approaches to handling the information, and it is common now to use object-oriented technologies to model bio-chemical data. On the other hand in the current stage relational data schema is more developed and robust. Moreover, major biological databases historically were developed and maintained in the form of formatted text files. Of course, current data explosion forces migration to more robust data management systems, but it is not reasonable to change the historic distribution format. As a result we have great heterogeneity among various databases and diversity of their distribution formats.

Several developers have identified the need to design database indexing and cross-referencing systems, which assist in the process of searching for entries in one database and cross-indexing them to another. The most important examples of these systems are SRS (*http://www.lionbio.co.uk*, ETZOLD et al., 1996), Entrez (*http://www.ncbi.nlm.nih.gov/Database/index.html*, SCHULER et al., 1996b) and DBGET (*http://www.genome.ad.jp/dbget/*

dbget.html, FUJIBUCHI et al., 1998), Atlas (*http://vms.mips.biochem.mpg.de/mips/programs/atlas. html*), Acnuc (*http://pbil.univ-lyon1.fr/databases/acnuc.html*, GOUY et al., 1985). The Sequence Retrieval System, or SRS, is one of the most successful approaches to this problem described in some detail in the next section of this chapter.

4 The SRS Approach

Started as a Sequence Retrieval System (ETZOLD and ARGOS, 1993; ETZOLD et al., 1996), SRS was originally aimed at facilitating access to biological sequence databases like EMBL Nucleotide Sequence Database stored in formatted text files. In fact, the format of such databases like EMBL or SWISS-PROT became a *de facto* standard for data distribution in the bioinformatics community. The text format is human readable, computer platform independent and there are a lot of tools to handle it. Self-descriptive XML format (*http://www.w3.org/XML/*) has advanced features, but it is still a text.

While RDBMS are highly advanced for data management, SRS has advantages as a retrieval system: First, it is faster by more than 1 or 2 orders of magnitude when retrieving whole entries from large databases with complex data schemas like EMBL. Second, it is less space storage demanding than RDBMS tables since it only retrieves fixed data. The average difference of 2–5 times is significant in the case of large databases like EMBL, which is about 30 Gb in flatfile format at present. Third, it is more scaleable with a number of databases, and it is reasonably easy to integrate new data with basic retrieval capabilities and extend it further to a more sophisticated data schema. Searchable links between databases and customizable data representation are original features of SRS.

Today it grew up in a powerful unified interface to over 400 different scientific databases. It provides capabilities to search multiple databases by shared attributes and to query across databases fast and efficiently. SRS has become an integration system for both data retrieval and data analysis applications. Originally SRS was developed at the EMBL and then later at the EBI. In 1999 LION Bioscience AG acquired it. Since then SRS has had undergone a major internal reconstruction and SRS6 was released as a licensed product that is freely available for academics. The EBI SRS server (*http://srs.ebi.ac.uk*) is a central resource for molecular biology data as well as a reference server for the latest developments in data integration.

4.1 Data Integration

The key feature of SRS is its unique object-oriented design. It uses meta-data to define a class of a database entry object and rules for text-parsing methods, coupled with the entry attributes. The fundamental idea is that you infer the defined data schema from the available data. For object definitions and recursive text parsing rules SRS uses its own scripting language Icarus.

4.2 Enforcing Uniformity

The integrating power of SRS benefits from sharing the definitions of conceptually equal attributes among different data sets. That allows multiple-database queries on common attributes. As described above the running time generated object of an entry gets its attribute values infered from the underlying data so that extracted information could be reformatted to enforce uniformity in data representation among different databases.

4.3 Linking

Data become more valuable in the context of other data. Besides enriching the original data by providing html linking, one of the original features of SRS is the ability to define indexed links between databases. These links reflect equal values of named entry attributes in two databases. It could be a link from an explicitly defined reference in DR (data reference) records in SWISS-PROT or an implicit link from SWISS-PROT to the ENZYME database by a corresponding EC (Enzyme Commission) number in the protein description. Once indexed, the links become bi-directional. They operate on sets of entries, can be weighted and can be combined with logical operators (AND, OR and NOT). This is similar to a table of relations in a relational database schema that allows querying of one table with conditions applied to others. The user can search not only the data contained in a particular database, but also in any conceptually related databases and then link back to the desired data. Using the linking graph, SRS makes it possible to link databases that do not contain direct references to each other (Fig. 1). Highly cross-linked data sets become a kind of domain knowledge base. This helps to perform queries like "give me all proteins that share InterPro domains with my protein" by linking from SWISS-PROT to InterPro and back to SWISS-PROT, or "give me all eukaryotic proteins for which the promoter is further characterized" by selecting only entries linked to the

EPD (Eukaryotic Promoter Database) from the current set.

4.4 Application Integration

Searching sequence databases is one of the most common tasks for any scientist with a newly discovered protein or nucleic acid sequence. That is used to determine or infer,

- if the sequence has been found and already exits in a database,
- the structure (secondary and tertiary),
- its function or chemical mechanism,
- the presence of an active site, ligand-binding site or reaction site,
- evolutionary relationships (homology).

Sequence database searching is different from a database query. Generally, sequence searching involves searching for a similar sequence in a database of sequences. By contrast, a query

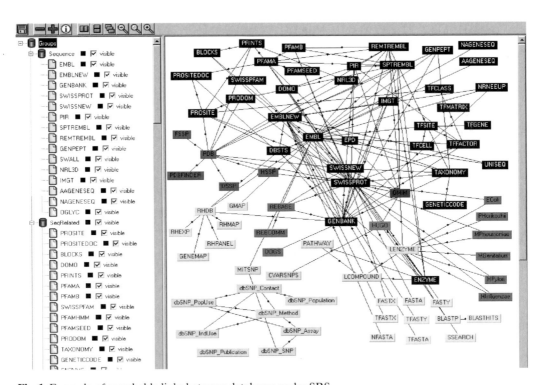

Fig. 1. Example of searchable links between databases under SRS.

involves searching for keywords or other text in the annotation associated with each sequence in a database.

The introduction of the biosequence object in SRS allowed the integration of various sequence analysis tools such as similarity search tool FASTA (PEARSON, 1990) or multiple alignment program CLUSTALW (THOMPSON et al., 1994). This integration allows treating the text output of these applications like any other text database. Linking to other databanks and user-defined data representations become then possible. Up to now more than 30 applications are already integrated into SRS and many others are in the pipeline. Expanding in this direction, SRS becomes not only a data retrieval system but also a data analysis application server (Fig. 2). Recent advances in application integration include different levels of user control over application parameters, support for different UNIX queuing systems (LSF, CODINE, DQS, NQS) and parallel threading. There is now also support for "user-owned data" (the user's own sequences), which make SRS a more comprehensive research tool.

5 A Case Study: The EBI SRS Server

The EBI SRS server plays an important role in EBI's mission to provide services in bioinformatics. It gives a flexible and up-to-date access to many major databases produced and maintained at the EBI and other institutions. The databases are grouped in specialized sections including nucleic acid and protein sequences, mapping data, macromolecular structure, sequence variations, protein domains and metabolic pathways (Tab. 1).

The EBI SRS server contains today more than 130 biological databases and integrates more than 10 applications. SRS is a constantly evolving system. New databases are being added, and the interfaces to the old ones are always being enhanced. This server is in high demand by the bioinformatics community. Currently, requests and queries on the system total more than 3 million genuine queries per month with a growth rate of more than 15% per month.

Applications in SRS

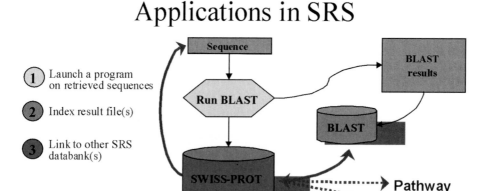

1 Launch a program on retrieved sequences

2 Index result file(s)

3 Link to other SRS databank(s)

"How many members of the TM4 family did I find ?"
"Did I find any enzyme s in the phenylanaline pathway?"
"Remove all viral sequences from my 'hit list'"

Fig. 2. The integration of applications in SRS has the advantage of treating the application output like any other database, which allows linking to other databanks and user-defined data representation.

Tab. 1. Some of the Databases Available through the EBI SRS Server (*http://srs.ebi.ac.uk*). Databases marked in bold are produced and maintained at the EBI. Short descriptions of each of the databases are available on the SRS database info pages html linked from the database names

Sequence

EMBL	**EMBLNEW**	**ENSEMBL**	
SWISSPROT	**SPTREMBL**	**TREMBLNEW**	**REMTREMBL**
SWALL	**IMGT**	**IMGTHLA**	

InterPro&Related

InterPro	PROSITE	PROSITEDOC	BLOCKS
PFAMA	PFAMB	PFAMHMM	PFAMSEED
PRINTS	**NICEDOM**	PRODOM	

SeqRelated

TAXONOMY	GENETICCODE	EPD	**HTG_QSCORE**
UTR	UTRSITE	**EMESTLIB**	

TransFac

TFSITE	TFFACTOR	TFCELL	TFCLASS
TFGENE	TFMATRIX		

Protein3DStruct

PDB	**DSSP**	**HSSP**	**FSSP**

Genome

HSAGENES	MOUSE2HUMAN	LOCUSLINK

Mapping

RHDB	**RHEXP**	**RHMAP**	**RHPANEL**
OMIMMAP			

Mutations

MUTRES	**MUTRESSTATUS**	OMIM	**OMIMALLELE**
OMIMOFFSET	**SWISSCHANGE**	**EMBLCHANGE**	**HUMUT**
HUMAN_MITBASE	**P53LINK**		

SNP

MITSNP	dbSNP_Contact	dbSNP_Method
dbSNP_Population	dbSNP_Publication	dbSNP_Assay
dbSNP_SNP	dbSNP_PopUse	dbSNP_IndUse
HGBASE	**HGBASE_SUBMITER**	**SNPLink**

Metabolic Pathways

PATHWAY	LENZYME	LCOMPOUND	BRENDA
EMP	MPW	UPATHWAY	UREACTION
UCOMPOUND	UIMAGEMAP	ENZYME	UENZYME

All SRS database parsers are available to external users and thus, the EBI SRS server plays an important role as a reference site for most other SRS servers. SRS has gained wide popularity and now there are more than 100 installations worldwide. To track the information available on publicly available databases on numerous SRS servers there is the "Database of Data Banks". It is based on a set of scripts that automatically gather information from SRS servers on the Internet and organizes these data into a searchable database (KREIL and ETZOLD, 1999).

5.1 Data Warehousing & SRS PRISMA

One of the hardest chores in maintaining an up-to-date SRS server is the constant hunting for new database releases and updates. Typically, the nightly update of the EBI SRS server consists of more than 1,000 processes. The recently introduced SRS PRISMA is a set of programs designed to automate this process. It integrates the monitoring for new data sets on remote servers, downloading and indexing. PRISMA can execute a user-defined number of parallel sessions in order to increase updating throughput and reduce the time it takes for users to be able to query the new data. Administratively, PRISMA combines parallel threads execution, automatic report generation with graphical diagrams, automated recovery and offline data processing, making it simple to quickly identify problems and take corrective actions.

The SRS server at the EBI uses extensively the capability of the system to prepare indices off-line. This feature of SRS6.x solves the problem of a database not being available for querying during the updating process. Although there is a drawback in terms of storage the mere fact that the database is always on-line outweighs this disadvantage.

6 Advanced Features & Recent Additions

6.1 Multiple Subentries

Data representation as a stream of entries in flat text implies restrictions to the underlying data schema. Since support for more advanced data schemas allows the resolution of more specific queries, SRS introduced subentries as logically independent concepts nested in the parent database entries. Probably the most commonly known examples of subentries are the elements of feature tables in sequence databases such as EMBL or SWISS-PROT. Other widely occurring cases are publication references. In SRS6, it is possible to define sev-

eral subentries per database. In the case of SWISS-PROT there are now several definitions for subentries corresponding to elements of the feature table, publication references and comments. A special purpose subentry, called "Counter", was introduced in order to make the number of links to other databanks and/or the number of certain features searchable. Using the "Counter" subentry it is possible to query for all proteins with exactly 7 transmembrane regions and with annotated similarities to receptors. The query can be easily constructed using the "extended query form" in the SRS web interface.

6.2 Virtual Data Fields

It is possible in SRS to define data fields that coupled with a method inferring "on-the-fly" new data from the original data. These could be the graphical visualization of protein domains and/or functional sites, links to external data sources or precompiled SRS queries. As an example, the "AllSeq" attribute of a PRO-DOM entry is the SRS query that leads to all SWISS-PROT proteins containing this PRO-DOM domain.

6.3 Composite Views

SRS allows the definition of composite views that dynamically link entries from the main query database to other related databases. These views display external data as if they were original database attributes.

An example is the visualization of InterPro Matches (InterPro domain composition of a protein sequence) using "SW_InterProMatches" available at the EBI SRS server. This view dynamically links protein sequences to the InterProMatches database, retrievs information of known InterPro signatures in the proteins and presents the data in a virtually composed graphical form (Fig. 3).

6.4 InterPro & XML Integration

As described earlier the InterPro data is distributed in XML format (*ftp://ftp.ebi.ac.uk/*

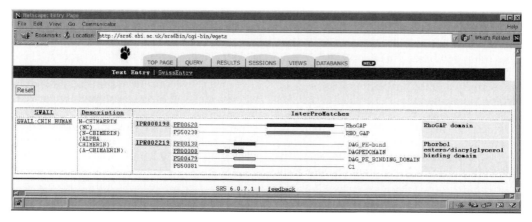

Fig. 3. Example showing the InterPro domain composition of a protein sequence. The view consists of the ID and description fields of a protein entry and linked InterProMatches data presented graphically.

pub/databases/interpro/). InterPro was the first XML formatted databank integrated in SRS and it represents an important milestone in the low-level integration of XML in SRS.

6.5 InterProScan

As an example of a data analysis application we present here InterProScan, which was recently implemented at the EBI. InterProScan is a wrapper on top of a set of applications for scanning protein sequences against InterPro member databases. It is implemented as a virtual application that launches underlying signature scanning applications in parallel mode and then presents their results in one view. Currently it is based on

(1) the FingerPRINTScan (SCORDIS et al., 1999) application that searches the PRINTS database for protein signatures
(2) ProfileScanner (pfscan) from the Pftools package for searching protein sequences against a collection of generalized profiles in PROSITE (*http://www.isrec.isb-sib.ch/software/ PFSCAN_form.html*)
(3) Ppsearch (FUCHS, 1994) for PROSITE pattern matching,
(4) HMMPfam from the HMMER package (*http://hmmer.wustl.edu*) or HMMS

implemented on a Decypher machine from TimeLogic that scans sequences against the Pfam collection of protein domain HMMs (Hidden Markov Models).

InterProScan provides an efficient way to analyze protein sequences for known domains and functional sites by launching the applications in parallel, parsing their output and combining the results at the level of unified attributes into one representation with graphical visualization of the matches (ZDOBNOV, unpublished results).

6.6 New Services Based on "SRS Objects"

LION Bioscience has made available with SRS6 some APIs to popular programming languages, namely C++, JAVA, PERL and PYTHON. The APIs allow the development of highly customized user interfaces, which can use "SRS Objects" for data retrieval, application launching and protected user sessions. This allows the creation of programs with specialized interfaces. We implemented InterProScan as an integrated SRS sequence analysis tool and as a web interface using the SRS Perl API (*http://www.ebi.ac.uk/interpro/ interproscan/ipsearch.html*). This client program generates interfaces to all InterPro related applications within SRS. It effectively uses SRS parsing of the results and SRS retrieval capabilities to look up related data from other databases. This approach is a compromise between the SRS inter-database linking integrity and the simplicity of the user interface, implementing "one-click-away" results. The provision of these APIs represents a big step in the integration of common languages with SRS, but it implies that the client program and the SRS server share the same file system (e.g., over NFS). Fortunately, SRS has a CORBA API as well, which allows development of truly distributed networked systems. For example, to enhance the searching capabilities of the simple interfaces for the CluSTr and InterPro databases stored in ORACLE, we use the SRS CORBA interface to extend the user query through an "all-text search" in linked databases under SRS.

6.7 Double Word Indexing

Many biological databases contain free text descriptions. The simplest indexing of all individual words in free text lacks the ability to reflect the word's semantic meaning and does not represent underlying concepts specifically enough. A recently introduced technique of indexing all consecutive pairs of words makes the querying of concepts buried deep in free text descriptions much more powerful without significant compromise on index size or search speed. As an example: the result of the query of "cytochrome c" is quite different from the query of "cytochrome" AND "c".

6.8 Bookmarklets

It is worth to mention the simple but very handy JavaScript interfaces to SRS that have also been developed recently (SRSQuick-Search). These have the advantage that they can be bookmarked as ordinary html links. In www parlance they are called Bookmark-Lets (*http://www.bookmarklets.com*). Modern browsers such as Netscape or Internet Explorer allow the user to rearrange their bookmarks so that they appear as buttons on the browser window from where the BookmarkLets can be

conveniently called at any time. The user can highlight one or more words on the current page and click on the SRSQuickSearch bookmarklet button to execute a query. These scripts are especially useful when customized for particular needs. To make users life easier we provide a set of the most popular pre-configured SRS bookmarklets as well as a tool to generate customized SRSQuickSearch bookmarklets. These scripts are used extensively by the curators at the EBI.

6.9 Simple Search

To simplify the user interface to SRS we introduced a number of simple web forms based on JavaScript code. These are shortcuts for simple queries. All the required code is in the page source and users are encouraged to take it from the EBI web pages and use it for particular local needs.

7 Final Remarks

The databases are still evolving. While the wealth of information in these databases is fast growing, there is a lot of molecular biology data still only available in the original publications. New advances in technology provide even faster means of generating data. It will remain a constant challenge to handle it efficiently as more discoveries are made.

8 References

APWEILER, R., BAIROCH, A. (1999), The Human Proteomics Initiative of SIB and EBI, *The Bioinformer* **5**.

APWEILER, R., BISWAS, M., FLEISCHMANN, W., KANAPIN, A., KARAVIDOPOULOU, Y. et al. (2001), Proteome Analysis Database: online application of InterPro and CluSTr for the functional classification of proteins in whole genomes, *Nucleic Acids Res.* **29** (1), 44–48.

APWEILER, R., GATEAU, A., CONTRINO, S., MARTIN, M. J., JUNKER, V. et al. (1997), Protein sequence annotation in the genome era: the annotation concept of SWISS-PROT + TREMBL, *Ismb* **5**, 33–43.

ASHBURNER, M., BALL, C. A., BLAKE, J. A., BOTSTEIN, D., BUTLER, H. et al. (2000), Gene ontology: tool for the unification of biology. The Gene Ontology Consortium, *Nature Genet.* **25**, 25–29.

ATTWOOD, T. K., CRONING, M. D., FLOWER, D. R., LEWIS, A. P., MABEY, J. E. et al. (2000), PRINTS-S: the database formerly known as PRINTS, *Nucleic Acids Res.* **28**, 225–227.

BAIROCH, A. (2000), The ENZYME database in 2000, *Nucleic Acids Res.* **28**, 304–305.

BAIROCH, A., APWEILER, R. (2000), The SWISS-PROT protein sequence database and its supplement TrEMBL in 2000, *Nucleic Acids Res.* **28**, 45–48.

BARKER, W. C., GARAVELLI, J. S. et al. (1999), The PIR-International Protein Sequence Database, *Nucleic Acids Res.* **27**, 39–43.

BATEMAN, A., BIRNEY, E. et al. (2000), The Pfam protein families database, *Nucleic Acids Res.* **28**, 263–266.

BERLYN, M. B., LETOVSKY, S. (1992), Genome-related datasets within the *E. coli* Genetic Stock Center database, *Nucleic Acids Res.* **20**, 6143–6151.

BLAKE, J. A., RICHARDSON, J. E. et al. (1999), The Mouse Genome Database (MGD): genetic and genomic information about the laboratory mouse. The Mouse Genome Database Group, *Nucleic Acids Res.* **27**, 95–98.

BUTLER, D. (2000), Ensembl gets a Wellcome boost, *Nature* **406**, 333.

CHERVITZ, S. A., HESTER, E. T. et al. (1999), Using the *Saccharomyces* Genome Database (SGD) for analysis of protein similarities and structure, *Nucleic Acids Res.* **27**, 74–78.

CORPET, F., SERVANT, F. et al. (2000), ProDom and ProDom-CG: tools for protein domain analysis and whole genome comparisons, *Nucleic Acids Res.* **28**, 267–269.

DAYHOFF, M. O. (1965), Computer aids to protein sequence determination, *J. Theor. Biol.* **8**, 97–112.

DAYHOFF, M. O., ORCUTT, B. C. (1979), Methods for identifying proteins by using partial sequences, *Proc. Natl. Acad. Sci. USA* **76**, 2170–2174.

DODGE, C., SCHNEIDER, R. et al. (1998), The HSSP database of protein structure–sequence alignments and family profiles, *Nucleic Acids Res.* **26**, 313–315.

ETZOLD, T., ARGOS, P. (1993), SRS – an indexing and retrieval tool for flat file data libraries, *Comput. Appl. Biosci.* **9**, 49–57.

ETZOLD, T., ULYANOV, A. et al. (1996), SRS: information retrieval system for molecular biology data banks, *Methods Enzymol.* **266**, 114–128.

FLEISCHMANN, W., MOLLER, S. et al. (1999), A novel method for automatic functional annotation of proteins, *Bioinformatics* **15**, 228–233.

FREZAL, J. (1998), Genatlas database, genes and development defects, *C. R. Acad. Sci. III* **321**, 805–817.

FUCHS, R. (1994), Predicting protein function: a versatile tool for the Apple Macintosh, *Comput. Appl. Biosci.* **10**, 171–178.

FUJIBUCHI, W., GOTO, S. et al. (1997), DBGET/LinkDB: an integrated database retrieval system, *Pac. Symp. Biocomput.* 1998, 683–694.

GOTO, S., NISHIOKA, T. et al. (2000), LIGAND: chemical database of enzyme reactions, *Nucleic Acids Res.* **28**, 380–382.

GOUY, M., GAUTIER, C. et al. (1985), ACNUC – a portable retrieval system for nucleic acid sequence databases: logical and physical designs and usage, *Comput. Appl. Biosci.* **1**, 167–172.

HAMOSH, A., SCOTT, A. F. et al. (2000), Online Mendelian Inheritance in Man (OMIM), *Hum. Mutat.* **15**, 57–61.

HENIKOFF, S., HENIKOFF, J. G. et al. (1999), Blocks + : a non-redundant database of protein alignment blocks derived from multiple compilations, *Bioinformatics* **15**, 471–479.

HODGES, P. E., MCKEE, A. H. et al. (1999), The Yeast Proteome Database (YPD): a model for the organization and presentation of genome-wide functional data, *Nucleic Acids Res.* **27**, 69–73.

HOFMANN, K., BUCHER, P. et al. (1999), The PROSITE database, its status in 1999, *Nucleic Acids Res.* **27**, 215–219.

HOOGLAND, C., SANCHEZ, J. C. et al. (2000), The 1999 SWISS-2DPAGE database update, *Nucleic Acids Res.* **28**, 286–288.

HORN, F., WEARE, J. et al. (1998), GPCRDB: an information system for G protein-coupled receptors, *Nucleic Acids Res.* **26**, 275–279.

KANEHISA, M., GOTO S. (2000), KEGG: kyoto encyclopedia of genes and genomes, *Nucleic Acids Res.* **28**, 27–30.

KARP, P. D., RILEY, M. et al. (2000), The EcoCyc and MetaCyc databases, *Nucleic Acids Res.* **28**, 56–59.

KORAB-LASKOWSKA, M., RIOUX, P. et al. (1998), The Organelle Genome Database Project (GOBASE), *Nucleic Acids Res.* **26**, 138–144.

KREIL, D. P., ETZOLD, T. (1999), DATABANKS – a catalogue database of molecular biology databases, *Trends Biochem. Sci.* **24**, 155–157.

KRIVENTSEVA, E. V., FLEISCHMANN, W., APWEILER, R. (2001), CluSTr: a database of Clusters of SWISS-PROT + TrEMBL proteins, *Nucleic Acids Res.* **29** (1), 33–36.

KROGER, M., WAHL, R. (1998), Compilation of DNA sequences of *Escherichia coli* K12: description of the interactive databases ECD and ECDC, *Nucleic Acids Res.* **26**, 46–49.

KUIKEN, C. L., FOLEY, F. B., HAHN, B., KORBER, B., MCCUTCHAN, F., MARX, P. A. et al. (Eds.) (1999),

Human Retroviruses and AIDS 1999: A Compilation and Analysis of Nucleic Acid and Amino Acid Sequences. Los Alamos National Laboratory, Los Alamos, NM.

LEFRANC, M. P., GIUDICELLI, V. et al. (1999), IMGT, the international ImMunoGeneTics database, *Nucleic Acids Res.* **27**, 209–212.

LEHVASLAIHO, H., STUPKA, E. et al. (2000), Sequence variation database project at the European Bioinformatics Institute, *Hum. Mutat.* **15**, 52–56.

LETOVSKY, S. I., COTTINGHAM, R. W. et al. (1998), GDB: the Human Genome Database, *Nucleic Acids Res.* **26**, 94–99.

MAIDAK, B. L., COLE, J. R. et al. (1999), A new version of the RDP (Ribosomal Database Project), *Nucleic Acids Res.* **27**, 171–173.

MEWES, H. W., FRISHMAN, D. et al. (2000), MIPS: a database for genomes and protein sequences, *Nucleic Acids Res.* **28**, 37–40.

MILLER, R. T., CHRISTOFFELS, A. G. et al. (1999), A comprehensive approach to clustering of expressed human gene sequence: the sequence tag alignment and consensus knowledge base, *Genome Res.* **9**, 1143–1155.

O'DONOVAN, C., MARTIN, M. J. et al. (1999), Removing redundancy in SWISS-PROT and TrEMBL, *Bioinformatics* **15**, 258–259.

OVERBEEK, R., LARSEN, N. et al. (2000), WIT: integrated system for high-throughput genome sequence analysis and metabolic reconstruction, *Nucleic Acids Res.* **28**, 123–125.

PEARSON, W. R. (1990), Rapid and sensitive sequence comparison with FASTP and FASTA, *Methods Enzymol.* **183**, 63–98.

PERIER, R. C., JUNIER, T. et al. (1999), The Eukaryotic Promoter Database (EPD): recent developments, *Nucleic Acids Res.* **27**, 307–309.

RAWLINGS, N. D., BARRETT, A. J. (2000), MEROPS: the peptidase database, *Nucleic Acids Res.* **28**, 323–325.

REARDON, E. M. (1999), Release 7.0 of Mendel database, *Trends Plant Sci.* **4**, 385.

REBHAN, M., CHALIFA-CASPI, V. et al. (1998), GeneCards: a novel functional genomics compendium with automated data mining and query reformulation support, *Bioinformatics* **14**, 656–664.

RHEE, S. Y., WENG, S. et al. (1999), Unified display of *Arabidopsis thaliana* physical maps from AtDB, the *A. thaliana* database, *Nucleic Acids Res.* **27**, 79–84.

ROBERTS, R. J., MACELIS, D. (2000), REBASE – restriction enzymes and methylases, *Nucleic Acids Res.* **28**, 306–307.

RUBIN, G. M., YANDELL, M. D. et al. (2000), Comparative genomics of the eukaryotes, *Science* **287**, 2204–2215.

SCHULER, G. D., BOGUSKI, M. S. et al. (1996a), A gene map of the human genome, *Science* **274**, 540–546.

SCHULER, G. D., EPSTEIN, J. A. et al. (1996b), Entrez: molecular biology database and retrieval system, *Methods Enzymol.* **266**, 141–162.

SCHULTZ, J., COPLEY, R. R. et al. (2000), SMART: a web-based tool for the study of genetically mobile domains, *Nucleic Acids Res.* **28**, 231–234.

SCORDIS, P., FLOWER, D. R. et al. (1999), FingerPRINTScan: intelligent searching of the PRINTS motif database, *Bioinformatics* **15**, 799–806.

STOESSER, G., TULI, M. A. et al. (1999), The EMBL Nucleotide Sequence Database, *Nucleic Acids Res.* **27**, 18–24.

SUSSMAN, J. L., LIN, D. et al. (1998), Protein Data Bank (PDB): database of three-dimensional structural information of biological macromolecules, *Acta Crystallogr. D. Biol. Crystallogr.* **54**, 1078–1084.

TATUSOV, R. L., GALPERIN, M. Y. et al. (2000), The COG database: a tool for genome-scale analysis of protein functions and evolution, *Nucleic Acids Res.* **28**, 33–36.

The FlyBase Consortium (GELBART, W. M., GROSBY, M. C., MATTHEWS, B., CHILLEMI, J., RUSSO TWOMBLY, S., EMMERT, D. et al.) (1999), The FlyBase database of the *Drosophila* Genome Projects and community literature, *Nucleic Acids Res.* **27**, 85–88.

The InterPro Consortium (APWEILER, R., ATTWOOD, T. K., BAIROCH, A., BATEMAN, A., BIRNEY, E. et al.) (in press), InterPro – An integrated documentation resource for protein families, domains and functional sites, *Nucleic Acids Res.* **29** (1), 37–40.

THOMPSON, J. D., HIGGINS, D. G. et al. (1994), CLUSTAL W: improving the sensitivity of progressive multiple sequence alignment through sequence weighting, position-specific gap penalties and weight matrix choice, *Nucleic Acids Res.* **22**, 4673–4680.

WESTERFIELD, M., DOERRY, E. et al. (1999), Zebrafish informatics and the ZFIN database, *Methods Cell. Biol.* **60**, 339–355.

WINGENDER, E., CHEN, X. et al. (2000), TRANSFAC: an integrated system for gene expression regulation, *Nucleic Acids Res.* **28**, 316–319.

13 Tools for DNA Technologies

PETER RICE

Cambridge, UK

1 Introduction

Other chapters in this book concentrate on the underlying technologies. This chapter covers the essentials of using bioinformatics tools to analyze DNA sequences. The examples are drawn from the new EMBOSS sequence analysis package (RICE et al., 2000), developed over the past few years by the members of the European Molecular Biology Network (EMBnet, *http://www.embnet.org/*).

The nature of DNA sequence data has changed significantly over the years since DNA sequencing became a common laboratory method. The earliest sequences covered just one gene of interest, either from a bacterial clone or from a cDNA clone. The sequences were finished to a reasonable standard, and at 1500 bases on average were an ideal size for analysis by computer. During the early 1990s, EST sequencing projects led to a vast increase in the number of short, single read sequences with high error rates both for single base substitutions and for single base insertions or deletions. These proved especially difficult for alignment methods. In recent years we have seen a massive growth in the release of large clones and complete genomes, so that the databases also contain many sequences of 100 kb or larger. These also give major problems to computer methods. This chapter will illustrate some of the solutions that have been applied to these new sequence analysis problems.

For DNA sequence analysis, the main methods can be divided into those based on alignment and those based on pattern matching.

2 Alignment Methods

Sequence alignment assumes there are one or more other sequences that can be compared to our starting sequence. The most obvious case is a search of the sequence databases, where we look for any other known sequence that has some similarity to ours, in other words one that can be aligned to it.

Having found other sequences that could be similar, we can align them to our starting sequence either one at a time (pairwise alignment) or all together (multiple alignment).

A graphical display of the matches between two sequences is often a good starting point, as this is the easiest way to see where there are potential regions of similarity. Most sequence analysis packages will have one or more programs to produce these plots, known by the generic name of "dot plots". In a dot plot, one sequence is plotted on each axis, and a dot is drawn where the two sequences "match". There are, of course, choices to be made about how a match is defined. The simplest is to pit a dot where a base in one sequence matches a base in the other sequence. However, with only 4 bases to choose from this automatically puts dots in 25% of the possible places and fills the plot so that any real matches become impossible to see.

The general rule for dot plots is to have about one dot for each base in the shorter sequence, so that real matches are obvious as lines on the plot. Some programs do this by counting several bases together, and putting a dot when enough of them match. With suitable scoring methods, this can work very well and show similarities between very distantly related sequences.

In the EMBOSS program dotmatcher, short regions (in the example below, 10 bases) are compared, and a dot is plotted when the number of matches exceeds a set score. A reasonable scoring system would require at least half of the residues to be identical. As EMBOSS scores +5 for each match, this would be a score of 25.

A particularly tricky problem, surprisingly, is comparing spliced cDNA sequences to genomic (unspliced) sequences. The reasons will become clearer when we consider pairwise alignment below. A graphical dot plot view, however, makes the problem much simpler. In the example, we use a window of 10 bases, enough to avoid "random" matches (unless there are simple repeats in both sequences), and insist on perfect matches a score of 50 over 10 bases). The 5 exons in the example sequences are then easy to see.

Example dotmatcher

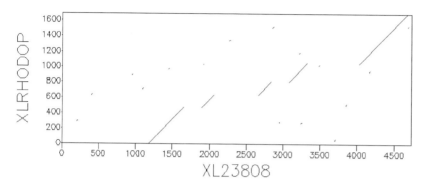

Dotmatcher: XL23808 vs XLRHODOP

(windowsize = 10, threshold = 50.00 04/06/00)

2.1 Pairwise Alignment

To see, base-by-base, how two sequences are related, we need to carry out a rigorous alignment of the two sequences, using string matching techniques from computer science.

The standard alignment method is that of dynamic programing. The name refers to the way in which the result is calculated automatically as the sequences are compared base by base. Scores for each comparison are stored in a table, like a spreadsheet, inside the program. These individual scores are then used to build an alignment score, stepping through the table from beginning to end. The most favorable alignment is simply calculated by tracing back from the highest value in the table.

There are two very similar methods, one for global alignment (NEEDLEMAN and WUNSCH, 1970) and another for local alignment (SMITH and WATERMAN, 1981). Of the two, the local alignment is generally preferred in bioinformatics as it will show the most significant alignment between the two sequences. In most practical cases only part of the sequences will be highly conserved. The global alignment forces the entire length of both sequences to be compared, and a long stretch of low or no similarity can prevent even a strong local high scoring match from being found.

A key part of these alignment methods is the way in which insertions and deletions

("gaps") in one or other sequence are scored. The favored method is to provide three score values to the program. The first is a comparison matrix which gives a single score for every possible match or mismatch between two bases, including if appropriate the nucleotide ambiguity codes. The second score is a penalty to be subtracted each time a gap is made in one sequence so that two other matching regions can be better aligned. The third score is a penalty to be subtracted each time a gap is extended by another base. Clearly the sizes of the gap penalty and the gap extension penalty are dependent on the score values in the comparison matrix. Mainly for this reason, most programs will hide the comparison matrix rom the user and prompt only for a pair of gap penalties.

In EMBOSS, the Needleman–Wunsch method is implemented in a program called needle. Sequence alignments for DNA are typically very long; the example below, purely for illustration, uses two short sequences.

In EMBOSS, the standard DNA comparison matrix has a simple score of $+5$ for a match between a base in each sequence, and -4 for a mismatch. The same scoring is used by the BLAST database search program. Typical gap penalties are 10 for opening a gap, so that it will be worth opening a gap to allow an extra 2 matches to be made, and 0.5 for a gap extension so that up to 10 gaps can be added to create one additional match.

Example needle

```
Global: SHORTA vs SHORTB

Score: 21.00
```

```
SHORTA          1     ATTACCACAT                    10
                      || |||||
SHORTB          1     ATA.CACAT                     8
```

The score is calculated as follows: 7 matches (marked by the vertical lines) score +5 each, a total of 35. One mismatch (T to A) scores −4. One gap scores −10, made up from −10 for the gap, and 0 for its length of 1. The resulting score is 21. We know from the dynamic programing method that this is the highest alignment score possible for these two sequences (and this scoring method), although it is possible that there could be more than one solution.

Although it is tempting to align the As at the starts of the sequences, this would produce a lower score because of the high penalty for adding another gap. This can be calculated as an exercise for the reader. The score would be 20.00. The gap penalty system is designed to allow only a relatively small number of gaps.

There are cases where we would prefer to allow a large number of gaps, but to keep them short. The most common is in the comparison of single sequencing reads where single base insertions or deletions are quite likely. We can achieve this result relatively easily by changing the gap penalties. For example, scoring −1 for starting a gap and −0.5 for each extra gap position. Now we find that aligning the As at the starts of the sequences is possible because the gap penalty is low enough to allow two gaps even in such short sequences.

```
Global: SHORTA vs SHORTB

Score: 38.00
```

```
SHORTA          1     ATTACCACAT                    10
                      | || |||||
SHORTB          1     A.TA.CACAT                    8
```

2.2 Local Alignment

For most DNA sequence comparisons, we are not really interested in aligning the full length of both sequences. Fortunately, a simple adjustment to the dynamic programing method allows us to find the best alignment anywhere within the two sequences. The program looks for the highest score anywhere in the results table, and traces through to the position where the score falls below zero. This method, first proposed by TEMPLE SMITH and MICHAEL WATERMAN, requires that mismatches are given a negative score, something which all alignment programs will include in their scoring methods.

The EMBOSS program for Smith–Waterman alignment is called water. For our short sequence example, the local alignment gives a higher score than the Needleman–Wunsch global method. The reason is simple: the mismatch at the start of the global alignment makes the score worse, and the local alignment stops when it traces back this far.

Example water

```
Local: SHORTA vs SHORTB
Score: 25.00
```

```
SHORTA          3       TACCACAT                            10
                        || |||||
SHORTB          2       TA.CACAT                            8
```

```
Local: SHORTA vs SHORTB
Score: 38.00
```

```
SHORTA          1       ATTACCACAT                          10
                        | || |||||
SHORTB          1       A.TA.CACAT                          8
```

2.3 Variations on Pairwise Alignment

Many sequence analysis packages offer only these simple global and local alignment methods. EMBOSS has some useful extensions which are easily programmed by making small changes to the standard programs. One, merger by GARY WILLIAMS from HGMP, merges two sequences by finding the highest scoring overlap.

Setting gap penalties appears complicated at first, but is really simple when broken down into simple rules.

- Large gap penalties will make it hard to include more than one matching segment in the highest scoring result, and will tend to produce short, local alignments.
- Small gap penalties will make it easy to include gaps to extend a match, and will tend to produce a global alignment even if only a short local match exists in practice.

Example merger

```
# LONGA position base        LONGB position base     Using
Global: LONGA vs LONGB
Score: 40.00
```

```
LONGA           1                               agagacatattactagata     19
                                                |||||||||
LONGB           1       tattgcgcagttgcagatcgcgagagacat                  30
```

- A low gap extension penalty will tend to produce a few very long gaps. A gap extension penalty of 0 is allowed, and can be useful for very long gaps such as introns when comparing a cDNA to a genomic sequence.
- A low gap open penalty will tend to produce a large number of very short gaps. A gap open penalty of 0 is allowed, and can be useful to allow many single base gaps when aligning a single read to a finished sequence.

2.4 Beyond Simple Alignment

Only one set of gap penalties is allowed at any one time. For more complicated situations, a specialized application is needed. For example, to align an EST sequence (many short gaps from poor quality sequence and a few long gaps from intron excision) to a genomic sequence can be done with the EMBOSS application est2genome. This program, by RICHARD MOTT at the Sanger Centre (MOTT, 1997), includes splice site consensus sequences and two different gap scoring systems for introns and exons.

Example est2genome

```
Note Best alignment is between forward est and forward genome, and splice
sites imply forward gene


Segment 469 100.0 1182 1650 XL23808    2   470 XLRHODOP X.laevis rhodopsin

Segment 169 100.0 1899 2067 XL23808  471   639 XLRHODOP X.laevis rhodopsin

Segment 166 100.0 2669 2834 XL23808  640   805 XLRHODOP X.laevis rhodopsin

Segment 240 100.0 3085 3324 XL23808  806  1045 XLRHODOP X.laevis rhodopsin

Segment 639 100.0 4030 4668 XL23808 1046  1684 XLRHODOP X.laevis rhodopsin


XL23808 vs XLRHODOP:


  XL23808    2065 caggtaaa.....tgcagatacatcccagagggaatgcaatgctcatgcg   2700

                  |||>>>>> 601 >>>>>||||||||||||||||||||||||||||||||||

  XLRHODOP    637 cag..............atacatcccagagggaatgcaatgctcatgcg    671


  XL23808    2701 gagtagactactacacactgaagcctgaggtcaacaatgaatcctttgtt   2750

                  ||||||||||||||||||||||||||||||||||||||||||||||||||

  XLRHODOP    672 gagtagactactacacactgaagcctgaggtcaacaatgaatcctttgtt    721


  XL23808    2751 atctacatgttcattgtccacttcaccattcccctgattgtcatcttctt   2800

                  ||||||||||||||||||||||||||||||||||||||||||||||||||

  XLRHODOP    722 atctacatgttcattgtccacttcaccattcccctgattgtcatcttctt    771


  XL23808    2801 ctgctatggtcgcctgctctgcactgtcaaagaggtgag.....catagg   3085

                  |||||||||||||||||||||||||||||||||||||||>>>>> 250 >>>>>|

  XLRHODOP    772 ctgctatggtcgcctgctctgcactgtcaaagag..............g    806
```

One problem with the dynamic programing method is that it does not scale to large sequences. It uses a comparison table to match each base of each sequence. For a typical bacterial gene or cDNA reading frame this could be 1,000 bases in each sequence, or 1,000,000 possible comparisons. As the sequences grow longer this quickly becomes too big a problem to fit in the computer"s memory or to be computed in a reasonable time.

For large sequences, such as BAC clones or complete genomes, some short cut is needed. One of the most common is to look only at localized ungapped alignments by comparing the two sequences in short stretches and then looking for ways to extend the match region but without allowing gaps so that the problem is far easier to compute. Where a gap appears in a long region of similarity the expectation is that there will be two ungapped matches reported. Even so, too many gaps will make matches impossible to detect.

A particularly rapid way to find local ungapped matches is to look for "words" in common between two sequences (WILBUR and LIPMAN, 1983). One sequence is converted to a list of all possible sequences of, for example, 6 bases and then each of these is searched for in the second sequence. Where there is a match, the neighboring bases can be checked to find whether the match can be extended. It is then very easy to look for clusters of high scores and to report the best matches. Sequences of up to complete bacterial genome size can be compared in a few seconds by this method. For increased speed and flexibility, some programs build a complete index of the words and look up the values instead of computing them each time the program runs. This can take a lot of disk space and a lot of time when the indices are calculated.

Multiple alignment is known to be a hard problem. Clearly, the dynamic programing method used for aligning two sequences is not practical for more. Just 3 sequences of 1,000 bases each would need 1,000,000,000 comparisons to be made and stored. Rather, the problem is usually broken down into a series of pairwise matches which are combined to produce a final alignment (THOMPSON et al., 1994).

2.5 Other Alignment Methods

To use all possible alignments, there are other methods. The most popular currently is to create a Hidden Markov Model (HMM) (DURBIN et al, 1998) which "learns" the characteristics of a set of sequences, including the most significant similarities between them. By aligning the original sequences to the HMM we can get a sequence alignment, and by searching a sequence database with the HMM we can search for new sequences to add to the alignment. HMMs have been very successfully used for protein domain analysis (Pfam) and have also been applied to gene prediction in bacteria as a pattern matching method (see below).

2.6 Sequence Comparison Methods

An alternative to full alignment of two sequences is the search for common patterns between them. One of the fastest to calculate is the occurrence of long subsequences ("k-tuples", or "words").

We start again with "dot plot" methods. The EMBOSS dottup program calculates word matches between two sequences and displays them on a "dot plot". The first step is to build a complete list of all the words in one sequence and make this into an indexed table. Then, for each word in the second sequence, a simple lookup in the table shows every match in the first sequence. With only a few thousand table lookups we have covered several million possible dot positions.

This method scales up exceptionally well, and can be used up to whole chromosome scale sequences.

Example dottup

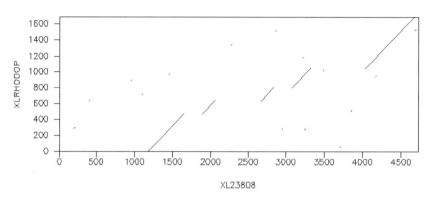

XL23808

Graphical views of the results are all very well, but we also need to have the results as sequence positions we can work with in other programs. Another EMBOSS program, wordmatch, uses the same method as dottup but instead reports the start positions in each sequence and the length.

Applied to our earlier cDNA/genomic sequence example we find that the exons are reported almost immediately.

The method does have some slight drawbacks, compared to the rigorous est2genome program. For example, if the next base after the exon is the same in each sequence then the match will be extended a little. Also, mismatches or insertions and deletions in the exons will produce two or more shorter matches, but est2genome needs much more time to produce its rigorous result.

2.7 Multiple Pairwise Comparisons

We can extend these rapid word-based methods to produce very fast comparisons of very long sequences. This is implemented in EMBOSS as the polydot program, which is the dottup method for self comparison of a set of sequences. These could be contigs from a fragment assembly project to check for overlaps (the original use for this program), or complete chromosomes to check for possible duplications.

In the example, we see the *Escherichia coli* lac operon (EMBL entry ECLAC) compared to the sequences of the four individual genes lacI, lacZ, lacY and lacA.

Example wordmatch

```
FINALLY length = 5
      XL23808    XLRHODOP Length
       4027        1043      642
       1182           2      471
       3083         804      242
       1898         470      170
       2666         637      170
```

Example polydot

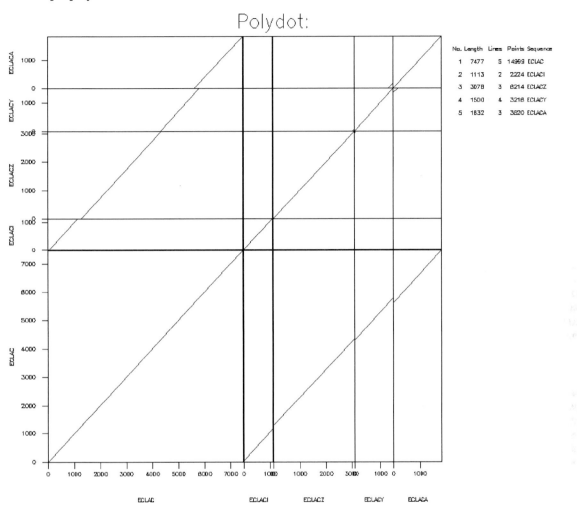

Polydot:

No.	Length	Lines	Points	Sequence
1	7477	5	14999	ECLAC
2	1113	2	2224	ECLACI
3	3078	3	8214	ECLACZ
4	1500	4	3218	ECLACY
5	1832	3	3820	ECLACA

Just like dottup and wordmatch, polydot has a companion program which produces a report of all the match positions. This program, seqmatchall, reports the alignment length, and the start and end positions in each sequence.

The matches (apart from self matches) are shown below. They include each of the individual gene sequences matched to the complete operon, plus two short matches between lacY and the flanking sequences.

Example seqmatchall

```
1113  49 1161 ECLAC 1 1113 ECLACI
3078  1287 4364 ECLAC 1 3078 ECLACZ
1500  4305 5804 ECLAC 1 1500 ECLACY
1832  5646 7477 ECLAC 1 1832 ECLACA
60   3019 3078 ECLACZ 1 60 ECLACY
159   1342 1500 ECLACY 1 159 ECLACA
```

2.8 Statistical Methods

Some methods look at statistical properties of sequences to classify them in various ways. One example in EMBOSS is the program complex (Pesole et al., 1994), contributed by the Italian EMBnet group (*http://www.it.embnet.org*), which calculates the complexity (randomness) of a set of sequences. In the example below, one of the selected sequences, HHTETRA, is highly repetitive and this is clearly reflected in the complexity statistic.

Example complex

```
Length of window : 100

jmin : 4

jmax : 6

step : 5

Execution without simulation
```

number of sequence	name of sequence	length of sequence	value of complexity
1	HS989235	495	0.7210
2	HSCAD5	3170	0.6921
3	HSD	781	0.6991
4	HSEGL1	3919	0.6618
5	HSFAU	518	0.6739
6	HSFOS	6210	0.6681
7	HSEF2	3075	0.6925
8	HSHT	1658	0.7314
9	HSTS1	18596	0.6668
10	HSNFG9	33760	0.6661
11	CEZK637	40699	0.6307
12	PDRHOD	1675	0.6201
13	ECLAC	7477	0.7137
14	ECLACA	1832	0.6916
15	ECLACI	1113	0.7480
16	ECLACY	1500	0.6801
17	ECLACZ	3078	0.7278
18	PAAMIB	1212	0.6596
19	PAAMIE	1065	0.6418
20	PAAMIR	2167	0.6562
21	PAAMIS	1130	0.6989
22	MMAM	366	0.7163

22	MMAM	366	0.7163
23	RNOPS	1493	0.6571
24	RNU68037	1218	0.6381
25	HHTETRA	1272	0.3114
26	XLRHODOP	1684	0.7193
27	XL23808	4734	0.7180

2.9 Consensus Methods

Consensus methods take a sequence alignment and calculate a consensus sequence which represents all the alignment members. These consensus sequences are less useful than the alignment or a weight matrix of Hidden Markov Model derived from it, but are often found in pattern databases.

Consensus sequences make use of base codes beyond the familiar A, C, G, and T. Each possible combination of bases has its own one letter code, and with a little practice it is relatively easy to learn them all. The full set of codes is listed in the Tab. 1. For the two-base alternatives some knowledge of the base chemistry if needed. Surprisingly, the three-base codes are extremely easy to learn though most biologists will not know them.

Tab. 1. Nucleotide Base Code Table

Code	Base(s)	Mnemonic
A	A	adenine
C	C	cytidine
G	G	cuanine
T	T or U	thymine
U	T or U	uracil (RNA equivalent of T)
R	A or G	pu-R-ine
Y	C or T/U	p-Y-rimidine
S	C or G	strong H-bonding
W	A or T/U	weak H-bonding
K	G or T/U	K-eto group
M	A or C	a-M-ino group
B	C or G or T/U	not A
D	A or G or T/U	not C
H	A or C or T/U	not G
V	A or C or G	not T or U
N	A or C or G or T/U	a-N-y base
. or -	Gap	

2.10 Simple Sequence Masking

The simplest sequence patterns to detect are often the ones that cause the most problems. Simple repeats, or runs of single bases, are usually removed before attempting to run a sequence database search because otherwise the highest scoring hits will be to other sequences with the same simple repeat and any functionally significant matches will be lost. The most common filter for DNA sequences is the "dust" program from NCBI.

2.11 Unusual Sequence Composition

Using the word-based methods familiar from the sequence comparison section above, we can easily identify the most common subsequences of any given length. Surprisingly, many genomes do show a strong bias in their composition of sequences ranging in size from 2 bases to 111 or more.

Example wordcount
The genome of *Escherichia coli* is known to have strong biases for certain 4 and 5 base sequences. The wordcount program reports the frequencies of all short sequences (words) of a given size. This example run on sequence ECUW87 from the EMBL database shows that the rarest 5 base sequences all contain the 4 base sequence ctag, and the most common include sequences close to CCAGG or CCTGG (MERKL et al., 1992)

```
ccagc 310

tggcg 295

gccag 281

cagcg 281
```

cagca 273

ctggc 268

ttttt 265

cgcca 263

cgccg 245

gctgg 239

aaaaa 234

tttgc 233

cgctg 233

ctagc 6

gctag 6

actag 3

ctagt 3

ctaga 2

cctag 2

ctagg 1

tctag 1

range of possible repeat sizes. In the example below, a run with a range of 2 to 10 bases reports a possible repeat size of 6 bases in a sequence.

Example equicktandem

```
339           191          935  6 124
```

The result of equicktandem does not tell us where the repeat was found, and does not guarantee that the repeat will be sufficiently conserved to be identified. However, the size value can be used by a further program etandem (DURBIN, R. personal communication) to identify the precise positions of exact or near exact repeat runs.

Example etandem

```
120          793          936  6  24  93.8 acccta
 90          283          420  6  23  84.8 taaccc
 38          432          485  6   9  90.7 ccctaa
 26          494          529  6   6  94.4 ccctaa
 24          568          597  6   5 100.0 aaccct
```

2.12 Repeat Identification

Other programs can find less obvious repeats, or can look for repeats in particular categories.

In EMBOSS, the problem of finding tandem (direct) repeats is simplified by breaking it into two steps.

The first program, equicktandem (DURBIN, R. personal communication), rapidly scans a sequence and reports possible similarities for a

Direct repeats are relatively easy to find, mainly because they appear one after another without gaps.

A further program, einverted (DURBIN, R. personal communication), looks for near perfect inverted repeats. This is a trickier problem, as the repeats can be very long and typically have a gap between the two repeated sequences. By limiting the size of the central gap we can significantly speed up the calculation.

Example einverted

```
Score 80: 44/51 ( 86%) matches, 2 gaps
   12246 ctcctgcctcag-cctccaagtagctgggattaca-gcatgtgccaccatgcc 12296
         ||||||  |||||  |  |||||   ||||||||||||  |||||  ||||||||  ||
   13938 gaggacagagtcagaaggtttcacgaccctaatgtccgtactcggtggtatgg 13886


Score 99: 53/65 ( 81%) matches, 1 gaps
   13884 tgggtatggtggctcatgcctgtaatcccagcactttggaagactgagacaggagcaattgcttga 13949
         |||||  |||||||   ||||||||||||||||      |||  ||  |||||  |||  ||  ||||||||||
   14692 acccacaccaccgtacacggacattagggtcgatggaccctccgactccgtcttc-ttaacgaact 14628
```

If we are less forgiving of mismatches and gaps, we can make a more exhaustive search for inverted repeats. The program palindrome offers such an alternative approach.

Example palindrome

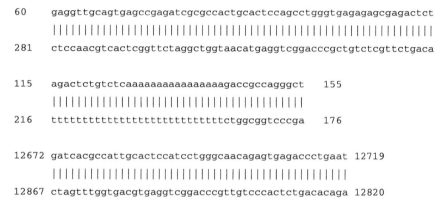

```
Palindromes of:  HSTS1

Sequence length is: 18596

Start at position: 1

End at position: 18596

Minimum length of Palindromes is: 40

Maximum length of Palindromes is: 100

Maximum gap between elements is: 100

Number of mismatches allowed in Palindrome: 10
```

```
Palindromes:
60      gaggttgcagtgagccgagatcgcgccactgcactccagcctgggtgagagagcgagactct

        ||||||||||||||||||||||||||||||||||||||||||||||||||||||||||||||

281     ctccaacgtcactcggttctaggctggtaacatgaggtcggacccgctgtctcgttctgaca


115     agactctgtctcaaaaaaaaaaaaaaaaagaccgccagggct    155

        ||||||||||||||||||||||||||||||||||||||||||

216     tttttttttttttttttttttttttttttctggcggtcccga    176


12672  gatcacgccattgcactccatcctgggcaacagagtgagaccctgaat 12719

        ||||||||||||||||||||||||||||||||||||||||||||||||

12867  ctagtttggtgacgtgaggtcggacccgttgtcccactctgacacaga 12820
```

2.13 Detection of Patterns in Sequences

Some applications will look for specific patterns in a sequence, and can be very useful when first analyzing a new sequence.

One of the simplest to program, and yet also one of the most effective for detecting unusual composition on large sequences, is the "chaos game representation" plot (JEFFREY, 1990). Although the plot looks complicated and difficult to understand, it is really very simple.

The example plot shows a typical result of a "chaos game" plot for human genomic DNA. Clearly, there is a pattern in the sequence although at first sight it is not obvious what the pattern could be.

The chaos game method makes everything clear. Imagine that the program is running on the sequence ATCCG. Now start in the center of the plot. The first base is A, so go half way to

the A corner and draw a dot. The next base is C so go half way from this position to the T corner and draw a dot. Now go half way to the C corner and draw a third dot. Then half way again to the C corner and draw a fourth dot. The next move, half way to the G corner, will land in the blank area on the plot.

The reason is simple. Moving "half way to the G corner" will always land in the top right quarter of the plot, no matter where you start. But if the previous base was a C you start in the top left quarter of the plot, and always move to the blank region.

In fact, any sequence that ends with CG will always produce a dot in this 1/16 area. The pattern is simply a result of the under-representation of the dinucleotide CG in the human genome.

The plot works on longer regions of unusual sequence composition. A plot of *Escherichia coli* genomic sequence will reveal small light and dark boxes, corresponding to the 4 and 5 base sequence biases resulting from "very short patch" repair mechanisms (MERKL et al, 1992).

Example chaos

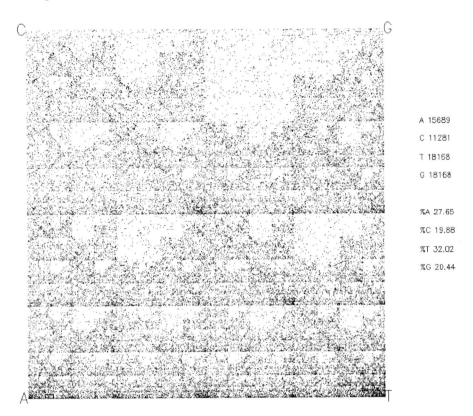

```
A  15689

C  11281

T  18168

G  18168


%A 27.65

%C 19.88

%T 32.02

%G 20.44
```

2.14 Physical Characteristics

Although accurate calculation of physical properties is not realistic as factors beyond the sequence are involved, it is possible to estimate the potential for bending in a DNA double helix. Various calculations have been proposed. One method is implemented in the EMBOSS program banana (GOODSELL and DICKERSON, 1994).

The program name, incidentally, is more than a simple joke on curvature. The consensus sequence for maximal bending is approximately AAAAA surrounded by 5 bases that are not A. From the DNA ambiguity codes in the table earlier, you can see that N is "any base" and B is "not-A". This means that the DNA sequence code "BANANA" matches the bent DNA consensus!

Example banana

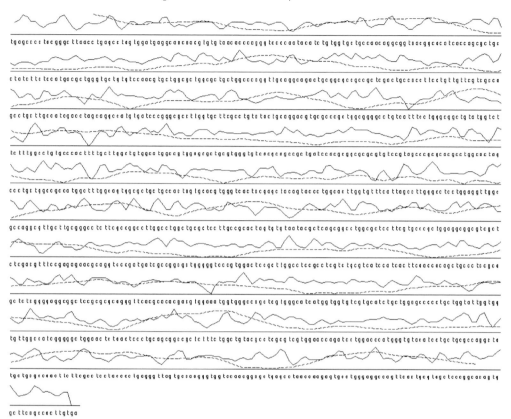

Bending and curvature plot of RNU68037

3 Detecting CpG Islands

The "chaos game" plot clearly shows under-representation of the dinucleotide sequence CG in the human genome. However, in some arts of the genome, especially in the 5′ ends of "housekeeping" genes, the sequence CG occurs far more frequently than in the remainder of the genome.

Rules have been derived for assigning such sequences as "CpG islands" (the p is simply the phosphate between the bases). The EMBOSS program cpgplot identifies CpG islands for the CPGISLE database project (LARSEN et al., 1992). The three plots show the ratio of observed to expected CG sequences over a sliding window, the C + G content over the same window (even a sequence that is 100% C and G will usually have rather fewer than the expected 25 CG dinucleotides). The third plot shows the region that meets the "CpG island" criteria.

Example cpgplot

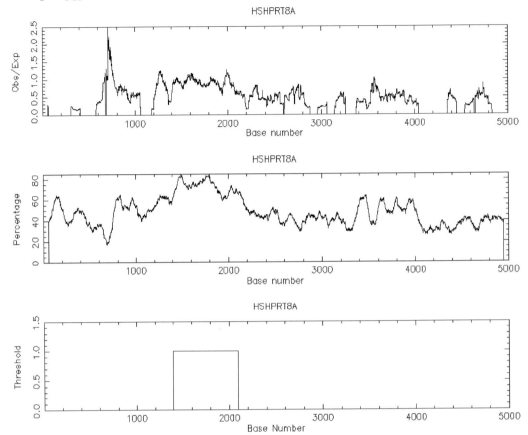

A second program, newcpgreport, generates the CPGISLE database entries, in an EMBL-like format, from sequences identified by cpgplot. The comments section (CC lines) of the entry also shows the CpG island criteria used.

Example newcpgreport

```
ID    HSHPRT8A    4900 BP.
XX
DE    CpG Island report.
XX
CC    Obs/Exp ratio > 0.60.
CC    % C + % G > 50.00.
CC    Length > 200.
XX
```

```
FH    Key                Location/Qualifiers
FT    CpG island         1448..2149
FT                       /size=702
FT                       /Sum C+G=515
FT                       /Percent CG=73.36
FT                       /ObsExp=0.86
FT    numislands         1
//
```

There are alternative approaches to CpG island identification. One used by another CpG island project is provided by the EMBOSS program newcpgseek (MICKLEM, G., personal communication) which calculates scores for potential CpG islands. The highest scoring match is the same, though slightly longer, as the region reported by cpgplot and newcpgreport.

Example newcpgseek

```
NEWCPGSEEK of HSHPRT8A from 1 to 5000
with score > 17


 Begin    End   Score      CpG   %CG   CG/GC
   737    871     64        11   63.0   0.00
  1003   1009     48         3  100.0   1.00
  1249   2163    814        96   71.0   0.00
  2262   2287     29         3   50.0   3.00
  2657   2661     32         2  100.0   2.00
  3420   3435     39         3   75.0   3.00
  3575   3589     22         2   66.7   1.00
  3638   3649     25         2   66.7   2.00
  3826   3843     37         3   77.8   1.00
  3920   3930     26         2   72.7   2.00

-------------------------------------------
```

Example fuzznuc

```
ECUW87      827 CTAG
ECUW87     1811 CTAG
ECUW87    32570 CTAG
ECUW87    40394 CTAG
ECUW87    49454 CTAG
ECUW87    57582 CTAG
ECUW87    60838 CTAG
ECUW87    68074 CTAG
ECUW87    69341 CTAG
ECUW87    75724 CTAG
ECUW87    88686 CTAG
ECUW87    95978 CTAG
```

Restriction sites and promoter consensus sequences.

3.1 Known Sequence Patterns

Nucleotide sequence patterns are used extensively where there are no similar sequences to compared to. Public databases are available for restriction enzyme target sites (REBASE) and transcription factor binding sites (TRANSFAC).

3.2 Data Mining with Sequence Patterns

Many sequence patterns remain to be discovered. Patterns can be specified in many ways, some of which are particularly difficult to identify in a computer program. Special difficulties are caused by allowing a wide range of possible gaps or a large number of mismatches.

The EMBOSS program fuzznuc allows the user to search for "fuzzy" patterns in DNA sequences. The patterns are specified in a similar way to patterns in the protein motif database PROSITE. Fuzznuc first examines the pattern, and then chooses the most appropriate string matching method from its extensive library.

The example run shows the results of a simple search of *Escherichia coli* genomic DNA for the under-represented tetranucleotide CTAG.

3.3 Restriction Mapping

A natural application for pattern searching in DNA sequences is to use the target sites of restriction enzymes to generate a fragment size map from sequence and to compare this with the expected fragment sizes from experiment.

Because there are many and varied uses for such a program, these methods typically offer an unusually large number of options for the user, including restricting the search to rare cutters (those with long specific target sites such as NotI), or looking only for enzymes that will cut the sequence once or twice.

When a program needs to be especially versatile, it is particularly useful to have access to the program's source code so that new functions can be added and existing functions can be modified. Such changes can easily save several days of tedious analysis.

Example restrict

```
# Restrict of PAAMIR from 1 to 2167
#
# Minimum cuts per enzyme: 1
# Maximum cuts per enzyme: 2
# Minimum length of recognition site: 6
# Sticky ends allowed
# DNA is linear
# No ambiguities allowed
# Number of hits: 31
```

# BaseNumber	Enzyme	Site	5'	3'
1	Acc65I	GGTACC	1	5
1	KpnI	GGTACC	5	1
57	BsgI	GTGCAG	37	35
80	BssSI	CACGAG	80	84
97	BsiWI	CGTACG	97	101
152	CciNI	GCGGCCGC	153	157
214	MluI	ACGCGT	214	218
231	Bse3DI	GCAATG	225	223
277	BspCI	CGATCG	280	278
374	BssSI	CACGAG	369	373
382	BspLU11I	ACATGT	382	386
500	BbeI	GGCGCC	504	500
500	Mly113I	GGCGCC	501	503
500	KasI	GGCGCC	500	504
591	Cfr42I	CCGCGG	594	592
648	PaeR7I	CTCGAG	648	652
671	BciVI	GTATCC	682	681
809	CciNI	GCGGCCGC	810	814
817	BspCI	CGATCG	820	818
886	Acc36I	ACCTGC	872	876
912	BanIII	ATCGAT	913	915
950	Bsp19I	CCATGG	950	954
1065	BsgI	GTGCAG	1086	1084
1167	BanIII	ATCGAT	1168	1170
1332	Acc36I	ACCTGC	1318	1322
1560	Cfr42I	CCGCGG	1563	1561
1602	BclI	TGATCA	1602	1606
1718	MroNI	GCCGGC	1718	1722
1765	Bsp19I	CCATGG	1765	1769
2016	BclI	TGATCA	2016	2020
2162	PaeR7I	CTCGAG	2162	2166

3.4 Codon Usage Analysis

Although more complicated methods for gene identification are covered elsewhere, it is worth reviewing some of the methods available in EMBOSS, especially for prokaryotic (i.e., unspliced) sequences.

Many methods depend on the strong bias in the use of alternative codons in true coding sequences (GRIBSKOV et al., 1984). This generally reflects the abundance of the respective tRNA species, so that the codons with the most abundant tRNAs will be used in preference, at least for genes that must be highly expressed under some conditions.

To apply such methods, we need to know the typical codon usage for a given species. Each has its own peculiar bias. In yeast *(Saccharomyces cerevisiae)* tables have been used successfully to distinguish low and highly expressed genes by their different codon usage patterns.

In the example below, which is sorted by amino acid, there is a clear bias in the codon choices for aspartate (D) where 75% of the codons will be GAT and only 25% will be the alternative GAC. For leucine (L) there is a strong preference for CTG or CTT and in this sample of genes a total avoidance of TTA.

Example cusp

```
# CUSP codon usage file
```

# Codon	Amino acid	Fract	/1000	Numb
GCA	A	0.200	31.812	23
GCC	A	0.200	31.812	23
GCG	A	0.296	47.026	34
GCT	A	0.304	48.409	35
TGC	C	0.500	17.981	13
TGT	C	0.500	17.981	13
GAC	D	0.250	6.916	5
GAT	D	0.750	20.747	15
GAA	E	0.692	12.448	9
GAG	E	0.308	5.533	4
TTC	F	0.750	8.299	6
TTT	F	0.250	2.766	2
GGA	G	0.187	19.364	14
GGC	G	0.293	30.429	22

GGG	G	0.213	22.130	16
GGT	G	0.307	31.812	23
CAC	H	0.400	8.299	6
CAT	H	0.600	12.448	9
ATA	I	0.273	4.149	3
ATC	I	0.545	8.299	6
ATT	I	0.182	2.766	2
AAA	K	0.750	8.299	6
AAG	K	0.250	2.766	2
CTA	L	0.130	4.149	3
CTC	L	0.130	4.149	3
CTG	L	0.304	9.682	7
CTT	L	0.391	12.448	9
TTA	L	0.000	0.000	0
TTG	L	0.043	1.383	1
ATG	M	1.000	6.916	5
AAC	N	0.750	4.149	3
AAT	N	0.250	1.383	1
CCA	P	0.329	34.578	25
CCC	P	0.066	6.916	5
CCG	P	0.355	37.344	27
CCT	P	0.250	26.279	19
CAA	Q	0.588	13.831	10
CAG	Q	0.412	9.682	.7
AGA	R	0.056	9.682	7
AGG	R	0.144	24.896	18
CGA	R	0.224	38.728	28
CGC	R	0.304	52.559	38
CGG	R	0.184	31.812	23
CGT	R	0.088	15.214	11
AGC	S	0.234	24.896	18
AGT	S	0.039	4.149	3
TCA	S	0.078	8.299	6
TCC	S	0.104	11.065	8
TCG	S	0.364	38.728	28
TCT	S	0.182	19.364	14
ACA	T	0.185	13.831	10
ACC	T	0.315	23.513	17
ACG	T	0.333	24.896	18
ACT	T	0.167	12.448	9
GTA	V	0.100	2.766	2
GTC	V	0.250	6.916	5
GTG	V	0.250	6.916	5
GTT	V	0.400	11.065	8
TGG	W	1.000	26.279	19
TAC	Y	0.750	4.149	3
TAT	Y	0.250	1.383	1
TAA	*	0.143	1.383	1
TAG	*	0.000	0.000	0
TGA	*	0.857	8.299	6

3.5 Plotting Open Reading Frames

One of the simplest methods, plotorf, displays the open reading frames (ORFs) in a sequence, defined as the longest sequences starting with a "start codon"(usually ATG) and ending with a stop codon. The longest ORFs are likely to be true genes, although care is needed in annotation. Although the stop codon is clearly fixed, there is no guarantee that the most distant start codon is the one actually used when the gene is expressed.

A further complication arises because the nature of amino acid composition and codon bias tends to reduce the number of possible stop codons on the opposite strand, and so it is common to see two open reading frames in opposite directions overlapping each other.

The plot below shows such a case for a sequence from *Pseudomonas aeruginosa* (LOWE et al., 1989), where the high G+C content in this organism further reduces the frequency of stop codons which are, of course, relatively AT-rich. The three true open reading frames can be shown by other methods to be the longest ones in the second and third panels of the plot.

Example plotorf

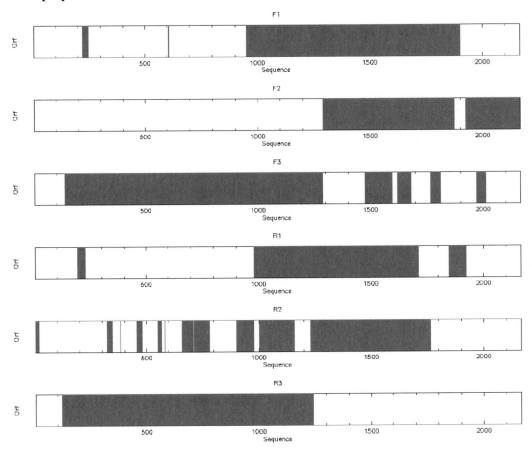

3.6 Codon Preference Statistics

To identify such genes, statistical methods are available which use either the expected codon usage table, or more crudely the biased base composition of the third (wobble) base position in each codon.

For codon usage, the EMBOSS program syco (synonymous codons) plots the Gribskov statistic (GRIBSKOV et al., 1984) over a sequence in each of the 3 possible reading frames. The example sequence is the same as the one in the plotorf example above.

Example syco

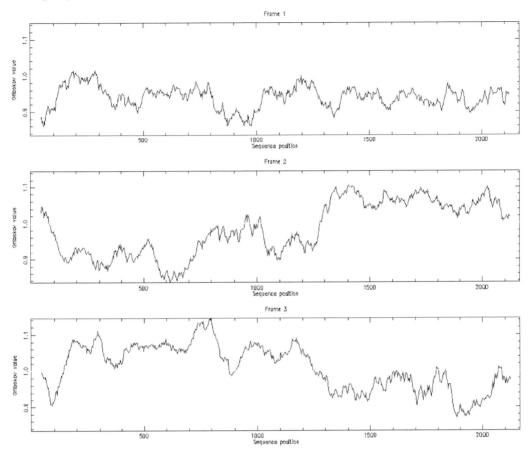

If a codon usage table is not available, or for some species where the codon bias is weak, statistical analysis of the third base position offers a practical alternative method.

The example below is for the same *Pseudomonas aeruginosa* sequence, where the high $G + C$ content leads to a specially strong third position bias reflected in the general rule "if any base can be C or G then it will be."

Example wobble

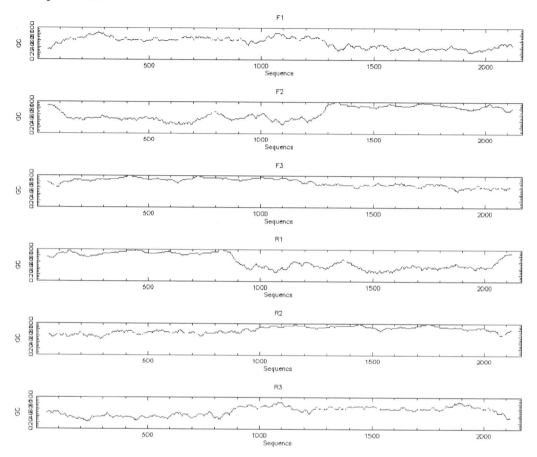

3.7 Reading Frame Statistics

Once a potential gene has been identified, various statistical methods are available to calculate the codon bias for that particular gene and to relate these to the possible expression level of the gene on the grounds that more highly expressed genes are generally found to have a more biased codon usage.

An example in EMBOSS is FRANK WRIGHT's "effective number of codons" statistic (WRIGHT, 1990), calculated by the program chips ("codon heuristics in protein-coding sequences")

Example chips

```
# CHIPS codon usage statistics


Nc = 32.951
```

In addition to using a codon usage table to analyze gene expression in one organism, these tables can be used to compare species. Closely related species will usually show similar codon frequencies, although there are exceptions caused by recent changes in GC content, and by recent horizontal transfer of DNA. These methods can be used, for example, to identify DNA or viral origin in bacterial genomes.

The program codcmp compares two codon usage tables, and calculates statistics to indicate their degree of difference.

Example codcmp

```
# CODCMP codon usage table comparison
# hum vs pseud.cusp

Sum Difference Squared = 1.832
Sum Difference        = 0.021
Codons not appearing  = 0
```

4 The Future for EMBOSS

The EMBOSS project is an open source effort, which means that all the source code is made freely available for everyone to share and develop further. The home of the project is now the United Kingdom's EMBnet node (*http://www.uk.embnet.org/*), at the Human Genome Mapping Project (HGMP) Resource Centre in Hinxton, next door to the Sanger Centre where the project started.

The programs were designed to be run as "commands" by simply typing the program name and answering a series of questions. It was always clear that for many users this would not be enough, and also that few users would in practice agree on the best approach to take.

The EMBOSS collaborators made plans to make all the programs available under as many different user interfaces as possible. Already in the first phase of the project (up to the end of the year 2000) we have added EMBOSS to Web interfaces such as Pise from Institut Pasteur, and W2H from the German Cancer Research Centre (DKFZ – the German EMBnet node *http://www.de.embnet.org/*). Web interfaces make the EMBOSS programs available to any user with a web browser, through access to a central service site that can install the databases and maintain the programs – something that EMBnet sites have many years of experience in providing to most of the major countries of the world.

We also have progress on including EMBOSS under other graphical user interfaces, and with other software packages, and we expect these will begin to appear during the coming few years.

The programs will also continue to develop. In the area of sequence analysis, we are rewriting the program output. As you can see from the examples in this chapter, the first release has a wide variety of output formats, reflecting current practice in each area. These are now changing to give a set of interchangeable report formats, which will make the processing of EMBOSS output and the interconnection of the programs much easier in the next release.

EMBOSS is also extending beyond pure sequence analysis, with groups interested in the fields of protein structure and population genetics (to give just two examples). With help from the bioinformatics community, EMBnet and beyond, the future s looking very interesting indeed.

5 References

DURBIN, R., EDDY, S., KROGH, A., MITCHISON, G. (1998), *Biological Sequence Analysis*, Cambridge: Cambridge University Press.

GOODSELL, D. S., DICKERSON, R. E. (1994), Bending and curvature calculations in B-DNA, *Nucleic Acids Res.* **22**, 5497–5503.

GRIBSKOV, M., DEVEREUX, J.M BURGESS, R. R. (1984), The codon preference plot: graphic analysis of protein coding sequences and prediction of gene expression, *Nucleic Acids Res.* **12**, 539–549.

JEFFREY, H. J. (1990), Chaos game representation of gene structure, *Nucleic Acids Res.* **18**, 2163–2170.

LARSEN, F., GUNDERSEN, G., LOPEZ, R., PRYDZ, H. (1992), CpG islands as gene markers in the human genome, *Genomics* **13**, 105–107.

LOWE N., RICE P. M., DREW R. E. (1989), Nucleotide sequence of the aliphatic amidase regulator gene of Pseudomonas aeruginosa, *FEBS Lett.* **246**, 39–43.

MERKL, R., KROEGER, M., RICE, P., FRITZ, H.-J. (1992), Statistical evaluation and biological interpretation of non-random abundance in the *E. coli* K-12 genome of tetra- and pentanucleotide sequences related to VSP DNA mismatch repair, *Nucleic Acids Res.* **20**, 1657–1662.

MOTT, R. F. (1997), Est-genome: a program to align spliced DNA sequences to unspliced genomic DNA, *Computer Appl. Biosci.* **13**, 477–478.

NEEDLEMAN, S. B., WUNSCH, C. D. (1970), A general method applicable to the search for similarities in the amino acid sequence of two proteins, *J. Mol. Biol.* **48**, 443–453.

PESOLE, G., ATTIMONELLI, M., SACCONE, C. (1994), Linguistic approaches to the analysis of sequence information, *Trends Biotechnol.* **12**, 401–408.

RICE, P., LONGDEN, I., BLEASBY, A. (2000), EM-BOSS: The European Molecular Biology Open Software Suite, *Trends Genet.* **16**, 276–277. (*http://www.uk.embnet.org/Software/EMBOSS/*)

SMITH, T. F., WATERMAN, M. S. (1981), Identification of common molecular subsequences, *J. Mol. Biol.* **147**, 195–197.

THOMPSON, J. D., HIGGINS, D. G., GIBSON, T. J. (1994), CLUSTAL W: improving the sensitivity of progressive multiple sequence alignment through sequence weighting, positions-specific gap penalties and weight matrix choice, *Nucleic Acids Res.* **22**, 4673–4680.

WILBUR, W. J., LIPMAN, D. J. (1983) Improved tools for biological sequence comparison, *Proc. Nat. Acad. Sci. USA* **80**, 726–730.

14 Tools for Protein Technologies

DAVID S. WISHART

Edmonton, Canada

1 Introduction

With the completion of the human genome now at hand, we will soon know the amino acid sequence of nearly every human protein. It is likely what within two more years, complete sets of protein sequences will be known for several important laboratory animals and key agricultural organisms (BRODER and VENTER, 2000; DUTT and LEE, 2000). The challenge over the coming decades will be to connect all those protein sequences with their respective actions and to translate that understanding into new approaches to manage or treat disease, to diagnose medical conditions, to monitor drug interactions, to improve crop yields, or to enhance the quality of our environment.

Key to translating this raw biological data to practical knowledge will be our ability to recognize or detect patterns that exist within these data (ATTWOOD, 2000). This is where bioinformatics comes in. Bioinformatics plays a vital role in all areas of proteomics (the study of proteins and their interactions) by providing the software tools that help sort, store, analyze, visualize, and extract important patterns from raw proteomic data. Computational tools such as correlational analysis, multiparametric fitting, dynamic programing, artificial intelligence, neural networks, and hidden Markov models are critical to teasing out many of the hidden patterns and relationships in sequence, 2D gel or mass spectrometric (MS) data. Complementing these tools are a growing array of queryable databases containing protein sequences, pre-calculated mass fragment data, 2D gel images, 3D structures, biochemical pathways, and functional sites that provide the critical "prior knowledge" necessary to extract additional information from unprocessed experimental data.

In previous chapters on protein and DNA technologies we have seen how the raw data on protein sequences, isoelectric points, and peptide mass fingerprints can be acquired. In this chapter we will see how these raw data can be transformed into useful biochemical knowledge. In particular, we will show how bioinformatics tools can be used to facilitate protein identification and characterization using 2D gel, MS, and unprocessed protein sequence data. This chapter will be divided into two sections. The first section will be concerned with describing the software tools and algorithms that can facilitate protein identification from 2D gels, mass spectrometric, or protein sequence data. The second section will describe the bioinformatics tools and databases that may be used to predict the functions, locations, and properties of proteins once they have been identified. Particular emphasis will be placed on describing freely available Web tools or software packages that have been published in the scientific literature.

2 Protein Identification

Unfortunately for us, proteins do not come with name tags. What's more, proteins like to hang out in crowds – usually with other proteins that look and behave almost identically. Indeed, the only way to uniquely identify a protein is to carefully separate it and painstakingly determine its sequence or precisely measure its mass. Consequently protein identification is an inherently difficult process that requires the close interplay between experimental and computational techniques (GEVAERT and VANDEKERCKHOVE, 2000). The experimental techniques provide the raw data while the computational techniques convert these raw data into a usable protein name or data bank accession number. These computational tools all rely on a common theme – i.e., they identify proteins by looking for close matches (in mass, in sequence, or in 2D gel position) to previously identified proteins. In this way protein identification is facilitated by making use of vast stores of previously accumulated knowledge about the 500,000 proteins already studied. In this section we will review three protein identification methods and their associated software tools:

(1) identification by 2D gel spot position,
(2) identification by mass spectrometry, and
(3) identification by sequence data.

2.1 Protein Identification from 2D Gels

As we have seen from Chapter 10, 2D gel electrophoresis allows for the precise and reproducible separation of up to 10,000 different proteins. The widespread use of 2D gels in functional genomics (i.e., proteomics) led to the development of some excellent software tools and a number of valuable 2D gel databases to facilitate protein identification and annotation. While most 2D gel analysis software is very image oriented, the fact that these packages can be used to measure physical properties (*p*I and MW) and identify proteins actually makes them a key part of the standard bioinformatics tool chest.

There are at least four major commercial programs for 2D gel analysis: Phoretix 2D, Investigator 2D, PDQuest, and Melanie 3 (Tab. 1). All four offer an impressive array of image manipulation facilities integrated into sophisticated graphical user interfaces. Three are specific to Windows platforms (Phoretix 2D, Investigator 2D, Melanie 3) while Melanie II and PDQuest runs on both Windows and MacOS. Some commercial packages, such as Melanie (Medical ELectrophoresis ANalysis Interactive Expert system) began as academic projects and have been under development for many years (APPEL et al., 1988, 1997). Most of these packages make use of machine learning, heuristic clustering, artificial intelligence, and high-level image manipulation techniques to support some very complex 2D gel analyses. Several commercial packages are typically sold as part of larger equipment purchases (2D gel systems with robotic gel cutters) and are closely tied to the major proteomics or 2D gel vendors.

Essentially all commercial packages offer an array of automated or manual spot manipulations including: spot detection, spot editing, spot normalization, spot filtering, spot quantitation, and spot annotation. This allows users to compare, quantify and archive 2D gel spots quickly and accurately. A task which is particu-

Tab. 1. Protein Identification Tools – Web Links

Tool/Database	Web Address
Flicker (2D gels)	*http://www.lecb.ncifcrf.gov/flicker/*
Phoretix 2D	*http://www.phoretix.com*
PDQuest/Melanie II	*http://www.proteomeworks.bio-rad.com/html/pdquest.html*
Melanie 3	*http://expasy.cbr.nrc.ca/melanie/*
Investigator 2D	*http://www.bioimage.com*
2DWG (2D databases)	*http://www.lecb.ncifcrf.gov/2dwgDB/*
WebGel	*http://www.lecb.ncifcrf.gov/webgel/*
SWISS-2DPAGE	*http://www.expasy.ch/*
E. coli 2D database	*http://pcsf.brcf.med.umich.edu/eco2dbase/*
Yeast 2D database	*http://www.ibgc.u-bordeaux2.fr/YPM/*
PeptIdent (MS Fingerprint)	*http://expasy.cbr.nrc.ca/tools/peptident.html*
Profound (MS Fingerprint)	*http://prowl.rockefeller.edu/cgi-bin/ProFound*
Mowse (MS Fingerprint)	*http://srs.hgmp.mrc.ac.uk/cgi-bin/mowse*
Mascot (MS Fingerprint)	*http://www.mascot.com*
turboSEQUEST (MS/MS)	*http://www.thermoquest.com/bioworks.html*
PepSea (MS Fingerprint)	*http://pepsea.protana.com/PA_PepSeaForm.html*
BLAST	*http://www.ncbi.nlm.nih.gov/BLAST/*
PSI-BLAST	*http://www.ncbi.nlm.nih.gov/blast/psiblast.cgi*
Swiss-Prot Database	*http://expasy.cbr.nrc.ca/sprot/*
Owl Database	*http://www.bioinf.man.ac.uk/dbbrowser/OWL/*
PIR Database	*http://www-nbrf.georgetown.edu/pirwww/pirhome.shtml*
GenBank Database	*http://www.ncbi.nlm.nih.gov*
Protein Data Bank	*http://www.rcsb.org*

larly important in monitoring changes in protein expression from gel to gel or experiment to experiment. In addition to individual spot manipulation, whole gel manipulations such as rotating, overlaying, referencing, "synthesizing", and averaging are typically supported in most commercial packages. This is done to facilitate inter-gel comparison and to calibrate gels to pI and molecular weight standards. Calibration is particularly important for 2D gels, if one wishes to extract accurate molecular weight or pI information for protein identification.

No matter how careful one is in casting or running a 2D gel, there is usually some inter-gel variability. Therefore, the ability to stretch or shrink certain gel regions (or even entire gels) is often necessary to permit direct comparisons. Techniques called spot matching, "landmarking", and image "warping" are offered by most programs to allow this kind of forced matching. Once this kind of image transformation has been completed most commercial packages allow additional gels may be overlaid, subtracted, alternately flashed (flickered), or color contrasted to identify significant changes or significant spots.

In addition to the commercial 2D gel packages, a particularly nice freeware package known as Flicker (LIPKIN and LEMKIN, 1980; LEMKIN and THORNWALL, 1999) is also available for 2D gel image analysis and comparison (Tab. 1). While not quite as sophisticated as image manipulation, Flicker has recently been translated to Java, making it a platform independent package that runs on any Java-enabled Web browser. Flicker is quite useful for transforming (warping, rotating, etc.) and visualizing pairs of 2D gels so that the gel of interest can be easily compared to a pre-existing gel obtained over the Web. Its name comes from the fact that the program allows gel images to be flipped on and off ("flickered") to facilitate visual comparison. When combined with other Web tools, such as WebGel and 2DWG (also developed by Lemkin), it is possible to create a very powerful suite for protein identification at essentially no cost.

Protein identification using 2D gels can be done any number of ways, be it through pI/MW measurements, Western blotting (if an antibody is known), or ^{32}P detection (if the protein of interest is known to be phosphorylated). However, the best method for protein identification is through visual database comparisons to previously annotated gels (CELIS et al., 1998; LEMKIN and THORNWALL, 1999; HOOGLAND et al., 2000). Over the past 25 years, thousands of 2D gels have been run on cell extracts of many different organisms and human tissues. A large number of these gels have been analyzed and their protein spots identified through microsequencing or mass spectrometry. These carefully annotated gels have been deposited into more than 30 different "Federated" 2D gel databases (such as SWISS-2D PAGE) with the intention that others who may be studying similar systems could use these standardized, annotated gel images to overlay with their own gels and rapidly identify proteins of interest.

For instance, suppose you have decided to study *Saccharomyces cerevisiae* under anaerobic conditions. By running a 2D gel of the proteins expressed under anaerobic conditions, you may save literally months of effort by comparing this gel with the fully annotated *S. cerevisiae* 2D gel (grown in aerobic conditions) found at *http://www.ibgc.u-bordeaux2. fr/YPM/*. Using a software package like Flicker or more sophisticated commercial packages, it should be possible to visually transform the two gels, overlay them and identify nearly 400 yeast proteins or protein fragments in less than an hour. Quantification of the differences in expression might take only a few more hours. Indeed, the intent of these federated 2D gel databases is to avoid costly or repetitive efforts that only lead to the re-identification of previously mapped or previously known proteins. The utility of 2D gel databases is bound to grow as more gels are collected and as more spots are progressively identified in various labs around the world. Indeed, one might optimistically predict that sometime in the near future, mass spectrometry and microsequencing will no longer be needed to routinely identify protein spots as all detectable spots will have been annotated and archived in a set of Web-accessible 2D gel databases.

However, much still remains to be done before this vision could become a reality. 2D gel spot patterns are highly dependent on the methods used to isolate and prepare the initial protein mixture (CELIS et al., 1998). Consequently, individuals wishing to do gel database comparisons must take into account such variables as the protein fraction that was isolated, how the sample was prepared, and how the gel was run. Even if sample preparation issues are eventually sorted out, continuing problems concerning 2D gel database maintenance and updates still persist. Indeed, most publicly available annotated 2D gels represent incomplete "best efforts" of a single graduate student rather than collective, sustained efforts arising from multiple laboratories. If the concept of 2D gel databases is going to succeed, it will need a well-funded central repository (like the NCBI or EBI) and open-minded funding agencies to support sustained gel annotation contributions from the whole scientific community.

In the future it is likely that other separation and display techniques such as, 2D HPLC, tandem capillary electrophoresis (see Chapter 11), and protein chips will gain greater prominence in functional proteomics. The resolution and separation reproducibility of these techniques suggests that similar database comparison methods (i.e., elution profile analysis) could eventually allow proteins to be identified without the need for MS or microsequencing analysis (LINK et al., 1999; YATES, 2000).

2.2 Protein Identification from Mass Spectrometry

Recent advances in mass spectrometry have led to a paradigm shift in the way peptides and proteins are identified (YATES, 2000). In particular, the introduction of "soft" ionization techniques (Electrospray and MALDI), coupled with substantial improvements in mass accuracy (5 ppm), resolution (MS/MS), and sensitivity (femtomoles) have made the rapid, high-throughput identification of peptides and proteins almost routine (DUTT and LEE, 2000). Key to making this paradigm shift possible has been the development of bioinformatics software that allows one to correlate biomolecular MS data directly with protein sequence databases. Two kinds of MS bioinformatics software exist:

(1) software for identifying proteins from peptide mass fingerprints and
(2) software for identifying peptides or proteins directly from uninterpreted tandem (MS/MS) mass spectra.

Peptide mass fingerprinting was developed in the early 1990s as a means to unambiguously identify proteins from proteolytic fragments (PAPPIN et al., 1993; YATES et al., 1993; MANN et al., 1993). Specifically, if a pure protein is digested with a protease that cuts at predictable locations (say trypsin), the result will be a peptide mixture containing a unique collection of between 10–50 different peptides, each with a different or characteristic mass. Running this mixture on a modern ESI or MALDI instrument will lead to an MS spectrum with dozens of peaks corresponding to the masses of each of these peptides. Because no two proteins are likely to share the same set of constituent peptides, this mixture is called a peptide mass fingerprint. By comparing the observed masses of the mixture with predicted peptide masses derived from all known protein sequences it is theoretically possible to identify the protein of interest (providing the protein has been previously sequenced). Specifically, in the course of performing a mass fingerprint search, database sequences are theoretically "cleaved" using known protease cutting rules, the resulting hypothetical peptide masses are calculated and the whole protein is ranked according to the number of exact (or near exact) cleavage fragment matches made to the observed set of peptide masses. The sequence with the highest number and quality of matches is usually selected as the most likely candidate.

There are nearly a dozen different types of mass fingerprinting software available for protein identification. While some are sold as commercial products, most are freely available over the Web (see Tab. 1 for a partial listing). Nearly all of the packages allow one to select a protein database (OWL, SWISS-PROT, or NCBI-nr), a source organism (to limit the search), a cleavage enzyme (trypsin is the most common), a cleavage tolerance (1 missed cleavage per peptide is usual), a mass tolerance (0.1 amu is typical), and a mass type (average or monoisotopic). Most of these values are pre-selected as defaults in the submission form and do not normally need to be changed. All packages expect users to enter a list of masses (with at least 2 decimal point accuracy) read from the MS spectrum prior to launching the search. On most days a Web search result can be returned within 10–20 s.

When a peptide mass is read off from an MS spectrum, it is important to remember that mono-isotopic masses (the mass of the most abundant isotope for a given peptide) are only accurately readable from 500–3,000 $m z^{-1}$. Peaks with mass-to-charge ratios above 3,000 generally correspond to an average mass and not to a mono-isotopic mass. Key to performing any successful peptide mass fingerprint search is to start with the most accurate masses possible. Internally calibrated mono-isotopic standards are essential. If one is very confident in the mass accuracy, then restricting the mass tolerance setting to less than 0.1 amu will generally improve the specificity of the search. Restricting the size of the database to search is also wise. In most cases the organism being studied is known, and so it is best to select only the portion of the protein database with protein sequences from the source organism (or very closely related organisms). It is not (yet) a good idea to search through translated EST databases as they have too many sequencing errors and contain only partial protein sequence information.

As a general rule one should try to use as many mass values as possible when performing an MS fingerprint analysis. An absolute minimum of 5, but more commonly 10 mass values should be entered to positively identify a protein. Typically the number of masses one enters should correspond to the molecular weight of the protein in kilodaltons (kDa). The need for so much mass data is primarily to compensate for the fact that experimental MS data is inherently "noisy". Indeed, it is not uncommon to have up to half of all predicted peptide peaks absent from any given MS fingerprint spectrum along with any number of additional peaks arising from contaminating proteins. Consequently, one is usually quite content to get coverage (the fraction of predicted peptide masses closely matching with observed peptide masses) of only 40–50%. It is very rare to see a perfect match or 100% coverage.

Because of the experimental noise associated with MS data, the analysis of peptide fingerprint searches is not always easy. Some of the common complications include:

(1) disappearance of key peaks due to non-specific ion suppression,
(2) appearance of extra peaks from protease autolysis,
(3) appearance of peaks from post-translational or artifactual chemical modification,
(4) appearance of peaks from non-specific cleavage, or from contaminating proteases, and
(5) appearance of peaks from contaminating impurities, contaminating homologs, or splice variants.

Because of these complications, the issue of how to score and rank peptide mass matches is actually quite critical to the performance and reliability of peptide mass fingerprint software. Most early fingerprinting programs used simple heuristic scoring schemes and arbitrary cut-offs to select candidate sequences. More recent programs such as Profound (ZHANG and CHAIT, 1995) use Bayesian statistics and posterior probabilities to rank database candidates. Some of the latest programs take at least some of the complications listed above into account and allow for secondary searches with so-called "orphan" masses. Mascot (PERKINS et al., 1999) is one program which uses a probabilistic model similar to an expectation or E-value to rank sequences. The use of probabilities allows a better estimation of significance (which guards against false positives). It also permits scores to be compared with other types of search algorithms (such as BLAST). Regardless of the advantages and disadvantages of individual programs, it is generally a good idea to run several different peptide mass fingerprinting programs and combine the results. This serves as a form of "signal averaging" and potentially reduces the occurrence of errors arising from algorithmic or database limitations in any single program.

Because peptide mass fingerprinting does not always work for unambiguous protein identification there has been increasing emphasis on using tandem mass spectrometers equipped with collision induced dissociation (CID) cells to provide more precise and interpretable peptide data. SEQUEST (ENG et al., 1994; YATES et al., 1996) and Mascot (PERKINS et al., 1999) are two software packages that can be used to analyze tandem mass data of peptide fragments. Both programs take uninterpreted tandem mass spectral data (i.e., the actual spectrum), perform sequence database searches, and identify probable peptides or protein matches. Typically these programs work by first scanning the protein databases for potential matches to the precursor peptide ion, then ranking the candidate peptides on the basis of their predicted similarity (ion continuity, intensity, etc.) to the observed fragment ion masses. After this screening step a model MS/MS spectrum for each candidate peptide is generated and then compared, scored, and ranked with the observed MS/MS spectrum using correlational or probabilistic analysis. As with peptide mass fingerprinting, similar kinds of information (database, source organism, mass tolerance, cleavage specificity, etc.) must be provided before running the programs. The only difference is that instead of typing in a list of masses, the user is expected to provide a spectral filename containing the digitized MS/MS spectrum. Overall, the reported performance of both programs is quite impressive (YATES et al., 1996; PERKINS et al., 1999).

It is likely that protein identification via mass spectral analysis will continue to grow in popularity and in importance. The wide availability of easy-to-use, freely available peptide mass fingerprinting software has made the entire protein identification process very accessible. Furthermore, as more protein sequence data is deposited into sequence data banks around the world, the utility of these database-driven techniques is expected to grow accordingly. While sequence databases continue to grow, mass spectrometer technology is also progressing rapidly. With continuing improvements in mass resolution (i.e., Fourier Transform Cyclotron Mass Spectrometers with 1 ppm resolution are now available) it is likely that peptide mass fingerprinting will become less common as only a single tryptic peptide will be sufficient to positively identify a protein (GOODLETT et al., 2000).

2.3 Protein Identification from Sequence Data

The most precise and accurate way of unambiguously identifying a protein is through its sequence. Historically proteins were identified by direct sequencing using painstakingly difficult chemical or enzymatic methods (Edman degradation, proteolytic digests). All that changed with the development of DNA sequencing techniques which proved to be faster, cheaper, and more robust (SANGER et al., 1977). Now more than 99% of all protein sequences deposited in databases such as OWL (BLEASBY et al., 1994), PIR (BARKER et al., 2000), SWISS-PROT + trEMBL (BAIROCH and APWEILER, 2000, and GenBank (BENSON et al., 2000) are derived directly from DNA sequence data. While complete sequence data is normally obtained via DNA sequencing, improvements in mass spectrometry and chemical microsequencing now allow for routine sequencing of short (10–20 residue) peptides from subpicomole quantities of protein (SHEVCHENKO et al., 1997). With the availability of several different rapid sequencing methods (MS/MS, chemical microsequencers, DNA sequencers, ladder sequencing, etc.) and the growing number of protein sequences (>500,000) and sequence databases, there is now increasing pressure to develop and use specific bioinformatics tools to facilitate protein identification from partial or homologous sequence data.

Protein identification via sequence analysis can be performed either through exact substring matches or through local sequence similarity to a database of known protein sequences (ref). Exact matching of short peptide sequences to known protein sequences is ideal for identifying proteins from partial sequence data (obtained via Edman microsequencing or tandem MS). This type of text matching to sequence data is currently supported by the OWL, SWISS-PROT, and PIR Web servers (but not GenBank!). Given the current size of the databases and the number of residues they contain, it is usually wise to sequence 7 or 8 residues to prevent the occurrence of false positives. Alternately, if some information about the protein mass, predicted pI, or source organism is available, only 4 or 5 residues need to be determined to guarantee a unique match. Note that exact string matching will only identify a protein if it is already contained in a sequence database.

While exact string matching is useful for certain types of protein identification problems, by far the most common method for protein identification is through "fuzzy matching" via sequence similarity. Unlike exact string matching, sequence similarity is a robust technique which allows proteins to be identified even if there are sequencing errors in either the query or database sequence. Furthermore, sequence similarity allows one to potentially identify or ascribe a function to a protein even if it is not contained in the database. In particular, the identification of a similar sequence (> 25% sequence identity to the query sequence) with a known function or name is usually sufficient to infer the function or name of an unknown protein (DOOLITTLE and BORK, 1993).

Sequence similarity is normally determined using database alignment algorithms wherein a query sequence is aligned and compared against all other sequences in a database. In many respects sequence alignment programs are just glorified spell checkers. Fundamentally there are two types of sequence alignment algorithms: dynamic programing methods (NEEDLEMAN and WUNSCH, 1970; SMITH and WATERMAN, 1981) and heuristic "fast" algorithms such as FASTA and BLAST (PEARSON and LIPMAN, 1988; ALTSCHUL et al., 1990). Both methods make use of amino acid substitution matrices such as PAM-250 and Blossum 62 (DAYHOFF et al., 1983; HENIKOFF and HENIKOFF, 1992) to score and assess pairwise sequence alignments. Dynamic programing methods are very slow N^2 type algorithms that are guaranteed to find the mathematically optimal alignment between any two sequences. On the other hand, heuristic methods such as FASTA and BLAST are much faster N-type algorithms that find short local alignments and attempt to string these local alignments into a longer global alignment. Heuristic algorithms make use of statistical models to rapidly assess the significance of any local alignments, making them particularly useful for biologists trying to understand the significance of their matches. Exact descriptions and detailed assessments of these algorithms are beyond the scope of this chapter, but suffice it to say that BLAST and its successors such as FASTA3 (PEARSON, 2000); BLAST2 and PSI-BLAST (ALTSCHUL et al., 1997) have probably become the most commonly used "high-end" tools in all of biology.

BLAST-type searches are generally available for all major protein databases through a variety of mirror sites and Web servers (see Tab. 1). Most servers offer a range of databases which can be "BLASTed". The largest and most complete database is GenBank's nonredundant (nr) protein database, which is largely equivalent to the translated EMBL (TREMBL) database. The second largest, and one that is frequently used in mass fingerprinting, is the OWL database. This non-redundant database is updated every two months. The Swiss-Prot database is the most completely annotated protein database, but does not contain the quantity of sequence data found in OWL or GenBank. The PIR database, which was started in the 1960s, is actually the oldest protein sequence database and contains many protein sequences determined through direct chemical or MS methods (which are typically not in GenBank records). Most of these protein databases can be freely downloaded by academics, but industrial users must pay a fee.

The new version of BLAST (BLAST2) offers several improvements over the original BLAST program, particularly in its ability to create longer, near global alignments from preliminary local alignments. However, in terms of generating global alignments, FASTA3 is probably the best program to use. BLAST2 and FASTA3 are particularly good at identifying sequence matches sharing between 25% to 100% identity with the query sequence. PSI-BLAST (position-specific iterated BLAST), on the other hand, is exceptionally good at identifying matches in the so-called twilight zone of between 15–25% sequence identity. PSI-BLAST can also identify higher scoring similarities with the same accuracy as BLAST2. The trick to using PSI-BLAST is to repeatedly press the "Iterate" button until the program indicates that it has converged. Apparently many first-time users of PSI-BLAST fail to realize this by running the program only once and coming away with little more than a regular BLAST2 output. Nevertheless, because of its near universal applicability, PSI-BLAST is probably the best all-round tool for protein identification from sequence analysis.

The stunning success that PSI-BLAST has had in "scraping the bottom of the barrel" in terms of its ability to identify sequence relationships is leading to increased efforts by bioinformaticians aimed at trying to develop methods to identify even more remote sequence similarities from database comparisons. This has led to the development of a number of techniques such as threading, neural network analysis, and Hidden Markov Modeling – all of which are aimed at extracting additional information hidden in the sequence databases. Many of these techniques are described in more detail in the following section.

3 Protein Property Prediction

Up to this point we have focused on how to identify a protein either from a spot on a gel, an MS fingerprint, or through DNA or protein sequencing. Once the identification problem has been solved, one is usually interested in finding out what this protein does and how/where it works. If a BLAST, PSI-BLAST, or PUBMED search turns up little in the way of useful information, it is still possible to employ a variety of bioinformatics tools to learn something directly from the protein's sequence. Indeed, as we shall see in the following pages, protein property prediction methods can often allow one to make a very good guess as to the function, location, structure, shape, solubility, and binding partners of a novel protein long before one has even lifted a test-tube.

3.1 Predicting Bulk Properties (*p*I, Absorptivity, MW)

While the amino acid sequence of a protein largely defines its structure and function, a protein's amino acid composition can also provide a great deal of information. Specifically,

amino acid composition can be used to predict a variety of bulk protein properties such as isoelectric point, UV absorptivity, molecular weight, radius of gyration, partial specific volume, solubility, and packing volume – all of which can be easily measured on commonly available instruments (gel electrophoresis systems, columns, mass spectrometers, UV spectrophotometers, amino acid analyzers, ultracentrifuges, etc.). Knowledge of these bulk properties can be particularly useful in cloning, expressing, isolating, purifying or characterizing any newly identified protein.

Many of these bulk properties can be calculated using simple formulas and commonly known parameters, some of which are presented in Tabs. 2 and 3. Typical ranges found in water-soluble globular proteins are also shown in Tab. 2. A large number of these calculations can also be performed with more comprehensive protein bioinformatics packages such as SEQSEE (WISHART et al., 1994, 2000) and ANTHEPROT (DELEAGE et al., 1988) as well as many commonly available commercial packages (GCG, LaserGene99, PepTool, VectorNTI).

3.2 Predicting Active Sites

As more and more protein sequences are being deposited into data banks, it is becoming increasingly obvious that certain amino acid residues remain highly conserved even among diverse members of protein families. These highly conserved sequence patterns are often called signature sequences and in many cases they define the active site of a protein. Because most signature patterns are relatively short (7–10 residues) this kind of sequence information is not easily detected from BLAST or FASTA searches. Consequently, it is always a good idea to scan against a signature sequences database (such as PROSITE) in an effort to detect additional information concerning a protein's structure, function, or activity.

Active site or signature sequence databases come in two varieties:

(1) pattern-based and
(2) profile-based.

Tab. 2. Formulas for Protein Property Prediction

Property	Formula	Typical Range
Molecular weight	$MW = \sum A_i * W_i + 18.01056$	N/A
Net charge (pI)	$Q = \sum A_i/(1 + 10**pH - pK_i)$	N/A
Molar absorptivity	$\varepsilon = (5690 \cdot \#W + 1280 \cdot \#Y)/MW$	N/A
Average hydrophobicity	$AH = \sum A_i * H_i$	($AH = -2.5 \pm 2.5$)
Hydrophobic ratio	$RH = \sum H(-)/\sum H(+)$	($RH = 1.3 \pm 0.5$)
Linear charge density	$\sigma = (\#K + \#R + \#D + \#E + \#H + 2)/N$	($\sigma = 0.25 \pm 0.5$)
Solubility	$\prod = RH + AH + \sigma$	($\prod = 1.6 \pm 0.5$)
Protein radius	$R = 3.875*(N**0.333)$	N/A
Partial specific volume	$PSV = \sum PS_i * W_i$	($PSV = 0.725 \pm 0.025$)
Packing volume	$VP = \sum A_i * V_i$	N/A
Accessible surface area	$ASA = 7.11*MW**0.718$	N/A
Unfolded ASA	$ASA(U) = \sum A_i * ASA_i$	N/A
Buried ASA	$ASA(B) = ASA(U) - ASA$	N/A
Interior volume	$V_{int} = \sum A_i * FB_i * V_i$	N/A
Exterior volume	$V_{ext} = VP - V_{int}$	N/A
Volume ratio	$VR = V_{ext}/V_{int}$	N/A
Fisher volume ratio	$FVR = [R**3/(R-4.0)**3] - 1$	N/A

MW: molecular weight in Da; A_i: number of amino acids of type i; W_i: molecular weight of amino acid type i; Q: charge; pK_i: pKa of amino acid type i; e: molar absorptivity; $\#W$: number of tryptophans; H_i: Kyte Doolittle hydropathy; $H(-)$: (hydrophilic residue hydropathy values; $H(+)$: hydrophobic residue hydropathy values; N: total number of residues in the protein; R: radius in Å; PS_i: partial specific volume of amino acid type i; V_i: volume in cubic Å of amino acid type i; ASA_i: accessible surface area in square Å of amino acid type i; FB_i: fraction buried of amino acid type i; see Tab. 3 for residue specific values of many of these parameters.

In general, pattern-based sequence motifs are the easiest to work with as they can be easily entered as simple regular expressions. Typically pattern-based sequence motifs are identified or confirmed through careful manual comparisons of multiply aligned proteins – most of which are known to have a specific function, active site, or binding site (BAIROCH, 1991). Profile-based signatures or "sequence profiles" are usually generated as a combination of amino acid and positional scoring matrices derived from multiple sequence alignments (GRIBSKOV et al., 1987). While sequence profiles are generally more robust than regular pattern expressions in identifying active sites, the effort required to prepare good sequence profiles has generally precluded their widespread adoption in the bioinformatics community.

The most extensive and best documented signature sequence database is PROSITE (HOFMANN et al., 1999). It currently has 1,386 signature sequences and sequence profiles as well as extensive bibliographic and statistical information on all of its signature data. PRO-

SITE can be accessed through a variety of commercial bioinformatics programs as well as a number of freely available Web servers (Tab. 4). The database itself can also be downloaded and run locally. PROSITE is by no means the only active site or signature sequence site database available. SEQSITE (WISHART et al., 1994) along with a number of smaller databases have also been published or made available over the years (HODGMAN, 1989; SETO et al., 1990; OGIWARA et al., 1992). More recently, a new kind of signature sequence database has started appearing in the literature. These databases are composed of automated compilations of multiply aligned sequence fingerprints or sequence blocks. PRINTS (ATTWOOD et al., 1999), BLOCKS (HENIKOFF et al., 1999), and the Pfam database (BATEMAN et al., 2000) – all of which have Web access – are examples of a few of these family signature databases. Although not quite as useful or fully annotated as PROSITE, they can be helpful in identifying sequence motifs, active sites, or signature sequences that may not be contained in PROSITE.

Tab. 3. Amino Acid Residue Properties (molecular weight [monoisotopic mass], frequency, pKa, absorbance at 280 nm, hydrophobicity, partial specific volume [ml/g], packing volume [Å^3], accessible surface area [Å^2], fraction buried)

AA	MW (Da)[a]	ν (%)	pKa	ε_{280}	Hphb[b]	PS[c]	Vol.[d]	ASA[e]	FB[f]
A	71.03712	8.80	–	–	1.8	0.748	88.6	115	0.38
C	103.00919	2.05	10.28	–	2.5	0.631	108.5	135	0.45
D	115.02695	5.91	3.65	–	−3.5	0.579	111.1	150	0.15
E	129.04260	5.89	4.25	–	−3.5	0.643	138.4	190	0.18
F	147.06842	3.76	–	–	2.8	0.774	189.9	210	0.50
G	57.02147	8.30	–	–	−0.4	0.632	60.1	75	0.36
H	137.05891	2.15	6.00	–	−3.2	0.670	153.2	195	0.17
I	113.08407	5.40	–	–	4.5	0.884	166.7	175	0.60
K	128.09497	6.20	10.53	–	−3.9	0.789	168.6	200	0.03
L	113.08407	8.09	–	–	3.8	0.884	166.7	170	0.45
M	131.04049	1.97	–	–	1.9	0.745	162.9	185	0.40
N	114.04293	4.58	–	–	−3.5	0.619	117.7	160	0.12
P	97.05277	4.48	–	–	−1.6	0.774	122.7	145	0.18
Q	128.05858	3.84	–	–	−3.5	0.674	143.9	180	0.07
R	156.10112	4.22	12.43	–	−4.5	0.666	173.4	225	0.01
S	87.03203	6.50	–	–	−0.8	0.613	89.0	115	0.22
T	101.04768	5.91	–	–	−0.7	0.689	116.1	140	0.23
V	99.06842	7.05	–	–	4.2	0.847	140.0	155	0.54
W	186.07932	1.39	–	5690	−0.9	0.734	227.8	255	0.27
Y	163.06333	3.52	–	1280	−1.3	0.712	193.6	230	0.15

[a] BEIMANN, 1990; [b] KYTE and DOOLITTLE, 1992; [c] ZAMAYATNIN, 1972; [d, e] RICHARDS, 1977; [f] CHOTHIA, 1976.

3.3 Predicting Modification Sites

Post-translational modification, such as proteolytic cleavages, phosphorylation, glycosylation, and myristilation can greatly affect the function and structure of proteins. The identification of these sites can be quite helpful in learning something about the function, preferred location, probable stability, and possible structure of a given protein. Furthermore, knowledge of the location or presence of these modification sites can assist in selecting expression systems and designing purification protocols. Most post-translational modifications occur at specific residues contained within well-defined sequence patterns. Consequently, many of these modification sites are contained in the PROSITE database and can be detected through a simple PROSITE scan. However, for glycosylation and phosphorylation, the sequence patterns are generally less-well defined and the use of simple sequence patterns can lead to many false positives and

false negatives. Neural networks trained on databases of known phosphorylation and glycosylation have been shown to have somewhat better specificity (>70%) in finding these hard-to-define sites (BLOM et al., 1999; HANSEN et al., 1998). NetPhos and NetOGlyc (Tab. 4) are examples of two Webservers that offer neural network identification of potential phosphorylation and O-glycosylation sites on proteins.

3.4 Finding Protein Interactions and Pathways

Sequence information, alone, can rarely provide sufficient information to indicate how or where a protein interacts with other proteins or where it sits within a given metabolic pathway. While detailed literature surveys and careful keyword searches can be of some help, there are now a number of freely available databases that allow one to query, visualize, and identify protein–protein interactions in the context of metabolic or signaling pathways. Some of these sites, including KEGG (Kyoto Encyclopedia of Genes and Genomes), STKE (Signal Transduction Knowledge Environment), and ProNet (Protein Network Database) have their URLs listed in Tab. 4. Most of these servers provide visual descriptors, textual information, and hyper-linked pointers to help understand the role a protein plays in a given pathway or interaction. Because many of these databases are quite new, they do not yet have the depth of coverage needed to make them broadly useful. Nevertheless, this kind of information can occasionally be quite revealing, particularly if one is trying to gain some "context" about why a particular protein is expressed, how it is being regulated, or where it is being found. In the not-to-distant future one can expect that more comprehensive protein interaction databases will appear and these will, in all likelihood, become just as important as sequence databases in helping biologists sort through the ever-growing mass of biological data being generated worldwide.

3.5 Predicting Location or Localization

Many proteins have sequence-based signals that will localize the protein to certain regions of the cell. For instance, proteins with transmembrane helices will end up in the lipid bilayer, proteins that are rich in positively charged residues will end up in the nucleus and proteins with specific signal peptides will be exported outside the cell (NIELSEN et al., 1999). Being able to identify a signaling or localization sequence can help in understanding the function, probable location, and biological context of a newly identified protein. This information can also be quite useful in designing cloning, purification, and separation protocols. Many of these signaling sequences are contained in the PROSITE database and so a simple PROSITE scan can be quite revealing. However, not all localization sequences are easily defined as simple PROSITE patterns. To cover those signaling sequences that are not so easily defined a protein sorting server has

been developed that makes use of a number of previously known sorting sequences and patterns to predict the location or localization of new sequences. Specifically, the PSORT server (Tab. 4) identifies probable sorting signals using Hidden Markov Models and nearest neighbor classifiers (HORTON and NAKAI, 1997).

In addition to the PSORT server, one may also employ a variety of other programs to identify those proteins that localize to the cell membrane (which account for up to 30% of all proteins). Transmembrane helix prediction historically was done using hydropathy or hydrophobicity plots (ENGELMAN et al., 1986; KYTE and DOOLITTLE, 1982). However, this graphical technique often proved to be inconsistent and unreliable. The introduction of neural network based approaches combined with multiple sequence alignments (or evolutionary information) has greatly improved the quality and reliability of transmembrane helix prediction (ROST et al., 1995). Indeed, transmembrane helices can often be identified with an accuracy exceeding 95%. This makes the identification of membrane proteins one of the more robust and reliable predictive methods in all of bioinformatics. Interestingly, while transmembrane helices are quite predictable, trans-membrane beta-sheets (as found in porins) are not. This problem continues to be one of the unmet challenges in membrane protein sequence analysis.

3.6 Predicting Stability, Globularity, and Shape

Prior to cloning or expressing a protein or protein fragment, obviously one would like to know whether it will be soluble, stable, and globular. While there are some crude predictors based on average hydrophobicity, localized hydrophobicity, hydrophobic ratios, charge density, and secondary structure, the prediction of protein solubility and expressibility is still on rather shaky ground (WILKINSON and HARRISON, 1991). Nevertheless, with more and more data being compiled and analyzed from various structural proteomics efforts around the globe, a clearer idea is being obtained about the key sequence/property features that determine the likelihood of successful, high-yield expression. One example of a protein sequence feature that determines stability is the so-called PEST sequence (RECHSTEINER and ROGERS, 1996). Eukaryotic proteins with intracellular half-lives of less than 2 h are often found to contain regions rich in proline, glutamic acid, serine, and threonine (P, E, S, and T). These PEST regions are generally flanked by clusters of positively charged amino acids. Therefore, the identification of PEST sequences in proteins can be a very important consideration in any protein expression project. Two PEST web servers (Tab. 4) are available to identify PEST sequences using a standard set of pattern-based rules. Similar information about intracellular protein lifetimes can be extracted using the N-end rule for protein ubiquitation (BACHMAIR et al., 1986).

The propensity of a protein to fold into a particular shape or to fold into a globule can also be predicted. Indeed, as far back as 1964 a very simple, but elegant procedure was developed to predict the shape and globularity of proteins based on simple packing rules, hydrophobicity, and amino acid volumes (FISHER, 1964). Specifically, if the calculated volume ratio for a given protein (see Tabs. 2 and 3) is slightly greater than the theoretical Fisher volume ratio, then the protein likely forms a soluble monomer. If the calculated volume ratio is much greater than the Fisher volume ratio, the protein likely does not form a compact globular structure (i.e., it is filamentous or unfolded). This procedure, with a few modifications, has been implemented in both the SEQSEE and PepTool software packages (WISHART et al., 1994, 2000). A more sophisticated approach to solubility/globularity prediction has been developed by Burkhard Rost. This technique uses evolutionary information, neural networks, and predicted protein accessibility to determine whether or not a protein will form a globular domain. The Web server (GLOBE) is located in the PredictProtein Web site listed in Tab. 4.

3.7 Predicting Protein Domains

Larger proteins will tend to fold into structures containing multiple domains. Typically domains are defined as contiguous stretches of 100–150 amino acids that have a globular fold or a function that is distinct from other parts of the protein. Many eukaryotic proteins are composed of multiple subunits or domains, each having a different function or a different fold. If one can identify (through sequence analysis) the location or presence of well-folded, well-defined domains, it is often possible to gain a greater understanding not only of the probable function, but also of the evolutionary history of that protein. Furthermore, the identification of domains can often allow one to "decompose" a protein into smaller parts to facilitate cloning and expression and to increase the likelihood that the protein, or parts of it, could be studied by X-ray or NMR techniques.

As with active-site identification, domain identification is typically performed using comparisons or alignments to known domains. Several databases now exist including: Domo (GRACY and ARGOS, 1998), Pfam (BATEMAN et al., 2000), and Prodom (CORPET et al., 2000) all of which have their URLs listed in Tab. 4. These represent compilations of protein domains that have been identified both manually and automatically through multiple sequence alignments and hierarchical clustering algorithms. All three databases can provide quick and reliable identification of most globular protein domains. However, some domains seem to defy routine identification. For instance, coiled-coil domains, with their non-descript sequence character and non-globular nature have had to have specialized software developed to aid in their identification (LUPAS, 1996). Several coiled-coil prediction services are also available over the Web (Tab. 4).

3.8 Predicting Secondary Structure

The primary structure (the sequence) determines both the secondary structure (helices and beta-strands) and the tertiary structure (the 3D fold) of a protein. In principle, if you know a protein's sequence, you should be able to predict both its structure and function. Secondary structure prediction offers a route to gaining some insight into the probable folds, expected domain structure, and possible functions of proteins (DELEAGE et al., 1997). Because secondary structure is more conserved than primary structure, secondary structure prediction may also be used to facilitate remote homolog detection or protein fold recognition (ROST et al., 1997).

Secondary structure prediction has been under development for more than 30 years. Consequently there are large number of different techniques available having widely varying degrees of accuracy and utility. The simplest approaches are statistical (CHOU and FASMAN, 1974) wherein intrinsic amino acid probabilities for being in helices and beta-strands are simply averaged over varying window sizes. Amino acid segments with the highest local scores for a particular secondary structure are then assigned to that structure. However, secondary structure generally depends on more than just averaged conformational preferences of individual amino acids. Sequence patterns, positional preferences, and pairwise interactions also play an important role. To account for these effects more sophisticated predictive approaches have had to be developed. These include information theoretic or Bayesian probabilistic approaches (GARNIER et al., 1978), stereochemical methods (LIM, 1974), nearest neighbor or database comparison techniques (LEVIN et al., 1986), and neural network approaches (QIAN and SEJNOWSKI, 1988). Typically these methods achieve a three-state (helix, sheet, coil) accuracy of between 55% (for the simplest statistical methods) to 65% (for the best neural network or nearest neighbor approaches) for water-soluble globular proteins.

A significant improvement in secondary structure prediction accuracy occurred in the early 1990s with the introduction of combined approaches that integrated multiple sequence alignments (i.e., evolutionary information) with neural network pattern recognition methods (ROST, 1996). This innovation allowed secondary structure prediction accuracy to improve to better than 72%. Similar efforts aimed at integrating evolutionary information with nearest neighbor approaches also led to comparable improvements in prediction accuracy (LEVIN, 1997). Most recently, the integration of better database searching methods (PSI-BLAST) has allowed protein secondary

prediction accuracy to approach 77% (JONES, 1999). Many of these "new and improved" secondary structure prediction methods are now freely available over the Web (see Tab. 4).

Given their wide availability and much improved reliability, secondary structure predictions should now be considered as an integral component to any standard protein sequence analysis. With steady, incremental improvements in prediction accuracy occurring every one or two years, it is likely that secondary structure prediction will soon achieve an accuracy in excess of 80%. At this level of accuracy it may be possible to use secondary structure predictions as starting points for 3D structure prediction.

3.9 Predicting 3D Folds (Threading)

Threading is a protein fold recognition or structure prediction technique that got its name because it conceptually resembles the method used to thread electrical cables through a conduit. Specifically threading involves placing or threading an amino acid sequence onto a database of different secondary or tertiary structures (pseudo-conduits). As the sequence is fed through each structure its fit or compatibility to that structure is evaluated using a heuristic potential. This evaluation may be done quickly using some empirical "energy" term or some measure of packing efficiency or secondary structure propensity. In this way it is possible to assess which protein sequences are compatible with the given backbone fold. While one would clearly expect that those sequences homologous to the original template sequence should fit best, it has been found that this approach can occasionally reveal that some seemingly unrelated sequences can also fit into previously known folds. One striking example of how successful threading can be in structure and function prediction can be found with the protein called leptin (the hormone responsible for obesity). In particular, standard threading techniques were able to show that leptin was a helical-like cytokine long before any confirmatory X-ray structure had been determined or biological receptors had been found (MADEJ et al., 1995).

Two approaches to threading exist. One is called 3D threading, which is classified as a distance-based method (DBM). The other is called 2D threading, which is classified as a prediction-based method (PBM). 3D threading was first described in the early 1980s (NOVOTNY and BRUCCOLERI, 1984) and later "rediscovered" about 10 years later (JONES et al., 1992; SIPPL and WEITCKUS, 1992; BRYANT and LAWRENCE, 1993) when the concept of heuristic potential functions matured. 3D threading uses distance-based or profile-based (BOWIE et al., 1991) energy functions and technically resembles the "pipe" threading description given earlier. In 3D threading, coordinates corresponding to the hypothesized protein fold are actually calculated and the energy functions evaluated on the basis of these 3D coordinates.

Just like 3D threading, 2D threading was first described in the mid 1980s (SHERIDAN et al., 1985) and then rediscovered in the mid 1990s (FISCHER and EISENBERG, 1996; RUSSEL et al., 1996; ROST et al., 1997) when the reliability of secondary structure predictions started improving. Rather than relying on 3D coordinates to evaluate the quality of a fold, 2D threading actually uses secondary structure (hence the name 2D) as the primary evaluation criterion. Indeed, 2D threading is based on the simple observation that secondary structure is more conserved than primary structure (sequence). Therefore, proteins that have lost detectable similarity at the sequence level, could still be expected to maintain some similarity at the secondary structure level.

Over the past few years 2D threading has matured so that secondary structure, solvent accessibility, and sequence information can now be used in the evaluation process. The advantage that 2D threading has over 3D threading is that all this structural information can be encoded into a 1D string of symbols (i.e., a pseudo-sequence). This allows one to use standard sequence comparison tools, like dynamic programing, to rapidly compare a query sequence and/or secondary structure to a database of sequences and secondary structures. Consequently, 2D threading is 10 to 100 times faster than the distance-based 3D threading approach and seems to give comparable (and in some cases, even better) results than 3D threading. The fact that the 2D threading algorithm is relatively simple to understand and to implement has led to the development of a number of freely available 2D threading servers (Tab. 4).

Typically the best 2D threading methods score between 30 and 40% when working with "minimal" databases. If the structural databases are expanded to include more than one related fold representative, the performance can be as high as 70–75%. As already mentioned 2D threading performs about as well as 3D threading, however, it is much faster and easier to implement. It is generally thought that, if 2D threading approaches could improve their secondary structure prediction accuracy and include more information about the "coil" state (such as approximate dihedral angles) then even further performance gains could be realized. Similarly, if the initial 2D threading predictions could be verified using 3D threading checks (post-threading analysis) and further tested by looking at other biochemical information (species of origin, known function, ligand contacts, binding partners) then additional improvements should be possible.

3.10 Comprehensive Commercial Packages

Commercial packages can offer an attractive alternative to many of the specialized or single-task analyses offered over the Web. In particular, these commercial tools integrate sophisticated graphical user interfaces (GUIs) with a wide range of well-tested analytical functions, databases, plotting tools, and well-maintained user-support systems to make most aspects of protein sequence analysis simpler, speedier, and "safer" than is possible with Web-based tools. However, commercial packages also have their disadvantages. Most are quite pricey ($1,000–$3,000) and most do not offer the range or even the currency of analytical functions available through the Web. Furthermore, most commercial packages are very platform-specific, meaning that they can only run on certain types of computers or operating systems. This is not a limitation for Web-based tools, as most are platform-independent.

Tab. 4. Protein Prediction Tools – Web Links

Tool/Database	Web Address
Compute pI/Mw	*http://www.expasy.ch/tools/pi_tool.html*
Scan ProSite	*http://www.expasy.ch/tools/scnpsite.html*
PROSITE database	*http://www.expasy.ch/prosite*
BLOCKS database	*http://www.blocks.fhcrc.org/*
PRINTS database	*http://bmbsgi11.leeds.ac.uk/bmb5dp/prints.html*
Pfam database	*http://pfam.wustl.edu/*
NetPhos server	*http://www.cbs.dtu.dk/services/NetPhos/*
NetOGlyc server	*http://www.cbs.dtu.dk/services/NetOGlyc/*
PredictProtein (PHDhtm)	*http://cubic.bioc.columbia.edu/predictprotein/*
TMpred (membrane helix)	*http://www.ch.embnet.org/software/TMPRED_form.html*
PSORT server	*http://psort.nibb.ac.jp/*
KEGG database	*http://www.genome.ad.jp/kegg/regulation.html*
STKE database	*http://www.stke.org*
ProNet database	*http://pronet.doubletwist.com*
PEST searcher	*http://www.icnet.uk/LRITu/projects/pest/*
PESTfind	*http://www.at.embnet.org/embnet/tools/bio/PESTfind/*
PredictProtein (GLOBE)	*http://cubic.bioc.columbia.edu/predictprotein/*
ProDom database	*http://protein.toulouse.inra.fr/prodom.html*
Domo database	*http://www.infobiogen.fr/services/domo/*
COILS (coil-coil prediction)	*http://www.ch.embnet.org/software/COILS_form.html*
PredictProtein (Coils)	*http://cubic.bioc.columbia.edu/predictprotein/*
PredictProtein (PHD)	*http://cubic.bioc.columbia.edu/predictprotein/*
PREDATOR (2° prediction)	*http://www.embl-heidelberg.de/cgi/predator_serv.pl*
PSIpred (2° prediction)	*http://insulin.brunel.ac.uk/psipred*
Fischer Method (Threading)	*http://www.cs.bgu.ac.il/~bioinbgu/query.html*
PredictProtein (TOPITS)	*http://dodo.cpmc.columbia.edu/predictprotein/*
GenThreader Server	*http://insulin.brunel.ac.uk/psipred/*
3DPSSM (Threading)	*http://www.bmm.icnet.uk/~3dpssm/*
ANTHEPROT	*http://pbil.ibcp.fr/ANTHEPROT/ie_sommaire.html*
SEQSEE	*http://www.pence.ualberta.ca/ftp/seqsee/seqsee.html*

The first commercial packages debuted in the mid 1980s and over the past 15 years they have evolved and improved considerably. Most packages integrate both protein and DNA sequence analysis into a single "suite", although some companies have opted to create separate protein-specific modules. Most commercial packages offer a fairly standard set of protein analysis tools including:

(1) bulk property calculations (molecular weight, pI),
(2) physicochemical property plotting (hydrophobicity, flexibility, hydrophobic moments),
(3) antigenicity prediction,
(4) sequence motif searching or identification,
(5) secondary structure prediction,
(6) database searching (internet or local),
(7) alignment and comparison tools (dot-plots or multiple alignment),
(8) plotting or publication tools, and
(9) multiformat (GenBank, EMBL, SWISS-PROT, PIR) sequence I/O.

It would be impossible to review all of the commercial packages here, but a brief summary of some of the more popular tools is given below.

Oxford Molecular Group (*www.oxmol.com*) offers perhaps the widest range of protein analysis tools including the GCG Wisconsin package (UNIX), Omiga (Windows), and MacVector (MacOS). The Wisconsin package along with its new graphical interfaces (Seq-

Web, SeqLab, and GCGlink to Omiga) is probably the most comprehensive and widely distributed bioinformatics package in the world. It offers an impressive array of standard tools as well as some very good motif detection routines (Meme) and sophisticated sequence profiling algorithms. With literally dozens of protein analysis tools, the GCG Wisconsin package is much more complete than either Omiga or MacVector. However, GCG's interface and user-friendliness still lag well behind industry standards.

DNAstar Inc. (*www.dnastar.com*) produces a highly acclaimed multi-component suite called LaserGene (MacOS, Windows). The protein sequence analysis module (Protean) has a well-conceived GUI and an array of sophisticated plotting and visualization tools. In addition to the usual analytical tools, the Protean module also offers some very innovative facilities for synthetic peptide design, linear structure display, and SDS PAGE gel simulation. Unfortunately, Protean does not offer the usual sequence comparison or alignment tools typically found in most comprehensive packages. Instead these must be obtained by purchasing a second stand-alone module called Megalign.

Informax Inc. (*www.informaxinc.com*) manufactures a very popular package called Vector NTI (Windows, MacOS). Although the emphasis is clearly on DNA analysis, the Vector NTI package also offers a modestly comprehensive set of protein tools presented in an easy-to-use GUI. Vector NTI supports most standard protein analytical functions, but sets itself apart from many other packages by offering integrated internet connectivity to Entrez, PubMed, and BLAST. Vector NTI also supports a nice interactive 3D molecular visualization tool as well as a number of interesting property evaluation functions to calculate free energies, polarity, refractivity, and sequence complexity.

BioTools Inc. (*www.biotools.com*), a relatively recent entry to this area, produces a comprehensive platform-independent protein analysis package called PepTool (Windows, MacOS, UNIX). As with most other commercial packages, PepTool supports all of the major protein analytical functions as well as offering a particularly broad range of statistical and property prediction tools. In addition to being the only package to offer universal platform compatibility, PepTool also has a particularly logical, easy-to-learn interface. Unfortunately, PepTool currently lacks the internet connectivity found in other packages and it does not yet support graphical export or overlays.

A question that is often asked by both novice and experienced users is: Should I choose expensive commercial packages or should I stick with freely available Web tools? There is no straightforward answer. If one is looking for the latest analytical tool or best-performing prediction algorithm, the Web is almost always the best place to find it. Similarly, if one is interested in doing rapid database comparisons or searches, the Web is still the place to go. However, if high quality plotting, graphing, or rendering is important (for papers or presentations) one often has to turn to commercial packages. Helical wheels, colored multiple alignments, overlaid or stacked plots, annotated graphs, graphical secondary structure assignments, etc. are examples of images that cannot yet be rendered (well) on the Web. For many, commercial biotech firms, security, uniformity and reliability are particularly important and so once again, commercial packages – with their resident databases, uniform interfaces and reliable user-support – are definitely the tools of choice.

Most recently, a "compromise" method for delivering bioinformatics services has emerged that seems to combine the advantages of both commercial and Web-based tools. Internet Service Providers (ISPs) and Application Service Providers (ASPs) are commercial vendors that sell or lease access to commercial quality bioinformatics packages and databases through the Web. ISPs and ASPs allow users get access to an array of secure, reliable, and fully supported software without the usual worries of platform incompatibility, local disk storage capacity, or internet traffic jams. These portals also allow a less expensive pay-as-you-go approach to licensing or accessing software. Some of the more popular sites include *www.ebioinformatics.com*, *www.doubletwist.com*, and *www.viaken.com*, with Viaken and Doubletwist offering subscription access to many of the commercial packages just discussed here. It will be interesting to see, if this novel approach to bioinformatics software delivery is widely adopted by the scientific and industrial research community.

Acknowledgement
The author wishes to acknowledge the Protein Engineering Network of Centres of Excellence (PENCE) for financial support.

4 References

ALTSCHUL, S. F., MADDEN, T. L., SCHAFFER, A. A., ZHANG, J., ZHANG, Z. et al. (1997), Gapped BLAST and PSI-BLAST: a new generation of protein database search programs, *Nucleic Acids Res.* **25**, 3389–3402.

ALTSCHUL, S. F., GISH, W., MILLER, W., MYERS, E. W., LIPMAN, D. J. (1990), Basic local alignment search tool, *J. Mol. Biol.* **215**, 403–410.

APPEL, R., HOCHSTRASSER, D., ROCH, C., FUNK, M., MULLER, A. F., PELLEGRINI, C. (1988), Automatic classification of two-dimensional gel electrophoresis pictures by heuristic clustering analysis: a step toward machine learning, *Electrophoresis* **9**, 136–142.

APPEL, R. D., VARGAS, J. R., PALAGI, P. M., WALTHER, D., HOCHSTRASSER, D. F. (1997), Melanie II – a third-generation software package for analysis of two-dimensional electrophoresis images, *Electrophoresis* **18**, 2735–2748.

ATTWOOD, T. K. (2000), The quest to deduce protein function from sequence: the role of pattern databases, *Int. J. Biochem. Cell Biol.* **32**, 139–155.

ATTWOOD, T. K., FLOWER, D. R., LEWIS, A. P., MABEY, J. E., MORGAN, S. R. et al. (1999), PRINTS prepares for the new millennium, *Nucleic Acids Res.* **27**, 220–225.

BACHMAIR, A., FINLEY, D., VARSHAVSKY, A. (1986), *In vivo* half-life of a protein is a function of its amino-terminal residue, *Science* **234**, 179–186.

BAIROCH, A. (1991), PROSITE: A dictionary of sites and patterns in proteins, *Nucleic Acids Res.* **19**, 2241–2245.

BAIROCH, A., APWEILER, R. (2000), The SWISS-PROT protein sequence database and its supplement TrEMBL in 2000, *Nucleic Acids Res.* **28**, 45–48.

BARKER, W. C., GARAVELLI, J. S., HUANG, H., McGARVEY, P. B., ORCUTT, B. C. et al. (2000), The protein information resource (PIR), *Nucleic Acids Res.* **28**, 41–44.

BATEMAN, A., BIRNEY, E., DURBIN, R., EDDY, S. R., HOWE, K. L., SONNHAMMER, E. L. (2000), The Pfam protein families database, *Nucleic Acids Res.* **28**, 263–266.

BENSON, D. A., KARSCH-MIZRACHI, I., LIPMAN, D. J., OSTELL, J., RAPP, B. A., WHEELER, D. L. (2000), GenBank, *Nucleic Acids Res.* **28**, 15–18.

BIEMANN, K. (1990), Appendix 6. Mass values for amino acid residues in peptides, in: *Methods in Enzymology* Vol. 193 (McCLOSKEY, J. A., Ed.), pp. 888. San Diego, CA: Academic Press.

BLEASBY, A. J., AKRIGG, D., ATTWOOD, T. K. (1994), OWL – a non-redundant composite protein sequence database, *Nucleic Acids Res.* **22**, 3574–3577.

BLOM, N., GAMMELTOFT, S., BRUNAK, S. (1999), Sequence- and structure-based prediction of eukaryotic protein phosphorylation sites, *J. Mol. Biol.* **294**, 1351–1362.

BOWIE, J. U., LUTHY, R., EISENBERG, D. (1991), A method to identify protein sequences that fold into a known 3-dimensional structure, *Science* **253**, 164–170.

BRODER, S., VENTER, J. C. (2000), Sequencing the entire genomes of free-living organisms: the foundation of pharmacology in the new millennium, *Annu. Rev. Pharmacol. Toxicol.* **40**, 97–132.

BRYANT, S. H., LAWRENCE, C. E. (1993), An empirical energy function for threading a protein sequence through a folding motif, *Proteins* **5**, 92–112.

CELIS, J. E., OSTERGAARD, M., JENSEN, N. A., GROMOVA, I., RASMUSSEN, H. H., GROMOV, P. (1998), Human and mouse proteomic databases: novel resources in the protein universe, *FEBS Lett.* **430**, 64–72.

CHOTHIA, C. (1976), The nature of the accessible and buried surfaces in proteins, *J. Mol. Biol.* **105**, 1–14.

CHOU, P. Y., FASMAN, G. D. (1974), Prediction of protein conformation, *Biochemistry* **13**, 222–245.

CORPET, F., SERVANT, F., GOUZY, J., KAHN, D. (2000), ProDom and ProDom-CG: tools for protein domain analysis and whole genome comparisons, *Nucleic Acids Res.* **28**, 267–269.

DAYHOFF, M. O., BARKER, W. C., HUNT, L. T. (1983), Establishing homologies in protein sequences, *Methods Enzymol.* **91**, 534–545.

DELEAGE, G., BLANCHET, C., GEOURJON, C. (1997), Protein structure prediction. Implications for the biologist, *Biochimie* **79**, 681–686.

DELEAGE, G., CLERC, F. F., ROUX, B., GAUTHERON, D. C. (1988), ANTHEPROT: a package for protein sequence analysis using a microcomputer, *Comput. Appl. Biosci.* **4**, 351–356.

DOOLITTLE, R. F., BORK, P. (1993), Evolutionarily mobile modules in proteins, *Sci. Am.* **269**, 50–56.

DUTT, M. J., LEE, K. H. (2000), Proteomic analysis, *Curr. Opin. Biotechnol.* **11**, 176–179.

ENG, J. K., McCORMACK, A. L., YATES, J. R. (1994), An approach to correlate tandem mass-spectral data of peptides with amino-acid sequences in a protein database, *J. Am. Soc. Mass Spect.* **5**, 976–989.

ENGELMAN, D. M., STEITZ, T. A., GOLDMAN, A. (1986), Identifying non-polar transbilayer helices in amino acid sequences of membrane proteins, *Annu. Rev. Biophys. Chem.* **15**, 321–353.

FISCHER, D., EISENBERG, D. (1996), Protein fold recognition using sequence-derived predictions, *Protein Sci.* **5**, 947–955.

FISHER, H. F. (1964), A limiting law relating the size and shape of protein molecules to their composition, *Proc. Natl. Acad. Sci. USA* **51**, 1285–1291.

GARNIER, J., OGUSTHORPE, D. J., ROBSON, B. (1978), Analysis of the accuracy and implementation of simple methods for predicting the secondary structure of globular proteins, *J. Mol. Biol.* **120**, 97–120.

GEVAERT, K., VANDEKERCKHOVE, J. (2000), Protein identification methods in proteomics, *Electrophoresis* **21**, 1145–1154.

GOODLETT, D. R., BRUCE, J. E., ANDERSON, G. A., RIST, B., PASA-TOLIC, L. et al. (2000), Protein identification with a single accurate mass of a cysteine-containing peptide and constrained database searching, *Anal. Chem.* **72**, 1112–1118.

GRACY, J., ARGOS, P. (1998), DOMO: a new database of aligned protein domains, *Trends Biochem. Sci.* **23**, 495–497.

GRIBSKOV, M., MCLACHLAN, A. D., EISENBERG, D. (1987), Profile analysis: detection of distantly related proteins, *Proc. Natl. Acad. Sci. USA* **84**, 4355–4358.

HANSEN, J. E., LUND, O., TOLSTRUP, N., GOOLEY, A. A., WILLIAMS, K. L., BRUNAK, S. (1998), NetOglyc: prediction of mucin type *O*-glycosylation sites based on sequence context and surface accessibility, *Glycoconj. J.* **15**, 115–130.

HENIKOFF, S., HENIKOFF, J. G. (1992), Amino acid substitution matrices from protein blocks, *Proc. Natl. Acad. Sci. USA* **89**, 10915–10919.

HENIKOFF, J. G., HENIKOFF, S., PIETROKOVSKI, S. (1999), Blocks +: a non-redundant database of protein alignment blocks derived from multiple compilations, *Biotransformatics* **15**, 471–479

HODGMAN, T. C. (1989), The elucidation of protein function by sequence motif analysis, *Comput. Applic. Biosci.* **5**, 1–13.

HOFMANN, K., BUCHER, P., FALQUET, L., BAIROCH, A. (1999), The PROSITE database, its status in 1999, *Nucleic Acids Res.* **27**, 215–219.

HOOGLAND, C., SANCHEZ, J. C., TONELLA, L., BINZ, P. A., BAIROCH, A. et al. (2000), The 1999 SWISS-2DPAGE database update, *Nucleic Acids Res.* **28**, 286–288.

HORTON, P., NAKAI, K. (1997), Better prediction of protein cellular localization sites with the *k* nearest neighbor classifier, *Intellig. Syst. Mol. Biol.* **5**, 147–152.

JONES, D. T. (1999), Protein secondary structure prediction based on position-specific scoring matrices, *J. Mol. Biol.* **292**, 195–202.

JONES, D. T., TAYLOR, W. R., THORNTON, J. M. (1992), A new approach to protein fold recognition, *Nature* **358**, 86–89.

KYTE, J., DOOLITTLE, R. F. (1982), A simple method for displaying the hydropathic character of a protein, *J. Mol. Biol.* **157**, 105–132.

LEE, B. K., RICHARDS, F. M. (1971), The interpretation of protein structures: estimation of static accessibility, *J. Mol. Biol.* **55**, 379–400.

LEMKIN, P. F., THORNWALL, G. (1999), Flicker image comparison of 2-D gel images for putative identification using the 2DWG meta-database, *Mol. Biotechnol.* **12**, 159–172.

LEVIN, J. M. (1997), Exploring the limits of nearest neighbour secondary structure prediction, *Protein Eng.* **10**, 771–776.

LEVIN, J. M., ROBSON, B., GARNIER, J. (1986), An algorithm for secondary structure determination in proteins based on sequence similarity, *FEBS Lett.* **205**, 303–308.

LIM, V. I. (1974), Algorithms for prediction of helices and beta-structural regions in globular proteins, *J. Mol. Biol.* **88**, 873–894.

LINK, A. J., ENG, J., SCHIELTZ, D. M., CARMACK, E., MIZE, G. J. et al. (1999), Direct analysis of protein complexes using mass spectrometry, *Nature Biotechnol.* **17**, 676–682.

LIPKIN, L. E., LEMKIN, P. F. (1980) Data-base techniques for multiple two-dimensional polyacrylamide gel electrophoresis analyses, *Clin. Chem.* **26**, 1403–1412.

LUPAS, A. (1996), Prediction and analysis of coiled-coil structures, *Methods Enzymol.* **266**, 513–525.

MADEJ, T., BOGUSKI, M. S., BRYANT, S. H. (1995), Threading analysis suggests that the obese gene product may be a helical cytokine, *FEBS Lett.* **373**, 13–18.

MANN, M., HOJRUP, P., ROEPSTORFF, P. (1993), Use of mass spectrometric molecular weight information to identify proteins in sequence databases, *Biol. Mass Spectrom.* **22**, 338–345.

NEEDLEMAN, S. B., WUNSCH, C. (1970), A general method applicable to the search for similarities in the amino acid sequence of two proteins, *J. Mol. Biol.* **48**, 443–453.

NIELSEN, H., BRUNAK, S., VON HEIJNE, G. (1999), Machine learning approaches for the prediction of signal peptides and other protein sorting signals, *Protein Eng.* **12**, 3–9.

NOVOTNY, J., BRUCCOLERI, R., KARPLUS, M. (1984), An analysis of incorrectly folded protein models. Implications for structure predictions, *J. Mol. Biol.* **177**, 787–818.

OGIWARA, A., UCHIYAMA, I., SETO, Y., KANEHISA, M. (1992), Construction of a dictionary of sequence motifs that characterize groups of related proteins, *Protein Eng.* **5**, 479–488.

PAPPIN, D. J. C., HOJRUP, P., BLEASBY, A. J. (1993), Rapid identification of proteins by peptide-mass fingerprinting, *Curr. Biol.* **3**, 327–332.

PEARSON, W. R. (2000), Flexible sequence similarity searching with the FASTA3 program package, *Methods Mol. Biol.* **132**, 185–219.

PEARSON, W. R., LIPMAN, D. J. (1988), Improved tools for biological sequence comparison, *Proc. Natl. Acad. Sci. USA* **85**, 2444–2448.

PERKINS, D. N., PAPPIN, D. J., CREASY, D. M., COTTRELL, J. S. (1999), Probability-based protein identification by searching sequence databases using mass spectrometry data, *Electrophoresis* **20**, 3551–3567.

QIAN, N., SEJNOWSKI, T. J. (1988), Predicting the secondary structure of globular proteins using neural network models, *J. Mol. Biol.* **202**, 865–884.

RECHSTEINER, M., ROGERS, S. W. (1996), PEST sequences and regulation by proteolysis, *Trends Biochem. Sci.* **21**, 267–271.

RICHARDS, F. M. (1977), Areas, volumes, packing and protein structure, *Annu. Rev. Biophys. Bioeng.* **6**, 151–175.

ROST, B. (1996), PHD: Predicting one-dimensional protein structure by profile-based neural networks, *Methods Enzymol.* **266**, 525–539.

ROST, B., CASADIO, R., FARISELLI, P., SANDER, C. (1995), Transmembrane helices predicted at 95% accuracy, *Protein Sci.* **4**, 521–533.

ROST, B., SCHNEIDER, R., SANDER, C. (1997), Protein fold recognition by prediction-based threading, *J. Mol. Biol.* **270**, 471–480.

RUSSEL, R. B., COPELY, R. R., BARTON, G. J. (1996), Protein fold recognition by mapping predicted secondary structures, *J. Mol. Biol.* **259**, 349–365.

SANGER, F., AIR, G. M., BARRELL, B. G., BROWN, N. L., COULSON, A. R. et al. (1977), Nucleotide sequence of bacteriophage phi X174 DNA, *Nature* **265**, 687–695.

SETO, Y., IKEUCHI, Y., KANEHISA, M. (1990), Fragment peptide library for classification and functional prediction for proteins, *Proteins: Struct. Funct. Genet.* **8**, 341–351.

SHERIDAN, R. P., DIXON, J. S., VENKATARAGHAVAN, R. (1985), Generating plausible protein folds by secondary structure similarity, *Int. J. Pept. Protein Res.* **25**, 132–143.

SHEVCHENKO, A., WILM, M., MANN, M. (1997), Peptide sequencing by mass spectrometry for homology searches and cloning of genes, *J. Protein Chem.* **16**, 481–490.

SIPPL, M. J., WEITCKUS, S. (1992), Detection of native-like models for amino acid sequences of unknown 3D structure, *Proteins* **13**, 258–271.

SMITH, T. F., WATERMAN, M. S. (1981), Identification of common molecular subsequences, *J. Mol. Biol.* **47**, 195–197.

WILKINSON, D. L., HARRISON, R. G. (1991), Predicting the solubility of recombinant proteins in *Escherichia coli*, *Biotechnology* **9**, 443–448.

WISHART, D. S., BOYKO, R. F., WILLARD, L., RICHARDS, F. M., SYKES, B. D. (1994), SEQSEE: a comprehensive program suite for protein sequence analysis, *Comput. Applic. Biosci.* **10**, 121–132.

WISHART, D. S., STOTHARD, P., VAN DOMSELAAR, G. H. (2000), PepTool and GeneTool: platform-independent tools for biological sequence analysis, *Methods Mol. Biol.* **132**, 93–113.

YATES, J. R. (2000), Mass spectrometry. From genomics to proteomics, *Trends Genet.* **16**, 5–8.

YATES, J. R., ENG, J. K., GLAUSER, K. R., BULRINGAME, A. L. (1996), Search of sequence databases with uninterpreted high-energy collision-induced dissociation spectra of peptides, *J. Am. Soc. Mass Spect.* **7**, 1089–1098.

YATES, J. R., SPEICHER, S., GRIFFIN, P. R., HUNKAPILLER, T. (1993), Peptide mass maps – a highly informative approach to protein identification, *Anal. Biochem.* **214**, 397–408.

ZAMAYATNIN, A. A. (1972), Protein volume in solution, *Progr. Biophys. Mol. Biol.* **24**, 107–123.

ZHANG, W., CHAIT, B. T. (1995), *Proc. 43rd ASMS Conf. Mass Spectrometry and Allied Topics*, Atlanta, GA.

15 Structure Information

MIROSLAW CYGLER

ALLAN MATTE

JOSEPH D. SCHRAG

Montreal, Canada

1 Introduction

The biological sciences begin the new millennium on the verge of a quantum leap in our understanding of the fundamental processes of life, their control and regulation, and interactions of living organisms – with one another and with their environment. This information explosion was initiated with the human genome sequencing project and has presently led to the determination of nucleic acid sequences of over 40 entire genomes (see *http://www.ncbi.nlm.nih.gov:80/entrez/*; *http://216.190.101.28/GOLD/*; KYRPIDES, 1999), the advanced work toward the sequencing of at least 200 other genomes (FRASER et al., 2000), near completion of human genome sequencing (MACILWAIN, 2000; BENTLEY, 2000), and rapid progress on sequencing several other mammalian genomes (WELLS and BROWN, 2000). There is now a large and growing information database from which to analyze similarities, differences, and relationships between various life forms. Comparative analysis of complete genome sequences will provide new insights into the minimum requirements for life and into mechanisms of adaptation and evolution of living organisms.

While the main thrust of research efforts in the last several years has focused on developing efficient DNA sequencing methods and determining the sequences of entire genomes (LEE and LEE, 2000), more recent efforts have shifted towards the next required level of knowledge – association of all of the identified/predicted genes with a specific biological function. Broad categorization of gene functions has been derived from comparative analysis of genome sequences (MARCOTTE et al., 1999). About 60% of the identified genes can be classified into COGs (clusters of orthologous groups) with annotated functions based on experimentally determined functions of some COG members (*http://ncbi.nlm.nih.gov/COG*; TATUSOV et al., 2000). Proteomics efforts (DUTT and LEE, 2000) are now aiming at rapid, large-scale determination of the cellular functions of proteins by direct analysis of proteins in cells and organelles, identification of protein expression, and post-translation modification patterns under various conditions using two-dimensional electrophoresis coupled with mass spectrometry (YATES III, 2000), microarray analysis of gene expression (TAO et al., 1999), expression profiling (RICHMOND et al., 1999), and by systematic gene knockout experiments on entire genomes, such as *Saccharomyces cerevisiae* (ROSS-MACDONALD et al., 1999).

Proteins are biological molecules performing their functions through interactions with other molecules, and these interactions occur at the level of atoms. Therefore, a full understanding of the molecular function of proteins requires information about their three-dimensional (3-D) structures at atomic or near-atomic resolution. This is not to imply that such a detailed view is sufficient to determine protein function, but rather that to understand how this function is accomplished, an atomic view of the interacting molecules is required. For example, the function of hemoglobin as an oxygen carrier was known long before its three-dimensional structure was determined, but the knowledge of the hemoglobin structure in various oxygenation states led to the understanding of the mechanisms by which this molecule binds and releases oxygen (PERUTZ et al., 1998). Similarly, an in depth understanding of an enzymatic reaction mechanism requires the knowledge of the precise disposition of active site residues relative to the substrate. The substrate specificity is defined by the type of residues contributing to and the overall shape of the substrate binding site. Specificity of molecular recognition between forms is the basic premise of signal transduction and specialized modules/domains have evolved to recognize specific motifs, e.g., phosphotyrosine (PAWSON, 1995). Specific recognition of defined DNA sequences by proteins of transcription (GAJIWALA and BURLEY, 2000) and DNA repair machinery (MOL et al., 1999) can only be understood by analyzing the three-dimensional structures of their complexes. One would, therefore, like to know the three-dimensional structures of all of the proteins coded for in the genomes.

A number of major initiatives aimed at automation and rapid determination of a large number of structures have begun. Despite the nearly exponential increase in the rate of structures deposited in the Protein Databank and the growing momentum toward large-scale structural determination, experimental determination of 3-D structures of all proteins coded for by the many thousands of genes identified in the sequenced genomes remains a long-term prospect. In the following paragraphs we describe ongoing and developing methods for maximizing the returns on available structural information.

2 Protein Fold

The information about the shape of the protein in its natural environment is ultimately encoded in its amino acid sequence. This implies that an understanding of the principles of protein folding would allow calculation of the three-dimensional structure of a protein from the sequence alone. The ability to do so would obviate the need for experimental determination of three-dimensional structures and would provide an atomic view of all the proteins encoded by entire genomes. Efforts in this direction have progressed through the years from prediction of secondary structural elements to homology modeling and more sophisticated approaches including threading

(MOULT, 1999). Large-scale computational approaches and the application of artificial intelligence (IBM Blue Gene project) are now being applied to structure calculation/prediction in an effort to achieve the ultimate goal of "sequence-to-structure". For the foreseeable future, however, experimental determination of three-dimensional structures will be required to provide the atomic level view of molecules.

The number of protein structures deposited within the Protein Databank (PDB, *http://www.rcsb.org/pdb/index.html*; BERNSTEIN et al., 1977; BERMAN et al., 2000) presently exceeds 13,000 and is growing rapidly (Fig. 1), yet this is only a very small fraction of the number of genes identified through the worldwide sequencing efforts. Therefore, the question of diversity of protein structures is central to the understanding of the "protein sequence ↔ 3-D structure" relationship. Had we found that each structure differs greatly from the others, we would have to entertain the prospects of determining the 3-D structure of each gene product in order to understand its action. Therefore, analysis of protein 3-D structures has been of great interest and provided us with the systematic view where the structures can be grouped into families or folds (MURZIN et al., 1995; ORENGO et al., 1997).

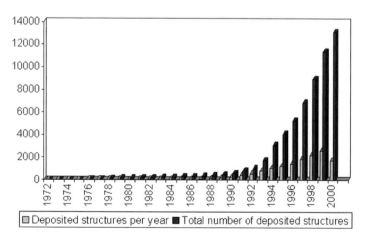

Fig. 1. Rate of increase of macromolecular structure deposition with the Protein Data Bank. Figure depicts number of entries per year from 1980 to 2000. Taken from analysis of the PDB archive.

The most general description of the three-dimensional structure of a protein is in terms of its fold, that is the topology of its polypeptide chain. This classification disregards local differences, and is mostly concerned with the secondary structural elements and their mutual disposition. The premise that proteins showing significant homology of their amino acid sequences share the same fold is by now well established. This statement generally holds even when the amino acid identity between the two sequences is as low as 20–25%. Many of the ~ 13,000 structures deposited in the Protein Databank represent the same protein complexed with different inhibitors, mutants, and complexes with other proteins. If we eliminate these duplicates and further consider only one representative of closely homologous proteins, say at the level greater than 40% sequence identity, we are left with only ~ 2,400 protein chains (FSSP, *http://www2. ebi.ac.uk/dali/fssp/*; HOLM and SANDER, 1997). This is a rather small subset of all proteins present in nature and, in particular, represents a very poor sampling of membrane-bound and transmembrane proteins. Nevertheless, comparative analysis of protein structures known to date has provided an extremely useful view of protein folding space. Several groups undertook classification of protein domains into folds and the resulting databases are easily available through the Internet (SCOP *http:// scop.mrc-lmb.cam.ac.uk/scop/*; CATH, *http:// www.biochem.ucl.ac.uk/bsm/cath/CATH.html*). With an increasing number of protein structures determined over the years it became evident that even when there is no recognizable sequence homology between proteins, they may have the same fold. Therefore, the number of folds is even smaller than the FSSP set. While the rate at which protein structures are being determined is increasing in a nearly exponential fashion, the rate at which the new folds are being identified among these structures is slowing down considerably. Presently, approximately 85% of the new structures deposited within PDB belong to an existing protein superfamily, as defined, e.g., by the CATH database (ORENGO et al., 1999a) and the question of how many different folds exist in nature becomes very relevant. The "structure ↔ fold" analyses indicate that the number of currently

characterized folds is between 540–580, depending on the defining criteria (HADLEY and JONES, 1999). The distribution of proteins among different folds is far from uniform. Almost a third of structural domains represented presently in the Protein Databank belong to one of only ten folds, known for this reason as superfolds (Fig. 2) (ORENGO et al., 1997). Each of these superfolds is associated with several different functions. On the other hand, most of the remaining folds correspond to a single homologous protein family and represent either a single or several related functions (ORENGO et al., 1999a). It has to be kept in mind that this distribution might be skewed by the available sample of protein structures, which represents to a large extent soluble proteins that have been relatively easy to crystallize or were amenable to NMR studies. The present view is that the number of folds existing in nature is rather limited and several arguments have been put forward to estimate the total number of folds to be somewhere between 1,000 and 5,000 (BRENNER et al., 1997). Thus, one of the fundamental goals of a large-scale structure determination effort is the identification of all naturally occurring folds and determining their number. The percentage of proteins with new folds among the newly determined protein structures decreases steadily with time. Therefore, the search for new folds will require improvement of methods for recognizing from sequence the proteins with potential new folds.

From a practical viewpoint (see homology modeling below) it is equally important to obtain a representative structure for every family of homologous proteins. This would allow the construction of models for all other proteins. The quality of such models would be related to the level of homology between the protein of interest and the one for which the structure is known.

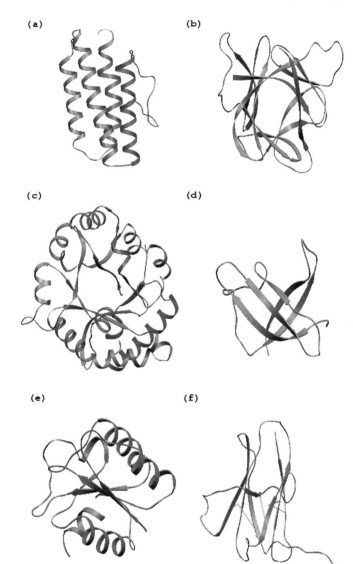

Fig. 2. Representative examples of protein superfolds. **a** UpDown fold (PDB code 2hmz), **b** trefoil (PDB code 1i1b), **c** TIM barrel (PDB code 7tim), **d** OB fold (PDB code 1mjc), **e** doubly wound (PDB code 3eca, chain A residues 213–326), **f** immunoglobulin fold (PDB code 3hhr, chain B residues 32–131) (SALEM et al., 1999).

3 Evolution and Three-Dimensional Structure

Comparison of the three-dimensional structures of distantly related proteins led to the conclusion that structure is maintained much longer during evolution than the amino acid sequence. Evolutionary connections between proteins can be derived either through homology in sequence or in structure (HOLM and SANDER, 1996a). While sequence similarity is often a satisfactory indicator of evolutionary relatedness on the short time scale, structural homology is often the only reliable indicator on a long time scale. Structural similarity is often maintained after all traces of sequence homology have vanished (ORENGO et al., 1999b). Databases of experimentally determined protein structures, such as FSSP (HOLM and SANDER, 1996b), SCOP (MURZIN et al., 1995), and CATH (ORENGO et al., 1997) allow an almost instantaneous view into these structural relationships. As protein domains are often the evolutionarily conserved fragments of proteins at both the sequence and structure levels, databases are organized based on domain classification. This is even more true now that such databases can be searched either with the sequence of a protein of unknown structure or with a newly determined protein structure. As one progresses from the level of "fold" to "superfamily" and finally "family", the structural similarity progressively increases. Fold comparison allows detection of very distant relationships and extends protein families. Some examples here include mammalian glycogen phosphorylase and DNA glycosyltransferase from bacteriophage T4 (HOLM and SANDER, 1995). Here, the substrate specificities and cellular functions of the proteins are dissimilar, although they have conserved structures.

Stability of the protein fold goes beyond divergence of the amino acid sequence. For several proteins with different folds, but having their N- and C-termini in close proximity, it was shown that cyclic permutations of amino acid sequence do not affect the fold. The permutations were created by a reshuffling of the gene in which a segment was removed from the 5′ end and attached to the 3′ end resulting in a protein with new N- and C-termini. Several examples include the TIM barrel (LUGER et al., 1989), SH3 domain (VIGUERA et al., 1995), T4 lysozyme (ZHANG et al., 1993), dihydrofolate reductase (IWAKURA et al., 2000), β-glucanase (HAHN et al., 1994). Such permutations are not only the result of our protein engineering activity. Homologous proteins from different organisms showing permutated arrangements have been observed in nature (HEINEMANN and HAHN, 1995).

4 Structure to Function, Structural Signatures

A recent study on the relationship between protein structure and function based on protein domains within the CATH database has revealed that the three-dimensional fold of enzymes does not correlate directly with specific catalytic functions (MARTIN et al., 1998; ORENGO et al., 1999b). That is, enzymes having the same enzyme commission (EC) number can have vastly different overall folds. This result, while seemingly counter-intuitive, makes the point that the function of a protein is as much a consequence of local structure, for example an enzyme active site, as the overall folding topology. An interesting result, however, was the apparent bias between the binding of a certain ligand (e.g., DNA, carbohydrate), and the structural class of the protein (mainly α, mainly β, or α/β (MARTIN et al., 1998).

The function of a protein is associated with its interaction with other molecules, be it other protein(s) or small molecules, e.g., substrate and cofactors, which result in formation of a complex between the protein and its functional partner. In the case of an enzyme, formation of such a complex is followed by a chemical reaction accelerated by a constellation of residues constituting an active site. Protein families that are identified based on sequence homology can often be characterized by a specific sequence pattern. Such patterns are, e.g., collected within the PROSITE database (HOFMANN et al., 1999) of the SwissProt database (BAIROCH and APWEILER, 2000). Spatial motifs within protein structures, once identified in one protein structure, can be used as templates to search the entire database of protein structures (WALLACE et al., 1997; RUSSELL et al., 1998; KLEYWEGT, 1999). These correspond to a three-dimensional substructure, a three-dimensional signature of the family. Usually, these signatures or motifs perform a specific function, e.g., nucleotide binding motif or Ser-His-Acid catalytic triad found in many proteins (WALLACE et al., 1996).

Closely similar structural motifs have been found in proteins whose sequences showed no detectable homology, either because of divergent evolution or because they appeared independently more than once during evolution. In particular, the PROSITE database has been analyzed in terms of protein structures having these sequence motifs, resulting in the compilation of a set of 3-D templates that can be used to analyze new structures for functional sites (KASUYA and THORNTON, 1999). Some protein superfolds consisting of analogous proteins (similar structure, but unrelated sequence) have been demonstrated to bind a ligand in the same spatial position within the structure (RUSSELL et al., 1998). Examples include the ligand-binding sites within TIM barrels, located at the C-terminal ends of β strands, and within proteins having Rossmann-like folds.

This type of information can be used to predict the ligand-binding site in a newly determined protein structure belonging to a previously characterized fold, even if the specific function of the protein is unknown. Therefore, the structure of a protein provides several venues for deriving the function. First, a search for known structural signatures may identify a specific active site and, thus, the type of catalyzed reaction. Otherwise, a binding motif for a specific substrate or cofactor may be found and this also defines the class of reactions. Second, analysis of the surface of an enzyme identifies the most likely location of the catalytic site. Therefore, computational methods could be used to find biological molecules that are most likely to bind in this region. These predictions can then be rapidly verified experimentally, e.g., by NMR methodology. Finally, the structure determination may reveal a molecule bound to the protein whose chemical identity is recognized from the electron density map (Fig. 3). In such a case, the nature of this molecule, be it a substrate, intermediate, or a cofactor, provides specific clues as to the possible function of the protein (ZAREMBINSKI et al., 1998).

5 Homology Modeling

Presently, three-dimensional structures are known for only a relatively small fraction of all

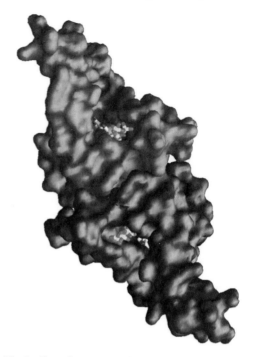

Fig. 3. Crystal structure of the MJ0577 protein from *Methanococcus jannaschii*. Each molecule of the dimer has a deep cleft on one side. A molecule of ATP and a Mn^{2+} ion were found in each cleft (PDB code 1mjh; ZAREMBINSKI et al., 1998) (see color plates p. 456).

proteins. If the entire content of the Protein Databank is reduced to nonredundant/nonhomologous proteins (say at less than 25% sequence identity), there are only ~2,400 independent chains left, unrelated or remotely related to each other (FSSP). Computational methods provide means to expand this number by building models of other proteins that show some level of homology to a protein with known 3-D structure (JONES, 2000). The reliability of such models depends on the level of sequence similarity between the two proteins.

The structure of a protein can be reliably modeled with the current methods only when its sequence is sufficiently similar to a sequence of a protein with known three-dimensional structure. This, so called homology model can be quite accurate, if the percent identity between the two sequences is high (>50%), in particular in places along the polypeptide chain where the conservation is highest, like the active site (THOMPSON et al., 1994). A homology model of such accuracy can be used to predict substrate specificity, design inhibitors, and analyze structure–activity relationships. Many successful examples of this approach are described in the literature, e.g., based on modeling, cathepsin X was correctly predicted to be a carboxypeptidase (NAGLER et al., 1999). The desirability of models with high accuracy is quite obvious.

When the sequences show less than 30% identity, the model is usually correct in gross features, but may be very inaccurate locally, especially if the two sequences can not be unequivocally aligned in some regions (TRAMONTANO, 1998; GUEX et al., 1999). In such cases, arriving at a correct alignment is crucial for the outcome of the modeling. Indeed, large errors usually occur due to a misalignment of the two sequences in a region of low homology. Better results are obtained when the alignment is based on multiple sequences and several methods were developed to perform such alignments (e.g., PSI-BLAST, ALTSCHUL et al., 1997; ClustalW, THOMPSON et al., 1994). Prediction of structures by threading and other methods (see below) can provide sufficiently good models to guide other functional studies, greatly expanding the number of proteins for which some structural information can be derived. Even though models built when there is only low sequence identity may have large local errors, they are of great value because they provide a three-dimensional global view of the molecule, albeit at a rather low resolution. In many instances such a view is sufficient for mapping residues proximal in space that are far apart in the sequence and to design further biological experiments with a three-dimensional view of the protein in mind. Such a model often provides a good basis for rationalization of site-directed mutagenesis results and provides a structural context for the analysis of naturally occurring mutations.

The genome-wide modeling exercise has been done for several bacterial genomes and resulted in models for a significant part of up to 1/3 of all proteins (SANCHEZ and SALI, 1998; M. PEITSCH, SwissModel Repository, *http://www.expasy.ch/swissmod/SM_3DCrunchhtml*; MARTI-RENOM et al., 2000). To maximize the benefit of modeling, the targets for structural genomics have to be chosen in such a way as to provide a wide coverage of protein sequence space in as uniform a fashion as possible. Since the accuracy of the model increases with the level of sequence identity, a high level of model accuracy will necessitate having experimentally determined structures for several moderately related representatives of each sequence family. It is likely that, with the further improvement of computational modeling methods, such models for proteins of a particular family, derived from a set of several related structures, may in the future alleviate the need for additional experimental structures for members of this family.

6 Fold Recognition by Threading

Structural similarities that have been found between apparently sequence-unrelated proteins showed that they could fold in a topologically similar manner. One of the earliest analyzed cases involved the unexpected similarity between domains of actin, hexokinase and heat-shock protein hsp70 (Fig. 4) (BOWIE et al., 1991; BORK et al., 1992). The attempts to develop *ab initio* methods for deriving protein fold directly from amino acid sequence have a long history, but they at best produce a very low resolution view of the protein with rather large root-mean-squares deviation from the experimentally determined structure.

As a result of these difficulties, a different approach was taken in the last few years, namely, the "inverse folding problem" has been addressed. The heart of this approach is the identification and alignment of a sequence to the fold with which it is most compatible, a process known as threading. This approach is quite promising and a number of algorithms have been developed to pursue this general idea. The first such algorithm, called the 3D profile method, was developed by EISENBERG and coworkers (BOWIE et al., 1991; LUTHY et al., 1992; ZHANG and EISENBERG, 1994). The protein sequence is threaded through the set

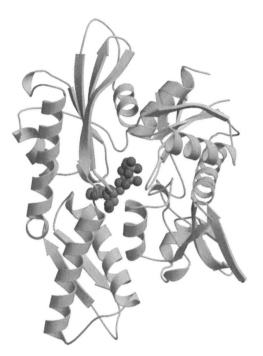

Fig. 4. Structural comparison of HSP70 β-and actin as an example of homologous protein structures. Shown in blue is the structure of actin complexed with ATP (PDB code 1hlu) and in green is Hsp70 complexed with ADP (PDB code 1hpm) (see color plates p. 456).

searching for conserved patterns of hydrophobicity (CALLEBAUT et al., 1997).

The progress of methods and algorithms toward the goal of predicting structure from sequence is periodically tested in the Critical Assessment of Techniques for Protein Structure Prediction (CASP) competition. Three such competitions have already taken place during which the sequences of 20–30 proteins whose structures have been solved, but not yet made publicly available, were provided to various laboratories specializing in structure prediction. The derived models were later compared with the crystal structures and the methods assessed. A clear improvement in predictive methods was noticed although they are still some way from producing high quality models. More exhaustive accounts of the results and a review of the various methods can be found in (VENCLOVAS et al., 1999; STERNBERG et al., 1999; CASP web site, *http://predictioncenter.llnl.gov/*).

7 Virtual Screening

Knowing the three-dimensional structure of a protein, and in particular the site of interaction with ligands, allows for computational screening of large libraries of compounds for their binding potential to a specific site on the protein surface. This is a complex computational problem due to the potential flexibility of the protein and the ligand, and has attracted significant attention for some time due to its practical importance (KIRKPATRICK et al., 1999). The driving force for a search for an efficient algorithm with good predictive powers is the pharmaceutical industry with its great need for selective and potent inhibitors of potential protein drug targets. The best known programs are DOCK and its progeny (OSHIRO et al., 1995), FlexX (RAREY et al., 1995), FLOGG (MILLER et al., 1994), DREAM + + (MAKINO et al., 1999) and several others. Most of these computational methods allow for flexibility of the ligand, but treat the protein as a rigid ensemble. Therefore, for these methods to be of value, the structure must correspond very closely to the conformation of the bound state.

of known template structures and for each position of the sequence along the chain the neighborhood of each residue is assessed by some measure derived from the statistical analysis of known structures. The results are scored for compatibility with the expected neighborhood for each residue and ordered according to a scoring function (ZHANG and EISENBERG, 1994). Several of the threading algorithms have been described in the literature and applied to the genome-wide prediction of protein folds (ROST et al., 1997; YADGARI et al., 1998; XU et al., 1998; JONES, 1999; PANCHENKO et al., 1999; RYCHLEWSKI et al., 1999; SANCHEZ et al., 2000). The fact that structural similarities are observed despite no apparent sequence similarity and the knowledge that a major determinant of protein folding is elimination of hydrophobic residues from the aqueous environment has led to three-dimensional structure prediction methods based on

The method can be applied equally well to homology-modeled protein structures, but the success of this approach depends critically on the quality of the homology model in the area of interest, e.g., substrate binding site, that is when the homology to a protein with known structure is quite high. There are many examples of successful application of molecular modeling to optimization of drug candidates and to identifying potential lead compounds (MARRONE et al., 1997). The success of the computational methods was clearly demonstrated in the development of inhibitors of HIV protease (WLODAWER and VONDRASEK, 1998).

One should expect that approaches of this type will play an important role in the near future when structural genomics projects will start producing 3-D structures of proteins of unknown functions. Virtual screening of each of these structures with a large set of biological small molecules contained within a living cell, i.e., substrates, cofactors, etc., might identify a small group of compounds compatible with the protein structure and provide a direction in the search for function. Analysis of surfaces of enzymes showed that in the majority of cases the substrate binding site was located in the largest cleft/cavity on the protein surface (LASKOWSKI et al., 1996). Therefore, for an unknown protein, the analysis of its surface can indicate if the protein has an enzymatic function and, if so, where the active site might be located.

8 Molecular Machines

Proteins act in concert with other proteins and many functions require formation of large multimeric protein complexes. For example, the translation and transcription machinery is composed of such complexes (polymerases, translation initiation complex, transcription initiation complex). Protein–protein interactions are essential for signaling, cell–cell recognition, and many other crucial processes. One of the most extraordinary achievements in structural biology has been the recent atomic resolution structure determination of the bacterial ribosome, the molecular machine responsible for translating mRNA into protein (BAN et al., 2000; SCHLUENZEN et al., 2000; WIMBERLY et al., 2000). These studies have revealed that the ribosome is in essence a ribozyme, with the binding and catalytic machinery consisting of rRNA rather than protein (NISSEN et al., 2000). Another example is the type-III secretion systems of enteric pathogenic bacteria, where protein–protein interactions between bacterial proteins, as well as between proteins from the pathogen and the host, are critical in pathogenesis (VALLANCE and FINLAY, 2000).

A variety of smaller proteins form complexes with other cellular proteins and utilize the energy derived from the hydrolysis of ATP to perform mechanical work. One example of this are the kinesin motor proteins, which transport organelles within cells, as well as perform chromosome segregation during meiosis and mitosis (KULL et al., 1996). There is structural similarity between the motor domain of kinesin and the catalytic domains of myosin and dynein, molecular motors which function during filament contraction in muscle. A second example is the F_1F_0 ATPase, which functions to pump protons across the membrane of bacteria, mitochondria and chloroplasts, thereby generating proton motive force at the expense of ATP hydrolysis. The first crystal structure of this group of proteins to be determined was that of the F_1-ATPase domain from beef heart mitochondria (ABRAHAMS et al., 1994). This protein consists of five subunits, having the stoichiometry $(\alpha)_3(\beta)_3\gamma\delta\varepsilon$. While there is little structural information about the structures of the γ, δ, and ε subunits, it is believed that rotation of the three $(\alpha\beta)$ dimer pairs occurs with respect to the γ and ε subunits in this motor, coupling ATP hydrolysis to proton translocation. A final example consists of the GroEL–GroES chaperonin complex, proteins required for the correct folding of many proteins in a variety of eubacteria. The crystal structure of GroEL–GroES–$(ADP)_7$ from *E. coli* has been determined, revealing large movements in GroEL as a consequence of GroES binding (XU et al., 1997). These movements increase the volume of the cavity within the protein complex that accommodates nascent polypeptide chains for folding, as well as making this cavity more hydrophilic in character. Hydrolysis of ATP results in a distinct conformational change in GroEL–GroES, which in turn facilitates protein folding by this complex.

It has become increasingly evident that many macromolecular interactions involve dynamic association, posttranslational modification, and conformational changes. In the next and very challenging stage, the emphasis in structural biology will shift from studying isolated molecules to investigation of molecular complexes. As the examples given above have shown, such data will be necessary to understand the next level of protein function, namely the transduction of signals between molecules. We are at an early stage in terms of dissecting the molecular details of protein–protein interactions as these relate to cell signaling and other processes. An understanding of the structural bases of the specificities of these interactions will contribute greatly to the development of therapeutics aimed at control of, and intervention into cellular processes.

9 *De novo* Protein Design

The availability of three-dimensional structures for a large number of proteins representing the entire spectrum of proteins existing in nature will provide a rich database for analyzing the principles governing the folding of a polypeptide chain. In addition, the information from NMR on the dynamics of the folding and/or unfolding process will be invaluable in understanding the physics of this process. What transpires from that is a potential for designer proteins, initially rather simple ones, but eventually of complexity matching those of naturally occurring proteins. The first successful attempts have been at building a rather simple structure, a four-helix bundle, and several such constructs had the desired properties (REGAN and DEGRADO, 1988; HECHT et al., 1990; SCHAFMEISTER et al., 1997). Other designs aimed at creating coiled-coil structures with two, three or four coils, in parallel or antiparallel fashion (Fig. 5) (NAUTIYAL et al., 1995; HARBURY et al., 1998).

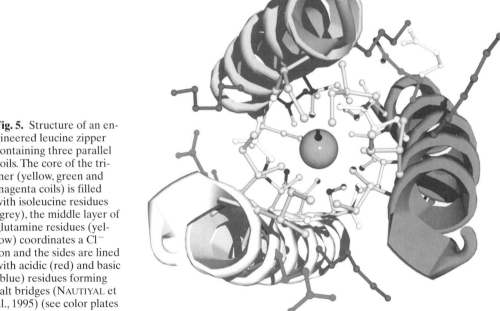

Fig. 5. Structure of an engineered leucine zipper containing three parallel coils. The core of the trimer (yellow, green and magenta coils) is filled with isoleucine residues (grey), the middle layer of glutamine residues (yellow) coordinates a Cl⁻ ion and the sides are lined with acidic (red) and basic (blue) residues forming salt bridges (NAUTIYAL et al., 1995) (see color plates p. 457).

Tab. 1. Web Sites of Interest for Macromolecular Structure Information

Site	Content	URL
Primary Databases		
PDB	protein structure repository	*http://www.rcsb.org/pdb/index.html*
PDBSum	summary of PDB entries	*http://www.biochem.ucl.ac.uk/bsm/pdbsum/*
NDB	nucleic acid structure repository	*http://ndbserver.rutgers.edu/*
NCBI Entrez	protein and DNA sequences	*http://www3.ncbi.nlm.nih.gov/Entrez/*
SWISS-PROT	annotated protein sequences	*http://www.expasy.ch/sprot/sprot-top.html*
PRODOM	protein domain families	*http://www.toulouse.inra.fr/prodom.html*
Modelling Databases		
ModBase	protein comparative model database	*http://pipe.rockefeller.edu/modbase/*
SWISS-MODEL	protein homology model database	*http://www.expasy.ch/swissmod/SWISS-MODEL.html*
Derived Databases		
FSSP	protein structure classification	*http://www.ebi.ac.uk/dali/fssp/*
CATH	protein structure classification	*http://www.biochem.ucl.ac.uk/bsm/cath/*
SCOP	protein structure classification	*http://scop.mrc-lmb.cam.ac.uk/scop/*
Enzyme Structures	structures of enzymes from PDB	*http://www.biochem.ucl.ac.uk/bsm/enzymes/*
3D_ALI	structure–sequence alignment	*http://www.embl-heidelberg.de/argos/ali/ali.html*
NCBI MMDB	structure alignment and visualization	*http://www.ncbi.nlm.nih.gov/Structure/MMDB/mmdb.shtml*
Structural Genomics Projects		
NIH Structural Genomics Resource	centralized resource	*http://www.structuralgenomics.org/*
PRESAGE	centralized resource	*http://presage.berkeley.edu/*
New York	project progress	*http://proteome.bnl.gov/*
CARB/TIGR	project progress	*http://structuralgenomics.org*
UC Berkeley	project progress	*http://www-kimgrp.lbl.gov/genomics/proteinlist.html*
Los Alamos/UCLA	project progress	*http://bdiv.lanl.gov/structure.htm*
NJCST/Rutgers	project progress	*http://www-nmr.cabm.rutgers.edu/structuralgenomics/*
Protein Structure Factory	German project based on human genes	*http://userpage.chemie.fu-berlin.de/~psf/*
Structure and Sequence Analysis Software		
DALI	comparison of 3D structures	*http://www2.ebi.ac.uk/dali/*
PhD	secondary structure prediction	*http://cubic.bioc.columbia.edu/predictprotein/*
WHATIF	many types of structure analysis	*http://www.cmbi.kun.nl/whatif/*
ExPASY	expert protein analysis system	*http://www.expasy.ch*
EBI	European Bioinformatics Institute	*http://www.ebi.ac.uk/*
PSIpred	geneTHREADER fold prediction software	*http://insulin.brunel.ac.uk/psipred/*

As was mentioned above, all our experience from analysis of protein structures and their sequences indicates that two naturally occurring proteins that share 50% sequence identity have the same fold. However, an elegant experiment in protein design showed that this is not necessarily the case. REGAN and coworkers (DALAL et al., 1997) have taken a protein whose structure contains predominantly β-sheets and replaced 50% of its amino acids according to rules that promote formation of α-helices. They were able to show that, indeed, the redesigned protein folds into a four-helix bundle, the fold drastically different from the original molecule. This experiment shows the potential of *de novo* design to go beyond the evolutionarily derived proteins. That natural proteins with much less sequence similarity maintain the same fold results from a non-random appearance of folds during evolution.

Several successful attempts have been undertaken to redesign the enzymatic activity or introduce a new enzymatic activity into an existing protein structure (SHAO and ARNOLD, 1996; BRAISTED et al., 1997; KELLNER et al., 1997). A novel approach was the concept of abzymes, antibodies with an enzymatic function introduced by design or selection (JACOBSEN and SCHULTZ, 1995). Significant progress has also been made in the design and introduction of metal binding site(s) into the protein (REGAN, 1993; MATTHEWS, 1995). The design of novel proteins is presently limited to a few well understood folds. However, with the accumulation of knowledge and better understanding of the principle of protein folding, design of more diverse folds will be attempted. It remains to be seen if proteins with folds not observed in nature can be successfully designed.

Biological systems and macromolecules are increasingly being recognized as useful and important components in the development of nanodevices (LOWE, 2000). Lessons learned from understanding the function, design, and construction of biological systems such as molecular motors and biopolymers promise to push forward the design and construction of miniaturized nanostructures and devices. Advances in this area will be greatly facilitated by the ability to rationally design specific properties and activities into proteins, polynucleotides, and polysaccharides.

10 References

ABRAHAMS, J. P., LESLIE, A. G., LUTTER, R., WALKER, J. E. (1994), Structure at 2.8 Å resolution of F1-ATPase from bovine heart mitochondria, *Nature* **370**, 621–628.

ALTSCHUL, S. F., MADDEN, T. L., SCHAFFER, A. A., ZHANG, J., ZHANG, Z. et al. (1997), Gapped BLAST and PSI-BLAST: a new generation of protein database search programs, *Nucleic Acids Res.* **25**, 3389–3402.

BAIROCH, A., APWEILER, R. (2000), The SWISS-PROT protein sequence database and its supplement TrEMBL in 2000, *Nucleic Acids Res.* **28**, 45–48.

BAN, N., NISSEN, P., HANSEN, J., MOORE, P. B., STEITZ, T. A. (2000), The complete atomic structure of the large ribosomal subunit at 2.4 Å resolution, *Science* **289**, 905–920.

BENTLEY, D. R. (2000), The Human Genome Project – an overview, *Med. Res. Rev.* **20**, 189–196.

BERMAN, H. M., WESTBROOK, J., FENG, Z., GILLILAND, G., BHAT, T. N. et al. (2000), The Protein Data Bank, *Nucleic Acids Res.* **28**, 235–242.

BERNSTEIN, F. C., KOETZLE, T. F., WILLIAMS, G. J., MEYER, E. E. J., BRICE, M. D. et al. (1977), The Protein Data Bank: a computer-based archival file for macromolecular structures, *J. Mol. Biol.* **112**, 535–542.

BORK, P., SANDER, C., VALENCIA, A. (1992), An ATPase domain common to prokaryotic cell cycle proteins, sugar kinases, actin, and hsp70 heat shock proteins, *Proc. Natl. Acad. Sci. USA* **89**, 7290–7294.

BOWIE, J. U., LUTHY, R., EISENBERG, D. (1991), A method to identify protein sequences that fold into a known three-dimensional structure, *Science* **253**, 164–170.

BRAISTED, A. C., JUDICE, J. K., WELLS, J. A. (1997), Synthesis of proteins by subtiligase, *Methods Enzymol.* **289**, 298–313.

BRENNER, S. E., CHOTHIA, C., HUBBARD, T. J. (1997), Population statistics of protein structures: lessons from structural classifications, *Curr. Opin. Struct. Biol.* **7**, 369–376.

CALLEBAUT, I., LABESSE, G., DURAND, P., POUPON, A., CANARD, L. et al. (1997), Deciphering protein sequence information through hydrophobic cluster analysis (HCA): current status and perspectives, *Cell Mol. Life Sci.* **53**, 621–645.

DALAL, S., BALASUBRAMANIAN, S., REGAN, L. (1997), Protein alchemy: changing beta-sheet into alpha-helix, *Nature Struct. Biol.* **4**, 548–552.

DUTT, M. J., LEE, K. H. (2000), Proteomic analysis, *Curr. Opin. Biotechnol.* **11**, 176–179.

FRASER, C. M., EISEN, J. A., SALZBERG, S. L. (2000),

Microbial genome sequencing, *Nature* **406**, 799–803.

GAJIWALA, K. S., BURLEY, S. K. (2000), Winged helix proteins, *Curr. Opin. Struct. Biol.* **10**, 110–116.

GUEX, N., DIEMAND, A., PEITSCH, M. C. (1999), Protein modelling for all, *Trends Biochem. Sci.* **24**, 364–367.

HADLEY, C., JONES, D. T. (1999), A systematic comparison of protein structure classifications: SCOP, CATH and FSSP, *Struct. Fold. Des.* **7**, 1099–1112.

HAHN, M., PIOTUKH, K., BORRISS, R., HEINEMANN, U. (1994), Native-like *in vivo* folding of a circularly permuted jellyroll protein shown by crystal structure analysis, *Proc. Natl. Acad. Sci. USA* **91**, 10417–10421.

HARBURY, P. B., PLECS, J. J., TIDOR, B., ALBER, T., KIM, P. S. (1998), High-resolution protein design with backbone freedom, *Science* **282**, 1462–1467.

HECHT, M. H., RICHARDSON, J. S., RICHARDSON, D. C., OGDEN, R. C. (1990), *De novo* design, expression, and characterization of Felix: a four-helix bundle protein of native-like sequence (published erratum appears in *Science* 1990 Aug 31; **249** (4972): 973), *Science* **249**, 884–891.

HEINEMANN, U., HAHN, M. (1995), Circular permutations of protein sequence: not so rare? *Trends Biochem. Sci.* **20**, 349–350.

HOFMANN, K., BUCHER, P., FALQUET, L., BAIROCH, A. (1999), The PROSITE database, its status in 1999, *Nucleic Acids Res.* **27**, 215–219.

HOLM, L., SANDER, C. (1995), Evolutionary link between glycogen phosphorylase and a DNA modifying enzyme, *EMBO J.* **14**, 1287–1293.

HOLM, L., SANDER, C. (1996a), Mapping the protein universe, *Science* **273**, 595–603.

HOLM, L., SANDER, C. (1996b), The FSSP database: fold classification based on structure–structure alignment of proteins, *Nucleic Acids Res.* **24**, 206–209.

HOLM, L., SANDER, C. (1997), Dali/FSSP classification of three-dimensional protein folds, *Nucleic Acids Res.* **25**, 231–234.

IWAKURA, M., NAKAMURA, T., YAMANE, C., MAKI, K. (2000), Systematic circular permutation of an entire protein reveals essential folding elements, *Nature Struct. Biol.* **7**, 580–585.

JACOBSEN, J. R., SCHULTZ, P. G. (1995), The scope of antibody catalysis, *Curr. Opin. Struct. Biol.* **5**, 818–824.

JONES, D. T. (1999), GenTHREADER: an efficient and reliable protein fold recognition method for genomic sequences, *J. Mol. Biol.* **287**, 797–815.

JONES, D. T. (2000), Protein structure prediction in the postgenomic era, *Curr. Opin. Struct. Biol.* **10**, 371–379.

KASUYA, A., THORNTON, J. M. (1999), Three-dimensional structure analysis of PROSITE patterns, *J. Mol. Biol.* **286**, 1673–1691.

KELLNER, D. G., MAVES, S. A., SLIGAR, S. G. (1997), Engineering cytochrome P450s for bioremediation, *Curr. Opin. Biotechnol.* **8**, 274–278.

KIRKPATRICK, D. L., WATSON, S., ULHAQ, S. (1999), Structure-based drug design: combinatorial chemistry and molecular modeling, *Comb. Chem. High Throughput Screen.* **2**, 211–221.

KLEYWEGT, G. J. (1999), Recognition of spatial motifs in protein structures, *J. Mol. Biol.* **285**, 1887–1897.

KULL, F. J., SABLIN, E. P., LAU, R., FLETTERICK, R. J., VALE, R. D. (1996), Crystal structure of the kinesin motor domain reveals a structural similarity to myosin, *Nature* **380**, 550–555.

KYRPIDES, N. C. (1999), Genomes OnLine Database (GOLD 1.0): a monitor of complete and ongoing genome projects world-wide, *Bioinformatics* **15**, 773–774.

LASKOWSKI, R. A., LUSCOMBE, N. M., SWINDELLS, M. B., THORNTON, J. M. (1996), Protein clefts in molecular recognition and function, *Protein Sci.* **5**, 2438–2452.

LEE, P. S., LEE, K. H. (2000), Genomic analysis, *Curr. Opin. Biotechnol.* **11**, 171–175.

LOWE, C. R. (2000), Nanobiotechnology: the fabrication and applications of chemical and biological nanostructures, *Curr. Opin. Struct. Biol.* **10**, 428–434.

LUGER, K., HOMMEL, U., HEROLD, M., HOFSTEENGE, J., KIRSCHNER, K. (1989), Correct folding of circularly permuted variants of a beta alpha barrel enzyme *in vivo*, *Science* **243**, 206–210.

LUTHY, R., BOWIE, J. U., EISENBERG, D. (1992), Assessment of protein models with three-dimensional profiles, *Nature* **356**, 83–85.

MACILWAIN, C. (2000), World leaders heap praise on human genome landmark, *Nature* **405**, 983–984.

MAKINO, S., EWING, T. J., KUNTZ, I. D. (1999), DREAM++: flexible docking program for virtual combinatorial libraries, *J. Comput. Aided Mol. Des.* **13**, 513–532.

MARCOTTE, E. M., PELLEGRINI, M., THOMPSON, M. J., YEATES, T. O., EISENBERG, D. (1999), A combined algorithm for genome-wide prediction of protein function, *Nature* **402**, 83–86.

MARRONE, T. J., BRIGGS, J. M., MCCAMMON, J. A. (1997), Structure-based drug design: computational advances, *Annu. Rev. Pharmacol. Toxicol.* **37**, 71–90.

MARTI-RENOM, M. A., STUART, A. C., FISER, A., SANCHEZ, R., MELO, F., SALI, A. (2000), Comparative protein structure modeling of genes and genomes, *Annu. Rev. Biophys. Biomol. Struct.* **29**, 291–325.

MARTIN, A. C., ORENGO, C. A., HUTCHINSON, E. G., JONES, S., KARMIRANTZOU, M. et al. (1998), Protein folds and functions, *Structure* **6**, 875–884.

MATTHEWS, D. J. (1995), Interfacial metal-binding

site design, *Curr. Opin. Biotechnol.* **6**, 419–424.

MILLER, M. D., KEARSLEY, S. K., UNDERWOOD, D. J., SHERIDAN, R. P. (1994), FLOG: a system to select "quasi-flexible" ligands complementary to a receptor of known three-dimensional structure, *J. Comput. Aided Mol. Des.* **8**, 153–174.

MOL, C. D., PARIKH, S. S., PUTNAM, C. D., LO, T. P., TAINER, J. A. (1999), DNA repair mechanisms for the recognition and removal of damaged DNA bases, *Annu. Rev. Biophys. Biomol. Struct.* **28**, 101–128.

MOULT, J. (1999), Predicting protein three-dimensional structure, *Curr. Opin. Biotechnol.* **10**, 583–588.

MURZIN, A. G., BRENNER, S. E., HUBBARD, T., CHOTHIA, C. (1995), SCOP: a structural classification of proteins database for the investigation of sequences and structures, *J. Mol. Biol.* **247**, 536–540.

NAGLER, D. K., ZHANG, R., TAM, W., SULEA, T., PURISIMA, E. O., MENARD, R. (1999), Human cathepsin X: A cysteine protease with unique carboxypeptidase activity, *Biochemistry* **38**, 12648–12654.

NAUTIYAL, S., WOOLFSON, D. N., KING, D. S., ALBER, T. (1995), A designed heterotrimeric coiled coil, *Biochemistry* **34**, 11645–11651.

NISSEN, P., HANSEN, J., BAN, N., MOORE, P. B., STEITZ, T. A. (2000), The structural basis of ribosome activity in peptide bond synthesis, *Science* **289**, 920–930.

ORENGO, C. A., MICHIE, A. D., JONES, S., JONES, D. T., SWINDELLS, M. B., THORNTON, J. M. (1997), CATH – a hierarchic classification of protein domain structures, *Structure* **5**, 1093–1108.

ORENGO, C. A., PEARL, F. M., BRAY, J. E., TODD, A. E., MARTIN, A. C. et al. (1999a), The CATH Database provides insights into protein structure/function relationships, *Nucleic Acids Res.* **27**, 275–279.

ORENGO, C. A., TODD, A. E., THORNTON, J. M. (1999b), From protein structure to function, *Curr. Opin. Struct. Biol.* **9**, 374–382.

OSHIRO, C. M., KUNTZ, I. D., DIXON, J. S. (1995), Flexible ligand docking using a genetic algorithm, *J. Comput. Aided Mol. Des.* **9**, 113–130.

PANCHENKO, A., MARCHLER-BAUER, A., BRYANT, S. H. (1999), Threading with explicit models for evolutionary conservation of structure and sequence, *Proteins* (Suppl.) **3**, 133–140.

PAWSON, T. (1995), Protein modules and signalling networks, *Nature* **373**, 573–580.

PERUTZ, M. F., WILKINSON, A. J., PAOLI, M., DODSON, G. G. (1998), The stereochemical mechanism of the cooperative effects in hemoglobin revisited, *Annu. Rev. Biophys. Biomol. Struct.* **27**, 1–34.

RAREY, M., KRAMER, B., LENGAUER, T. (1995), Time-efficient docking of flexible ligands into active sites of proteins, *Ismb.* **3**, 300–308.

REGAN, L. (1993), The design of metal-binding sites in proteins, *Annu. Rev. Biophys. Biomol. Struct.* **22**, 257–287.

REGAN, L., DEGRADO, W. F. (1988), Characterization of a helical protein designed from first principles, *Science* **241**, 976–978.

RICHMOND, C. S., GLASNER, J. D., MAU, R., JIN, H., BLATTNER, F. R. (1999), Genome-wide expression profiling in *Escherichia coli* K-12, *Nucleic Acids Res.* **27**, 3821–3835.

ROSS-MACDONALD, P., COELHO, P. S., ROEMER, T., AGARWAL, S., KUMAR, A. et al. (1999), Large-scale analysis of the yeast genome by transposon tagging and gene disruption, *Nature* **402**, 413–418.

ROST, B., SCHNEIDER, R., SANDER, C. (1997), Protein fold recognition by prediction-based threading, *J. Mol. Biol.* **270**, 471–480.

RUSSELL, R. B., SASIENI, P. D., STERNBERG, M. J. E. (1998), Supersites within superfolds. Binding site similarity in the absence of homology, *J. Mol. Biol.* **282**, 903–918.

RYCHLEWSKI, L., ZHANG, B., GODZIK, A. (1999), Functional insights from structural predictions: analysis of the *Escherichia coli* genome, *Protein Sci.* **8**, 614–624.

SALEM, G. M., HUTCHINSON, E. G., ORENGO, C. A., THORNTON, J. M. (1999), Correlation of observed fold frequency with the occurrence of local structural motifs, *J. Mol. Biol.* **287**, 969–981.

SANCHEZ, R., PIEPER, U., MIRKOVIC, N., DE BAKKER, P. I., WITTENSTEIN, E., SALI, A. (2000), MODBASE, a database of annotated comparative protein structure models, *Nucleic Acids Res.* **28**, 250–253.

SANCHEZ, R. SALI, A. (1998), Large-scale protein structure modeling of the *Saccharomyces cerevisiae* genome, *Proc. Natl. Acad. Sci. USA* **95**, 13597–13602.

SCHAFMEISTER, C. E., LAPÒRTE, S. L., MIERCKE, L. J., STROUD, R. M. (1997), A designed four helix bundle protein with native-like structure, *Nature Struct. Biol.* **4**, 1039–1046.

SCHLUENZEN, F., TOCILJ, A., ZARIVACH, R., HARMS, J., GLUEHMANN, M. et al. (2000), Structure of functionally activated small ribosomal subunit at 3.3 angstroms resolution, *Cell* **102**, 615–623.

SHAO, Z., ARNOLD, F. H. (1996), Engineering new functions and altering existing functions, *Curr. Opin. Struct. Biol.* **6**, 513–518.

STERNBERG, M. J., BATES, P. A., KELLEY, L. A., MACCALLUM, R. M. (1999), Progress in protein structure prediction: assessment of CASP3, *Curr. Opin. Struct. Biol.* **9**, 368–373.

TAO, H., BAUSCH, C., RICHMOND, C., BLATTNER, F. R., CONWAY, T. (1999), Functional genomics: expression analysis of *Escherichia coli* growing on minimal and rich media, *J. Bacteriol.* **181**, 6425–6440.

TATUSOV, R. L., GALPERIN, M. Y., NATALE, D. A., KOONIN, E. V. (2000), The COG database: a tool for genome-scale analysis of protein functions and evolution, *Nucleic Acids Res.* **28**, 33–36.

THOMPSON, J. D., HIGGINS, D. G., GIBSON, T. J. (1994), CLUSTAL W: improving the sensitivity of progressive multiple sequence alignment through sequence weighting, position-specific gap penalties and weight matrix choice, *Nucleic Acids Res.* **22**, 4673–4680.

TRAMONTANO, A. (1998), Homology modeling with low sequence identity, *Methods* **14**, 293–300.

VALLANCE, B. A., FINLAY, B. B. (2000), Exploitation of host cells by enteropathogenic *Escherichia coli*, *Proc. Natl. Acad. Sci. USA* **97**, 8799–8806.

VENCLOVAS, C., ZEMLA, A., FIDELIS, K., MOULT, J. (1999), Some measures of comparative performance in the three CASPs, *Proteins* (Suppl.) **3**, 231–237.

VIGUERA, A. R., BLANCO, F. J., SERRANO, L. (1995), The order of secondary structure elements does not determine the structure of a protein but does affect its folding kinetics, *J. Mol. Biol.* **247**, 670–681.

WALLACE, A. C., BORKAKOTI, N., THORNTON, J. M. (1997), TESS: a geometric hashing algorithm for deriving 3D coordinate templates for searching structural databases. Application to enzyme active sites, *Protein Sci.* **6**, 2308–2323.

WALLACE, A. C., LASKOWSKI, R. A., THORNTON, J. M. (1996), Derivation of 3D coordinate templates for searching structural databases: application to Ser-His-Asp catalytic triads in the serine proteinases and lipases, *Protein Sci.* **5**, 1001–1013.

WELLS, C., BROWN, S. D. (2000), Genomics meets genetics: towards a mutant map of the mouse, *Mamm. Genome* **11**, 472–477.

WIMBERLY, B. T., BRODERSEN, D. E., CLEMONS, W. M., Jr., MORGAN-WARREN, R. J., CARTER, A. P. et al. (2000), Structure of the 30S ribosomal subunit, *Nature* **407**, 327–339.

WLODAWER, A., VONDRASEK, J. (1998), Inhibitors of HIV-1 protease: a major success of structure-assisted drug design, *Annu. Rev. Biophys. Biomol. Struct.* **27**, 249–284.

XU, Y., XU, D., UBERBACHER, E. C. (1998), An efficient computational method for globally optimal threading, *J. Comput. Biol.* **5**, 597–614.

XU, Z., HORWICH, A. L., SIGLER, P. B. (1997), The crystal structure of the asymmetric GroEL-GroES-(ADP)7 chaperonin complex, *Nature* **388**, 741–750.

YADGARI, J., AMIR, A., UNGER, R. (1998), Genetic algorithms for protein threading, *Ismb.* **6**, 193–202.

YATES, J. R., III (2000), Mass spectrometry. From genomics to proteomics, *Trends Genet.* **16**, 5–8.

ZAREMBINSKI, T. I., HUNG, L. W., MUELLER-DIECKMANN, H. J., KIM, K. K., YOKOTA, H. et al. (1998), Structure-based assignment of the biochemical function of a hypothetical protein: a test case of structural genomics, *Proc. Natl. Acad. Sci. USA* **95**, 15189–15193.

ZHANG, K. Y., EISENBERG, D. (1994), The three-dimensional profile method using residue preference as a continuous function of residue environment, *Protein Sci.* **3**, 687–695.

ZHANG, T., BERTELSEN, E., BENVEGNU, D., ALBER, T. (1993), Circular permutation of T4 lysozyme, *Biochemistry* **32**, 12311–12318.

16 Automated Genome Annotation and Comparative Genomics

Shuba Gopal

Terry Gaasterland

New York, NY, USA

1 Introduction

The goal of genome annotation is to extract biologically relevant and scientifically interesting knowledge from raw genomic sequence data and to integrate it with previously existing genome-scale information. The primary emphasis of genome annotation is the assignment of biological functions to as many predicted coding sequences as possible. In addition, a complete annotation of a genome includes information regarding gene location and organization, the transcripts and products of those genes, and the regulation and control of expression, translation, and degradation. Annotation must also in some way present the manner in which genes and their products work together in cellular processes.

Annotation is done at two levels, initially at the DNA sequence level and then at the level of predicted protein sequences. DNA sequence analysis estimates boundaries between coding and non-coding sequence, identifies DNA features associated with gene structures, translates protein-coding genes into protein sequence, and characterizes conditions under which different forms of a gene may be expressed. Protein sequence analysis takes annotation a step further. Protein level analysis helps to predict function, to associate cellular processes with genome organization, and ties back to DNA level analyses by assisting in refining predicted gene structures.

Once the individual coding sequences are discerned, genome annotation constructs systems of genes and gene products by combining knowledge gained from sequence analysis with knowledge from other data sources. For example, pathway databases provide information to associate individual genes predicted in an organism with protein-level networks in the same and other organisms. Even in the absence of pathway information or global similarity to proteins of known function, individual regions of the predicted protein can hint at putative function. Leader domains target proteins for secretion through the cell membrane or transport through a nuclear, mitochondrial, or chloroplast membrane, for instance. Internal protein domains can indicate whether the protein is likely to be involved

in signaling. Other protein properties indicate whether the protein is likely to be an enzyme, an intracellular globular protein, or a transmembrane protein. Protein sequence motifs can suggest regions that may be modified or play key roles in regulation and activation. The presence of phosphorylation, glycosylation, cleavage or splice motifs can often support functional assignment and placing a given protein within the context of a larger network or pathway (ALBERTS et al., 1994).

Information about genes that are expressed together under different conditions further serves to associate individual gene products with cellular subsystems. Co-expression data helps to confirm that genes do indeed participate in common cell processes. More importantly, co-expression data indicate relationships between genes beyond those that can be deduced from known sequence and pathway information (CALIFANO et al., 2000).

Genome annotation is the process of uncovering all of these features at the DNA and protein sequence levels and integrating them into a systematic view of the molecular machinery of the organism. The very first challenge in annotating a genome is separating the information coding portions of a genomic sequence from the non-coding sequences. In all organisms, coding regions are differentiated from neighboring non-coding regions by specific features. Detecting these features is essential to transforming sequence data into knowledge of a genome. Once a gene is identified or predicted, the next step is to identify putative functions, possible homologs in other organisms and within the genome, and to postulate its role in the life of the organism. These tasks require the use of multiple bioinformatics tools. Gene finding tools assist in identifying genes, and the analysis of gene products via homology bases searches and other methods allows identification of functions and roles in the larger cellular context (GAASTERLAND et al., 2000; GOPAL et al., 2001).

This chapter discusses the process of annotation from the initial gene finding to the comprehensive analyses of pathways and networks. We will also highlight the importance and limitations of using complementary DNA (cDNA) and expressed sequence tag (EST) data for annotation of eukaryotic genome sequence data. Finally, we present two examples of genome re-annotation, *Escherichia coli* and *Drosophila melanogaster*, to illustrate nuances of genome annotation and how available bioinformatics tools facilitate this process.

2 The Process of Genome Annotation

Genome sequence annotation is an iterative process. It requires finding putative coding regions, confirming that these regions are truly coding regions, identifying what each encodes, and using the available evidence to refine the coding regions. Each of these steps involves collection of data from specific software tools that identify genomic or protein features. All of the steps must be controlled by a meta-level process that decides what tools to run, what input to give to each tool, and how to synthesize the output from all the tools into coherent reports about proteins and protein systems. There are four essential data collection steps: (1) parsing genes from the genome, (2) homology searches, (3) organization and evaluation of the data, and (4) the generation of an annotation (automatic or human-directed) based on the evidence collected (GAASTERLAND and SENSEN, 1996). The final step remains human-intensive, but no thorough evaluation of a genome could (or should) be accomplished in an entirely automated fashion.

Pre-processing genomic sequence data is the first step in annotation of a eukaryotic genome. First, the genomic sequence must be split into overlapping contiguous sequences (contigs) whose length is acceptable as input to genomic sequence analysis tools. Next, repeat regions are identified. Masked and unmasked contigs are sent through gene finding tools to generate exons and extract spliced coding sequences. Gene finding remains as much an art as a science, and the computational challenges will be discussed in more detail shortly. However, the putative coding regions are determined, the annotation process begins in earnest at the next step, with the search for a functional assignment for each of these predicted coding regions (GAASTERLAND and SENSEN, 1996).

Each coding sequence is compared with sequences in the public databases at both the DNA and protein levels. Expressed sequence tags, full-length and partial cDNA sequences, and known protein sequences are compared to the predicted DNA sequence, while protein sequences translated from each coding sequence are searched for patterns that indicate functional and structural features. The results of these homology-based searches are used in developing a comprehensive picture of the putative function for the coding region (GAASTERLAND et al., 2000).

Evidence collected from these methods and other bioinformatics tools provides the means for generating a logical model of the structure of each gene and a possible functional assignment for the protein encoded by that gene. The gene model includes predicted promoters, transcription factor binding sites, terminator sequences, 5′ and 3′ UTR, poly-adenylation sites, intron boundaries, and, when possible, alternative splicing of exons. At the protein level, evidence is used in predicting putative binding, post-translational modification sites, and, when possible, subcellular localization.

These individual evaluations and function assignments are then used in more global evaluations of the genome annotation. At this stage, one can assess the completeness of known metabolic or signaling pathways, for instance. If a given pathway is incomplete, it might indicate that a region of the genome was not analyzed, or that some evidence was overlooked. It may also hint at unusual biology or alternative pathways in that particular organism. These meta-level analyses build upon the initial annotation and develop a genome-scale knowledge base available for others to evaluate and use in their own annotation efforts. A summary of the annotation process is shown in Fig. 1.

2.1 Separating Coding Sequence from Non-Coding Sequence

The initial annotation step is to find likely coding regions with gene finding tools trained for the organism. Any series of non-stop co-

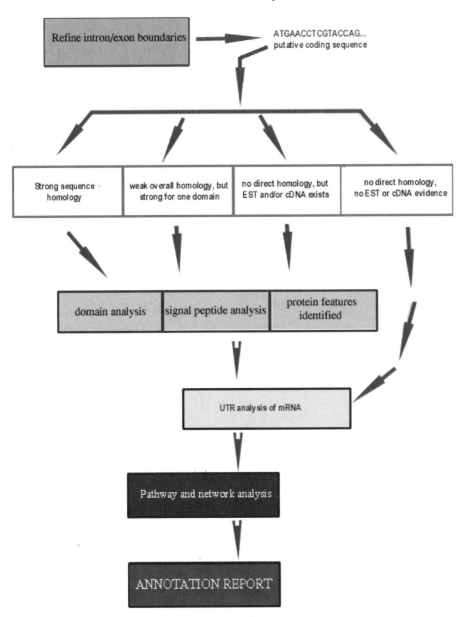

Fig. 1. Summary of the genome annotation process. Colored boxes indicate areas where computational tools and methods assist in processing and interpretation.

dons can be translated computationally into an amino acid sequence. If such a region, called an open reading frame or ORF, begins a gene, it starts with a start codon. The first translated codon in a eukaryotic gene encodes methio-nine (ATG); in prokaryotic genes, the first co-don is usually ATG and sometimes TTG or GTG (ALBERTS et al., 1994). Gene finding tools use a variety of algorithms and computa-tional techniques to assess the overall frequency of each possible codon in known genes of a particular genome and downgrade or upgrade translatable regions according to codon usage. For eukaryotic genomes, other information such as splice site patterns, average length of introns, and presence of poly-adenylation sites help to discriminate coding from non-coding regions (BURGE and KARLIN, 1997; DELCHER et al., 1999).

In prokaryotes, ORFs can be easily defined syntactically; they are the regions of the genome that contain a start codon, a series of translatable codons and a stop codon. In addition, one can specify certain hallmarks of "true" coding regions. For instance, most ORFs contain the highly conserved Shine–Delgarno sequences for improved translation of the transcript. In addition, the percentage of rare codons tends to be relatively low in genuine coding regions. These and other criteria can be used to produce a fairly robust ORF-finding tool (DELCHER et al., 1999; GAASTERLAND and SENSEN, 1996).

As with prokaryotes, eukaryotic genomes contain signals that delineate genuine coding regions from the non-coding regions of the genome. However, the simplistic definitions of an ORF that are satisfactory in prokaryotes almost always fail to capture true coding regions for eukaryotic genes because of the presence of introns. The presence of introns makes gene prediction computationally much more complicated because the actual coding sequence is split across multiple open reading frames (ALBERTS et al., 1994; BURGE and KARLIN, 1997).

Tab. 1 compares the predictions made by three well-known gene prediction tools, GENSCAN from BURGE and KARLIN, HMMGENE from ANDERSON et al., FGENES and FGENESH from the Sanger Centre's VICTOR SOLOVYEV (SOLOVYEV, 1999). Included is a BLASTx search (translation in all 6 frames of the nucleotide query and search against the non-redundant protein database) of a 100 kbp test region of eukaryotic genomic DNA.

As becomes obvious from this simple comparison, while all the tools are within a "ball-park figure" of the number of likely coding regions in this test sequence, the final numbers do vary somewhat from tool to tool. GENSCAN predicted the fewest exons, although FGENESH was not far behind. However,

Tab. 1. Gene Predictions by Several Publicly Available Gene Finding Tools. Test sequence was a 100 kbp region from the left arm of chromosome II of *Drosophila melanogaster* (sequence as released by Celera Genomics, Inc. on 3/24/2000)

Tool	Number of Predicted Genes	Number of Predicted Exons
BLASTx	–	45[a]
GENSCAN	12	57
FGENESH	14	58
HMMGENE	14	62

[a] Confirmed coding sequences by way of high homology to a sequence in the protein database (see text for details).

GENSCAN produced a more parsimonious set of gene structure predictions. Indeed, as the documentation for FGENESH notes, this tool is best at finding the individual exons, but does less well at stringing any given set of exons into a coherent gene structure (SOLOVYEV, personal communication and documentation, 1999). HMMGENE did the most poorly of the three in this very simplistic comparison, although a brief perusal of the actual predictions suggests that it has a bias toward finding several very small exons (between 50 bp and 150 bp) around what the other tools identify as a single, larger initiating exon. Regardless of such variations, what this simple comparison does demonstrate is that gene prediction in eukaryotes is still as much an art as it is a science.

Some regions of the gene structure can be more easily confirmed than others. Initial and terminal exons and intron–exon boundaries may be confirmed by matches with sequences known to be expressed in the organism. Alignments with known 3′ and 5′ ends of expressed sequence tags (ESTs) from cDNA clones anchor ends of genes and indicate alternative ends of transcription for the same gene. The use of ESTs in gene refinement is discussed further in the sections that follow. Other features within the DNA sequence help to signal which coding regions are contained within a single transcript. For example, in eukaryotic genomes, poly-adenylation sites after a terminal exon and transcription initiation sites upstream of the first coding exon indicate transcription boundaries. In prokaryotic genomes, a series of coding regions may be transcribed together as an operon and then each translated as independent amino acid sequences after transcription (ALBERTS et al., 1994; BURGE and KARLIN, 1997; DELCHER et al., 1999).

2.2 Characterizing Non-Coding Regions of the Genome

Even if the coding region of a gene is correctly predicted, additional information may be required to predict the actual transcript and protein product. This is especially true for higher eukaryotic genomes, where much more post-transcriptional and translational modification occurs in the process of generating a functional protein from the genomic DNA sequence of the gene. A short list of such modifications includes alternative splicing at the DNA and RNA levels, RNA editing, post-translational modifications to the proteins, and many other variations on these themes. All these manipulations of a coding sequence must be included in the computational analysis of a eukaryotic gene, if it is to be of any use to the biologist (PAVY et al., 1999; SMIT, 1999).

The non-coding regions of eukaryotic genomes are also rich with variation. Non-coding regions in eukaryotes play significant roles in the biology of the organism, containing regulatory regions with promoters and transcription factor binding sites. Less obvious and harder to identify computationally are non-coding regions with structural features that may be valuable for overall regulation of a genomic sequence, such as histone binding sites and larger structures including centromeres and telomeres (ALBERTS et al., 1994).

Tab. 2 contains a sampling of the computational tools for the prediction and classification of these regions. The first program, bend.It, attempts to predict the curvature of DNA, and of specific DNA motifs. These can be valuable in predicting the likelihood that a given motif does bind a protein, for instance (MUNTEANU et al., 1998). The structure of RNA, as predicted by mfold, can be used in predicting post-transcriptional modifications, including likely alternative splicing of the transcript, putative sites of RNA editing and other modifications (ZUKER, 2000). In addition, more specific tools are becoming available, including tools such as Recon, which identify specific sites where histones bind to package DNA into nucleosomes (LEVITSKY et al., 2000).

Finally, there are non-coding regions with no known biological function. While these may be genuine "junk", the flotsam of eons of evolution, they must nevertheless be identified and categorized, so as to provide a more comprehensive view of the genome. Within this category fall insertion elements and transposons, long, short, and tandem repeats, and pseudogenes (SMIT, 1999; ALBERTS et al., 1994).

2.3 DNA and Protein Level Analysis of Coding Regions

Once coding sequences are identified and extracted from the genome, evidence for each coding sequence (CDS) is collected from a variety of sequence homology and motif search tools. The output from each tool is integrated into a generic relational format. The most common tools currently employed for homology searches include BLAST, FASTA, and BLAIZE searches against the public nucleotide and non-redundant protein databases. Searches of the dbEST expressed sequence tag

Tab. 2. Tools for Analyzing Non-Coding Regions of the Genome. A sampling of tools that are currently available for analyzing non-coding regions of sequences, both at the genomic and transcript level

Tool	Input Type	Predictions
bend.It (MUNTEANU et -al., 1998)	DNA genomic sequence	curvature of DNA
mfold (ZUKER, 2000)	DNA or RNA sequence	secondary structure
Recon (LEVITSKY et al., 2000)	DNA sequence	nucleosome binding sites

(EST) database provide valuable information to confirm the transcription and splicing of predicted open reading frames even in the absence of protein or direct nucleotide homologies. EST searches can be done at the DNA level with BLASTn or FASTA or at the protein level via tBLASTn, tBLASTx or FASTAx. Protein motif identification can be done with searches against the PROSITE, pFAM, and BLOCKS databases with PROSEARCH and the pFAM and BLOCKS search software. Finally, a survey of RNA genes can be done using tRNAscan-SE to identify tRNA genes (LOWE and EDDY, 1997) and BLASTn against the rRNA subset of Genbank to identify rRNA genes. Tab. 3 shows the most commonly used software tools used to collect supporting evidence during genome annotation.

If protein or DNA level sequence homologies are sufficiently strong and informative, their functional information can be transferred to the genomic query sequence as a putative functional annotation. Alternatively, function can be postulated based on the collective evidence of other non-homology-based analyses. Tools that inspect the protein for structural and subcellular localization features help to determine the cellular location and role of the protein. Tab. 4 lists some instances of tools (by no means a complete list!) that identify specific protein features valuable for annotation purposes.

If a portion of the protein sequence matches a domain of a protein of known three-dimensional (3D) structure, a 3D homology model can be constructed for the protein. A 3D model serves to confirm or overrule function assignments and to gain further insight into how the protein may bind with other molecules. Structure prediction from homology is complemented by attempts at *ab initio* structure prediction which is done without prior knowledge of homologous protein structures. Tab. 5 lists a few structure prediction programs and the methods they use for generating predictions.

While homology-based evidence is a valuable method for assigning function to a predicted coding region, it does have its shortcomings. A review of genomes published to date demonstrates the most significant of these: of the approximately 4,200 genes annotated in *E. coli*, nearly 40% lack any functional annotation (MONICA RILEY, personal communication). In organisms such as the anciently diverged classes of eukaryotes, the number of ORFs with effective functional designations

Tab. 3. Commonly Used Tools in Sequence Homology Searches. Tools are listed in approximate order of the time taken to complete one query against the relevant database derived from benchmark tests on SUN cpus)

Tool	Query Sequence Type	Target Database
prosearch	amino acid	PROSITE database
blocks	amino acid	BLOCKS database (at SwissProt)
psi-BLAST	amino acid	PDB (structural models database)
pfam	amino acid	Pfam HMM database (collection of HMM models for sequence motifs)
BLASTp	amino acid	nr (non-redundant protein database at NCBI)
BLAIZE	amino acid	SwissProt database
BLASTn	nucleotide	nt (nucleotide database at NCBI)
BLASTnEST	nucleotide	est (dbEST, EST database at NCBI)
tBLASTn	amino acid	six-frame translation of nt
tBLASTnEST	amino acid	six-frame translation of est
BLASTx	six-frame translation of nucleotide sequence	nr (non-redundant protein database at NCBI)
tBLASTx	six-frame translation of nucleotide sequence	six-frame translation of nt

Tab. 4. Tools for Protein Feature Identification. Some instances of protein features that can be analyzed computationally and some tools that can provide information regarding the features

Protein Feature	Examples of Available Tools	References
Secondary structure	PREDATOR SAM T98/T99	FRISHMAN and ARGOS (1996) KARPLUS et al. (1998)
Signals for subcellular localization and protein modifications	PSORT/PSORT II SignalP	NAKAI and HORTON (1999) NIELSEN et al. (1997)
Transmembrane domains	TopPred2 DAS PredictProtein	GUNNER (1992) CSERZO et al. (1997) ROST (1996)
Motif analysis tools	GIBBS PRATT MEME AlignAce	LAWRENCE et al. (1993) JONASSEN et al. (1995) BAILEY and GRIBSKOV (1996) ROTH et al. (1998)

Tab. 5. 3D Structure Prediction Tools. A sample of tools that generate 3D structure predictions given a protein sequence as input

Tool	Comments
ICM (ABAGYAN and TOTROV, 1994)	allows for structure modeling and ligand-binding predictions with virtual ligand screening
123D + (ALEXANDROV et al., 1995)	structure prediction using threading on homologous proteins
WAM (REES et al., 1996)	antibody structure prediction
Modeller/ModBase (SANCHEZ and SALI, 1997)	threading of sequence onto homologous proteins with solved structure to generate structure prediction
GROMACS (SANSOM et al., 1998)	entire package of structure modeling tools with a variety of strategies and algorithms available

is even lower (MCKEAN et al., 1999; MYLER et al., 1999 and preliminary results). Therefore, homology-based searching will not provide as comprehensive a view of the genome as would be desired.

Yet, even in cases where strict homology-based annotation is impossible, accumulated evidence from several tools can sometimes provide as useful an annotation. The key to developing supportive evidence of the "authenticity" of a predicted coding region is in identifying the hallmarks of a true eukaryotic gene. While these sequence motifs and structures are not as highly conserved from eukaryote to eukaryote, some are conductive to computa-

tional analysis and identification. The primary challenge for coding regions with relatively poor homology-based evidence is confirming the validity of the prediction. For eukaryotes, this begins with a review of the predicted intron/exon boundaries (GAASTERLAND et al., 2000).

As already discussed, promoter elements are a singularly strong suggestion that a predicted coding region really does produce a transcript and a downstream protein product. However, the identification of promoter elements has proved an elusive goal, in part because the sequence motifs are not highly conserved, even between genes sharing common regulatory processes (ALBERTS et al., 1994). Nevertheless, numerous tools exist that, to some extent, are able to provide promoter site predictions as shown in Tab. 6.

The 5′ and 3′ untranslated regions (UTRs) of the transcript can provide additional support for evaluating predicted sequences. These regions appear to play a central role in mediating the efficiency of translation, the stability of the mRNA and may play a role in the regulation of the gene's expression. If a predicted coding sequence contains appropriate signals for mRNA processing in the 5′ and 3′ UTRs (untranslated regions), this gives greater credence to the prediction. In certain eukaryotes, 5′ and 3′ UTRs contain highly conserved motifs related to these processes, and may be tractable to computational analysis (ALBERTS et al., 1994; LEE and PLOEG, 1997).

2.4 Gene Refinement

The most difficult genome annotation task is to determine exact boundaries when separating coding sequence from non-coding. In most eukaryotic organisms, boundary predictions are complicated by the fact that multiple coding regions often comprise the complete gene encoding a protein, and different combinations occur under different conditions. In prokaryotic organisms and some microbial eukaryotes, each open reading frame generally corresponds to one coding region, but the start of the actual coding region can be ambiguous (BURGE and KARLIN, 1997; DELCHER et al., 1999).

If gene boundary signals are weak, ambiguous, or not detectable, distinct genes may be merged, exons can be assigned to the wrong gene, and single genes can be split into multiple genes. As each predicted gene product is compared to protein sequences, motifs and 3D structures, the alignments help to refine boundaries between coding and non-coding regions. Of particular value in these evaluations are alignments to ESTs. For example, if an EST spans two exons (Fig. 2a), the EST sequence constrains possible intron splice sites. Similarly, a protein sequence can be reverse translated to all possible combinations of codons and mapped to the genomic sequence to refine boundaries (Fig. 2b). If the alignment of a predicted gene with a cDNA sequence, protein sequence, or 3D structure contains large gaps (Fig. 2c), a gap in the predicted gene indicates a missing exon; a gap in the protein indicates an extra exon (GAASTERLAND et al., 2000).

If just one portion of a predicted gene aligns with an entire protein (Fig. 3a), it may be appropriate to split the gene into two, especially if the remaining portion aligns with a different complete protein. Likewise, if different portions of the same cDNA or protein sequence align with neighboring predicted genes on the

Tab. 6. Promoter Prediction Tools. A sampling of some promoter prediction tools that span the gamut from core promoter predictions to *de novo* promoter prediction

Tool	Comments
Promoter Scan (PRESTRIDGE, 1995)	only for use on mammalian sequences, searches for hallmarks of a Pol II promoter
Neural Net Prediction Program (NNPP) (REESE and EECKMAN, 1995)	uses neural networks to do *de novo* promoter identification
CorePromoter (ZHANG, 1998)	finds the transcription start site and TATAA box (i.e., the "core" promoter region)
Promoter 2.0 (KNUDSEN, 1999)	only designed for use with vertebrate genomic sequence
GrailExp 2.0 (HYATT et al., 2000)	primarily for gene prediction, but attempts to identify core promoter region as well

Figure 2a. EST and cDNA alignments help refine exon boundaries

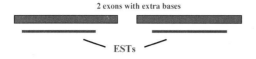

Figure 2b. Protein sequences help to refine intron/exon boundaries

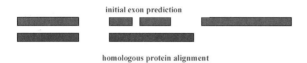

Figure 2c. Missing exons and extra exons create gaps in the alignment

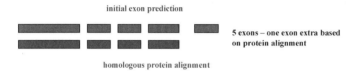

Fig. 2. Using ESTs and cDNAs helps to refine gene boundaries.

same coding strand, it may be appropriate to merge the genes, especially if the junction splits the protein in the middle of a structural domain or known binding site (GAASTERLAND et al., 2000; SCHULER, 1997).

The decision to re-define a gene's predicted structure involves many considerations. A key concern is that an apparently fused gene is in fact the actual gene structure rather than a mis-prediction. Gene fusion can and often does occur over evolutionary time as two organisms become more phylogenetically distant. For instance, the *E. coli* gene *thrA* generates a bifunctional protein, encoding both threonine synthetase and arginine dehydrogenase functionality in the same protein (BLATTNER et al., 1997). Alternatively, domains of proteins can shuffle and merge with domains from other proteins, as illustrated by the SH2 domain of many eukaryotic signaling proteins (KURIYAN and DARNELL, 1999).

Consequently, the decision whether to split a gene that aligns successively with two proteins or whether to fuse two genes that match successive regions of one protein becomes increasingly ambiguous as the organisms con-

tributing to the alignment grow phylogenetically more distant. Orthologous proteins can occur as fused domains in one organism, within an operon in another organism, or as multi-gene heteromers in yet another organism (see Fig. 3 for examples of such issues). These considerations make it difficult to automate the evaluation of gene boundaries, which remains a human-intensive task in annotation (GAASTERLAND et al., 2000).

2.5 Alternative Splicing and the Use of ESTs

In eukaryotic organisms, after transcription, the mRNA for a gene may be alternatively spliced. This phenomenon adds to the complexities of determining intron–exon boundaries. However, alternative splicing can be deduced if sufficient EST evidence is available for the predicted gene. Consequently, the EST information generated by cDNA sequencing projects is critical to interpret a eukaryotic genome completely (GOPAL et al., 2001).

Figure 3a. An alignment of part of one gene with 1 protein indicates a possible for splitting the gene

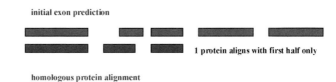

initial exon prediction

1 protein aligns with first half only

homologous protein alignment

Figure 3b. If one gene aligns with two distinct proteins, it is possible that the gene should be split into two genes unless there are fused orthologs of the protein in some other organism.

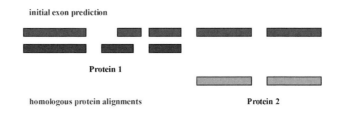

initial exon prediction

Protein 1

homologous protein alignments   Protein 2

Figure 3c. Protein 1 and Protein 2 are co-transcribed in another organism

Figure 3d. Protein 1 and Protein 2 are fused in a third organism

Figure 3e. Protein 1 and Protein 2 are completely separate in yet another organism

Fig. 3. Homologous p alignments can also refine boundaries and suggest phylogenetic and evolutionary history.

The use of EST information in gene structure prediction and refinement is a clear instance of the importance of integrating experimentally derived information with computational techniques and methods. To fully understand the importance of EST alignments, it is worthwhile to understand how ESTs are generated. Most EST generation projects are done in a staggered manner, in part because the time and effort required for a complete EST sequencing project is at least akin to that of sequencing the genome again (GILL and SANSEAU, 2000).

The first generation of cDNA sequencing projects that complement whole-genome projects aim to create as much EST data as possible for that genome. An EST sequencing project samples libraries of cDNA molecules prepared from a variety of tissues. An ideal databank of 5′ and 3′ ESTs would provide evidence for every gene used by an organism. However, some genes are rarely expressed or are uniquely expressed in a small number of tissues. Some proteins such as actin are present at high concentrations in every cell. Consequently, first pass EST data tends to be incomplete.

There are two means of improving the completeness of EST data sets. First, as EST sequencing efforts reach into normalized libraries, in which high-frequency cDNA molecules are minimized, rarer transcripts can be identified and sequenced. Libraries from tissues rich in rare transcripts provide another source for complete coverage of the organism's transcriptome (GILL and SANSEAU, 2000).

Large amounts of EST data are deposited in the major public databases every day. The vast quantities of ESTs are particularly valuable in evaluating gene structure predictions. ESTs from the same gene can reinforce and correct each other. If ESTs from different ends of the same gene are long enough to overlap, together they provide far more information than each EST alone. Multiple ESTs for the same gene might indicate alternative splicings or other modifications to the transcripts. One way to take advantage of these collections of ESTs for predicted genes is to use computational approaches in clustering ESTs by sequence similarity or homology to known genes. This allows each predicted gene to be compared against an array of EST sequences, enabling more effective and informative annotations (SCHULER et al., 1996; BOGUSKI and SCHULER, 1995).

The general approach to cluster ESTs is to identify known repeats and compute pairwise alignments between all ESTs from an organism and then apply constrained transitive closure to alignment pairs. One constraint is to ignore alignments within repeat regions. Another is to require that no more than one EST continue beyond either end of the alignment. Constraints on percent identity are more delicate: setting a high stringency on the percent of identical bases in a pairwise alignment may eliminate true duplicates, yet may also eliminate two distinct ESTs that share a region of homology. Too low a stringency may cluster ESTs from many different genes instead. Technical difficulties such as chimeric clones that combine different parts of different genes complicate clusters further, but can be minimized by constraints that apply to the entire cluster (SCHULER, 1997).

While EST clusters may be useful for some instances, there are certain issues that must be kept in mind. Unlike most published genomic sequences, EST sequences are fraught with error. Since ESTs are usually single-pass reads from sequencing machines, there is no easy way to ensure the accuracy of the read. In addition, genes may be alternatively spliced in different tissues or undergo modifications in some tissues, but not all. Finally, ESTs from paralogous genes or recently duplicated genomic regions may contain very few, but very significant, differences (ALBERTS et al., 1994). Teasing out the differences between sequencing error, polymorphisms and alternative splicing is almost impossible to accomplish in a consistent or reliable manner. As a result, efforts to cluster ESTs into sets of overlapping sequences and associate those clusters with genes in genomes are valuable, but insufficient.

A better source of data for confirming gene predictions is the second generation of EST/cDNA sequencing projects. These seek to generate full-length sequences of cDNA clones and thereby obviate the EST clustering problem. Complete sequences of full-length cDNA molecules cloned from a variety of tissues provide reliable information about both alternative splicing and intron–exon boundaries. As clones are selected for sequencing, physical constraints increase the likelihood that a clone contains a complete gene transcript. Such constraints include requiring the presence in the molecule of the 5' protein cap complex that initiates translation and of the 3' poly-A tail that indicates that mRNA splicing is complete. While clusters of ESTs are difficult to mine for alternative splicing, clusters of full-length cDNA sequences with poly-A tails immediately indicate alternatively spliced genes (GILL and SANSEAU, 2000).

As more genomes enter this phase of sequencing and larger cDNA libraries are built, the issues surrounding gene structure prediction are likely to become less pressing. In the meantime, refinements on many predicted genes must be deduced from EST and protein evidence. While not always accurate, these do provide a framework on which to base a gene prediction and the associated annotation and functional assignment (GAASTERLAND et al., 2000).

3 Genome Annotation Examples

The annotations of genomes involves using many of the bioinformatics tools and computational methods that are currently available. Two examples follow, one prokaryotic and one eukaryotic, to illustrate the specific challenges and resources available today for genome annotation. In both cases, we used the MAGPIE annotation system, which provides a graphic interface for users to explore the genomic sequence space and assists in automated data collection, analysis and annotation.

3.1 Example 1: Re-Annotating *Escherichia coli* – A New Look at a "Classic" Bacterial Genome

Three years after the initial release of the completed, fully annotated K-12 *E. coli* genome (BLATTNER et al., 1997), we re-analyzed the entire K-12 genome. The original annotation lacked a functional assignment for approximately 40% of all predicted ORFs. Since then, the sequencing efforts of dozens of other organisms have driven an exponential increase in the amount of sequence data available for comparison. This has greatly enriched the search for homologies between predicted coding sequences. In addition, experimental work in the intervening years has provided new information concerning the functions of many previously unannotated ORFs.

In preparing the MAGPIE project of *E. coli*, we chose to divide the genome into 100,000 bp segments with 10,000 bp overlaps. A quick review of the previously annotated genes from *E. coli* suggested that the longest gene was about 7,000 bp in length (BLATTNER et al., 1997 and annotated gene list from NCBI), so our choice of 10,000 bp ensured that no known gene would be split across two 100 kbp segments. Each segment became a MAGPIE "group", a logical subset of the total sequence data.

For the actual coding sequence prediction, the first phase of any genome annotation, we took a three-step approach. First, we incorporated GLIMMER 2.0, available from The Institute for Genome Research (TIGR), as the first gene prediction tool. GLIMMER uses statistical methods to develop a model of a "gene" for each organism, and then evaluates each potential coding region against the model. Any coding regions with sufficiently high scores are marked as putative ORFs for further consideration. This strategy allows GLIMMER to capture about 97% of all genes in a genome (DELCHER et al., 1999). We chose to run GLIMMER with a minimum ORF length cutoff of 80 nt (approximately 26 aa), as some regulatory proteins in *E. coli* are known to be very small (BLATTNER et al., 1997; ROBERT MAU, personal communication).

Because GLIMMER selectively identifies ORFs that match a statistical model of a gene for the organism, GLIMMER may miss genes that were laterally transferred or acquired more recently from other genomes. These genes may not yet have fully integrated into the genome, and may lack the hallmarks of a "native" gene. Given the frequency with which lateral transfer may occur in bacterial genomes (as suggested in DOOLITTLE, 1999), we felt that GLIMMER predictions alone would be insufficient for gene prediction. Therefore, we chose to combine the GLIMMER predictions with those of the syntactic gene finding tool available in MAGPIE.

MAGPIE uses Spliceorf, a syntactic gene prediction tool developed for use specifically with the system (GAASTERLAND and SENSEN, 1996). Spliceorf identifies stop codons and then "backtracks" to the farthest acceptable start codon in frame and defines this as an "open reading frame" (ORF). The syntactic approach is fairly effective for low- to medium-G+C content organisms, including *E. coli* which has a G+C content of 52% (determined from the MAGPIE analysis and supported by previous literature).

However, Spliceorf does overpredict genes and does not exclude ORFs that overlap each other on alternative reading frames. Because of the syntactic nature of Spliceorf's gene prediction, overprediction is a common issue. In addition, as one lowers the cutoff limit for length of an acceptable open reading frame, there is an exponential increase in the number of putative ORFs that Spliceorf will find. The number of genes predicted by Spliceorf with a cutoff of 80 nt, equivalent to the GLIMMER cutoff, would have predicted somewhere in the neighborhood of 20,000 genes for *E. coli*. Therefore, we set the lower bound on ORF length to 300 nt (100 aa) for the Spliceorf search, which gave us a preliminary set of about 8,000 putative coding regions.

By comparing Spliceorf predictions to GLIMMER predictions, we were able to create a non-redundant set of genes produced by each gene finding technique. Essentially, all Spliceorf predictions with an overlapping GLIMMER prediction were replaced by the GLIMMER prediction, whereas any Spliceorf predictions that had no overlapping GLIM-MER predictions were retained. In this fashion, we were able to refine the Spliceorf predictions, while retaining putative ORFs that could have been transferred to the *E. coli* genome fairly recently in its evolutionary history.

A final revision of this ORF list involved comparing each Spliceorf + GLIMMER-predicted ORF with the annotated Genbank set from the original BLATTNER et al. annotation. For every ORF predicted by either of these methods, we compared the stop position with the annotated version, if it existed. In cases where the stops were identical, but alternative starts were found, the longer ORF, whether from the original annotation or the Spliceorf/GLIMMER prediction was retained. In cases where the Spliceorf/GLIMMER prediction was longer than the annotated version, retaining the longer ORF allowed us to identify any ORFs that might have been previously mis-annotated. On the other hand, by retaining the longer version from the previous annotation, we corrected for errors in the GLIM-MER or Spliceorf prediction. However, where an annotated gene overlapped with, but did not share a common stop codon, we chose to retain both the Spliceorf/GLIMMER prediction and the annotated version.

This final comparison with the previous annotation also allowed us to capture many genes below the threshold cutoffs of both GLIMMER and Spliceorf predictions. For instance, the regulatory protein of the threonine operon is 23 amino acids in length (see MAG-PIE project, ORF ec_G0001), and while present in the original annotation, would not have been in either the GLIMMER or the Spliceorf set of predicted genes.

The final number of predicted genes from this tri-partite review of predictions was 5,527 genes compared with the original annotation of 4,405 genes (BLATTNER et al., 1997). To keep track of all ORFs, we chose a numbering system that numbered coding regions from 1 to "N" across the entire genome, irrespective of the individual group or groups that a coding sequence might lie in. This enabled us to do analyses across the entire genome without being hindered by the artificial division of the sequence into 100 kbp groups.

The second step of the annotation process involved data collection via homology-based searches. We chose to search the NCBI nucleotide (nt) and non-redundant protein (nr) databases as well as motif searches via prosearch. We also included searches against the predicted proteins of over 40 completed genomes, including the previously annotated *E. coli* set. We divided this last search into three databases containing bacterial genomes (BLASTpB), archeal genomes (BLASTpA), and eukaryotic genomes (essentially *C. elegans* and yeast, BLASTpE). This allowed for a more fine-tuned evaluation of homology across the three main branches of the phylogenetic tree.

Gene prediction and homology-based searching are only two of the steps toward a comprehensive genome annotation. The final step is a review of the information and the selection of a functional annotation for each predicted coding region. MAGPIE now utilizes a fully automated system for annotating a genome based on evidence from homology searches. The system, known as HERON, is still in development and currently evaluates the description lines for each high-scoring matching sequence to a query region. By following a protocol that emulates the human's decision-making process, the script attempts to select the "best" annotation possible from the description lines available. It is still assumed that a human reviewer will evaluate these automated selections, but in most cases, the automated annotation closely matches what a human annotator might have chosen for the ORF.

HERON works best when there are several complete or nearly complete hits to protein databases and seems to select an annotation that matches previous annotations closely. Therefore, the HERON script dramatically reduces the time required to review the evidence for each ORF in a genome. As will become apparent in the following example, this is crucial for rapid and effective annotation of eukaryotic genomes.

3.2 Example 2: Re-Annotating *Drosophila melanogaster* – Improving on Prediction and Annotation in a "Recent" Eukaryotic Genome

The challenges of annotating an eukaryotic genome are quite different from those of a prokaryotic genome as discussed earlier. Therefore, the strategy for eukaryotic annotation that we chose reflects our concern in ensuring the most complete annotation possible for the *Drosophila melanogaster* genome.

Eukaryotic genome capacities in MAGPIE are encoded as a suite of associated modules, collectively known as EGRET (for Eukaryotic Genome Rapid Evaluation Tools). Gene finding in EGRET occurs via GENSCAN, a gene identification program designed for use on eukaryotic organisms including human, *Arabidopsis thaliana* and others. GENSCAN uses a probabilistic model for the identification of the most likely "parse" of a gene, and attempts to *ab initio* predict exon–intron boundaries within these putative genes (BURGE and KARLIN, 1997). Output from GENSCAN is parsed into pre-existing formats and from thence onward, data collection proceeds as described above in the prokaryotic annotation strategy.

The process of developing a MAGPIE project around the very recently released sequence data (ADAMS et al., 2000) involved several steps. As a way to make the information processing simpler, we chose to create a MAGPIE project for each arm of the two large chromosomes (II and III), and a project for each of the other chromosomes (IV, X, and "U" for unassigned). We had 7 projects in all, each with about 25 Mbp of genomic sequence data.

Within the projects, the sequence was split in a manner similar to the *E. coli* project, into 100 kbp contigs with 10 kbp overlaps. We then ran GENSCAN as the primary gene prediction tool.

Since eukaryotic gene prediction is error-prone, an additional step is used to review the intron–exon boundaries that GENSCAN predicts. The first-pass data collection is followed by refinement either via manual intervention or the use of tools that can effectively map a protein sequence back to its genomic sequence. A recently developed tool, Genewise, is designed to automate the process of refining intron–exon boundaries given a protein sequence and its corresponding genomic sequence. Genewise maps the protein back to the genomic sequence, determining the exact intron–exon boundary in the process (BIRNEY and COPLEY, 1999). Genewise builds on statistical models of the Hidden Markov variety to be able to effectively map a protein back to its putative genomic sequence. Intron–exon boundaries are predicted based on a series of statistical calculations from known protein–genomic sequence sets, but can be calculated *ab initio* if desired as well (BIRNEY and COPLEY, 1999). This technique is the most conclusive mapping of intron–exon boundaries, but is limited in that complete protein or cDNA sequence is required. In contrast, the current manual review of evidence supporting intron–exon boundaries can be attempted even when evidence is relatively sparse and limited to EST (expressed sequence tags) or other minimal evidence.

GENSCAN was run on the *Drosophila* genome with the parameters suggested by the authors (BURGE, personal communication) which included using the intron–exon statistical model developed for human sequences. The first pass prediction provided approximately 22,290 coding sequences, but this number includes redundancies in the 10 kbp overlap regions between groups. The non-redundant set included 21,165 predicted coding sequences distributed across the four sequenced chromosomes, a significant increase in predicted coding sequences from the published set of 13,601 genes (ADAMS et al., 2000).

After gene prediction with GENSCAN, 10 tools were chosen for searches against the public databases. In addition to the standard BLASTn search against the nucleotide database and BLASTp search agains the non-redundant protein database (both at NCBI), we chose several additional searches. To determine which of our predicted coding sequences most closely matched the ADAMS et al. annotated genes, we ran a BLASTp and a BLASTn search against the annotated set provided by ADAMS et al. To capture possible paralogs and gene families within the genome, we also searched against the complete genomic sequence at both the protein and nucleotide levels. Finally, to identify genes that may have been missed during the annotation of ADAMS et al. because of their stringent screens for gene identification (ADAMS et al., 2000), we ran a search of each coding sequence (CDS) translated in the predicted reading frame against the EST database at NCBI translated in all 6 frames.

This final search, a tBLASTn search of the EST database, proved the largest bottleneck for rapid data processing. The version of the EST database used (as of April 2, 2000) had 3.6 million sequences in it with a total base pair count of just over 2 gigabases (2,000,000,000 bases). Each search against this database takes, on average, about 2.5 min to complete. Given the 21,165 sequences that had to be processed, the tBLASTnEST data collection took nearly 2 weeks running full-time on 12 SUN cpus. One of the future challenges we will face with annotation will be precisely this: as the sequence databases double in size over the next 6 months to a year, we will need ever larger amounts of computer memory to parse the databases for searching. Nevertheless, it is the tBLASTnEST data that has been the most effective in identifying potentially interesting genes that were not annotated by ADAMS et al.

HERON enables a first pass fully automated genome annotation to include putative function assignments for predicted coding regions, thereby allowing users to pursue questions of biological interest that might arise even at this stage. In the second stage of analysis, we will review each coding sequence's predicted exon–intron structure and the corroborating evidence for these predictions. In the past, this has been done entirely by manual intervention, but the development of Genewise,

a tool that can map protein sequence to genomic sequence will be used to automate the first level of review.

4 Conclusions

Genome annotation requires a spectrum of bioinformatics tools, each tuned to the genome of interest. Annotation is a tri-partite procedure. First, the information coding portions of a genomic sequence must be separated from the non-coding sequences. In all organisms, coding regions are differentiated from neighboring non-coding regions by specific features. Detecting these features is essential to transforming sequence data into knowledge of a genome. Once a gene is identified or predicted, the next step is to identify putative functions, possible homologs in other organisms and within the genome, and to postulate its role in the life of the organism. The final step in the process of annotating a genome generates a comprehensive collection of data related to each identified gene in the context of the genome.

The real challenge in genome annotation lies not in the collection or organization of data, but in the translation of that data into easily accessible forms for review by human users. A solid, carefully conducted annotation of a genome is merely the first step to developing intra- and inter-genomic analyses of coding sequences. In addition to understanding the specific function of a gene, users should be able to investigate the ways in which any one gene interacts with the rest of the genomic complement of genes and products. In order to do this, one must have a means of querying the collected data in different ways. In the examples above we presented the MAGPIE system, an integrated annotation tool that allows users to explore the genome space via a graphic interface. Effective automation of genome annotation will depend on the development of such systems that can integrate the multiple computational and analysis tools and their output for easy review by a human user.

5 References

ABAGYAN, R. A., TOTROV, M. M. (1994), Biased probability Monte Carlo conformational searches and electrostatic calculations for peptides and proteins, *J. Mol. Biol.* **235**, 983–1002.

ADAMS, M. D. and the sequencing teams at BDGP and Celera Genomics, Inc (2000), The genome Sequence of *Drosophila melanogaster*, *Science* **287**, 2185–2195.

ALBERTS, B., BRAY, D., LEWIS, J., RAFF, M., ROBERTS, K., WATSON, J. D. (1995), *Molecular Biology of the Cell*. New York: Garland Publishing.

ALEXANDROV, N. N., NUSSINOV, R., ZIMMER, R. M. (1995), Fast protein fold recognition via sequence to structure alignment and contact capacity potentials, *Proc. Pacific Symp. Biocomputing '96* (HUNTER, L., KLEIN, T. E., Eds.), pp. 53–72. Singapore: World Scientific Publishing.

ALTSCHUL, S., MADDEN, T., SCHAFFER, A., ZHANG, J., MILLER, W., LIPMAN, D. (1997), Gapped BLAST and PSI-BLAST: a new generation of protein database search programs, *Nucleic Acids Res.* **25**, 3389–3402.

BAILEY, T. L., GRIBSKOV, M. (1996), The megaprior heuristic for discovering protein sequence patterns, *Proc. 4th Int. Conf. Intelligent Systems for Molecular Biology*, pp. 15–24. Menlo Park, CA: AAAI Press.

BAIROCH, A. (1992), PROSITE: A dictionary of sites and patterns in proteins, *Nucleic Acids Res.* **20**, 2013–2018.

BAIROCH, A., APWEILER, R. (1999), The SWISS-PROT protein sequence data bank and its supplement TrEMBL in 1999, *Nucleic Acids Res.* **27**, 49–54.

BATEMAN, A., BIRNEY, E., DURBIN, R., EDDY, S., FINN, R., SONNHAMMER, E. (1999), Pfam 3.1: 1313 multiple alignments and profile HMMs match the majority of proteins, *Nucleic Acids Res.* **27**, 260–262.

BIRNEY, E., COPLEY, R. (1999), *Documentation for Genewise and Associated Package*. Sanger Centre, Cambridge, UK.

BLAST Documentation (1999), National Center for Biotechnology Information (NCBI), Bethesda, MD.

BLATTNER, F. R., PLUNKETT, G. 3rd, BLOCH, C. A., PERNA, N. T., BURLAND, V. et al. (1997), The complete genome sequence of *Escherichia coli* K-12, *Science* **277**, 1453–1474.

BOGUSKI, M. S., SCHULER, G. D. (1995), ESTablishing a human transcript map, *Nature Genet.* **10**, 369–371.

BURGE, C. (1997), *Documentation for Command Line Version of GENSCAN*. Department of Mathematics, Stanford University, CA.

BURGE, C., KARLIN, S. (1997), Prediction of complete gene structures in human genomic DNA, *J. Mol. Biol.* **268**, 78–94.

CALIFANO, A., STOLOVITZKY, G., TU, Y. (2000), Analysis of gene expression microarrays for phenotype classification, *Proc. 8th Int. Conf. Intelligent Systems for Molecular Biology* (ALTMAN, R., BAILEY, T. L., BOURNE, P., GRIBSKOV, M., LENGAUER, T. et al., Eds.), pp. 75–85. Menlo Park, CA: AAAI Press.

CSERZO, M., WALLIN, E., SIMON, I., VON HEIJNE, G., ELOFSSON, A. (1997), Prediction of transmembrane alpha-helices in prokaryotic membrane proteins: the Dense Alignment Surface method, *Protein Eng.* **10**, 673–676.

DELCHER, A. L., HARMON, D., KASIF, S., WHITE, O., SALZBERG, S. L. (1999), Improved microbial gene identification with GLIMMER, *Nucleic Acids Res.* **27**, 4636–4641.

DOOLITTLE, W. F. (1999), Lateral genomics, *Trends Cell Biol.* **9**, M5–8.

FRISHMAN, D., ARGOS, P. (1996), Incorporation of long-distance interactions into a secondary structure prediction algorithm, *Protein Eng.* **9** (2), 133–142.

GAASTERLAND, T., SENSEN, C. W. (1996), Fully automated genome analysis that reflects user needs and preferences – a detailed introduction to the MAGPIE system architecture, *Biochimie* **78**, 4.

GAASTERLAND, T., SCZYRBA, A., THOMAS, E., AYTEKIN-KURBAN, G., GORDON, P., SENSEN, C. W. (2000), MAGPIE annotation of the 2.91 MB *Drosophila melanogaster* ADH Region, *Genome Res.* **10**, 502–510.

GILL, R.W., SANSEAU, P. (2000), Rapid *in silico* cloning of genes using expressed sequence tags (ESTs), *Biotechnol. Ann. Rev.* **5**, 25–44.

GOPAL, S., SCHROEDER, M., PIEPER, U., SCZYRBA, A., AYTEKIN-KURBAN, G. et al. (2001), Homology-based annotation yields 1042 new candidate genes in the *Drosophila melanogaster* genome, *Nature Genet.* **27** (3), 337–340.

HENIKOFF, J. G., PIETROVSKI, S., MCCALLUM, C. M., HENIKOFF, S. (2000), Block-based methods for detecting protein homology, *Electrophoresis* **21** (9), 1700–1706.

HYATT, D., SNODDY, J., SCHMOYER, D., CHEN, G., FISCHER, K. et al. (2000), *Improved analysis and annotation tools for whole-genome computational annotation and analysis: GRAIL-EXP genome analysis toolkit and related analysis tools*, presented at Genome Sequencing and Biology Meeting, May 2000. *Proc. Genome Sequencing and Biology Meeting*, Santa Fe, NM, May 2000.

JONASSEN, I., COLLINS, J. F., HIGGINS, D. G. (1995), Finding flexible patterns in unaligned protein sequences, *Protein Sci.* **4**, 1587–1595.

KARPLUS, K., BARRETT, C., HUGHEY, R. (1998), Hidden Markov Models for detecting remote protein homologies, *Bioinformatics* **14**, 846–856.

KNUDSEN, S. (1999), Promoter2.0: for the recognition of PolII promoter sequences, *Bioinformatics* **15**, 356–361.

KURIYAN, J., DARNELL, J. E., Jr. (1999), Protein recognition – An SH2 domain in disguise, *Nature* **398**, 22–23, 25.

LAWRENCE, C. E., ALTSCHUL, S. F., BOGUSKI, M. S., LIU, J. S., NEUWALD, A. F., WOOTTON, J. C. (1993), Detecting subtle sequence signals: A Gibbs sampling strategy for multiple alignment, *Science* **262**, 208– 214.

LEVITSKY, V. G., KATOKHIN, A. V., KOLCHANOV, N. A. (2000), Computational technologies, *Novosibirsk* (special issue) **5**, 41–47.

LOWE, T., EDDY, S. E. (1997), tRNAscan-SE: a program for improved detection of transfer RNA genes in genomic sequence, *Nucleic Acids Res.* **25**, 955–964.

MUNTEANU, M. G., VLAHOVICEK, K., PARTHASARATY, S., SIMON, I., PONGOR, S. (1998), Rod models of DNA: sequence-dependent anisotropic elastic modeling of local bending phenomena, *Trends Biochem. Sci.* **23**, 341–346.

MYLER, P., AUDELMAN, L. and their sequencing teams (1999), *Leishmania major* Friedlin chromosome I has an unusual distribution of protein-coding genes, *Proc. Natl. Acad. Sci. USA* **96**, 2902–2906.

PAVY, N., ROMBAUTS, S., DEHAIS, P., MATHE, C., RAMANA, D. V. V. et al. (1999), Evaluation of gene prediction software using a genomic data set: application to *Arabidopsis thaliana* sequences, *Bioinformatics* **15**, 887–899.

PEARSON, W. (1990), Rapid and sensitive comparison with FASTA and FASTP, *Methods Enzymol.* **183**, 63–98.

PRESTRIDGE, D. S. (1995), Prediction of Pol II promoter sequences using transcription factor binding sites, *J. Mol. Biol.* **249**, 923–932

ROST, B. (1996), PHD: predicting one-dimensional protein structure by profile based neural networks, *Methods Enzymol.* **266**, 525–539.

ROTH, F. R., HUGHES, J. D., ESTEP, P. E., CHURCH, G. M. (1998), Finding DNA regulatory motifs within unaligned non-coding sequences clustered by whole-genome mRNA quantitation, *Nature Biotechnol.* **16**, 939–945.

SANCHEZ, R., SALI, A. (1997), Advances in comparative protein-structure modeling, *Curr. Opin. Struct. Biol.* **7**, 206–214.

SANSOM, M. S. P., KERR, I. D., LAW, R., DAVISON, L., TIELEMAN, D. P. (1998), Modelling the packing of transmembrane helices: Application to aquaporin 1, *Biochem. Soc. Trans.* **26**, 509–515.

SCHULER, G. D. (1997), Pieces of the puzzle: expressed sequence tags and the catalog of human genes, *J. Mol. Med.* **75**, 694–698.

SCHULER G. D., BOGUSKI, M. S., STEWART, E. A., STEIN, L. D., GYAPAY, G. et al. (1996), A gene map of the human genome, *Science* **274**, 540–546.

SMIT, A. F. (1999), Interspersed repeats and other mementos of transposable elements in mammalian genomes, *Curr. Opin. Genet. Devel.* **9**, 657–663.

SOLOVYEV, V. (1999), *Documentation for FGENES and FGENESH*. Sanger Centre, Cambridge, UK.

VON HEIJNE, G. (1992), Membrane protein structure prediction, hydrophobicity analysis and the positive-inside rule, *J. Mol. Biol.* **225**, 487–494.

ZHANG, M. Q. (1998), Identification of human gene core-promters *in silico*, to appear in the Bioinformatics special issue of *Genome Res.*, **8** (3), 319–326.

ZUKER, M. (2000), Calculating nucleic acid secondary structure, *Curr. Opin. Struct. Biol.* **10**, 303–310.

17 Genomic Data Representation through Images – MAGPIE as an Example

PAUL GORDON

Halifax, Canada

TERRY GAASTERLAND

New York, NY, USA

CHRISTOPH W. SENSEN

Calgary, Canada

1 Introduction

Graphical display systems for complex data, which can be used to analyze what would otherwise be overwhelming amounts of data, are becoming increasingly important in many scientific fields. Molecular biology and genomics are key areas for this development, because the size and number of genomic databases and the number of analysis tools is continually increasing. A typical microbial genome has up to 1,000 open reading frames (ORFs) per megabase of genomic sequence. If just 10 tools were used to analyze such a genome, and each tool showed hits against 100 database entries (which is not an uncommon number), about one million pieces of evidence would need to be recorded and mapped along the megabase of genomic sequence. When dealing with this amount of information, the saying "A picture is worth a thousand words" is an understatement!

In 1996, GAASTERLAND and SENSEN introduced a system for the automated analysis of biological sequences, called MAGPIE (Multipurpose Automated Genome Invesigation Environment) (GAASTERLAND and SENSEN, 1996). This tool-integration system executes and then integrates the results from multiple bioinformatics tools into an easily interpretable form. It has been used for the analysis of complete genomes, genomic DNA fragments, ESTs, proteins, and protein fragments. Initially, all MAGPIE output was in tabular format, but the need for rich visual representations of genomic analyses became evident early on. MAGPIE summarizes information about evidence supporting functional assignments for genes, information about regulatory elements in genomic sequences, including promoters, terminators and Shine–Dalgarno sequences, metabolic pathways, and the phylogenetic origin of genes. MAGPIE sorts and ranks the evidence by strength and is able to display the level of confidence associated with database search results.

MAGPIE was the first genome analysis and annotation system to add graphical representations to the results. Based on continual user feedback from various installations, the images have been refined over time, allowing annotators to process the quantity of relevant information quickly and efficiently. MAGPIE has one of the most comprehensive graphical capabilities available, thus we will use MAGPIE as the example for genomic annotation that is supported by imaging. We describe the meaning of the various images, the algorithms used to create them, and the data types from which they are derived.

The MAGPIE image types reflect the fact that sequence data is stored and presented in a hierarchical manner. A MAGPIE project generally consists of, but is not limited to, related sequences generated for a particular organism. To facilitate browsing and data maintenance, sequences are organized into logical groups. For example, all sequences from a single clone are normally placed in one group, because they will be joined as sequencing progresses. The images presented in this text are from the *Sulfolobus solfataricus* P2 sequencing project (CHARLEBOIS et al., 1996). This project was chosen as the example because it includes all of the graphical representations that can be produced by MAGPIE. The complete *Sulfolobus* MAGPIE project is available at *http://www.cbr.nrc.ca/sulfhome*.

MAGPIE images can be classified into three categories: representation of evidence, summary of genomic features, and biochemical assay simulation, which supports the design of follow-up experiments. We discuss how these images aid the researcher in genome sequencing and annotation through pattern recognition and information filtering and how they can be used to support the validation of genomic data.

2 The Graphical System

The MAGPIE user interface, which was initially implemented with a Web-based display that supported text and table output, has been developed over time into a Web-based graphical display system. As described previously (GAASTERLAND and SENSEN, 1996), the tools used in MAGPIE include, but are not limited to: the FastA (PEARSON and LIPMAN, 1988) family of programs (including the *ssearch* Smith–Waterman implementation, and protein fragment analysis tools *fastf* and *tfastf*), the BLAST (ALTSCHUL et al., 1997) family of programs (ungapped, gapped, and Position Specific Iterative), Blitz (*http://www.ebi.ac.uk/bic_sw/*), BLOCKS (HE-NIKOFF et al., 1999), ProSearch (KOLAKOWSKI et al., 1992), Genscan (BURGE and KARLIN, 1997), Glimmer (SALZBERG et al., 1998), and GeneMark (BORODOVSKY and MCININCH, 1993). To link image representations to alignments, individual tool "hits" and other related information, the original tool outputs (e.g., BLAST or BLOCKS responses) are stored as HTML files after data processing. Hit length and scores are extracted during the response processing using dedicated parsers. The hits are sorted into user-defined confidence levels.

The components of the modular computer code used for MAGPIE input and output are modules for Web standards written in the Perl5 (*http://www.perl.com*) programing language: HTML (Hypertext Markup Language), CGI (Common Gateway Interface), GIF (Graphics Interchange Format), and PNG (Portable Network Graphics). The text reporting system is based on a combination of pre-computed HTML pages and CGI programs producing HTML dynamically. A graphics library module called GD.pm (*http://stein.cshl.org/www/soft-ware/GD*), which is dynamically patched into the Perl5 system, is used to generate the MAGPIE graphics. GD.pm provides functionality for drawing on a two-dimensional canvas. The canvas can be translated into a browser-readable form such as PNG. Lines, basic geometric shapes, and arbitrary polygons can be drawn and color-filled using GD.pm. The drawing functions, when applied to linearly encoded data such as DNA or proteins, lend themselves to particular succinct representations. Rulers, along which features are positioned, are drawn as straight lines. Simple ticks (small lines perpendicular to the ruler) generally represent position-specific features that occur frequently, e.g., stop codons. Unique polygons that occupy more space are used for position-specific features, which occur less frequently, e.g., promoter sites. Data that cover a range, e.g., open reading frames, are displayed as boxes. These boxes may be subdivided when additional information needs to be encoded.

MAGPIE's graphics fulfill three main needs in a genome project once data is collected: display of the genomic features at various levels of detail in the genomic context, evaluation the evidence supporting a feature's annotation, and quality control. Four types of information are used to generate images in MAGPIE: user preferences, sequence information, data from tool outputs as well as analysis results, and manual user annotations (verifications). User-based annotations are stored as text files, usually created via the Web interface, but they can also be imported from files using standards such as GenBank flat format or the General Feature Format (*http://www.sanger.ac.uk/ Software/formats/GFF/*). User-configurable visualization parameters (e.g., the bases-per-pixel scale and the maximum image width) are stored in plain text configuration files, similar to the previously described (GAASTERLAND and SENSEN, 1996) configuration files for confidence level criteria and other configurable parts of MAGPIE. By default, images in MAGPIE are defined with a maximum width of 1,000 pixels. This allows landscape mode printing of the images on $8 \times 11''$ paper.

While most data in MAGPIE is hierarchical and stored as text files, cross-referencing of equivalent sequence identifiers is done using binary Gnu Database Manager (GDBM) files. This exception to the text-file storage allows the rapid location of a current analysis report via an identifier for a previous version of the respective sequence (version control).

The graphical reporting system in MAGPIE has two distinct user-configurable modes: static or dynamic, respectively. In the static graphical mode, all images are pre-computed for viewing after the analysis is finished. This requires considerable disk space, but it is computationally and temporally efficient when the sequence and analyses do not change much, e.g., when a completely finished genome is analyzed. In the dynamic mode, images are created on demand, using the data extracted from the analysis. Although this requires more *ad hoc* computation, it is appropriate when the underlying sequences or the analyses are frequently updated, for example in the case of an ongoing genome sequencing project.

Two key features for viewing data in context are the hierarchical representation of the data and consistent display idioms. Idioms such as bolder coloring for stronger evidence (darker text, brighter hues, respectively), and using red and blue to indicate the forward or reverse DNA strand respectively, pervade the images and allow complex information to be encoded in the graphics without cluttering them. Good idioms need only be learned once (COOPER, 1995). For example, by surveying the annotation and evidence strength status from Fig. 6, as well as the overlaps, it becomes evident how the clone relates to its genomic neighbors, and how much information has been gathered about the non-redundant part of the sequence.

Consistent color use improves the delivery of information in the MAGPIE images. Blue bars and borders indicate information located on the positive DNA strand (forward strand), while red bars and borders represent information located on the negative DNA strand. Black generally indicates information that is not strand specific. Coloration of analysis data is specified in a user-definable color preference file. Consistent coloring can group the evidence from similar tools by color range. For example and shown later in the text, FastA (PEARSON and LIPMAN, 1988) hits against an EST collection will always be shown in a particular green hue, and FastA hits against a protein database may be colored in a different green hue.

Shading is used throughout the MAGPIE interface to denote the confidence level of the evidence. Stronger evidence is always displayed in darker shades. For example, description text in reports is black when the evidence is good, gray when it is moderate, and white when it is only marginally useful. As shown below, this holds true for the graphical evidence displays as well. It is easy to filter out information related to potential genomic function by following this simple concept. These representational consistencies also reduce visual clutter. Information is implicitly conveyed in the color instead of requiring explicit depiction or labeling of the displayed features.

Three key features of the individual ORF display are succinctness, pattern display, and data linking. The succinct representation of the tool responses is essential to allow the annotator a quick survey of the salient information. Some of the information, such as the subject

description and discriminant score, remains in textual form within the images, while details such as the location of the hits are graphically mapped onto the images. The exact positioning of hit patterns is important to the annotator, because it can determine the relevance of the match. The comparative match display is maximized in MAGPIE by displaying results from tools, which are based on similar algorithms, data sets and evidence types (i.e., amino acid, DNA, and motif), atop each other.

Even though a succinct representation of the responses can be very useful, it is equally important to be able to access the original responses, database entries, associated metabolic pathways and other related information, thus most image types contain configurable hyperlinks to Web-based information, including SRS-6, ExPASy, or the NCBI Web services.

Finally, simulation images assist the wet-lab researcher during the verification process. MAGPIE analyses, like all results generated by automated genome analysis and annotation systems, are only computer models, which often need verification through a biochemical experiment.

2.1 The Hierarchical MAGPIE Display System

In the following paragraphs, we introduce the various graphical displays that are implemented in MAGPIE. The MAGPIE hierarchy is reflected in the set of graphical images. The resolution of the images increases over several levels until an almost single base pair resolution is reached. Fig. 1 shows the hierarchical connection between the different images. Depending on the state of the analyzed sequence, not all images are present; some images are mutually exclusive.

2.1.1 Whole Project View

MAGPIE can track sequencing efforts from single sequence reads to fully assembled clones and genomes. Based on mapping information, which identifies the relationship of clones in the sequencing project, MAGPIE can automatically generate and display non-

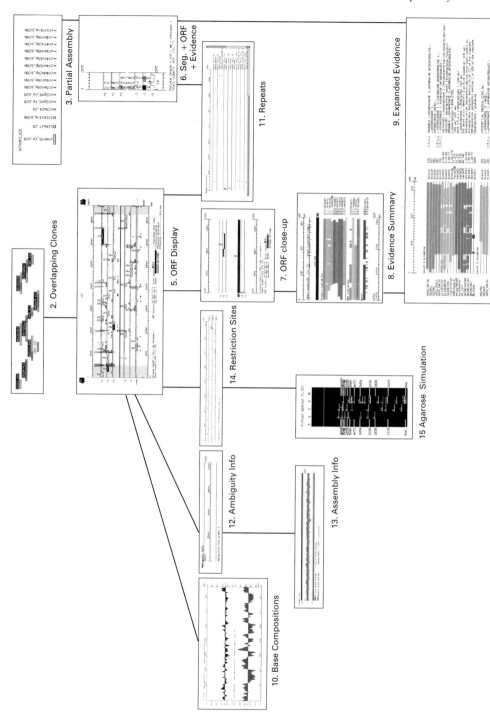

Fig. 1. Image hierarchy. From top to bottom, more detail is shown about smaller subsequences. Where connecting lines exist, the images are either juxtaposed or click-through. The images are labeled according to their figure numbers.

Fig. 2. Overlapping clone cluster in a Magpie project. Each sequence is hyperlinked to its Magpie report. All of the sequences are filled in with green, denoting finished sequence, and shaded where redundant. Red-outlined sequences are reverse complemented in the assembly. White letter labels denote the presence of annotations in the sequence (see color plates p. 458).

Fig. 3. Partially assembled fragments of BAC b07zd03_b28. Fragments are sorted by size and hyperlinked to their respective Magpie reports. The yellow fill indicates that sequences are in the linking state (see color plates p. 458).

redundant sequence(s) and the non-redundant gene set. Figs. 2 and 3 show examples of the images that display the summary "whole project view" page of the *Sulfolobus solfataricus* genome. Acting as a starting point for the annotator, this single page contains hyperlinked images representing all of the contiguous sequences (contigs) in the MAGPIE project. Contigs are drawn to scale, and color-filled according to their MAGPIE state. The states used in the *Sulfolobus* MAGPIE project are primary, linking, polishing and finished respectively, but the user for other projects could define other states. The colors for the states can be set in the user preference files. Users can also define in which of the states the sequence is resolved well enough so that MAGPIE can assemble larger contigs from individual clones. In the example of the *Sulfolobus* MAGPIE project, this would be sequence in state "polishing". Overlapping contigs are appropriately positioned in the image, and the areas of overlap are grayed out to denote redundancy. In keeping with the color usage described earlier, blue outlines denote that the contigs are in their normal (forward) orientation, while those outlined in red were reverse complemented in order to fit into the genome assembly. White text on a black background denotes the presence of manual annotation (verifica-

tion) on the labeled contig. In Fig. 2, it is discernable by the black label text and gray fill that clone l910_127 is the only clone without annotation. This clone was not annotated because it is completely redundant.

Even though overlaps between MAGPIE contigs are calculated in order to remove redundancies, MAGPIE is not meant to be a full-fledged assembly engine. For contigs to be considered overlapping, the user must specify that two clones are neighbors. This is information usually known from the clone-mapping phase or derived from self-identity searches in MAGPIE. The user specification avoids spurious assemblies that may be taken for granted. The extent and orientation of the sequence overlap is determined by running a FastA similarity search. Based on the percent similarity and length criteria set for the project, the overlap is either accepted or rejected. If the overlaps do not occur at the very ends of the contigs, the match is also rejected. This helps to avoid linking contigs based on repetitive regions. Based on the one-to-one neighbor information, larger contigs are formed using the following logic:

(1) let S be a set of sets, each containing a single contig
(2) let N be the set of neighbor relationships

(3) while N is not the empty set
    a) pick a relationship R(C1,C2) from N
    b) $N = N - R$
    c) Find S1, the set in S to which the
       contig C1 belongs
    d) Find S2, the set in S to which the
       contig C2 belongs
    e) $S = (S - S1 - S2) \cup (S1 \cup S2)$

The exclusive sets in S are implemented in an array format. This format provides constant-time union of sets and logarithmic-order time set membership determination. A consensus sequence for each contig set in S is determined. When conflicts between overlapping contigs occur, the better resolved sequence of the contig in a more complete state takes precedence. The ORFs on the consensus sequence are identified and a table of ORF equivalencies across all the contigs is created subsequently.

At step 1, each contig is in its own set. Because the data set consists of non-redundant contigs, the sets in S are by default disjoint. Steps 2, 3a and 3b iterate through all known connections. Two sets, which are determined in steps 3c and 3d can be combined when a contig from the first set is connected to one from the second set, which is valid because of the transitive property of contig connections. In step 3e, the two now connected contigs are removed from S and replaced with a combined set. The resulting sets of disjoint contigs are called "supercontigs" in MAGPIE, because they may consist of more than one clone.

The overlap and equivalency information is used in the images described below to propagate ORF information across equivalent sequence feature displays. The algorithm has been used in the *Sulfolobus solfataricus* P2 genome project to successfully assemble 110 bac-, cosmid- and lambda clones into a single non-redundant contig.

2.1.2 Coding Region Displays

A number of different image types represent the ORF evidence with increasing levels of detail. Sometimes it is necessary to see the actual evidence, e.g., for decision making during the manual annotation (verification) process. At other times the larger context of the ORFs is more useful, in this case, a detailed display containing all MAGPIE evidence would be overcrowded. These needs necessitate multiple representations of the same data. The varying levels of evidence abstraction are the most powerful part of the MAGPIE graphical environment.

A user may specify a wide variety of tools to be run against all contigs in a particular MAGPIE state. Tools used by MAGPIE can also produce periodically updated results, adding to the dynamic nature of the evidence. A user may wish to store all of the MAGPIE-generated images or create them on demand. This depends on available disk space, CPU power, and the frequency at which contigs and tool outputs are updated.

As shown in Fig. 4, the same code is used to generate both images and their image maps. When the CGI is called, it generates the necessary HTML content on the Web page by setting the image maker command-line arguments to print the image to standard output. Standard output is redirected to a port on the server, and the image URL points to that same port. The CGI then redirects standard error to the former standard output. It includes (as opposed to launching as a separate process) the image-making script with the appropriate arguments. When the script prints the image to the standard output stream instead of to a file, it is configured to print the image map to standard error. The stderr data stream is now redirected to the client's HTML page, therefore, no temporary files are created, even though two output streams are used simultaneously.

All scripts that generate a graphical representation of a contig and its ORFs may generate more than one image for the sequence. If the combined contig length and scale factor exceeds the maximum image width, the image is split into multiple "panes". The number and size of the images must be determined before any drawing takes place. This allows the drawing canvases to be allocated in the program. The information is also used to create the required number of image references in the HTML pages. The height of the image is fixed, because it depends entirely on fixed parameters. The image width is variable because sequences are represented horizontally. The

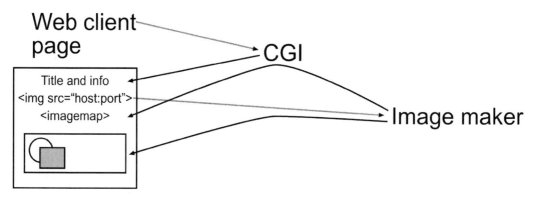

Fig. 4. Fileless HTML and image generation. Dashes represent client requests. The upper curved line represents the standard error output stream of the image maker redirected through the CGI standard output stream. The lower curved line represents the former's standard output redirected to the open port specified by CGI.

width is a function of the sequence length times the scale factor, plus constant elements such as border padding. When multiple panes are required, all but the last image have the maximum width. If the last image had ten or less pixels, it is merged into the preceding image. This slightly exceeds the maximum permitted width, but avoids an unintelligibly small sequence display.

2.1.3 Contiguous Sequence with ORF Evidence

Fig. 5 displays a sequence and its features with the highest degree of data abstraction. Images of the type shown in Fig. 5 summarize the evidence against all of the ORFs in a contig in all six open reading frames. They also show additional genomic features, which may be located around coding regions: promoters, terminators, and stop codons. Links to the corresponding ORF reports are provided via the imagemap. ORFs are labeled sequentially from left to right. In our example from the *Sulfolobus* project, the 100 amino acid residue cutoff is stated in the lower left-hand corner of Fig. 4. Inter-ORF regions are analyzed separately by MAGPIE using a set of scoring criteria, which is different from the one defined for large-ORF regions. The goal of the inter-ORF analysis is to identify small coding regions

(e.g., small proteins and RNA-coding regions). The names of the potential small coding regions meeting the user-defined criteria are denoted with the prefix "n". Clicking on a displayed small coding region brings up a screen that displays the evidence as in shown in Fig. 9. The user must confirm that the small sequence segment is indeed a coding region. Once confirmed, these identifiers of the small ORF are displayed with the "s" prefix. This naming convention differentiates types of small ORFs without the need for renaming ORFs when small ORFs are verified to be coding.

There are several aspects to ORF coloration. Gray-shaded boxes indicate ORF suppression when users deem particular ORFs to be non-coding. This may be when the ORF is more than a certain percentage inside another ORF (typically completely contained in an ORF on the opposite DNA strand), or the ORF shows an unusual amino acid composition. MAGPIE can be configured to automatically suppress ORFs for either reason, or for lack of evidence. ORFs with Xs (so called Saint Andrew's Crosses) through them have been annotated. A white background denotes that the assigned function is "putative", "hypothetical", or "uncharacterized". All three words stand for unknown function. These functional assignments (or lack thereof) may be carried over from similar ORFs in other genomes. The equivalencies used were determined in the creation of Figs. 2 and 3. Outline

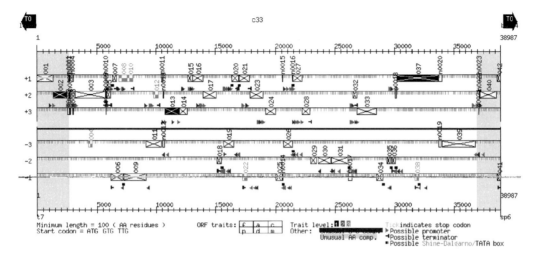

Fig. 5. Contiguous sequence with open reading frames displayed. Boxes on the six reading frame lines represent possible genes. The boxes are "x"ed when annotated. Light "x"ed boxes have annotations described as hypothetical or uncharacterized. Sub-boxes in unannotated genes indicate composition characteristics, plus the best level of protein, DNA and motif database hits. Grayed-out genes have been suppressed. Background shading and hyperlinked arrows in the corners indicate neighboring sequence overlaps. Boxes with labels that start in "n" are possible genes shorter than the specified minimum length (see color plates p. 458).

colors for ORFs starting with ATG, GTG and TTG are blue, green, and red, respectively. The different colors are only used, if they are defined by the user to represent the valid start codons for the organism. When the ORF starts upstream of the current contig and the start codon is unknown, the outline is black. Black also indicates the use of alternative start codons, which may occur in some organisms.

ORFs that are not validated by a user are split into three by two isometric blocks. These blocks can be colored to indicate the presence and strength of certain evidence. This is indicated in the "ORF traits" section of the Fig. 4 legend. The blocks in the upper half denote calculated sequence characteristics. Blocks "f" and "a" indicate on/off traits. A blue "f" block denotes that the codon usage in the ORF is within 10% of frequencies observed for this organism. A blue "a" block denotes that the purine $(A+G)$ composition of this ORF is greater than 50%. This is known to be an indicator for a good coding likelihood in many prokaryotes (CHARLEBOIS et al., 1996). For organisms with a $G+C\%$ greater or smaller than 50%, the "c" block in the upper right corner of

the block represents another indicator for coding sequence; $G+C\%$ codon compensation. The calculation of this parameter is based on the third (and to a lesser degree second) position codon wobble. Compensation at confidence level 1 occurs when the combined frequency of $G+C\%$ compensation is highest in the third base of the codons. For level 2, the second base $G+C\%$ compensation is the highest, followed by the compensation for the third base of the codons.

The blocks in the lower half denote database search results and the level of confidence in three levels, level one indicating the strongest evidence. The colors for the three levels are blue, cyan, and gray respectively. The lower half trait levels are determined by comparing extracted similarity analysis scores with the user-specified criteria. The "p" block indicates the highest level of protein similarity found through the database searches. The "d" block indicates the highest level of DNA similarity found. The "m" block indicates the best level of sequence motifs found (e.g., scored Prosite hits). After learning the representation scheme, the annotator can quickly see the nature and strength of coding indicators for all ORFs in a sequence.

Other indicators for transcription include Shine–Dalgarno motifs (SHINE and DALGARNO, 1974), promoters and terminators. These features are displayed in the appropriate reading frame as small black rectangles, green triangles and red sideways T's, respectively. In keeping with the representational consistency, the candidate with the highest score for each of these features around any ORF is colored in a darker shade. Shine–Dalgarno motifs are found by matching a user-defined subsequence, which represents the reverse complement of the 3' end of the organism's 16s rRNA molecule. Promoter and terminator searching has so far been implemented for archaeal DNA sequences (GORDON and SENSEN, unpublished data).

Stop codons are marked with orange ticks within the reading frames. The location of stop codons is determined while the image is being created as follows: In each forward translation frame, the search for the next stop codon begins at the first in-frame base represented by the next pixel, thus increasing the calculation efficiency without sacrificing information in the display. For example, if a stop codon is found at base 23, and each pixel represents 50 bases, the search for the next stop codon in that frame starts at base 51. This is because 51 is the first in-frame triplet in the next pixel, representing the [50, 100] range. By not repeatedly drawing stop ticks in the same pixel, no wasted rendering effort is made. Another display shortcut is to search for the reverse complements of the stop codons on the forward strand in the same manner when rendering the negative DNA strand. Finding stop codon reverse complements saves the effort of reverse complementing the whole sequence and inverting the information again for graphical rendering.

On the ends of the lower ruler, labels can be added to indicate information about the priming sites, which are flanking the insert of the clone. In Fig. 5 it is shown that the insert is in sp6 to t7 orientation. The information about the ends of a clone is stored in the configuration file that defines the clone relationship.

The image also displays the overlaps between contigs, which were determined dur-

ing the generation of Figs. 2 and 3. Overlaps are denoted with arrows in the upper left and right hand corners of the image. The extent and orientation of the overlaps is indicated by the shading of the background between the upper and lower rulers in the blue (for forward orientation of the neighbor) or red (for reverse complement orientation of the neighbor).

2.1.4 Contiguous Sequence with Evidence

Fig. 6 shows an example of a contiguous sequence display with evidence. Images of this type are similar in layout to the image shown in Fig. 5, reducing the user's learning curve. This type of image is used when the sequence is in a primary or linking state, i.e., when multiple sequencing errors and ambiguities still may exist in the sequence. The key difference between Figs. 5 and 6 is that evidence in Fig. 6 is not abstracted into ORF traits. Evidence is displayed at its absolute position on the DNA strand. Fig. 6 demonstrates the usefulness of this display. The cyan (level 2) and blue (level 1) evidence is on the –2 DNA strand, and a

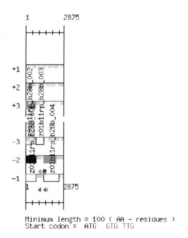

Fig. 6. Contiguous sequence with open reading frames and evidence displayed. Evidence and ORFs are displayed in their frames and locations. This facilitates easy recognition of frameshifts and partial genes. Other characteristics are the same as in Fig. 5 (see color plates p. 459).

frameshift in the 3′ end of the sequence is likely, because of the neighboring ORFs on the –3 strand. Higher ranked evidence always appears above lower ranked evidence. In this way, the more pertinent information is displayed in the limited screen area available. The evidence markings are hyperlinked to the pattern matches and alignments, which they represent. The ORFs are linked to a Fig. 9 view of the evidence that falls within the same boundaries on the same DNA strand.

These images can also be used to find previously unrecognized assembly overlaps. To identify potential overlaps, a MAGPIE sequence database search of each contig against all contigs in the project is performed, and self-matches of contigs are suppressed in the subsequent analysis. Any matches potentially mating two contigs are then visible. This kind of information would not be represented in Fig. 5.

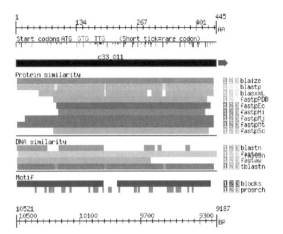

Fig. 8. Evidence summary. Evidence is grouped by type, and displayed as one line for all hits from a tool. Better evidence is darker and closer in the foreground (see color plates p. 459).

2.1.5 ORF Close-Up

The main purpose of Fig. 7 is to display a close-up of the ORF and surrounding features. Links to the overlapping ORFs are provided, so that the user can check whether the inner ORF or the outer ORF is an artifact, and to provide indications for frame shifts, which might result in slightly overlapping ORFs such as ORF number 009 in the example. The color-

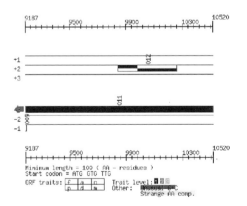

Fig. 7. ORF close-up. Displays the exact start and stop coordinates, and hyperlinked overlapping ORFs. Gene labels and trait sub-boxes are potentially easier to read than in Fig. 5 (see color plates p. 459).

ation of the ORFs is analogous to Fig. 5, this is useful for smaller ORFs which otherwise could be difficult to read. The green arrow indicates the orientation of the ORF. ORFs on the negative DNA strand are reverse-complemented in Figs. 8 and 9 to display all evidence in 5′-3′ orientation, thus they may be displayed in different orientations than Fig. 7.

Because of the simplicity of this image, its width is fixed. Unlike the situation in most other images, the scale factor is a function of the sequence length divided by the fixed width. The rulers have major markings every 100 pixels. For example in Fig. 7, the scale is (9900–9500)/100, or 4 bases/pixel.

2.1.6 Analysis Tools Summary

Fig. 8 contains a summary of the location of evidence from all tools along an ORF. This provides an overview of which parts of the ORF have evidence. The image has a fixed width, and it is usually shown next to the similarly sized Fig. 7. Its height is dependent on the number of tools that yielded valuable responses. This requires that all of the evidence information is loaded into memory before the canvas allocation and drawing of the image is performed.

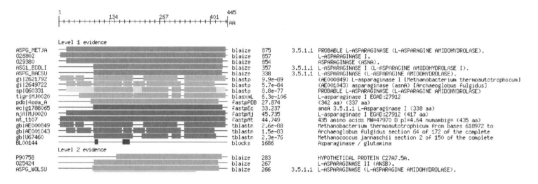

Fig. 9. Expanded evidence. Evidence is sorted by level, tool, score, and length. The first column links to the database ID of the similar sequence. The second displays the similarity location. It is linked to the original tool report. The third names the tool used. The fourth displays the tool's scoring of the match. The fifth displays the Enzyme Commission Number, hyperlinked to further enzyme information. The last column displays the matching sequence description (see color plates p. 460).

The graphic can also be useful in determining the location of the real start codon. One can rule out the first start codon under certain conditions by using the rare and start codon indicators at the top of the image, combined with the fact that supporting evidence may only exist from a certain start codon onwards. By default, rare codons are the ones that normally constitute less than 10% of the encoding of a particular amino acid for the particular organism. Rare codons are colored according to the color scheme for start codons. Shine–Dalgarno sequences are denoted with a black rectangle near the start codon indicators. Similar to Fig. 6, highly ranked evidence is placed on top so that best results are always shown for any region of the ORF.

2.1.7 Expanded Tool Summary

Fig. 9 displays in detail all of the evidence accumulated for the ORF during the MAGPIE analysis. The top ruler indicates the length of the translated ORF in amino acids. The bottom ruler indicates the position of the ORF within the contig. The rulers are numbered from right to left when the ORF is on the reverse-complement DNA strand. In this way the evidence is always presented in the same direction as the ORF translation. The evidence is separated into those database entries that

have at least a single level one hit, at least one level two hit, and others. The evidence coloring and order of placement is identical to that in Fig. 5.

This kind of image can be created for any subsequence. It is also used to display evidence in the confirmation pages for small inter-ORF features. For every database subject, the hit score is shown in the third text column, and the database subject description in the last column, thus consistently high scores and consistent descriptions are easy to spot.

The three types of representational display link to more in-depth information. The first linked data in text form are the accession number for the database subjects that hit against the query sequence in the first text column. The accession numbers are linked to the original database entries, e.g., GenBank or EMBL in accordance with a link-configuration file. For example, at NRC, the MAGPIE links connect to information provided by the Sequence Retrieval System (*http://www.lionbio.co.uk*) of the Canadian Bioinformatics Resource (*http://www.cbr.nrc.ca*). Other sites may configure these links to point to Entrez (*http://ncbi.nlm.nih.gov/Entrez*) at NCBI. If an Enzyme Commission (EC) number is associated with the database subject, the number is placed in the fourth text column and linked to a MAGPIE page listing the metabolic pathways in which the enzyme occurs.

The last but most informative linked component is the colored match coverage. The quality and types of evidence are clear because of the different colors assigned to the respective tools and respective confidence levels. The color differentiation can also be used to display other differences. For example, if protein-level BLAST analyses are performed as separate tools against eukaryotes, archaebacteria, and bacteria, and the resulting responses are colored using slightly different hues, the commonality of the gene is visually evident. Placing the mouse over the area with similarity causes a message to appear on the browser's status line. The message contains the exact interval of the similarity on both the query and subject sequences. The similarity display is hyperlinked to the original data in the text response.

The positioning of evidence rows in this image is more complex than in Fig. 8. The logic behind this display is as follows:

(1) Separate the evidence into sets where the tool and database id for the matches are the same.
(2) Separate the tool/id hit sets into three sets where the top level match in the tool/id set is either 1, 2, or 3.
(3) Order in a descending manner the tool/id match sets within each top match-level set by the user-specified tool rankings.
(4) Order tool/id sets within a tool ranking by score. If all scores are greater than one, order in a descending manner, otherwise order in an ascending manner (e.g., expected random probability scores).
(5) Within a score, rank in descending order tool/id sets by the total length of ORF intervals they cover, effectively giving longer hits higher priority.
(6) Within a length, sort alphabetically by hit description.
(7) Within a description, sort alphabetically by database identifier.

This fine level of sorting ensures predictability for the user. Evidence is sorted in terms of relevance and lexical ordering from top to bottom. In practice, the sorting is quite fast because a differentiation is usually made between tool/id sets in step 4. Steps one and two are only executed once. The system uses Perl's built-in sort, which is an implementation of the all-purpose quicksort (HOARE and QUICKSORT, 1974) algorithm. All of the information about database search tools, scores and hit lengths is kept in hash tables for quick reference during sorting. The speed benefit of hash table lookup outweighs its space cost. MAGPIE is usually run on servers with the capacity to execute multiple analyses in parallel, therefore, short-term requirements for large amounts of main memory are usually dealt with easily. The total height of the image can be calculated only after the number of tool/id sets and their ordering is determined. The image width is determined by keeping track of the longest value in each of columns while the evidence is loaded. Drawing can only begin after the height and width are determined.

2.1.8 Base Composition

Images displaying characteristics of the whole DNA sequences, e.g., the Base Composition or the Assembly Coverage Figures are usually drawn to the same scale as Fig. 4. They can be juxtaposed on top of each other to view them in context.

Fig. 10 displays two base composition distributions along the DNA sequence. In both graphs the colored region shows the actual base composition, which is determined by sliding a window along the forward DNA strand.

The $G+C\%$ graph is configured with a mean (as indicated by a horizontal line) equal to the average of the complete genomic sequence. Denoted on the left scale in the example, the *Sulfolobus solfataricus P2* genome average is 35%. The graph allows the rapid detection of areas of unusual base composition. Such aberrations may for example indicate the presence of transposable elements or other genomic anomalies. In the chosen example though, there is no great variability, indicating a low likelihood for the occurrence of transposable elements in this region of the genome. The 34.1% average base composition for this sequence is denoted on the right hand scale.

As previously mentioned, the red purine $(A+G\%)$ composition graph can be used for

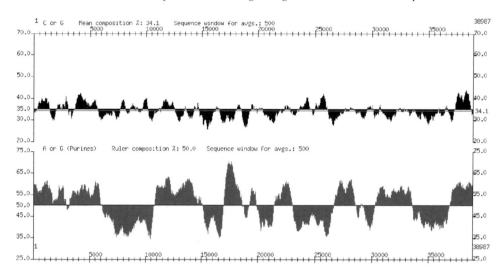

Fig. 10. Base compositions. Average A + G and G + C compositions are calculated using a sliding window of 500 bases. The moving average is displayed as a filled graph, both above and below the centerline average. Unusual G + C may indicate the presence of transposable elements. Majority A + G indicates coding strand in many organisms (see color plates p. 460).

many species to predict the strand containing the coding sequence. Lined up against Fig. 5, ORFs that are most likely non-coding can be detected. When the purine composition is greater than 50%, the coding ORFs are likely on the positive strand. They are likely to be on the negative strand when the composition is much below 50%.

The composition percentages are smoothed out by calculating averages with a sliding window of 500 base pairs. When each pixel represents 50 base pairs on the scale, and the window for composition averaging is 500, we can use the previous 5 pixels' (50/pixel 5pixels = 250 bases) and the next five pixels' totals for each plotted pixel column to calculate the current average. At each pixel column, we add a new total and discard the total for the first pixel column in the sliding window. Calculating the average at any location requires only averaging 10 numbers. Otherwise, in our example, 500 would need to be averaged at the sequence ends where the look-ahead and memory about the values for previous columns do not exist. To avoid invading whitespace, the average peak is truncated when it is outside of the user-defined scale ranges.

2.1.9 Sequence Repeats

Fig. 11 indicates the portions of the sequence that are repeated in the project. MAGPIE calculates families of repeated sequences sharing a minimum number of contiguous bases. By default, the minimum number is 20. Repeats are sorted into families of matching subsequences, which are further sorted by size in descending order. The matching sequence name is in the left right hand column, while the location and size of the match are displayed as filled boxes under the ruler. Red, blue, and green boxes represent forward, reverse complement, and complement matches.

When many repeats occur in a sequence, the images are very tall. In these instances, the dimensions may exceed the maximum image dimensions that can be shown by the browser. The stored image is loaded into a Java applet with scroll bars in order to overcome this browser limitation. A faster, memory intensive repeats search has recently replaced the original repeats finder. The new repeats finder exports the repeats information to enable specialized viewing of the data by Java applets accepting a special data format.

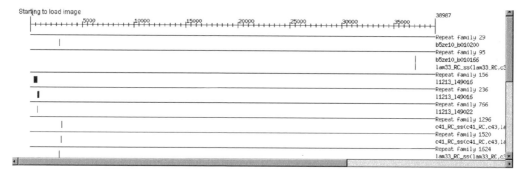

Fig. 11. Sequence repeats. The image has scrollbars (as part of an applet) because of its large dimensions. Repeats are sorted into families of matching subsequences, which are further sorted by size in descending order. The matching sequence name is in the left part of the right hand column, while the location and size of the match are displayed as filled boxes under the ruler. Red, blue and green boxes represent forward, reverse complement, and complement matches (see color plates p. 461).

2.1.10 Sequence Ambiguities

Fig. 12 displays the location of ambiguous bases in a contig. This image can be used in the polishing stage of the DNA sequencing project. It is usually viewed atop Fig. 13, which provides the assembly context. The generation of this image requires assembly information from the Staden package (STADEN et al., 1998). The scale in the center denotes the number of base pairs in the sequence. Red vertical ticks on the upper line represent ambiguities, where there is only positive strand coverage. Blue ticks on the lower bar represent ambiguities when only the negative strand is sequenced. When an ambiguity exists in a region of double stranded coverage, the tick appears on the center scale. As an exception to the consistent use of color, this tick is colored red for better visibility. The total number of ambiguities is dis-

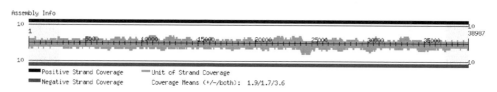

Fig. 12. Ambiguity information. This graph starts at base 50,001 because it is the second pane for a 71,519 base sequence limited to 50,000 bases per image. If the ambiguous base has been sequences on the forward, reverse, or both strands, the tick is displayed on the top, bottom, or center line respectively. The displayed ambiguities total is for the pane, not the sequence as a whole (see color plates p. 461).

Fig. 13. Assembly information. Histograms of average positive and negative strand assembly coverage are above and below the centerline. Breaks in the blue and red bars indicate gaps in the positive and negative strand coverage. Genomic neighbors are indicated by background shading as in Fig. 5 (see color plates p. 461).

played in the lower-left corner. In our example, the second pane of the display is shown. The number of ambiguities (two) is valid for this pane only; there may be additional ambiguities in the first pane.

2.1.11 Sequence Strand Assembly Coverage

The image shown in Fig. 13 summarizes the quality of the sequence assembly for emerging genomes in MAGPIE. MAGPIE can extract assembly information from the output of the Staden package assembler *gap4*. As in Fig. 5, background shading denotes the extent and orientation of the neighboring contigs.

Average coverage multiplicity is the average number of times any base of the contig has been sequenced. This number is separately calculated for both strands. The two values and the total for the contig are displayed at bottom-center of the image. The green area in the center of the image can be interpreted as two separate histograms, the one above the ruler quantifying the average coverage on the positive strand, while the one below the center ruler quantifies the average coverage on the negative strand. The histogram bars are truncated if greater than ten-fold coverage is reached. The histograms are used to determine the reliability of the data. This is useful where frame shifts or miscalled bases are suspected.

Further resolution of poorly covered regions is provided through the continuity of the large horizontal blue and red bars. A gap will appear in the blue or red bar if even a single base pair has not been sequenced on the forward or reverse strand, respectively. This allows the user to see how much DNA sequence polishing is required to double-strand the entire sequence assembly. The displayed overlap with other project sequences is once again useful. Gaps in the current sequence's assembly may be less worrisome in regions overlapping with other contigs.

In order to create this image, the assembly information is read in chunks. The chunk size is equal to the scale factor for the image (50 bases/pixel by default). The base pairs on each strand are mapped onto a blank template string of 50 bases. Blanks in the template indicate gaps.

The average sequencing coverage for that pixel is calculated at the same time. The sequence averages are calculated by summing up the pixel totals. This provides major savings over the much larger individual base totals.

2.1.12 Restriction Enzyme Fragmentation

Fig. 13 displays the location of restriction enzyme cuts on the insert. The MAGPIE user can define the set of restriction enzymes. The cloning vector and the orientation of the insert in the clone can be specified when the contig is added to the MAGPIE project. This information is taken into account during the fragment calculation. In the figure, cut locations are denoted as vertical ticks on each enzyme's lines. The fragments are numbered from the 5′ to the 3′ end, including the vector sequence (which is not shown in this display). The vector is always oriented to the 5′ end (i.e., left end) of the insert. Fragment numbers in green represent parts of the vector-free insert. Fragments labeled in yellow contain parts of the vector sequence. Fragments that contain only vector are not displayed in this figure, because the ruler only includes the range of the insert. Such fragments appear only in the agarose gel simulation described below. The numbering of fragments in the Fig. 10 display does not always start at number one, because of the out-of-sight vector fragments. The example clone is from a *Hind III* restricted library. The enzyme cuts the sequence at the very start and at the very end of the insert (i.e., there are no yellow-labeled fragments).

2.1.13 Agarose Gel Simulation

Fig. 15 is a computer simulation of an agarose gel displaying the same the restriction digests as Fig. 14. The width of the image is based on the number of restriction enzymes. The height depends on the user-configurable size of the agarose plate. Given that we know the theoretical fragment lengths, the hypothetical

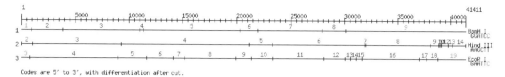

Fig. 14. Restriction sites. Ticks correspond to the location of cut sites for the restriction enzymes listed in the right hand margin. The fragments produced are labeled 5′ to 3′, with the undisplayed cloning vector on the 5′ end. Fragments containing vector have yellow labels, otherwise they are green. The HindIII line has restriction sites at the ends of the sequence because the clone library was HindIII restricted. By consequence, it has no yellow labeled fragments (see color plates p. 462).

Fig. 15. Agarose gel simulation. Fragment lanes and labels correspond to those in Fig. 14. The fragment migration is calculated using the specified standard marker "M" lane migrations. Fragments containing vector are colored yellow. Fragments composed entirely of vector are colored red. Where fragments are very close, luminescence is more intense and labels are offset for readability (see color plates p. 462).

where m is the mobility in the agarose gel and L the fragment length.

The constants from the least-squares fit are m_0, L_0, and c. Like most of the settings for the graphical displays in MAGPIE, the marker mobility data are specified in a configuration file. The regression is performed using the traditional summation method as opposed to the matrix multiplication method. This method was chosen because of the limited efficiency of array manipulation in Perl5.

Describing the features of the gel from top to bottom, the agarose percentage is displayed. This number is specified in the marker mobility configuration. The outside lanes marked "M" are the marker lanes. The inside lanes are numbered left to right in the top to bottom order of the restriction enzymes of Fig. 14. The horizontal gray line represents the location of the wells. The fragments in the marker lanes have their lengths displayed in the image margins. The bands of marker lane fragments are solidly colored to ensure they can be clearly observed. The bands in the enzyme lanes appear slightly diffused in order to more closely resemble the appearance of physical gels. These bands are also distinguishable from undifferentiated bands, which are more solid and overall brighter in color. When bands are close to each other in a lane, some labels for fragment numbers are offset in an attempt to maximize readability. As an example of these distinctions, observe fragments 16, 8, and 11 in lane 3 (an *EcoR I* digestion). Fragment 16 is isolated, has a diffused band, and a left justified label. On top of one another, fragments 8 and 11 are given a larger bright area. This highlights their overlap even though they occupy the same amount of space as fragment 16. The

fragment migration distances are calculated using the reciprocal method (SCHAFFER and SEDEROFF, 1981). This is used instead of the less accurate logarithmic scale based on the sequence length. The reciprocal method calculates mobility as the inverse of the length. The exact relationship is governed by constants calculated from a least-squares fit of the marker mobility data. The reciprocal formula is:

$$(m - m_0)(L - L_0) = c$$

label for number 11 is offset to the right of number 8.

Fragments containing or entirely composed of vector sequence are also displayed. This is done in order to remain true to the physical manifestation. Partial vector fragments are colored yellow. Full vector fragments are colored red. This labeling is done so that fragments containing vector are not accidentally isolated from the real agarose gels. As would be expected, in the *Hind III* lane (number 2) there is a single fragment, number 1, which contains the entire vector.

3 Conclusions and Open Issues

The images shown in this text are extremely useful as a support tool for genome annotation. Without the images, the efficient scanning of genomic evidence would be much harder and in many cases probably impossible. More than 20 genome projects have used MAGPIE to date for the annotation of "their" genome.

Yet, the system cannot satisfy the need for more complex data queries, which will be a theme for at least the next decade. For example, most prokaryotic genomes are circular in nature, a fact that is not displayed in any of the MAGPIE graphics. While the administrator can configure particular features of the MAGPIE graphics, there is a limit to the flexibility. For example, a graphical display of a query like: "Display all the tRNA coding genes and the tRNA-synthetase coding genes in a genome and show potential relations between the two gene sets" cannot be served by the current MAGPIE environment. Refinement of the displays and the addition of input forms for eukaryotic features (GAASTERLAND et al., 2000) in MAGPIE will continue as more analysis and genomic annotation is done.

Acknowledgement

The authors wish to thank Dr. JOHN P. VAN DER MEER, Director of Research at IMB for critically reading the manuscript.

4 References

ALTSCHUL, S. F., MADDEN, T. L., SCHÄFFER, A. A., ZHANG, J., ZHANG, Z. et al. (1997), Gapped BLAST and PSI-BLAST: a new generation of protein database search programs, *Nucleic Acids Res.* **25**, 3389–3402.

BORODOVSKY, M., MCININCH, J. (1993), GeneMark: Parallel gene recognition for both DNA strands, *Computers & Chemistry* **17**, 123–133.

BURGE, C., KARLIN, S. (1997), Prediction of complete gene structures in human genomic DNA, *J. Mol. Biol.* **268**, 78–94.

CHARLEBOIS R. L., GAASTERLAND T., RAGAN M. A., DOOLITTLE W. F., SENSEN, C. W. (1996), The *Sulfolobus solfataricus* P2 genome project, *FEBS Lett.* **389**, 88–91.

COOPER, A. (1995), *About Face: The Essentials of User Interface Design*. Foster City, CA: IDG Books Worldwide.

GAASTERLAND T., SENSEN, C. W. (1996), Fully automated genome analysis that reflects user needs and preferences – A detailed introduction to the MAGPIE system architecture, *Biochimie* **78**, 302–310.

GAASTERLAND, T., SCZYRBA, A., THOMAS, E., AYTEKIN-KURBAN G., GORDON, P., SENSEN, C. W. (2000), MAGPIE/EGRET Annotation of the 2.9 Mb *Drosophila melanogaster* ADH region, *Genome Research* **10**, 502–510.

GORDON, P., SENSEN, C. W. (1999), Bluejay: A browser for linear units in Java, in: *Proc. 13th Ann. Int. Symp. High Performance Computing Systems and Applications*, pp. 183–194.

HENIKOFF S., HENIKOFF, J. G., PIETROKOVSKI, S. (1999), Blocks +: A non-redundant database of protein alignment blocks derived from multiple compilations, *Bioinformatics* **15**, 471–479.

HOARE, C. A. R. (1962), Quicksort, *Computer J.* **5**, 10–15.

KOLAKOWSKI, L. F., Jr., LEUNISSEN J. A. M., SMITH, J. E. (1992), ProSearch: fast searching of protein sequences with regular expression patterns related to protein structure and function, *Biotechniques* **13**, 919–921.

PEARSON, W. R., LIPMAN, D. J. (1988), Improved tools for biological sequence comparison, *Proceedings of the National Academy of Science* **85**, 2444–2448.

SALZBERG, S., DELCHER, A., KASIF, S., WHITE, O. (1998), Microbial gene identification using interpolated Markov models, *Nucleic Acids Res.* **26**, 544–548.

SCHAFFER, H. E., SEDEROFF, R. R. (1981), Least squares fit of DNA fragment length to gel mobility, *Anal. Biochem.* **115**, 113–122.

SHINE, J., DALGARNO, L. (1974), The 3′-terminal sequence of *Escherichia coli* 16S ribosomal RNA: complementarity to nonsense triplets and ribosome binding sites, *Proc. Nat. Acad. Sci.* **71**, 1342–1346.

STADEN, R., BEAL, K. F., BONFIELD, J. K. (1998), The Staden Package, *Computer Methods Mol. Biol.* **132**, 115–130.

18 Interaction Databases

GARY D. BADER

CHRISTOPHER W. V. HOGUE

Toronto, Canada

1 Introduction

Given theoretical estimates of between 40,000 to 100,000 human genes, it is apparent that only a minority of these genes encode conventional metabolic enzymes or transcription–translation apparatus. Complete genomic sequencing and new protein discovery is uncovering large numbers of complex multi-domain proteins, containing interacting modules, especially in metazoans and vertebrates. The complexity of the DNA blueprint is augmented in an exponential fashion when one considers the possibility that these proteins could bind to several other biomolecules simultaneously. These "interaction networks" form conventional signaling cascades, classical metabolic pathways, transcription activation complexes, vesicle mechanisms, cellular growth and differentiation systems, indeed all of the systems that make cells work.

The ultimate manifestation of gene function is through intermolecular interactions. It is impossible to disentangle the mechanistic description of the function of a biomolecule from a description of other molecules with which it interacts. We think that one of the best forms of the annotation of the function of a gene, from the perspective of a machine-readable archive, would be information linking molecular interactions together, an *interaction database*. Interactions, as the manifestation of function, and interaction databases are going to be critical components as we move towards a complete functional description of molecular life. Interaction databases are essential to the future of bioinformatics as a new science. In this chapter we consider broadly, what can be achieved through integration of current interaction information into a common framework, and examine a number of databases that contain what we consider to be interaction information.

2 Scientific Foundations of Molecular Interaction Information

Interaction information is based on the experimental observation of a specific molecular interaction between two or more molecules. For the purposes of this discussion, we speak of natural, biological molecules spanning the entire range of biochemistry, including proteins, nucleic acids, carbohydrates and small molecules, both organic and inorganic. Interaction information is an inference that two or more molecules have a preferred specific molecular interaction with one another, within a living organism, and that inference is based on experimental evidence using conventional experimental molecular and cell biology techniques.

The number of types of experiments that can provide this evidence is large. One can broadly define primary interaction experiments as being based on direct observation of a measurable phenomenon based on bringing two molecules directly together in an experimental system. This may be *in vivo*, such as a yeast two-hybrid assay, or *in vitro* such as in a fluorescence polarization experiment using purified reagents in a cuvette. Another type of experimental information is based on genetic evidence. For example, a tandem gene knockout in an organism may cause a certain phenotype to appear such as a growth defect. There is evidence that the two genes combine their effects to cause the outcome of the observed phenotype, and that may imply a molecular interaction between the two gene products, but it is indirect and possibly dependent on other genes. Nonetheless, all this information is important in helping us to understand the networks of interacting molecules and how life arises and is maintained in the organism by the interplay of interacting molecules. By deciding globally to store primary interaction data into a common machine-readable archive, such as is done now for gene sequence and molecular structure information, we would have a tremendous resource for research biology and data retrieval.

Interaction databases ideally should contain information that is in the form of a correlated pair or group of molecules, some link to the experimental evidence that led to the interaction, and machine-readable information about what experimental interaction parameters are known. For example, did the interacting molecules undergo a chemical change? Was the binding reversible? What are the kinetic and thermodynamic parameters, if they were indeed measured in the experiment? Were the forms of the molecules in the experiment wild-type, or variants? What are the binding sites on the molecules?

3 The Graph Abstraction for Interaction Databases

Consider the collection of molecules in a cell as a graph. Each molecule is a vertex, and each interaction is an edge. Classical bioinformatics databases hold the "vertex" part of this graph. Protein sequence, DNA sequence and small molecule chemistry databases, collectively, hold molecules, which are the vertices of the interaction networks inside the cell. In contrast, the ideal interaction database will hold the "edge" information – which two molecules come together, under what cellular conditions, location and stage, how do they interact, and what happens to them in the course of the interaction. We refer to this concept as the graph-theory abstraction for interaction databases, and it is a powerful data abstraction as it simplifies the underlying concepts and allows one to apply algorithms that are well understood to the larger problems of data mining and visualization.

By decomposing interaction data into pairwise interactions that form graphs, there is a broad range of computer science methods that can be used to analyze, mine and understand this information. Having a clear picture of a general graph abstraction for interaction databases is the key to the integration of data into a universal framework.

We have already written about our effort, the Biomolecular Interaction Network Database (BIND) (BADER and HOGUE, 2000), which seeks to create a database of interaction information around a generalized graph theory abstraction of interaction data. We describe here some of the concepts that we found essential for the overall integration of interaction information resulting from our research.

4 Why Contemplate Integration of Interaction Data?

In our own effort to build the BIND data model, we pursued a prototyping approach very different from the way most biological databases are created. We sought a comprehensive data model that allowed interaction information to be represented in a machine-readable format, spanning all types of molecular interactions, including protein, RNA, DNA and small molecules and the biochemical reactions they are involved in. The BIND data specification was created following the NCBI ASN.1 architectural model (OSTELL and KANS, 1998) and the NCBI software development toolkit for implementing early versions of the BIND database and its tools. We focused on spending time designing the data model for BIND, contemplating the way we would store molecular interaction information and information about molecular mechanisms, from inferences as broad as a genetic experiment, to as precise as the atomic level of details found in a crystal structure of an interacting complex. Our hypothesis was: Is there a plausible universal description of molecular interactions and mechanisms that can suffice to drive whole-cell visualization, simulation and data retrieval services? We wanted to be certain that it was sufficient to describe the richness of molecular interaction information and molecular mechanisms. In doing so, we asked ourselves and others: What data should be represented? What abstractions should we use? How can we describe interactions together with chemical alterations to the interacting molecules? While the outcome of this hypothesis testing exercise is described in detail and embodied in the BIND specification (*www.binddb.org*), we focus here on the question: Besides the scientific literature, and new high-throughput interaction experiments, what sources of information are there that can help populate a general interaction database?

5 A Requirement for more Detailed Abstractions

Molecular interaction data must be represented by an abstraction so that computations may be carried out and data maintained in a machine-readable archive. This is a simple idea with an analogy in biological sequence information. Recall that biopolymer molecules, DNA and proteins, are abstracted for the computer as strings of letters. This information tells us nothing about conformation or structure, just of composition and biopolymer sequence. The IUPAC single letter code for DNA and for amino acids are abstractions that do contain sufficient information to reconstruct chemical bonding information, provided that a standard form of the biopolymer is being represented, and not a phosphorylated, methylated or otherwise modified form.

One cannot imagine a database of interactions of all the biomolecules in a cell without first having an enumeration of the contents of the cell, the biomolecular "parts list". Sequence databases partially fulfill this requirement. We say partially, because sequence information alone is not sufficient to encode precise information about post-translational activation of biopolymers, as we stated, phosphorylation or methylation for example. In order to encode exact information about biomolecules, one must have the capacity to describe the biopolymer both as sequence and as an atom-and-bonds representation, known as the *chemical graph*.

A chemical graph description of a biomolecule is an essential concept to grasp. It is sufficient to recreate the atoms, bonds and chirality of the molecule, but without specifying the exact location of the atoms in 3-D space. In other words, a chemical graph is an atomic structure without coordinate information. Chemical graph data abstraction exists within the NCBI MMDB data specification and database of molecular structure information (WANG et al., 2000). The MMDB structure specification is the only example of a chemical graph-based structure abstraction, and a complete chemical structure may be encoded in MMDB without knowing a single X, Y, Z atom coordinate. Neither the PDB or the newer mmCIF structure file format have a chemical graph data structure that can describe the complete chemistry of a molecule without atomic coordinate information (BERMAN et al., 2000).

Sequence alphabet abstractions have been invaluable in bioinformatics, having enabled all computer-based sequence analysis. This would have been very difficult to compute had an exact database of atoms and bonds making up each biopolymer sequence been chosen as the abstraction. While this information might bog down sequence comparison, it is required for a more precise record of the chemical state of a biopolymer following post-translational changes. These chemical states, once accessible through a precise database query, are important to have recorded as they form the control points for uncounted pathways and mechanisms for cellular regulation.

Abstractions are rarely applicable universally for all kinds of computations. However, as computing power increases, the abstractions we require can be expanded to fulfill the requirements of more kinds of computations. So far, there has been resistance to expand the abstractions of sequence information to more complete descriptions like a chemical graph, but it is clear these will be required to describe large and important parts of molecular biology such as phosphorylation, carbohydrate or lipid modification, and other post-translational changes upon which so many molecular mechanisms depend.

We can begin to contemplate interaction databases now, because we have demonstrated that computer infrastructure can keep up with genomic information. However, the representational models we select need to be carefully chosen in order that they not preclude a computation that might otherwise be required in the future. Indeed it may be time to find *the most complete description for molecular information as one can imagine*. With adequate standard data representations for molecules that are unambiguous for the purposes of general computation, specifying sequences, structures and small molecule chemistry, we may be ready to move ahead with the wholesale annotation of molecular function in a very complete fashion. Without it, our machine-readable descriptions will be ambiguous and we will be very limited in the precision with which we can contemplate performing simulation, visualization and data mining with biological information.

6 An Interaction Database as a Framework for a Cellular CAD System

In order to achieve the goal of a computing and software system that can achieve whole-cell simulation, we must build something that looks very much like a CAD (Computer Aided Design) system. CAD systems are used in engineering, e.g., in the design of electronic circuitry. In biology, such a system could be used for the representation and possible design of cellular circuitry. Unlike engineering, the biological CAD system we contemplate would often be used "backwards" as a reverse-engineering tool to understand the complexity of cellular life. This CAD system would have a detailed representation of biochemistry sufficient to allow one to draw from the system a snapshot of a living cell first in the form of data, to be either presented to a simulation engine, to data mining tools or to a visualization system. In engineering, CAD systems are, in fact, database-driven software systems, and the utility of a particular CAD system is proportional to the content of its database of "parts". Likewise we must focus on fulfilling a biomolecular interaction database with complete information about the parts we need for a cellular CAD system. However, in order to get to such a CAD system, bioinformaticists must start behaving more like engineers, and less like biologists or computer scientists. Integrated efforts, software systems and databases are the keys to success in any endeavor as grand as trying to make a detailed computer simulation of how the molecules and events take place inside a living cell on a computer. Federated databases with highly latent network interconnections and imprecise data models will not suffice for the uses we contemplate here. Interaction information, as it comprises a network of information, is best consolidated into a single system in order to drive the data demands of high performance computing required for whole-cell simulation, visualization and data mining. Overall, such a system requires a formal data model for molecular interactions that provides a good abstraction of the data with precise computability, and can store the full complexity of the information. Only with such a system does interaction data become manageable. The emergence of a standard will allow diverse groups to collaborate and work towards their common goals more efficiently.

7 BIND – The Biomolecular Interaction Network Database

We have designed and implemented a database to store biomolecular interactions possessing the attributes of an interaction database discussed above. BIND is a web-based database system that is based on a data model written in ASN.1 (Abstract Syntax Notation – *http://asn1.elibel.tm.fr/*). ASN.1 is a hierarchical data description language used by the NCBI to describe all of the data in PubMed, GenBank, MMDB and other NCBI resources. ASN.1 is also used extensively in air traffic control systems, international telecommunications and Internet security schemes. The advantages of ASN.1 include being strongly typed, computer-readable, having an elegant binary encoding scheme that saves space and CPU resources when transmitting data and being fully interoperable across platforms. Disadvantages are that commercial ASN.1 tools are very expensive and that the ASN.1 standard process is closed. The NCBI, however, provides a public domain cross-platform software development toolkit written in the C language to deal with the NCBI data model and with ASN.1. The toolkit can read an ASN.1 defined data model and generate C code that allows automatic reading (parsing), writing and management of ASN.1 objects. Recently, the toolkit also added support for the automatic translation of ASN.1 defined objects to and from XML as well as the automatic generation of an XML DTD for an ASN.1 data specification. A C++ version of the toolkit is under development and should be available soon after this book is published. This powerful data description language and toolkit allows us to circumvent the large and time-consuming problem in Bioinformatics of parsing primary databases. With the toolkit, parsing is automatic. The use of ASN.1 also allows the BIND specification to use mature NCBI data types for biological sequence, structure and publications.

Recently, the XML (Extensible Markup Language – *http://www.w3.org/XML/*) language has gained popularity for data description. XML matches ASN.1 in its ease of use, although it does not provide strong types. For instance, ASN.1 recognizes integers and can validate them, while XML treats numerical data as text. The advantages of XML are its open nature and familiarity, since it is similar to HTML. Many tools use XML these days, although there are currently no free code-generation and rapid application development tools. XML also wastes space and does not have a binary encoding scheme. An XML message will be many times larger than a binary encoded ASN.1 message.

The BIND data specification describes biomolecular interaction, molecular complex and molecular pathway data. Both genetic and physical interactions can be stored.

8 Other Molecular Interaction Databases

As we have outlined the requirements for a variety of future endeavors in bioinformatics based on interaction databases, we look at the current databases and their encoding of various kinds of interaction information. Most interaction data resides in the literature, in unstructured text, tables and figures in countless papers in molecular and cellular biology. It is impossible to retrieve information from this archive by computers or natural language query methods with the accuracy required by scientists. Several efforts have been put forward to create small databases of information that we classify now as interaction information, although it is unlikely that an idea of a generalized interaction database was in the mind of the creators of these databases. Most of these small databases have very select niches of interaction information, e.g., the restriction enzyme database REBase (ROBERTS and MACELIS, 2000), maintained by New England Biolabs. It is instructive to point out that a very widely known database like REBase falls into the category of a molecular interaction database. It satisfies the "edge" criterion of the graph abstraction of interaction data. REBase contains information that links a protein or gene that is a restriction enzyme to a molecular locus of DNA sequence that is specifically cleaved by the enzyme. Hence it is, in our parlance, a database of protein–DNA interaction, and as such it is as a very valuable source of molecular interaction information.

9 Examples of Interaction Databases

In our review of databases, we have examined both the literature and the Internet and here we report a current list of databases that contain interaction information covering proteins, DNA, RNA and small molecules. The number of projects indicates the importance of this data. However, the variety of formats, data architectures and license agreements is a daunting challenge to integration of this information into a common architecture. One can classify databases according to whether they are linked back to primary experimental data in the literature, or are secondary sources of information based on review articles or are the knowledge of expert curators. The databases based on primary information are few in this list, yet are among the most valuable.

In our review, we highlight whether or not these databases have data in a machine-readable form. Many databases are packed with information, but the information is entered in such a way that it cannot be unambiguously matched to other databases. For example, some databases are missing key data descriptors like sequence accession numbers, CAS chemical compound numbers, MEDLINE identifiers for publication references, or unambiguous taxonomy information when data from multiple organisms is present. It is critical that these projects move towards sound database principles when describing data such that it may be computed upon unambiguously and precisely.

Finally, where possible, we indicate the primary reference for the database and license terms of the database, if applicable, to academic and industrial users of the data.

Aminoacyl-tRNA synthetase Database
URL: *http://rose.man.poznan.pl/aars/index.html*
Ref: SZYMANSKI and BARCISZEWSKI (2000)
Contains aminoacyl-tRNA synthetase (AARS) sequences for many organisms. This database is simply a sequence collection, but collated pairs of AARS + tRNA can be used to create RNA–protein interaction records. The database is available freely over the web.

Biocarta
URL: *http://www.biocarta.com/*
Biocarta is a commercial venture whose purpose is to provide manually created clickable pathway maps for signal transduction as a resource to the scientific community. The presence of a standard set of symbols to represent various different protein components of pathways make the pathway maps clear and easy to understand. Proteins are linked to many different primary databases including PubMed,

GenBank, OMIM, Unigene (WHEELER et al., 2000), KEGG, SwissProt (BAIROCH and AP-WEILER, 2000) and Genecard. Genes may be sponsored by companies and links are present to commercially available reagents. Biocarta invites volunteer users to supply pathways as figures, and Biocarta then creates clickable linked maps and makes them available on the web. The data model is not public and the database has not been published in peer reviewed literature.

Biocatalysis/Biodegradation Database
URL: *http://www.labmed.umn.edu/umbbd/*
Ref: JOHNSON (1977)
Contains microbial biocatalytic reactions and biodegradation pathways primarily for xenobiotic, chemical compounds: currently about 100 pathways, over 650 reactions, 600 compounds, 400 enzymes and 250 microorganisms are represented. The data model is chemical reaction-based with a graph abstraction for pathways. The graph abstraction allows the "Generate a pathway starting from this reaction" function. PDB files for some of the small molecules are available. Graphics (clickable GIFs) are available for the various pathways. The work is funded by several organizations and is free to all users. Data is entered on a volunteer basis and records contain literary references to PubMed.

BRENDA
URL: *http://www.brenda.uni-koeln.de/*
BRENDA is a database of enzymes. It is based on EC number and contains much information about each particular enzyme including reaction and specificity, enzyme structure, post-translational modification, isolation/preparation, stability and cross references to structure databanks. Information about the chemical reaction is extensive, but is in free-text form and is not completely computer-readable. The database is copyright and is free to academics. Commercial users must obtain a license.

BRITE (Biomolecular Reaction pathways for Information Transfer and Expression)
URL: *http://www.genome.ad.jp/brite/brite.html*
BRITE is a database of binary relations based on the KEGG system. It contains pro-

tein–protein interactions, enzyme–enzyme relations from KEGG, sequence similarity, expression similarity and positional correlations of genes on the genome. The database mentions that it is based on graph theory, but no path finding tools are present. BRITE contains some cell cycle controlling pathways that have now been incorporated into KEGG.

COMPEL (Composite Regulatory Elements)
URL: *http://compel.bionet.nsc.ru/*
Ref: KEL-MARGOULIS et al. (2000)

Contains protein–DNA and protein–protein interactions for Composite Regulatory Elements (CREs) including the positions on the DNA that the protein binds to. The database provides links to TRANSFAC. The data model is text-based and does not use a graph theory abstraction. COMPEL 3.0 in January 1999, contained 178 composite elements.

CSNDB (Cell Signaling Networks Database)
URL: *http://geo.nihs.go.jp/csndb/*
Ref: TAKAI-IGARASHI et al. (1998)

CSNDB contains cell signaling pathway information in *homo sapiens*. It has a data model that is specific only to cell signaling. It is based on both interactions and reactions and stores information mainly as unstructured text. An extensive graph theory abstraction is present. It is probably one of the first databases to use a simple graph theory abstraction since its first publication in 1998. It contains the most elaborate pathway finder using shortest path algorithms. It can limit the graph to a specific organ and can mask sub-trees for this feature. The data model is quite good, but is *ad hoc*. Fields have been added as they are needed and the system is not general. It contains interesting pharmacological fields for drugs, such as IC50. The database can represent proteins, complexes and small molecules. It is linked to Medline and TRANSFAC. Recently, TRANSFAC has imported the CSNDB to seed its TRANSPATH database of regulatory pathways that link with transcription factors. It contains an extensive license agreement that limits corporate use. Free to academics. Funded by the Japanese National Institute of Health Sciences.

Curagen Pathcalling
URL: *http://curatools.curagen.com/*

The commercial Curagen Pathcalling program uses information from the Stanley Fields lab high-throughput yeast two-hybrid screen of the yeast genome along with other yeast protein–protein interactions from the literature. It contains only protein–protein interactions. Pathcalling uses a graph theory abstraction which allows the use of a Java applet to visually navigate the database. Each protein may be linked to SaccDB (SGD) (BALL et al., 2000), GenBank or SwissProt. Because it is proprietary, the database does not make any of its information, software or data model available.

DIP – Database of Interacting Proteins
URL: *http://dip.doe-mbi.ucla.edu*
Ref: XENARIOS et al. (2000)

The DIP project is an example of a nascent interaction database. The DIP database stores only protein–protein interactions. It is based on an binary interaction scheme for representing interactions and uses a graph abstraction for its tools. A visual navigation tool is present. DIP does not use a formal grammar for its data specification. The DIP data model allows the description of the interacting proteins, the experimental methods used to determine the interaction, the dissociation constant, the amino acid residue ranges of the interaction site and references for the interaction. DIP contains over 3,500 protein–protein interactions representing about 80 different organisms. Academic users may register to download the database for free if they agree to the click-through license. Commercial users must contact DIP for a license.

DPInteract
URL: *http://arep.med.harvard.edu/dpinteract/*
Ref: ROBISON et al. (1998)

DPInteract is a curated relational database of *E. coli* DNA binding proteins and their target genes. It provides BLASTN searching for DNA and has links to SwissProt, EcoCyc, Medline and Prosite (HOFMANN et al., 1999). The database is text based with a limited data specification. This database is not trying to be general. It is single-purpose. Interesting position specific matrices are available to describe the

DNA binding motif. Records are organized by structure family (e.g., helix-turn-helix family proteins). Updating of the database continued from 1993–1997 and has now stopped. The database is copyright, but is freely available over the web.

Ecocyc (and Metacyc)
URL: *http://ecocyc.doubletwist.com/*
Ref: KARP et al. (1999)

Ecocyc is a private database (freely available to academics) that contains metabolic and signaling pathways from *E. coli*. EcoCyc is one of the oldest interaction databases. It is based on a sound object-oriented data model. Chemical reactions are used to describe the data, which makes sense, since the main goal of EcoCyc is to catalog metabolic pathways from *E. coli*. It is currently being retrofitted to deal with cell signaling pathways, although data is still described using a chemical reaction scheme. The fields of this database are mostly free text-based. All types of molecules from small molecules to molecular complexes may be represented. The data model is not based on a chemical graph, however, and atomic level detail is not present. EcoCyc uses a graph abstraction model that has allowed pathway tools to be written. Ecocyc contains interactions of proteins with proteins and small molecules. MetaCyc contains EcoCyc and also pathways from some other non-eukaryotic organisms.

EMP
URL: *http://wit.mcs.anl.gov/EMP/*
Ref: SELKOV et al. (1996)

EMP is an enzyme database that is chemical reaction-based. It stores information as detailed as chemical reaction and k_m. Fields are stored as semi-structured text which may allow some of the database to be easily machine-readable. The database is part of the WIT project and is now more easily accessed from the WIT system. Some source code is available for the project and the database is freely available over the web.

ENZYME
URL: *http://www.expasy.ch/enzyme/*
Ref: BAIROCH (2000)

This database contains enzyme, substrate, product and cofactor information for over 3,700 enzymes. It has been a crucial resource for metabolic databases including EcoCyc. It is chemical reaction-based. This database can be translated to an interaction model by breaking down the chemical reactions into substrate–enzyme, product–enzyme and cofactor–enzyme groups. ENZYME links to BRENDA, EMP/PUMA, WIT and KEGG. The database is free and is run by the not-for-profit Swiss Institute of Bioinformatics. There are no restrictions on its use by any institutions as long as its content is in no way modified.

FIMM (Functional Molecular Immunology)
URL: *http://sdmc.krdl.org.sg:8080/fimm/*
Ref: SCHONBACH et al. (2000)

The FIMM database contains information about functional immunology. It is primarily not an interaction database, but it contains information about major histocompatibility complex (MHC)/human leukocyte antigen-(HLA)-associated peptides. The database contains over 1,450 peptides and 875 HLA records at time of writing. It is linked to GenBank, SwissProt, MHCPEP, OMIM, and PubMed, among others. This data provides protein–peptide interaction records that are important immunologically. The database is provided "as-is" by Kent Ridge Digital Labs in Singapore.

Database of Ribosomal Cross-links (DRC)
URL: *http://www.mpimg-berlin-dahlem.mpg.de/~ag_ribo/ag_brimacombe/drc*
Ref: BARANOV et al. (1999)

This database keeps a collection of all published cross-linking data in the *E. coli* ribosome. This is a database of hand-curated dBASE IV files with a web interface (last updated March 7th, 1998). The possibilities for machine parsing the database seem limited since the field data is non-standardized and meant to be human-readable only.

GeNet (Genetic networks)
URL: *http://www.csa.ru/Inst/gorb_dep/inbios/genet/genet.htm*

GeNet curates genetic networks for a few example species. It provides Java visualization tools for the genetic networks. The database contains extensive information about each example network in free-text form. This database is not machine-readable, although it is a good resource for genetic interaction information.

GeneNet (Genetic networks)

URL: *http://wwwmgs.bionet.nsc.ru/systems/mgl/genenet/*
Ref: KOLPAKOV et al. (1998)

GeneNet describes genetic networks from gene through cell to organism level using a chemical reaction-based formalism, i.e., substrates, entities affecting course of reaction and products. The database is based on a formal object-oriented data model. GeneNet contains 30 gene network diagrams and over 1,000 genetic interactions (termed relations in GeneNet) from varied organisms including human. The database is current and is regularly updated. Visual tools are present for examining the data, but are plagued by network latency problems that can prevent complete loading.

HIV Molecular Immunology Database

URL: *http://hiv-web.lanl.gov/immunology/index.html*
Ref: BETTE KORBER (1998)

The primary purpose of this database is not to store interactions. It contains data about binding events between HIV and the immune system including HIV epitope and antibody binding sites which could provide data for an interaction database. This information is freely available from the FTP site of the database.

HOX Pro

URL: *http://www.iephb.nw.ru/hoxpro* and *http://www.mssm.edu/molbio/hoxpro*
Ref: SPIROV et al. (2000)

The main purpose of this database is to provide a curated human readable resource for homeobox genes. It also stores extensive information about genetic networks of homeobox genes in a few model organisms. Clickable picture and a Java applet are available to visualize the networks. The visualization system is the same one used for GeNet.

InBase (The Intein Database)

URL: *http://www.neb.com/neb/inteins.html*
Ref: PERLER (2000)

The main purpose of this database is to be a curated resource for protein splicing. The database contains good descriptions of intein proteins (self-catalytic proteins) which are good examples of intramolecular interactions. The database records are present in a machine-readable format. Each record could be used by an interaction database to generate intramolecular interaction records containing chemical reaction description.

Interact

URL: *http://bc38.sbc.man.ac.uk/interactpr.htm*
Ref: EILBECK et al. (1999)

Interact is an object-oriented database for protein–protein interactions. It has a formal data-model that describes interactions, molecular complexes and genetic interactions. It stores information about experimental methods and is based on an object-oriented description of proteins and genes. The database does not provide other details about the interaction and the underlying description of genes and proteins is simplified compared to that of GenBank. The database is not yet available over the web, but the object-oriented design approach has been described in the literature. The database contains over 1,000 interactions.

Kyoto Encyclopedia of Genes and Genomes (KEGG)

URL: *http://www.genome.ad.jp/kegg/*
Ref: KANEHISA and GOTO (2000)

KEGG represents most of the known metabolic pathways and some of the regulatory pathways as graphical diagrams that are manually drawn and updated. Each of the metabolic pathway drawings are intended to represent all chemically feasible pathways for a given system. As such, these pathways are abstractions onto which enzymes and substrates from specific organisms can be mapped. The KEGG does not explicitly represent specific biomolecular interactions, however, the pathway representations are a valuable source of information for someone assembling pathway information from interaction records. The database is machine-readable, except for the pathway diagrams. Each enzyme entry contains a substrate and a product field that can be used to translate between the chemical reaction description scheme and a binary interaction scheme. The KEGG project distributes all database freely for academics via FTP. KEGG is one of the best freely available resources of metabolic and small molecule information.

MDB (Metalloprotein Database)
URL: *http://metallo.scripps.edu/*

MDB contains the metal-binding sites from entries in the PDB database. The database is based on open source software and is freely available. The data is present down to the atomic level of detail. An extensive Java applet is available to query and examine the data in detail.

MHCPEP
URL: *http://wehih.wehi.edu.au/mhcpep/*
Ref: BRUSIC et al. (1998)

MHCPEP is a database comprising over 13,000 peptide sequences known to bind MHC molecules compiled from the literature and from direct submissions. It has not been updated since mid-1998. While this database is not a typical interaction database, it provides peptide–protein interaction information relevant to immunology. The database is freely available via FTP in a text based machine-readable format.

MIPS Yeast Genome Database
URL: *http://www.mips.biochem.mpg.de/proj/yeast/*
Ref: MEWES et al. (2000)

The MIPS Yeast Genome Database (MYGD) presents a comprehensive database which summarizes the current knowledge regarding the more than 6,000 ORFs encoded by the yeast genome. This database is similar to SGD and YPD and is not primarily an interaction database. The MIPS center, however, makes available large tables for direct protein–protein interactions as well as genetic interactions in yeast free for download at *http://www.mips.biochem.mpg.de/proj/yeast/tables/interaction/index.html*. Each interaction contains an experimental method used and usually contains a literature reference. Manually created clickable pathway maps are also available for various metabolic and regulatory pathways in yeast. The MIPS Yeast Genome Database uses a relational model, but most fields use unstructured text. For example, the experimental method used to determine the interaction field is unstructured and the same experimental type may be represented in many different ways. This makes the database difficult to parse with a computer, but MYGD is an extremely useful resource for yeast protein–protein interaction information.

MMDB – Molecular Modeling Database
URL: *http://www.ncbi.nlm.nih.gov/Structure/*
Ref: WANG et al. (2000)

This database is an NCBI resource that contains all of the data in the PDB database in ASN.1 form. The MMDB validates all PDB file information and describes all atomic level detail data explicitly and in a formal machine-readable manner. While this database is not an interaction database, it does contain atomic level detail of molecular interactions present in some records that describe molecular complexes. Sequence linkage is improved and MMDB is easily accessed by machine-readable methods which can obtain information about molecular interactions. MMDB is in the public domain and all software and data is freely available to academics or corporations.

ORDB (Olfactory Receptor Database)
URL: *http://ycmi.med.yale.edu/senselab/ordb/*
Ref: SKOUFOS et al. (2000)

The ORDB is primarily a database of sequences of olfactory receptor proteins. It contains a section on small molecule ligands that bind to olfactory receptors. About 80 ligand–protein interactions are present in the database.

ooTFD (Object Oriented Transcription Factors Database)
URL: *http://www.ifti.org/*
Ref: GHOSH (2000)

The ooTFD contains information on transcription factors from various organisms including transcription factor binding sites on DNA and transcription factor molecular complex information. This means that it contains protein–DNA and protein–protein interactions. The database is based on a formal machine-readable object-oriented format and is available in numerous forms. The database contains thousands of sites and transcription factors and is freely available (including software) from *http://ncbi.nlm.nih.gov/repository/TFD/*.

PhosphoBase

URL: *http://www.cbs.dtu.dk/databases/ PhosphoBase/*

Ref: KREEGIPUU et al. (1999)

This database contains information on kinases and phosphorylation sites. The phosphorylation sites are stored along with kinetics information and references for each kinase. While this is not an interaction database directly, good information is present about protein–protein interactions involved in cell signaling and their chemistry.

PFBP (Protein Function and Biochemical Networks)

URL: *http://www.ebi.ac.uk/research/pfmp/*

The aim of the PFBP is to describe metabolism, gene regulation, transport and signal transduction. PFBP is based on a formal object-oriented data model that will be integrated with CORBA. The database is chemical reaction based and has started by describing metabolic pathways only. PFBP uses a graph abstraction for the interaction data and can describe chemical reactions and pathways. The database has an extensive web site describing it, but is not yet publicly available over the web.

ProNet (Myriad)

URL: *http://pronet.doubletwist.com/*

This commercial database provides protein–protein interaction information to the public from Myriad Genetics proprietary high-throughput yeast two-hybrid system for human proteins and from published literature. Each protein record describes interacting proteins and a Java applet is available to navigate the database. The database stores only protein interaction information with links to primary sequence databases and PubMed. It uses a graph abstraction to display the interactions. The database is fully proprietary and has not been published.

REBASE – restriction enzymes and methylases

URL: *http://rebase.neb.com*

Ref: ROBERTS and MACELIS (2000)

REBASE is a comprehensive database of information about restriction enzymes and related proteins, such as methylases. While it is not an interaction database, restriction enzymes and methylases take part in specific DNA–protein interactions. REBASE describes the enzyme and the recognition site. Useful links are present to commercially available enzymes. REBASE is freely available in many different formats to the academic community.

RegulonDB

URL: *http://www.cifn.unam.mx/Computational_Biology/regulondb/*

Ref: SALGADO et al. (2000)

RegulonDB is mainly an *E. coli* operon database, although it does contain protein–DNA interaction (e.g., ribosome binding sites) and protein complexes. The database is free for non-commercial use. Commercial users require a license.

SELEX_DB

URL: *http://wwwmgs.bionet.nsc.ru/mgs/ systems/selex/*

Ref: PONOMARENKO et al. (2000)

SELEX_DB is a curated resource that stores experimental data for functional site sequences obtained by using SELEX-like random sequence pool technologies to study interactions. The database contains DNA–protein interactions. It is available over the web and the records are in a machine readable flat-file format.

STKE (Signal Transduction Knowledge Environment)

URL: *http://www.stke.org/*

STKE is a curated resource for signal transduction. It provides a manually created clickable picture of various signal transduction pathways linked to a primary database. The data model is based on an upstream and downstream components view, which is a graph abstraction. Database fields are unstructured and are thus not machine-readable. STKE is free with registration until January 2001. Otherwise it is available via a paid subscription to *Science* magazine.

SPAD (Signaling Pathway Database)

URL: *http://www.grt.kyushu-u.ac.jp/eny-doc/*

SPAD provides clickable image maps for a handful of pathways. Clicking on an element of the pathway diagram links to sequence information of the protein or gene. Protein-protein and protein–DNA interactions are covered with respect to signal transduction. The database does not have a formal data model. SPAD has not been updated since 1998, but still gives useful overviews of the pathways it contains.

SoyBase
URL: *http://ars-genome.cornell.edu/cgi-bin/WebAce/webace?db=soybase& query=find+Reaction_or_pathway*
SoyBase is an ACEDB database that contains information about the soybean, including metabolism. Metabolic pathways are based on chemical reactions. SoyBase contains 545 automatically generated clickable diagrams of metabolic pathways covering 996 enzymes and 771 metabolites. SoyBase is based on a formal machine readable data model and is available over the web.

SYFPEITHI
URL: *http://www.uni-tuebingen.de/uni/kxi/*
Ref: RAMMENSEE et al. (1999)
SYFPEITHI is a database of MHC ligands and peptide motifs. It contains about 200 peptide sequences known to bind class I and class II MHC molecules. All entries have been compiled from the literature. While this database is not a typical interaction database, it provides peptide–protein interaction information relevant to immunology.

TRANSFAC
URL: *http://transfac.gbf.de/TRANSFAC/*
Ref: WINGENDER et al. (2000)
TRANSFAC is a database of transcription factors containing genomic binding sites and DNA-binding profiles. As such, it is not a typical interaction database, but it does contain protein–DNA interactions. TRANSFAC is freely available to academics for download via FTP and is based on a formal relational database model.

TRANSPATH
URL: *http://transfac.gbf.de/TRANSFAC/*
Ref: WINGENDER et al. (2000)
TRANSPATH is an effort underway at TRANSFAC to link regulatory pathways to transcription factors. The database is based on

a chemical reaction view of interactions and contains a strong graph abstraction. Graph algorithms have been implemented to navigate the data. The data can describe regulatory pathways, their components and the cellular locations of those components. It can store information about various species. TRANSPATH includes all of the data from the CSNDB and it is obvious that TRANSPATH is using graph theory ideas from the CSNDB. TRANSPATH is free for academic users and can be downloaded in machine readable XML format.

TRRD – Transcription Regulatory Regions Database
URL: *http://wwwmgs.bionet.nsc.ru/mgs/ dbases/trrd4/*
Ref: KOLCHANOV et al. (2000)
TRRD contains information about regulatory regions including over 3,600 transcription factor binding sites (DNA–protein interactions). It is available via an SRS database interface freely over the web.

WIT (What Is There?)
URL: *http://wit.mcs.anl.gov/WIT2*
Ref: OVERBEEK et al. (2000)
WIT is a database project whose purpose is to be able to reconstruct metabolic pathways in newly sequenced genomes by comparing predicted proteins with proteins in known metabolic networks. Metabolic networks are stored in a chemical reaction based scheme with a graph abstraction.

YPD (Yeast Proteome Database) and WormPD
URL: *http://www.proteome.com/databases/ index.html*
Ref: COSTANZO et al. (2000)
This proprietary commercial curated database effort by Proteome Inc. contains extensive information about all proteins in yeast and worm. Among these are data about protein interactions, molecular complexes and sub-cellular location. Most of the database fields are free form text, but there is enough structure in the data model to make it amenable to machine-reading of protein–protein interaction information. YPD is freely available to academics via the web.

Acknowledgements

GARY BADER is a graduate student in the Department of Biochemistry, University of Toronto. The BIND project is supported by the CIHR (formerly MRC). The authors thank IAN DONALDSON and CHERYL WOLTING for their constructive comments regarding this manuscript.

10 References

BADER, G. D., HOGUE, C. W. (2000), BIND – a data specification for storing and describing biomolecular interactions, molecular complexes and pathways, *Bioinformatics* **16**, 465–477.

BAIROCH, A. (2000), The ENZYME database in 2000, *Nucleic Acids Res.* **28**, 304–305.

BAIROCH, A., APWEILER, R. (2000), The SWISS-PROT protein sequence database and its supplement TrEMBL in 2000, *Nucleic Acids Res.* **28**, 45–48.

BALL, C. A., DOLINSKI, K., DWIGHT, S. S., HARRIS, M. A., ISSEL-TARVER, L. et al. (2000), Integrating functional genomic information into the *Saccharomyces* genome database, *Nucleic Acids Res.* **28**, 77–80.

BARANOV, P. V., KUBARENKO, A. V., GURVICH, O. L., SHAMOLINA, T. A., BRIMACOMBE, R. (1999), The Database of Ribosomal Cross-links: an update, *Nucleic Acids Res.* **27**, 184–185.

BERMAN, H. M., WESTBROOK, J., FENG, Z., GILLILAND, G., BHAT, T. N. et al. (2000), The Protein Data Bank, *Nucleic Acids Res.* **28**, 235–242.

BETTE KORBER, C. B. B. H. R. K. J. M. B. W. (1998), *HIV Molecular Immunology Database 1998.* Theoretical Biology and Biophysics Group, Los Alamos National Laboratory, Los Alamos, NM.

BRUSIC, V., RUDY, G., HARRISON, L. C. (1998), MHC-PEP, a database of MHC-binding peptides: update 1997, *Nucleic Acids Res.* **26**, 368–371.

COSTANZO, M. C., HOGAN, J. D., CUSICK, M. E., DAVIS, B. P., FANCHER, A. M. et al. (2000), The yeast proteome database (YPD) and *Caenorhabditis elegans* proteane database (WormPD): comprehensive resources for the organization and comparison of model organism protein information, *Nuclei acids Res.* **28**, 73–76.

EILBECK, K., BRASS, A., PATON, N., HODGMAN, C. (1999), INTERACT: an object oriented protein–protein interaction database, *Ismb.* 87–94 (Abstract).

GHOSH, D. (2000), Object-oriented transcription factors database (ooTFD), *Nucleic Acids Res.* **28**, 308–310.

HOFMANN, K., BUCHER, P., FALQUET, L., BAIROCH, A. (1999), The PROSITE database, its status in 1999, *Nucleic Acids Res.* **27**, 215–219.

JOHNSON, D. B. (1977), Efficient algorithms for shortest paths in sparse networks, *JACM* **24**, 1–13.

KANEHISA, M., GOTO, S. (2000), KEGG: kyoto encyclopedia of genes and genomes, *Nucleic Acids Res.* **28**, 27–30.

KARP, P. D., RILEY, M., PALEY, S. M., PELLEGRINI-TOOLE, A., KRUMMENACKER, M. (1999), Eco Cyc: encyclopedia of *Escherichia coli* genes and metabolism, *Nucleic Acids Res.* **27**, 55–58.

KEL-MARGOULIS, O. V., ROMASHCHENKO, A. G., KOLCHANOV, N. A., WINGENDER, E., KEL, A. E. (2000), COMPEL: a database on composite regulatory elements providing combinatorial transcriptional regulation, *Nucleic Acids Res.* **28**, 311–315.

KOLCHANOV, N. A., PODKOLODNAYA, O. A., ANANKO, E. A., IGNATIEVA, E. V., STEPANENKO, I. L. et al. (2000), Transcription regulatory regions database (TRRD): its status in 2000, *Nucleic Acids Res.* **28**, 298–301.

KOLPAKOV, F. A., ANANKO, E. A., KOLESOV, G. B., KOLCHANOV, N. A. (1998), GeneNet: a gene network database and its automated visualization, *Bioinformatics* **14**, 529–537.

KREEGIPUU, A., BLOM, N., BRUNAK, S. (1999), PhosphoBase, a database of phosphorylation sites: release 2.0, *Nucleic Acids Res.* **27**, 237–239.

MEWES, H. W., FRISHMAN, D., GRUBER, C., GEIER, B., HAASE, D. et al. (2000), MIPS: a database for genomes and protein sequences, *Nucleic Acids Res.* **28**, 37–40.

OSTELL, J., KANS, J. A. (1998), in: *Bioinformatics, a Practical Guide to the Analysis of Genes and Proteins* (BAXEVANIS, A. D., OUELLETTE, B. F., Eds.), pp. 121–144. New York: John Wiley & Sons.

OVERBEEK, R., LARSEN, N., PUSCH, G. D., D'SOUZA, M., SELKOV, E. J. et al. (2000), WIT: integrated system for high-throughput genome sequence analysis and metabolic reconstruction, *Nucleic Acids Res.* **28**, 123–125.

PERLER, F. B. (2000), InBase, the Intein Database, *Nucleic Acids Res.* **28**, 344–345.

PONOMARENKO, J. V., ORLOVA, G. V., PONOMARENKO, M. P., LAVRYUSHEV, S. V., FROLOV, A. S. et al. (2000), SELEX_DB: an activated database on selected randomized DNA/RNA sequences addressed to genomic sequence annotation, *Nucleic Acids Res.* **28**, 205–208.

RAMMENSEE, H., BACHMANN, J., EMMERICH, N. P., BACHOR, O. A., STEVANOVIC, S. (1999), SYFPEITHI: database for MHC ligands and peptide motifs, *Immunogenetics* **50**, 213–219.

ROBERTS, R. J., MACELIS, D. (2000), REBASE – restriction enzymes and methylases, *Nucleic Acids Res.* **28**, 306–307.

ROBISON, K., MCGUIRE, A. M., CHURCH, G. M. (1998), A comprehensive library of DNA-binding site matrices for 55 proteins applied to the complete *Escherichia coli* K-12 genome, *J. Mol. Biol.* **284**, 241–254.

SALGADO, H., SANTOS-ZAVALETA, A., GAMA-CASTRO, S., MILLAN-ZARATE, D., BLATTNER, F. R., COLLADO-VIDES, J. (2000), RegulonDB (version 3.0): transcriptional regulation and operon organization in *Escherichia coli* K-12, *Nucleic Acids Res.* **28**, 65–67.

SCHONBACH, C., KOH, J. L., SHENG, X., WONG, L., BRUSIC, V. (2000), FIMM, a database of functional molecular immunology, *Nucleic Acids Res.* **28**, 222–224.

SELKOV, E., BASMANOVA, S., GAASTERLAND, T., GORYANIN, I., GRETCHKIN, Y. et al. (1996), The metabolic pathway collection from EMP: the enzymes and metabolic pathways database, *Nucleic Acids Res.* **24**, 26–28.

SKOUFOS, E., MARENCO, L., NADKARNI, P. M., MILLER, P. L., SHEPHERD, G. M. (2000), Olfactory receptor database: a sensory chemoreceptor resource, *Nucleic Acids Res.* **28**, 341–343.

SPIROV, A. V., BOWLER, T., REINITZ, J. (2000), HOX Pro: a specialized database for clusters and networks of homeobox genes, *Nucleic Acids Res.* **28**, 337–340.

SZYMANSKI, M., BARCISZEWSKI, J. (2000), Aminoacyl-tRNA synthetases database Y2K, *Nucleic Acids Res.* **28**, 326–328.

TAKAI-IGARASHI, T., NADAOKA, Y., KAMINUMA, T. (1998), A database for cell signaling networks, *J. Comput. Biol.* **5**, 747–754.

WANG, Y., ADDESS, K. J., GEER, L., MADEJ, T., MARCHLER-BAUER, A., ZIMMERMAN, D., BRYANT, S. H. (2000), MMDB: 3D structure data in Entrez, *Nucleic Acids Res.* **28**, 243–245.

WHEELER, D. L., CHAPPEY, C., LASH, A. E., LEIPE, D. D., MADDEN, T. L. et al. (2000), Database resources of the National Center for Biotechnology Information, *Nucleic Acids Res.* **28**, 10–14.

WINGENDER, E., CHEN, X., HEHL, R., KARAS, H., LIEBICH, I. et al. (2000), TRANSFAC: an integrated system for gene expression regulation, *Nucleic Acids Res.* **28**, 316–319.

XENARIOS, I., RICE, D. W., SALWINSKI, L., BARON, M. K., MARCOTTE, E. M., EISENBERG, D. (2000), DIP: the database of interacting proteins, *Nucleic Acids Res.* **28**, 289–291.

Ethical, Legal and Social Issues

19 Ethical Aspects of Genome Research and Banking

BARTHA MARIA KNOPPERS

Montreal, Canada

1 Introduction

Genetic research is increasingly used to cover a wide range of research activities. These activities extend from classical research into diseases following Mendelian patterns, to the search for genetic risk factors in common diseases, to the more recent interest in pharmacogenomics and finally, to the actual need for studies of normal genetic variation across entire populations. This all encompassing nature of the term genetic research would not be so problematic were it not for the fact that corresponding distinctions (if necessary) may not in fact be applied in the ethical norms applied to evaluate such research. In order to address this issue, we need to understand the ethical aspects of the different types of genetic research. Beginning then with a cursory overview of the types of genetic research (Sect. 2), we will then proceed to an introduction to the ethics norms of research in general (Sect. 3), before analyzing their further elaboration in the area of genetic research (Sect. 4). Particular attention will then be paid to the problems raised by DNA banking (Sect. 5), with the conclusion focussing on the issue of ownership of the samples in an increasingly commercial environment (Sect. 6). Finally, the term "banking" will be used to cover all stored tissue samples used in genetic research whether obtained following medical care or specifically for research.

2 Types of Genetic Research

Incredible progress has been achieved in our ability to discover and develop diagnostic tests for hereditary, single gene disorders with calculable mathematical precision to say nothing of a known degree of morbidity and mortality. The same progress has not been made in the treatment of these conditions. They are, however, prime candidates for gene therapy research. Often inherited not only through families but also following racial and ethnic lines, these latter features together with the quasi-certainty of expression have led to the development of ethical guidelines and legislation sensitive to both potential discrimination and to the possibility of stigmatization by association (COLLINS, 1999) (see Sect. 4).

Understanding of the role of genetic factors in common conditions such as, e.g., hypertension, cancer, and diabetes is more complex. Other than perhaps certain rare forms of these conditions that follow familial patterns, their expression is often determined by the interplay of environmental, socioeconomic, cultural and other influences. This poses interdisciplinary challenges for ethics review to say nothing of determining the appropriateness of legislation in this area.

Pharmacogenomics is seeking to understand the role of genetic variation (polymorphisms) in individual response (e.g., toxicity, efficacy, dosage, etc.) and requires the expansion of epidemiology studies to entire populations (whether ill, at-risk, or, not) so as to establish normal, genetic diversity. While anthropological, demographic and surveillance research was hitherto free from "genetic" taint, the same is not true of the study of population genetics. Interestingly most population studies of genetic variation do not require personal, identifying medical information but rather seek to use anonymized DNA samples (ROSES, 2000) (see Sect. 5).

Across this spectrum then, from certainty, to probabilistic percentages in common diseases, to individualized susceptibility, to the anonymized sample, the possibility of applying uniform ethical criteria is unlikely. The same difficulties may not be present, however, in the application of the larger ethical framework governing biomedical research generally.

3 Research Ethics

The existence of a myriad of rules of conduct concerning the protection of human research subjects has prompted the National Bioethics Advisory Committee (NBAC of the United States) to make international harmonization a priority (3rd Global Summit of National Bioethics Commissions, Imperial College, London/England, Sept. 20–21, 2000). Globalization, the explosion of new technologies, the North–South divide, sensitivity to differing cultural and religious worldviews and to the lessons learned from the biodiversity debate, make such harmonization difficult but not impossible. The real test may well be that of ensuring that not only the public sector but also the private sector (which is the largest source of funding), abide by such a future international approach. The other challenge relates to an endemic problem, that of proper, ongoing oversight.

Following the adoption of the *Nuremberg Code* (1946–1949) and later of the *Helsinki Declaration* (World Medical Association, 1964, 1975, 1983, 1989, 1996, 2000), the main tenets of research ethics are both integrated into the biomedical world and yet evolving. The most common elements include respect for privacy and autonomy through the process of informed consent and choice, the right to withdraw, and the protection of the vulnerable.

The last decade has seen the emergence of new issues and additional elements such as: community consent, commercialization, statutory regulation of clinical trials, benefit-sharing, inclusiveness, and equitable access to research trials and benefits. There is also a much greater specificity in that particular areas or groups of persons are singled out such as those suffering from mental disorders, HIV/AIDS, and the disabled. Moreover, frameworks are being or have been developed for particular areas such as organ transplantation, reproductive technologies, or tissue banking, to name but a few (LE BRIS et al., 1997).

The adoption of the European *Convention on Biomedicine and Human Rights* (Council of Europe, 1997) illustrates the difficulty, however, of finding common principles and positions when technologies are already well entrenched and different countries have adopted legislation. For example, no agreement could be reached in the *Convention* on embryo research, an area where guidance is again required now that stem cell and therapeutic cloning techniques are offering new breakthroughs. Indeed, September 7, 2000, the European Parliament narrowly passed a resolution (237 vs. 230 votes with 43 abstentions) con-

demning the deliberate creation of embryos for therapeutic cloning (European Parliament, 2000). If this difficulty in finding consensus continues, the same will hold no doubt in the actual and future elaboration of the specific procedural protocols pursuant to the *Convention*.

The *Convention* is notable, however, in its broadening of the inclusion criteria governing incompetent adults and children. Indeed, rather than excluding them from biomedical research in the absence of direct benefit, the *Convention* would permit inclusion with the consent of the legal representative even if the benefits were only indirect, that is, for persons of the same age or condition (art. 17). This evolution in biomedical ethics bears examination in the field of bioethics and genetics (LE BRIS et al., 1997).

4 "Genethics"

At the international level, UNESCO adopted the *Universal Declaration on the Human Genome and Human Rights* in 1997 (United Nations Educational, Scientific and Cultural Organization, 1997). The *Declaration* is prospective in nature and embraces the concepts of human dignity and diversity of the genome as the common heritage of humanity, of non-commodification, of the need for international solidarity, and of concern over technologies such as germ-line interventions that could affect future generations (art. 24). It specifically prohibits human reproductive cloning (art. 11). This *Declaration* then, comes at the beginning of a technology and hopefully will serve to prospectively guide national approaches, thus ensuring a minimum of harmonization.

Also anticipatory in nature and 10 years in the making, the 1998 European *Directive on the Legal Protection of Biotechnological Inventions* (Council of the European Union, 1998) is not only a clarification (if not ratification) of existing trends but also innovates. The *Directive* reaffirms the non-patentability of human genes in their natural state and under the umbrella of public policy (an ethical filter also found in the *European Patent Conven-*

tion) prohibits techniques such as human cloning and germ-line intervention [art. 5(2)]. The preamble ("recital"), while not having legal force, is the first legal instrument to require that a patent application for an invention using human biological material must be "from a person who has had the opportunity of expressing [a] free and informed consent thereto, in accordance with national law" (para. 26). This means that at a minimum, participants in genetic research and banking must be notified of the possibility of eventual commercialization. In the absence of "national law", however, its impact will be weakened.

It is interesting to note that both international and regional instruments are strengthening barriers to access by third parties (e.g., insurers and employers). Notable in this regard is the European *Convention on Human Rights and Biomedicine* (Council of Europe, 1997) mentioned earlier which in limiting genetic testing to health purposes (art. 12) effectively limits requests for testing by insurers and employers.

A significant development is the creation of both a right not to know under the *Convention on Human Rights and Biomedicine* [art. 10(2)], and yet, a new exception for professional disclosure to at-risk family members for serious or preventable conditions where the patient or research participant refuses to do so. This is the position of the 1997 *Proposed International Guidelines* of WHO (World Health Organization, 1997), of the 1998 HUGO *Statement on DNA Sampling: Control and Access* (Human Genome Organisation, 1999a) and of the European Society of Human Genetics (2000). This is interesting in the banking context since ongoing access to banked samples (unless anonymized) could create a similar ongoing obligation for the researcher-banker as new tests become available.

Finally, another change in international "genethics" is the attempt to move away from traditional, categorical, wholesale prohibitions in the area of cloning and germ-line therapy. Yet unfortunately, while the International Bioethics Committee of UNESCO in its penultimate draft had agreed to keep the *Universal Declaration on the Human Genome and Human Rights* (United Nations Educational, Scientific and Cultural Organization, 1997)

free from the mention of any specific technology, the aim being to guarantee its viability over time and its universality as well as to strengthen the impact of concepts such as human dignity and diversity if justification for prohibitions were needed. Nevertheless, the governmental representatives convened to approve the Committee's final draft sought (political?) refuge in inserting "technique-specific" prohibitions in the *Declaration* with regard to human cloning and germ-line therapy as mentioned earlier. It bears noting that the WHO in both its 1997 *Proposed International Guidelines* (World Health Organization, 1997), its 1999 *Draft Guidelines on Bioethics* (World Health Organization, 1999), and its resolution on *Ethical, Scientific and Social Implications of Cloning in Human Health* (World Health Organization, 1998) distinguishes between the different types of cloning. Both WHO and HUGO (Human Genome Organisation, 1999b) prohibit human reproductive cloning but encourage relevant research in the field of therapeutic cloning and stem cell research.

This is instructive in the banking context where as we shall see, former, similar wholesale proclamations about the DNA as "person" or as "property" have ultimately proved secondary to the need to ensure personal control whatever the legal qualification and this without impact as regards commercialization.

5 DNA Banking

The last 10 years have seen tremendous upheaval and uncertainty in the world of DNA banking and research. Indeed, 1995 saw the hitherto unfettered access by researchers to archived samples come to a halt with the report of a NIH study group on informed consent for genetic research on stored tissue samples suggesting that the proof of consent to research was required even for those samples already stored during routine medical care (Clayton et al., 1995). While generally, the ethical and legal norms governing banking had been moving towards a more informed choice approach with options in the case of samples provided in the research context *per se*, the implementation of this approach would effectively have halted the largest "source" of DNA samples for genetic research to say nothing of epidemiological or public health research (even if the latter wished to use only anonymized samples). This conservative position was followed by a myriad of contradictory positions around the world (KNOPPERS et al., 1998).

Five years later, in May 2000, the UK's Royal College of Physicians Committee on Ethical Issues in Medicine published its recommendation on *Research Based on Archived Information and Samples* (Royal College of Physicians, 1999a) and the circle was closed. The College does not consider it necessary to obtain specific consent for:

- The retrospective use of existing medical records for analysis of disease prevalence, clinical features, prognosis, response to treatment, etc.
- The use of biological samples that have been previously taken during the course of medical diagnosis or treatment, at autopsy, or for research, and are in excess of requirement for their original purpose, e.g., "left over" portions of blood samples or tissue biopsies (Royal College of Physicians, 1999b).

Thus, according to this most recent report, irrespective of whether the person is still alive or has consented or not to the research in question, subject to certain conditions, medical research using biological samples may be conducted without the express consent of the individual patients or research subjects. Nevertheless, the material must be anonymized at the earliest possible stage consistent with obtaining the information necessary for the research. The minimum level of anonymization is that which precludes identification of individuals from the output of the research (Royal College of Physicians, 1999b).

Where does this position stand relative to international norms or to that of other countries? To answer that question, we will examine the varying responses in the time period of 1995–2000 with respect to samples already archived that were obtained during medical care or from autopsies, samples provided specifically for research, and finally, samples obtained for research but where other research is now proposed.

It should be mentioned at the outset that perhaps more confusing than the plethora of contradictory positions is that of the terminology used. Only terms such as "identified", or, "nominative", or, "personally identified" are understandable by all. In contrast, "identifiable" or "traceable" is used interchangeably with the term "coded", and, the term "anonymous" (i.e., never had any identifiers such as with specimens found in archeological digs), is often confused with "anonymized".

For the purpose of clarity, we will use the term "anonymized" (e.g., originally identified or coded/identifiable/traceable but now stripped except for some clinical or demographic data), and, the term "coded" (e.g., identifiable only through breaking the unique code given the sample in lieu of personal identifiers). We will examine international (Sect. 5.1) and regional (Sect. 5.2) positions on abandoned or research samples before turning to particular countries (Sect. 5.3).

5.1 International

Most international statements and guidelines on the ethics of genetic research do not address the specific issue of archived samples originating from medical care, the context of medical care being largely left to individual countries.

One notable exception is the 1998 *Statement on DNA Sampling: Control and Access* of the Ethics Committee of the Human Genome Organisation (HUGO) (Human Genome Organization, 1999a, rec. 2). The very mission of the Committee is to provide such guidance. Like the Royal College of the United Kingdom (Royal College of Physicians, 1999a) the HUGO Ethics Committee holds that:

"Routine samples, obtained during medical care and stored may be used for research if: there is general notification of such a policy, the patient has not objected, and the sample to be used by the researcher has been coded or anonymized. Routine samples obtained during medical care and stored before notification of such a policy may be used for research if the sample has been anonymized prior to use."

WHO's 1997 *Proposed International Guidelines* did not take a position on leftover or "abandoned" samples except to say that "specimens that could be useful to families in the future should be saved and should be available" (World Health Organization, 1997, Tab. 10, guideline 10).

The relative absence of international guidelines specific to possible research uses for "left over" samples is regrettable for many reasons. The first, as already mentioned, is the need for large scale epidemiology, for the study of population variations (genetics), and for general health surveillance (an often forgotten obligation of the State). Such studies would be greatly facilitated if abandoned, anonymized samples were made available. The second is the application by default of the rules governing samples obtained for specific research projects, or, of the rules of consent to genetic research generally. The third is the extreme difficulty, if not impossibility, of fulfilling the ethical obligation of international collaboration due to the lack of international guidance and harmonization. This last deficiency is further exacerbated by the confusion surrounding the use of different terminology to describe the samples as mentioned earlier.

Turning to samples obtained specifically for research purposes, only two international documents have addressed the issue, HUGO and the WHO. As we will see, until very recently, both were in stark opposition to the more conservative national positions. Indeed, WHO's 1997 *Proposed International Guidelines* (World Health Organization, 1997, Tab. 10, guideline 10) maintains that "a blanket informed consent that would allow use of a sample in future projects is the most efficient approach" (Tab. 10). This is somewhat tempered by the assertion that "genetic samples from individuals must be handled with respect, should be taken only after the consent is obtained, and, should be used only as stated in the consent document" (p. 4). Other than the general need to preserve confidentiality, no distinction is made between coded or anonymized samples for research purposes.

Due to its mandate, the 1998 HUGO *Statement on DNA Sampling: Control and Access* (Human Genome Organisation, 1999a) specifically addresses the issue and holds that:

"Research samples obtained with consent and stored may be used for other research if, there is general notification of such a policy, the participant has not yet objected, and the sample to be used by the researcher has been coded or anonymized. For the use of research samples obtained before notification of a policy, these samples may be used for other research if the sample has been coded or anonymized prior to use." (rec. 3).

While consent to specific research is a *sine qua non*, both international bodies do not require an explicit consent for other uses. As just seen, HUGO would require notification and the opportunity for objection as well as mandating anonymization if such prior notification did not take place.

5.2 Regional

Other than upholding the need for informed consent for all medical interventions including research, at the regional level, there is very little guidance on genetic research with regard to either archived samples left over after medical care or research samples.

Article 22 of the 1997 Council of Europe's *Convention on Human Rights and Biomedicine* (Council of Europe, 1997) maintains that:

"When in the course of an intervention any part of a human body is removed, it may be stored and used for a purpose other than that for which it was removed only if this is done in conformity with appropriate information and consent procedures."

In the mean time, what is "appropriate" depends on national positions. The Council of Europe is currently preparing a protocol to the *Convention* specifically on genetic research. Thus, perhaps some guidance on the thorny issue of the use of archived abandoned samples, or, on research samples will be forth-coming (see also the European Society of Human Genetics, 2000, the Society is preparing a position paper on sampling).

5.3 National

The majority of countries still do not distinguish between archived and research samples or have positions on the issue of other uses. Thus, unless a new and explicit consent is obtained, neither abandoned samples taken during medical care or research samples can be used for other purposes than those outlined in the protocol.

Before addressing the topic of other uses of research samples, it bears mentioning that the issue has not arisen in the context of leftover samples from routine care in that generally, any research use would require a specific consent unless the sample is anonymized (KNOP-PERS et al., 1998). For example, the Health Council of the Netherlands, in its report on the *Proper Use of Human Tissue* states:

> "If residual material is to be used for purposes of which patients are unaware,
> then – the argument runs – they should at least be informed and given the opportunity to object." (Health Research Council of the Netherlands, 1994).

Taking notice of the fact that obviously, consent is required for the actual obtaining of the sample in medical care or research, it is only in the last year that some national jurisdictions have distinguished between obtaining consent at the time of sampling for research and the issue of other uses. Generally, they are becoming less stringent in always requiring an explicit consent for further uses. To take but a few examples, Australia's 1999 *National Statement on Ethical Conduct in Research Involving Humans* (National Health and Medical Research Council, 1999a), "normally" requires a new consent from donors of archived samples (princ. 15.7). Yet, the possibility of waiver by an Ethics Committee for the obtaining of another consent is foreseen in the context of research samples (princ. 15.6).

Indeed, "[A]n HREC [Human Research Ethics Committee] may sometimes waive, with or without conditions, the requirement of consent. In determining whether consent may be waived or waived subject to conditions, an HREC may take into account:

- The nature of any existing consent relating to the collection and storage of the sample;
- The justification presented for seeking waiver of consent including the extent to which it is impossible or difficult or intrusive to obtain specific consent;
- The proposed arrangements to protect privacy including the extent to which it is possible to de-identify the sample;
- The extent to which the proposed research poses a risk to the privacy or well being of the individual;
- Whether the research proposal is an extension of, or closely related to, a previously approved research project;
- The possibility of commercial exploitation of derivatives of the sample; and
- Relevant statutory provisions." (princ. 15.8) (National Health and Medical Research Council, 1999a).

Similarly, Japan also seems to be moving in this direction, in that the Bioethics Committee of the Council for Science and Technology in its *Fundamental Principles of Research on the Human Genome* (Council for Science and Technology, 2000) mentions that "[i]f a participant consents to provide a research sample for a genome analysis in a particular research project and, at the same time, anticipates and consents to the use of the same sample in other genome analyses or related medical research, the research sample may be used for 'studies aimed at other purposes'" (princ. 8.1.a).

Finally, it bears mentioning the Council of Regional Networks for Genetic Services (Council of Regional Networks for Genetic Services, 1997) in the USA did not exclude blanket consent when it stated: "[...] Any deliberate act of the medical profession to separate entire specimens from identifiers may be viewed as usurping the patient's/subject's right to determine subsequence uses for tissue. Consent forms should provide options of blanket consent (waiving the right to be asked for further specific consent), as well as the option to limit their uses".

In short, on the national level, three positions typify the move away from the strict rule of requiring a new consent for other uses of research samples. The first is that of requiring ethics review when foreseeing the possibility of either anonymizing or coding the sample without going back to the source provided there is only minimal risk and confidentiality is ensured (National Bioethics Advisory Commission, 1999, rec. 9f; Medical Research Council, 1999; National Health and Medical Research Council, 1999b). The second requires ethics review but samples must always be anonymized (Health Research Council of the Netherlands, 1994; p. 88; American Society of Human Genetics, 1996; Medical Research Council of Canada, Natural Science and Engineering Research Council of Canada, Social Science and Humanities Research Council of Canada, 1998) and, the third eschews the automatic exclusion of "blanket consents" to future research. Indeed, a majority of members of the National Bioethics Commission of the United States would allow the use of "coded materials" for any kind of future study without further specification as to what kind of research, or the need for further consent, or even anonymization (National Bioethics Advisory Commission, 1999, rec. 9). Coding raises, however, other issues such as that of recontact should subsequent findings become clinically significant.

The advantage of coded samples is that clinical data can be added over time and so scientifically they remain viable. The disadvantage for researchers over time is that at a certain point in time the combination of research and clinical knowledge will become significant enough to have medical importance in the situation where prevention or treatment is available. NBAC has recommended that in this "exceptional" circumstance recontact and disclosure should occur (National Bioethics Advisory Commission, 1999, rec. 14).

In the same vein, on the issue of access by relatives to such information, Japan's Bioethics Committee of the Council for Science and Technology holds that "in case the genetic information obtained by research may lead to

an interpretation that a portion of the genetic characteristics of the participant is or, is supposed to be, connected to the etiology of a disease, this interpretation may be disclosed to his/her blood relatives following authorization by the Ethics Committee only if a preventive measure or a cure has already been established for the disease in question." (princ. 15.2) (Council for Science and Technology, 2000, this is similar to the position of the Human Genome Organisation, 1999a, rec. 5: "[S]pecial considerations should be made for access by immediate relatives. Where there is high risk of having or transmitting a serious disorder and prevention or treatment is available, immediate relatives should have access to stored DNA for the purpose of learning their own status. These exceptional circumstances should be made generally known at both the institutional level and in the research relationship.").

The scientific advantage of coded samples has to be weighed against the potential ongoing obligations that may emerge. Even if such potential obligations could be foreclosed in part by asking research participants in advance whether they would want to be recontacted or not in the event of medically significant findings, what is the longevity or validity of an anticipatory "yes" or "no"? No doubt, the courts will settle this latter question but in the meantime, the option should be presented. If not, automatic communication of at-risk information to participants may run afoul of the emerging right not to know and yet, the failure to do so, of an emerging duty to warn!

To conclude this section on banking, the following comments can be made with regard to the issue of other uses without obtaining another explicit consent:

(1) the wholesale prohibition against both blanket consent to future unspecified uses of research samples and against the use of leftover samples from medical care without a specific consent is increasingly nuanced (and may be on the wane);

(2) there is a need to re-examine the automatic anonymization of samples as the expedient solution to ethical and legal quandaries;

(3) a distinction should be drawn between refusal of access to third parties such as insurers or employers and the legitimate needs for communication to blood relatives; and

(4) discussion is required on the issue of recontact and communication of results in the situation of other research that yields medically relevant information.

It goes without saying that underlying these difficult choices is the ultimate question: to whom does the DNA belong in this commercialized research environment?

6 Ownership

Intimately linked to the issue of ownership is that of the legal status of human genetic material. Even though this issue is one of principle, surprisingly, different legal status – person or property – has not had a concomitant impact on the ultimate issue, that of control of access and use by others.

At the international level, there is an increasing recognition that at the level of the species, the human genome is the common heritage of humanity (for example see United Nations Educational, Scientific and Cultural Organization, 1997, art. 1; KNOPPERS et al., 1998). Contrary to common misunderstanding, it means that at the collective level, like space and the sea, no appropriation is possible by nation states. Other characteristics of this approach include peaceful and responsible international stewardship with a view to future generations and equitable access. In the absence of a binding international treaty (UNESCO's *Declaration* and the WHO's and HUGO's positions being only proclamatory in nature), it remains to be seen if this concept will come to legally binding fruition.

This position, however, is particularly important in that it serves to place new sequences that fail to meet the strict conditions of patenting into the public domain. While patenting is not the subject of this analysis, a strictly personal property approach to DNA samples, would theoretically require a specific personal

consent to eventual patenting. Yet, likewise giving the DNA sample the status of "person" also mandates obtaining consent, or at a minimum notification of patenting, as already mentioned under the 1998 European *Directive* (Council of the European Union, 1998). At the international level then, this position in favor of both the common heritage approach at the level of the collective human genome and that of personal control over individual samples and information has slowly been consolidated. Indeed, the last few years have seen the emergence of a new concept in the international arena, that of benefit sharing. This approach, largely sponsored by HUGO but gradually taking hold in industry, mandates recognition of the participation and contribution of participating populations and communities. Founded on notions of justice and equity, it upholds the common heritage approach but encourages "giving-back" by profit-making entities such as, e.g., contributions to the healthcare infrastructure (Human Genome Organisation, 2000).

Turning to the regional level, the "gift" language of a decade ago, that was replaced with "source", "owner", and "subject" has returned (see, e.g., European Society of Human Genetics, 2000). Lest there be any misgiving, a gift implies the complete transfer of any property or personal rights a person may have. The individual would also give up rights to a share of the profits derived from any commercial application. The language of gift is not found in international instruments, the former emphasizing the common heritage concept (Human Genome Organisation, 1996) or the notion of "general property" or "public domain" (German Society of Human Genetics, 1997), thus obviating the issue of status but excluding private ownership and concentrating on "shared goods".

The European *Convention* (Council of Europe, 1997) mirroring both UNESCO and WHO, limits itself to prohibiting financial gain by stating: "The human body and its parts shall not as such, give rise to financial gain" (art. 21). The *Convention* does, however, maintain that: "When in the course of an intervention any part of a human body is removed, it may be stored and used for a purpose other than that for which it was removed, only if this is done in conformity with appropriate information and consent procedures" (art. 22). It is not known whether the term "intervention" includes research but it goes without saying that if an "abandoned" sample obtained during routine care requires a specific consent for other purposes, including one could presume, eventual commercialization, the same would hold for samples specifically obtained for research.

At the national level, it should be stated at the outset that payment to a research participant for time and inconvenience or cost recovery by the researcher or institution (both being minimal in the case of DNA sampling), neither affords the status of property to a sample nor undermines the notion of gift. Furthermore, the notion of gift, while obviously involving transfer, may not necessarily create immediate property rights in the researcher. Indeed, in the absence of intellectual property which could be afforded to any invention, increasingly we will see that the researcher-"banker" is described as a "custodian". This is both a real and symbolic statement. Real, in that the current complex, public-private funding of research involves multiple economic partners in any eventual profits from patenting. Symbolic, in that the researchers involved may be bench scientists or clinician-researchers and so both may be simple guardians and fiduciaries of the samples for the research participants or patients and their families.

Even in Iceland, with its controversial presumed consent to the storage and use of health data, the recent *Act on Biobanks* (Iceland Minister of Health and Social Security, 2000) extends this notion of "non-ownership" to any company licensed by the Government to do research on accompanying biological samples: "The licensee shall not be counted as the owner of the biological samples, but has rights over them, with the limitations laid down by law, and is responsible for their handling being consistent with the provisions of this Act and of government directives based on it. The licensee may thus not pass the biological samples to another party, nor use them as collateral for financial liabilities, and they are not subject to attachment for debt".

The language of "donation" of human genetic material was particularly prevalent in the countries of civilian tradition (see, e.g., Net-

work of Applied Genetic Medicine, 2000) but has also been adopted in common law jurisdictions. Indeed, the recent MRC interim ethical guidelines on *Human Tissue and Biological Samples for Research* have placed the onus on the custodian of a tissue collection to manage access (Sect. 3.2) (Medical Research Council, 1999). While the MRC recommends that tissue samples donated for research be treated as gifts (Sect. 2.1), the definition of custodianship "implies some property rights over the samples but also some responsibility for safeguarding the interests of the donor" (...) (Medical Research Council, 1999).

Likewise, even those American states that have adopted the *Genetic Privacy Act* (KNOPPERS et al., 1998) have not done so with original articles on the property rights of the "source". Theoretically, the implementation of this approach would have given every "source-owner" an opportunity to bargain for a percentage of eventual profits (if any). The result of all of this debate as well as of increased commercialization of genetic research, is that most consent forms now inform research participants that their sample, or products derived from it, may be commercialized and that they will not be entitled to a share of any eventual profits (CARDINAL et al., unpublished data). Ultimately, it is usually universities, research institutes and/or commercial entities that maintain "biobanks" and share in any profits that may ensue (KNOPPERS, 1999).

7 Conclusion

Genetic research is moving to the forefront of the bioethics debate. This is due in part to public interest in the role of genetic factors in common diseases and also to the possibility of tailoring drugs to individual genetic susceptibility. Ethical frameworks will have to make a corresponding shift from an emphasis on monogenic diseases and the stigma they carry to the "normalization" of genetic information in common diseases. This is all the more important in that the study of normal genetic variation (diversity) will require large population banks. A corresponding "normalization" of the treatment of DNA samples and genetic information as medical information with increased protection will also be welcome.

For now, two issues have served to attract attention to the ethical issues surrounding genetic research and DNA sampling – consent to sampling and commercialization. We have seen that the issue of consent is characterized and stratified by the origin of the sample (medical or research) and by the type of information accompanying the sample as well as the issue of other research uses. The debate on sampling is moving towards a recognition of the need to distinguish between coded and anonymized samples. The trend to favor the latter with its lower risk of possible socioeconomic discrimination may well be short-lived. This is due to the fact that increasingly, if medical and research information is better protected generally, participants themselves may want to be "coded" and followed-up over time and be offered that choice. Furthermore, the anonymized samples themselves may lose their scientific utility over time considering the absence of ongoing clinical information. Researchers may also come to favor coding when the issue of responsibility for recontact is clarified.

On the issue of commercialization of research, while some clarification has been forthcoming in that raw sequences with no specific or substantial utility are seen as being in the public domain and not patentable *per se*. The issue of benefit-sharing raises the possibility of balancing legitimate returns on investment (profit-making) with concerns with equity and justice for participating families, communities, and populations. Influence on consent to sampling has been largely limited to ensuring a clear renunciation of any interest in potential intellectual property by the research participant. The next step may well be to also clarify the role of the researcher, the university (if applicable) and industry. The possibility for conflicts of interest are real and actual where the researcher is not only a clinician but the custodian of the sample and has a financial interest in the research.

As we move from the gene map to gene function, there is a need to understand normal genetic variation and diversity. This will require the participation of large populations. The

lessons learned in the last decade with respect to the need to not only respect personal values and choices in the control of and access to DNA samples in genetic research but also to communicate clearly its goals, should serve to direct the next decade. Transparency and ongoing communication of any change in the direction of the research will do much to ensure public trust in the noble goals of genetic research.

8 References

American Society of Human Genetics (1996), American Society of Human Genetics Statement on Informed Consent for Genetic Research, *Am. J. Hum. Genet.* **59**, 471.

CLAYTON, E. W., STEINBERG, K. K., KHOURY, M. J. et al. (1995), Informed Consent for Genetic Research on Stored Tissue Samples, *JAMA* **274**, 1786.

COLLINS, F. S. (1999), Shattuck Lecture – Medical and Societal Consequences of the Human Genome Project (July 1, 1999), *N. Engl. J. Med.* **341**, 28.

Council for Science and Technology (2000), Bioethics Committee, *Fundamental Principles of Research on the Human Genome,* Japan, June 14, 2000, Official site of the Council for Science and Technology, *http://www.sta.go.jp/shimon/cst/rinri/pri00614.html* (accessed October 2, 2000).

Council of Europe (1997), *Convention for the Protection of Human Rights and Dignity of the Human Being with Regard to the Application of Biology and Medicine: Convention on Human Rights and Biomedicine* (1997), *Int. Dig. Health Leg.* **48**, 99, Official site of the Council of Europe, *http://www.coe.fr/eng/legaltxt/164e.htm* (accessed October 2, 2000).

Council of Regional Network for Genetic Services (1997), *Issues in the Use of Archived Specimens for Genetic Research, Points to Consider,* Albany, New York (January 1997).

Council of the European Union (1998), Directive 98/44/EC of the European Parliament and of the Council of 6 July 1999 on the Legal Protection of Biotechnological Inventions (July 30, 1998) L 213 *Official Journal of the European Communities,* p. 13, Official site of the European Union, *http://europa.eu.int/eur-lex/en/lif/dat/1998/en_398L0044.html* (accessed October 2, 2000).

European Parliament (2000), *European Parliament Resolution on Human Cloning,* Document B5-

0710, 0753 and 0753/2000, Official site of the European Parliament, *http://www.europarl.eu.int/dg3/sdp/pointses/en/ps000904_ens.htm#28* (accessed October 2, 2000).

European Society of Human Genetics (Public and Professional Policy Committee) (2000), *Population Genetic Screening Program, Recommendations of the European Society of Human Genetics*, July 2000, Official site of the European Society of Human Genetics, *http://www.eshg.org/Recommendations%20genetic%20screening.pdf* (accessed October 2, 2000).

German Society of Human Genetics (Committee for Public Relations and Ethical Issues) (1997), *Statement on the Patenting of Human Genes and DNA Sequences*, Germany, 1997, art. 2.

Health Research Council of the Netherlands (1994), *Proper Use of Human Tissue*, The Hague.

Human Genome Organisation (1996), Statement on the Principled Conduct of Genetic Research (May 1996), *Genome Digest* **3**, 2, princ. 1.

Human Genome Organisation (1999a), Statement on DNA Sampling: Control and Access, *Genome Digest* **6**, 8, Official site of HUGO, *http://www.gene.ucl.ac.uk/hugo/sampling.html* (accessed October 2, 2000).

Human Genome Organisation (1999b), Statement on Cloning, *Eubios J. Asian Int. Bioethics* **9**, 70, *http://www.biol.tsukuba.ac.jp/~macer/hugoclone.html* (accessed October 9, 2000).

Human Genome Organisation (2000), *Statement on Benefit-Sharing*, Vancouver, BC, April 9, 2000, Official site of HUGO, *http://www.gene.ucl.ac.uk/hugo/benefit.html* (accessed October 2, 2000).

Iceland Minister of Health and Social Security (2000), *Act on Biobanks*, Iceland, May 13, 2000, art. 10.

KNOPPERS, B. M. (1999), Status, Sale and Patenting of Human Genetic Material: An International Survey, *Nature Genet.* **22**, 23–26.

KNOPPERS, B. M., HIRTLE, M., LORMEAU, S., LABERGE, C., LAFLAMME, M. (1998), Control of DNA samples and information, *Genomics* **50**, 385.

LE BRIS, S., KNOPPERS, B. M., LUTHER, L. (1997), International Bioethics, Human Genetics and Normativity, *Houston Law Rev.* **33**, 1363.

Medical Research Council of Canada, Natural Science and Engineering Research Council of Canada, Social Science and Humanities Research Council of Canada (1998), *Tri-Council Policy Statement – Ethical Conduct for Research Involving Humans*, Canada, August 1998, Official site of the Tri-Council, *http://www.nserc.ca/programs/ethics/english/ethics-e.pdf* (date accessed October 2, 2000, art. 10.3b).

Medical Research Council (1999), *Human Tissue and Biological Samples for Use in Research: Interim and Ethical Guidelines Issued by the Medical Research Council*, London, November, 1999, Official site of the MRC, *http://www.mrc.ac.uk/tissue_gde.pdf* (accessed October 2, 2000), sect. 2.6 et 3.3.

National Bioethics Advisory Commission (1999), *Research Involving Human Biological Materials: Ethical Issues and Policy Guidance Vol. I, Report and Recommendations*, Rockville, MD, August 1999, Official site of NBAC, *http://www.bioethics.gov/hbm.pdf* (accessed October 2, 2000).

National Health and Medical Research Council (1999a), *National Statement on Ethical Conduct in Research Involving Humans*, Australia, 1999, Official site of the NHMRC, *http://www.health.gov.au:80/nhmrc/publicat/pdf/e35.pdf* (accessed October 2, 2000).

National Health and Medical Research Council (1999b), *Guidelines for Genetic Registers and Associated Genetic Material*, Australia, November 1999, Official site of the NHMRC, *http://www.health.gov.au/nhmrc/publicat/pdf/e14.pdf* (accessed October 2, 2000), Chap. 4.1 (o).

Network of Applied Genetic Medicine (2000), *Statement of Principles: Human Genomic Research*, Montréal, Québec, April 2000, Official site of the RMGA, *http://www.rmga.qc.ca/doc/principes_en_2000.html* (accessed October 2, 2000).

ROSES, A. D. (2000), Pharmacogenetics and the Practice of Medicine, *Nature* **405**, 857.

Royal College of Physicians (1999a), Recommendations from the Royal College of Physicians Committee on Ethical Issues in Medicine, *J. R. C. Phys. London* **33**, 264.

Royal College of Physicians (1999b), Recommendations from the Royal College of Physicians Committee on Ethical Issues in Medicine, *J. R. C. Phys. London* **33**, 265.

The Nuremberg Code (1946–1949), *Trials of the War Criminals before the Nuremberg Military Tribunals under Control Council Law No. 10*, October 1946–April 1949. Washington, DC: U.S. Government Printing Office.

United Nations Educational, Scientific and Cultural Organization (International Bioethics Committee) (1997), *Universal Declaration on the Human Genome and Human Rights*, Paris, November 11, 1997, Official site of the UNESCO, *http://www.unesco.org/ibc/uk/genome/projet/index.html* (accessed October 2, 2000).

World Health Organization (1997), *Proposed International Guidelines on Ethical Issues in Medical Genetic and Genetic Services*, Geneva, December 15 and 16, 1997, Official site of the World Health Organization, *http:///www.who.int/ncd/hgn/hgn-ethic.htm* (accessed October 2, 2000).

World Health Organization (1998), *Resolution WHA51.10 on Ethical, Scientific and Social Implications of Cloning in Human Health*, Geneva,

May 16, 1998, Official site of the WHO, *http://www.who.int/wha-1998/pdf98/ear10.pdf* (accessed October 2, 2000).

World Health Organization (1999), *Draft World Health Organization (WHO) Guidelines on Bioethics*, May 1999, *http://www.helix.nature.com/wcs/b23a.html* (accessed October 5, 2000).

World Medical Association (1964, 1975, 1983, 1989, 1996), Declaration of Helsinki, *Recommendations Guiding Physicians in Biomedical Research Involving Human Subjects*, World Medical Assembly, 1964, 1975, 1983, 1989, 1996, 2000).

20 Genomics – Five Years from Now

CHRISTOPH W. SENSEN

Calgary, Canada

1 Introduction

Genomics is a very fast evolving field, making it hard just to keep up with all the developments that are emerging. Predicting five years into the future is essentially impossible, but based on the current trends, we can at least try to project some future developments. The final chapter of this book is an attempt to do that, and, as with all predictions, may well generate a degree of controversy.

To forecast what might happen in the next five years, it is worthwhile to look into the past and extrapolate from the pace of previous developments. Reviewing some of the other chapters in this book, it becomes evident that nothing in the field of Genomics is really new; Genomics only takes a much larger-scale, automated and integrated approach to biochemistry, biology, and molecular medicine. Almost all of the technologies and techniques were in use before the advent of Genomics, only on a smaller scale and with less automation.

Two developments were crucial for the emergence of Genomics. The first development was laser technology, which allowed environmentally friendly versions of several existing technologies. For example, the use of radioactivity for DNA sequencing has all but been abandoned in favor of fluorescent biochemistries. Lasers have "invaded" many aspects of the molecular biology laboratory, they are part of automated DNA sequencers, high-density DNA array scanners, MALDI mass spectrometers, and confocal microscopes, to name a few.

The second technology was the development of the computer. Computers and the Internet have played a major role in the development of Genomics. All major machines in a molecular biology laboratory are connected to computers, often the data collection system is directly coupled to a laser-based detector.

At the same time, data exchange in the Genomics world is almost completely computer-based. "It is on the Web" is now a common notion and the interface that is provided through Web browsers is considered the major work environment for many scientific analyses. Genomic databases are shared through the Web, to the point that all new data are entered into Web-accessible databases on a daily basis, and prior to publication. Many large-scale projects are now conducted as international collaborations, there is even a new word for this: collaboratory.

A typical genome project involves many individuals. This is a dramatic change from the pre-genomic era when most molecular biological publications contained from two to five authors. The large size of the laboratories involved in Genomics research has resulted in new modes of operation, which in many cases are more or less along the lines of a factory operation. Tasks are distributed in a defined way, allowing few degrees of freedom for an individual. We can easily predict that this trend will continue, making single-author publications more or less a "thing of the past".

2 The Evolution of the Hardware

Hardware for Genomics has been developed at an astonishing rate. It is impossible to predict details of future machine developments, but we see several trends where new approaches may emerge in the near future. The following paragraphs contain speculations about some of these trends.

2.1 DNA Sequencing as an Example

As an example of the potential for future development, we would like to look at automated DNA sequencing. While it was quite good to obtain 1,000–2,000 base pairs (bp) of raw sequence per day from a radioactive DNA sequencing gel, today's capillary sequencers can produce up to 300,000 bp per machine in the same time frame. A combination of robotics, which allows up to 6 automated machine loads per day, enhancements in biochemistry that allow the even labeling of DNA fragments, and automation in data processing has resulted in this increased throughput. Radioactive sequencing gels all had to be analyzed more or less manually, while today's data can be assembled automatically, allowing the researcher to spend most of the time on data analysis. DNA sequencing laboratories are well on the way towards resembling the "Ford Model-T factory".

We can anticipate that another 4-fold capacity increase in DNA sequencing machines will happen when the current 96-capillary formats are replaced with 384-well formats. In addition to an increase in the number of capillaries per machine, there is certainly also room for improvement in the DNA separation technology. The DNA sequencing reaction covers at least 2,000 bp, while the detection systems can only use between 500 and 1,200 bp. By using more sophisticated separation strategies and better detection systems, there could be another 2- to 4-fold increase in DNA sequence production on a single device, without any changes to the biochemistry.

Further enhancements to DNA sequencing can be predicted, if the shortcomings of Acrylamide-based separation can be overcome. The DNA polymerases used for DNA sequencing today are selected for efficiency in the first 2,000 bp. If a separation technology with a much longer separation range could be established, it is easy to foresee that DNA sequencing could see another order-of-magnitude increase, because the biochemistry could then be adjusted to accommodate the new technical possibilities. Routine future DNA sequencing reactions could yield readouts many thousands of base pairs long.

2.2 General Trends

The DNA sequencing example above highlights some general trends that we can see ongoing right now. Some of these general trends are discussed in more detail in the paragraphs below and in much more detail in other chapters in this book. In Germany, we have a saying though: "Das Bessere ist der Feind des Guten", which translates as: "The better is the enemy of the good". There will be a continuation of gradual updates and upgrades that result in dramatic enhancements of performance, as well as completely new approaches

that outperform any known technology by orders of magnitude. While general laboratory technology, e.g., spectrophotometers, liquid chromatography equipment and the like have a useful lifetime of many years, this likely will not be true for some of the Genomics-related hardware, which will need replacement at a much faster pace because of the unforeseeable jumps in technology development. This probably will not pose a major threat or problem to Genomics research, as only approximately 10–15% of the total cost of a Genome project are related to hardware. Even with the current level of automation, most of the expenses in a Genome project are related to expendables and human labor. In future, the human component will be even further reduced through increased levels of automation.

2.3 Existing Hardware will be Enhanced for More Throughput

Almost none of the existing machines used in Genomics are stretched to their physical limits. Single photon detection systems, which can increase the detection sensitivity of Genomics hardware dramatically, and bring the resolution to the theoretical physical limit, higher density data processing (many devices still detect only on the 8-bit or 16-bit level), and faster data processing strategies can be used to enhance current technology. Certain systems, e.g., DNA sequencers and high-density DNA arrays can be scaled by adding more capillaries or spots to a unit. Separation-based machines (e.g., sequencing gels or protein gels) have seen some improvements in capability over time. They may be replaced by other strategies that allow much better separation. The crucial make or break for the establishment of a completely new technology will always be the cost factor. Even if a new concept proved to be technically superior, it still had to compete on production cost. At the end of the day, it does not really matter how data is produced, as long as the data quality is comparable.

2.4 The PC Style Computers that Run Most of the Current Hardware will be Replaced with Web-Based Computers

One of the biggest problems in molecular biology and Genomics laboratories today is the fact that most machines are operated with the help of a PC-style computer. Currently, there are almost as many different operating systems in use (from MacOS to OS/2 to Windows) as there are machines and today it is not unusual to see all of these operating systems in the same laboratory. This situation causes several problems. The PC-style computers age quickly and require high levels of system maintenance. For example, it was quite difficult to network many of the original ABI DNA sequencing machines using TCP/IP networks, which have now become standard in most laboratories, requiring major computer upgrades to achieve the integration. A typical sequencing machine, for example, might last 10 years, making it necessary to replace the controlling PC two to three times to keep up with the pace of development.

To address the problems of PC-style computers, we predict that client–server models will replace the stand-alone systems currently in use. Thin clients that collect the data and post them to the Web will be complemented by platform-independent analysis software that can be executed on large servers or the latest workstations. An early example of this approach is the LiCor 4200 Global system (see Chapter 8, this volume). Data collection on the LiCor 4200 Global system is performed by a Linux-based Netwinder thin client, which operates an Apache Web server for data access. Users can control and monitor this kind of machine from any Web browser (e.g., Netscape or Internet Explorer). The analysis software for the LiCor 4200 is written in the Java programing language, making it platform-independent, thus allowing execution on any Java-enabled platform.

2.4.1 Integration of Machinery will become Tighter

In the past, laboratories have operated many devices that could not directly "talk" to each other. Incompatible operating systems and the lack of data exchange possibilities prevented a high degree of automation. We see attempts to change this, machines from different manufacturers are starting to "speak the same language". Interestingly, this does not necessarily mean that all machines understand a common, standardized data format, instead they are capable of exporting and importing results generated by other devices. For example, spectrophotometers have become capable of exporting data sheets that can be imported into pipetting robots, saving time in the setup of PCR reactions. Sequence assembly programs can "pick" DNA sequencing primers and export the lists to oligonucleotide synthesizers, saving the time of retyping them before oligonucleotide synthesis. Laboratories are starting to build more and more "manufacturing belts" for Genomics and logically, the trend for integration and data exchange will continue.

2.4.2 More and More Biological and Medical Machinery will be "Genomisized"

The initial "Genomics Toolkit" mainly consisted of machines for DNA sequencing. The next add-on of major machinery were proteomics-related mass spectrometers, followed by machinery for expression studies, including DNA high density arrayers and DNA chip readers. As more and more aspects of biological, biochemical, biophysical, and medical research get automated, we can expect more and more of the mostly manual devices used today to be automated and developed further for high throughput. With the complete blueprints of organisms at hand, it is logical that the current set of Genomics tools will be expanded by *in vivo* and *in vitro* studies of organelles, cells, organs, and organisms. Microscopy equipment, and imaging equipment in general, will be part of the Genomics toolkit of the future. Physiological research will be more and more automated to achieve a level of throughput that allows one to study the global biological system rather than a single aspect. Structural aspects of molecules will become more and more important to Genomics research, protein crystallization factories may be emerging soon that allow a dramatic increase in the throughput of protein structure determination. The latter will be a prerequisite for efficient use of synchrotron facilities for protein structure determination.

The lessons learned by automating procedures in the molecular biology laboratory will certainly be applied in the integration and automation of more and more biological techniques. While there have been some attempts to create fully automated laboratories, we predict that there will always be a human factor in the laboratory. Manned space flight, even though considered dangerous and costly, and thus not feasible, is still a major factor today and we expect that high-tech Genomics laboratories will develop under similar constraints.

3 Genomic Data and Data Handling

Bioinformatics is the "glue" that connects the various Genomics experiments. The role of computer-based analysis and modeling cannot be overestimated. Currently, more than 100 genomics-related databases exist. Many of these databases are updated daily and growing exponentially. New databases are being created at an enormous pace, as more and different experiments get added to the collected works of Genomics. It is completely impossible to host all genomic data on personal computers or workstations; thus, there is a requirement for high-performance computing environments in every serious genome research effort. We predict that in the future, the computational infrastructure for most bioinformatics laboratories will be organized as a client–server model. This will address the performance issue that comes with the exponentially growing data environment, while controlling the computer maintenance aspects and thus the major cost factor of the computational infrastructure.

To date, computer chip development has been according to Moore's law (which predicts that the computing power of a CPU doubles approximately every 18 months) and Genomic data have been produced at a rate that could be accommodated with the current computer advancements. It is conceivable that the pace of computer development and Genomic data production will in the future loose this synchronization, causing problems with the amount of data that needs to be handled. The likelihood of this seems small, however, as there exist other large-scale databases that are much more extensive than all of the Genomic data in consideration (e.g., astronomical databases or weather data).

In addition to scaling current computational environments, more and more dedicated hardware is being developed that can assist Genomics research. The goal for future bioinformatics environments should be to provide real-time analysis environments. We are certainly far away from this goal today, but more systems such as the Paracel GeneMatcher, which accelerates database searches by a factor of 1,000, will enter the market and revolutionize the computational environment.

Networking between Genomics laboratories is a basic requirement for the success of any Genome project, thus biological and Genomics applications have been a major application domain for the development of advanced networking strategies. For example, the Canadian Bioinformatics Resource, CBR-RBC is the major test bed for new networking developments in Canada (see *http://www.cbr. nrc.ca*). The trend to establish advanced networking connections in Genomics laboratories will continue and ultimately all Genomics laboratories will be connected to broad-band systems. Distributed computational infrastructure has been established in several genome projects and efforts are underway to create nation-wide bioinformatics networks in several countries, including countries in Europe as well as Canada.

Almost all of the current databases are initially organized as ASCII flat files, with no relational database infrastructure to support them. Data access is currently handled through Web-based database integration systems such as ENTREZ and SRS (see Chapter 12, this volume), or tool integration systems such as MAGPIE or PEDANT (see Chapter 16, this volume). In both cases, reformatting of original data into standardized HTML files creates the illusion for the user that they are dealing with a single database. HTML has been accepted as the major working environment for biologists because the Web browsers (e.g., Netscape and Internet Explorer) allow a platform-independent graphical view of the results of Genomic analyses.

The use of HTML for biological and medical data is certainly very limited, as HTML was originally designed for text files, rather than data files. Several approaches are currently being pursued to create a more standardized approach to Genomics-related data. The most promising candidate seems to be the XML language. XML is an extendable Web language. New data types and display modes can easily be introduced into the system. Many of the current ASCII flat file databases are being provided additionally in XML format, even MEDLINE will soon offer a downloadable XML version. Web browsers of the future will be XML-compatible, allowing multiple views of the same data without the need for the reformatting of the dataset or reprograming the display interface.

The bioinformatics tools of the future will integrate as many different biological data types as possible into coherent interfaces and allow online research that provides answers to complex queries in real-time environments. Computer models of biological systems will become sophisticated enough to conduct meaningful "in silico" biology. This will certainly help to reduce the cost related with Genomics research and assist in the design of smarter wet-lab experiments.

4 Next Generation Genomics Laboratories

4.1 The Toolset of the Future

The first generation of Genomics laboratories focused on the establishment of "sequencing factories" that could produce genomic sequence in the most efficient manner. Initially, many of the sequencing projects, e.g., the European yeast genome project, were set up in a distributed fashion. This proved to be unworkable, as the entire project depended on the slowest partner. Thus, most recent projects involve relatively few partners of equal capabilities. This has lead to the development of very large laboratories in industry and academia that are capable of producing up to hundreds of megabases of finished sequence per year. Examples of such development are the Sanger Centre in Cambridge, UK, the DNA sequencing laboratory at the Washington University in St. Louis, the formation of TIGR in Maryland, and companies such as Incyte and Celera.

The exceptions to the rule that genomic DNA sequencing is conducted in large-scale laboratories are projects in developing countries, which are just entering the field. Projects in these countries are sometimes conducted in a setup similar to the European yeast project, but even here a concentration eventually takes place that restricts Genomics support to the more advanced laboratories. The trend to larger and larger DNA sequencing laboratories will certainly not stop, because the greatest efficiency in DNA sequencing can be achieved in large settings. It can be expected that most, if not all, DNA sequencing in future will be contracted to large-scale laboratories, similar to film-processing laboratories in the photographic sector before the advent of digital photography.

Over time, it has become apparent that much of the DNA sequence that is being produced initially stays meaningless. Many of the potential genes that are identified through gene searching algorithms do not match any entries in the public databases, thus no function can be assigned to them. Without functional assignment, the true goal of any Genome project, which is to understand how genomes are organized, and how the organism functions, cannot be achieved. It is also not possible to protect any intellectual property derived from genomic sequence, if the function of the molecule is not known (see Chapter 19, this volume). The logical consequence is that many Genome laboratories today are trying to diversify and broaden their toolset.

Proteomics was the first addition to the genomic toolset and is also a very rapidly evolving field. Today, many expressed proteins can be identified via the 2D-gel mass spectrometry route (see Chapter 10, this volume), but there are still many limitations that make it impossible to obtain and examine a complete proteome. Proteins that exist in very small quantities in a cell, as well as proteins that are rarely expressed can elude the current detection methods. It can be anticipated that new proteomics-related techniques and approaches will be introduced at a fast pace. Protein chips and new separation techniques that will bypass the 2D-gel systems will be introduced and will help to advance protein studies dramatically. At the same time, current technologies will constantly be improved. As an example, we now see the first MS-MS-ToF systems coupled to a MALDI front-end entering the market. It can be expected that within the next 5 years, the large-scale study of proteins will reach the same level as the large-scale study of DNA molecules today.

Expression studies (macro arrays, high-density arrays, chips) have become a very popular addition to the genomic toolset (see Chapter 9, this volume). Theoretically, this methodology allows screening of the entire genome for activity under a wide variety of conditions. At this point, there are still more publications about technical aspects of DNA arrays or DNA chips than applications of the technology, but this situation is changing. The major bottleneck for this technology is currently the computer analysis component, which has not yet caught up with the technology. We expect that it will be seamless in the future to go from an RNA extraction to the functional analysis of the DNA array or DNA chip experiment. While today's chip readers are about the size of a medium-sized photocopier, we expect that integrated handheld devices will be

developed over time, which allow field testing with diagnostic chips, returning results instantaneously.

Certain organisms, like yeast, *Caenorhabditis elegans*, and mice are being subjected to intense knockout studies in order to gain insight into the function of as many genes as possible. This technology is not without pitfalls, as the knockout of many genes does not cause any "visible" effect, while the knockout of other genes is lethal to the organism, which causes great difficulty in the interpretation of the function of these genes. Moreover, most organismal characteristics are derived from the actions of several genes. We predict that this technology will not be applied widely outside the current model systems, because of the cost and the sometimes uninterpretable results.

For many years, researchers have attempted to use structural information to deduce the function of proteins (see Chapter 15, this volume). Mostly, these attempts have been unsuccessful, yet we predict that there will be a better chance for the deduction of function from structure in the future because then the entire dataset (many complete genomes, and all biologically relevant folds) will be known. This knowledge can then be used to predict the function of a similar gene with much more accuracy. "Structure mining" projects are currently under way that are attempting to determine all biologically relevant structures, and even though this process is tedious and slow, it is likely that within the next 5–10 years these efforts will be successful.

While it will certainly be possible to identify the function of many genes using the technologies mentioned above, existing technologies in other fields will need to be applied to genomics in innovative ways, and entirely new technologies will have to be added to the existing biological, biochemical, and medical tools, to gain a complete understanding of the function of all genes in the context of the living organism.

Imaging technologies are a prime candidate for systems that will have to join the current tool set in order to gain three-dimensional data and time-related information about processes in cells, organs, and organisms. Automation has started to "invade" the microscopy sector, assisted by laser technologies that allow micromanipulations unheard of 20 years ago. Today, automated microinjection, optical tweezers, and confocal microscopes that can be switched to high-speed imaging, shooting thousands of images per second are reality. Flow cytometry is now capable of screening hundreds of thousands of cells per minute, sorting candidate cells with desired features in an axenic manner. This equipment will soon be introduced to Genomics to allow the location of elements in a cell, monitor them over time, and create the input for more realistic computer models of life (virtual cell).

4.2 Laboratory Organization

There will likely be three types of Genomics laboratories in the future: (1) large-scale factories dedicated to certain technologies, (2) integrated Genomics laboratories, (3) laboratories that coordinate data production rather than produce data.

(1) Large-scale factories will continue to exist. The advantage of lower production cost that comes with the scale and integration level of these operations will justify their existence for a long time to come. It is crucial to understand that most of these factories will do most of their business based on contracts with other parties. The terms of these contracts (i.e., who owns the rights to the data), has been and will be the issue that determines the success of such a factory.

(2) Most of the genomics laboratories of the future will be tightly integrated units that employ as many different technologies as possible on a large scale in order to get a "complete" picture that helps to build more precise models of how cells function. This is the logical consequence of the lessons learned from first generation Genomics laboratories. The development into this direction is probably best described with LEROY HOODS term "Systems Biology". The integrated laboratories will not necessarily have to produce any data on a large scale, they might, for example, collaborate with factory-style operations for large-scale data collection in most of their projects. The crucial aspect of an integrated laboratory is the computer setup. The bioinformatics infrastructure of a "Systems Biology" laboratory

has to be quite large and capable of handling and integrating many different data types, some of them being produced at astonishing rates. "Systems Biology" laboratories will thus employ large bioinformatics development teams that can provide custom software solutions to address research and development needs.

(3) We envision the emergence of a new type of laboratory, without any capability to produce data, but rather a coordination office with data analysis capability. This kind of laboratory will outsource all wet-lab activities to third parties, allowing for very low overheads and high degrees of flexibility as such a laboratory can always draw on the latest technology by choosing the right partners. Many startup companies have begun to work this way, at least for part of their activities, and large pharmaceutical companies have outsourced many of their Genomics-related activities in the last 5 years.

The risk in this case is almost completely on the side of the data producer. This is an area where Governments will have to involve themselves in Genome research and development, as many of the risks in high technology environments are hard to calculate and will probably not initially be taken by industry. The countries most involved in Genomics today all have large Government programs to address this need and almost certainly this involvement will have to continue for the foreseeable future.

5 Genome Projects of the Future

As said before, the goal of Genomics research is to obtain the "blue print" of an organism in order to understand how it is organized and how it functions. To achieve this goal, Genome projects of the future will have to become truly integrated. Connecting all the different bits and pieces available in many locations around the world is certainly one of the major challenges for future Genomics research. In order to facilitate this, data will have to be available instantaneously. In many cases, DNA sequence is already produced under the

"Bermuda agreement", which calls for posting of new data to the Web immediately after data generation.

Complete openness before publication is certainly a new paradigm in biological and medical research, but the notion that only a complex data analysis leads to new insights and that any particular data type in itself remains meaningless without the connection to all the other data types is becoming more and more established.

Genome research will focus increasingly on biological questions. Almost 15 years after the first plastid genomes were completely sequenced in Japan, there are still certain genes in plastids, which have no identified function. This can certainly be attributed to the fact that much of the plastid genome research stopped when these genomes were released, because of the false notion that now "the work was done". The real work starts with the completion of the genomic sequencing and it might never stop because of the complexity of biological systems.

A key factor for the continuation of Genomics research will be the public perception and level of acceptance of this kind of research. To date, scientists have not been very successful in two key areas, the education of the public about ongoing activities and the evaluation of basic science using ethical criteria. Companies have been perceived as creating "Frankenfoods" that are potentially harmful rather than better products that serve mankind. Molecular biology experiments are very different from atomic research, as they can be conducted almost anywhere with little effort, and thus it is quite hard to exercise control. Future genome projects will have to employ higher standards in the evaluation of ethical aspects of the research. A better education of the public is absolutely necessary to generate the consensus needed for continued support.

The openness that is asked for between scientists in order to address the goals of Genomics research should be a starting point for the relations with the general public. If there is free access to the data by everyone, public control can be exercised more readily than in a secretive environment. Much of today's Genomics research is funded with tax dollars, which is an easily forgotten, but very important fact. The openness that is being asked for should certainly not go as far as to inflict on the protection of intellectual property, but the general approaches to Genomics science should be common knowledge.

6 Epilog

One of the most frequently asked questions in Genomics is: "How long is Genome sequencing going to last?" Genome sequencing will be around for a very long time to come. The sequencing process has become so affordable that it is feasible to completely characterize a genome to answer very few initial scientific questions. The genomic data are now typically shared with the entire scientific community. Unlike many other biological data, the genomic sequence is a fixed item, new technology will not help to improve it any more, and a complete genome of good quality is, therefore, the true end product of the sequencing process.

Several million different species exist on earth today, and each individual organism has a genotype that is different from any other individual. Comparative Genomics will lead to many new insights into the organization of life. The knowledge about the diversity of life will lead to new products and cures for diseases at an unprecedented pace. The need to characterize genomic data will, therefore, be almost endless.

In a related research field, studying the phylogeny of species, many thousand 18S rDNA sequences have been obtained to date, even though the basic dataset that could answer most questions had been generated more than 10 years ago. Today, probably more 18S rDNA sequences get produced in a week than in all of 1990. While it took many years to produce the first 5S rDNA sequence, today this technology is standard in any sequencing laboratory around the world.

The almost endless possibilities of Genomics research are very exciting for all of us involved in this kind of research. This keeps us going through the enormous efforts that it takes to complete any large-scale genome project.

Index

Color Plates

Chapter 2

B.

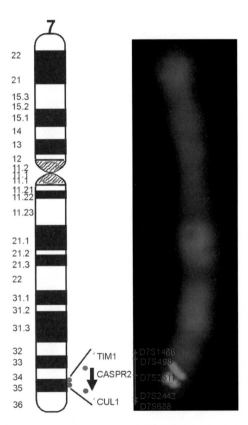

Fig. 5. Integrated physical, genetic and DNA sequence map of human chromosome 7q35. (**A**) The physical order of DNA markers (microsatellites, STSs, ESTs, and genes) as determined by their presence or absence in YAC and BAC clones, is shown. CASPR2 was determined to be one of the largest genes in the human genome spanning 2.3 Mb (K. NAKABAYASHI and S. W. SCHERER, unpublished data). The representation of "working draft" DNA sequence alone compared and combined with "whole genome shotgun" (Celera contigs) data is shown. (**B**) A BAC from exon 1 (pink) and another from exon 25 (green) of CASPR2 were hybridized using FISH to map the gene to chromosome 7q35.

Chapter 15

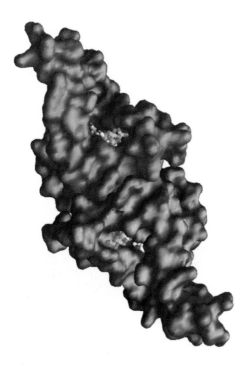

Fig. 3. Crystal structure of the MJ0577 protein from *Methanococcus jannaschii*. Each molecule of the dimer has a deep cleft on one side. A molecule of ATP and a Mn^{2+} ion were found in each cleft (PDB code 1mjh; ZAREMBINSKI et al., 1998).

Fig. 4. Structural comparison of HSP70 β-and actin as an example of homologous protein structures. Shown in blue is the structure of actin complexed with ATP (PDB code 1hlu) and in green is Hsp70 complexed with ADP (PDB code 1hpm).

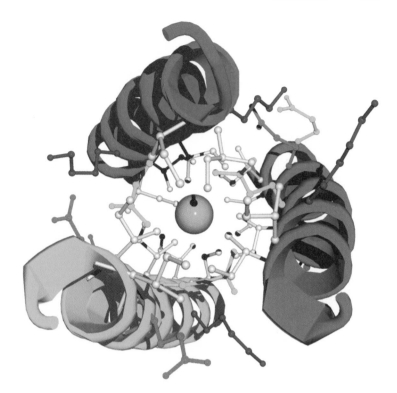

Fig. 5. Structure of an engineered leucine zipper containing three parallel coils. The core of the trimer (yellow, green and magenta coils) is filled with isoleucine residues (grey), the middle layer of glutamine residues (yellow) coordinates a Cl^- ion and the sides are lined with acidic (red) and basic (blue) residues forming salt bridges (NAUTIYAL et al., 1995).

Chapter 17

Fig. 2. Overlapping clone cluster in a Magpie project. Each sequence is hyperlinked to its Magpie report. All of the sequences are filled in with green, denoting finished sequence, and shaded where redundant. Red-outlined sequences are reverse complemented in the assembly. White letter labels denote the presence of annotations in the sequence.

Fig. 3. Partially assembled fragments of BAC b07zd03_b28. Fragments are sorted by size and hyperlinked to their respective Magpie reports. The yellow fill indicates that sequences are in the linking state.

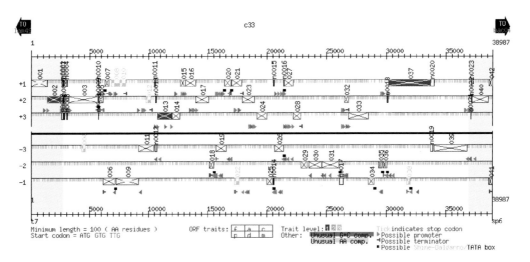

Fig. 5. Contiguous sequence with open reading frames displayed. Boxes on the six reading frame lines represent possible genes. The boxes are "x"ed when annotated. Light "x"ed boxes have annotations described as hypothetical or uncharacterized. Sub-boxes in unannotated genes indicate composition characteristics, plus the best level of protein, DNA and motif database hits. Grayed-out genes have been suppressed. Background shading and hyperlinked arrows in the corners indicate neighboring sequence overlaps. Boxes with labels that start in "n" are possible genes shorter than the specified minimum length.

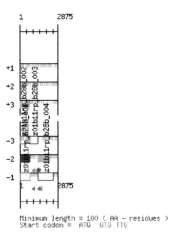

Fig. 6. Contiguous sequence with open reading frames and evidence displayed. Evidence and ORFs are displayed in their frames and locations. This facilitates easy recognition of frameshifts and partial genes. Other characteristics are the same as in Fig. 5.

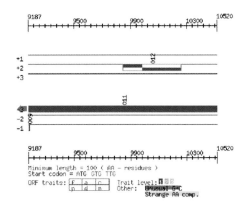

Fig. 7. ORF close-up. Displays the exact start and stop coordinates, and hyperlinked overlapping ORFs. Gene labels and trait sub-boxes are potentially easier to read than in Fig. 5.

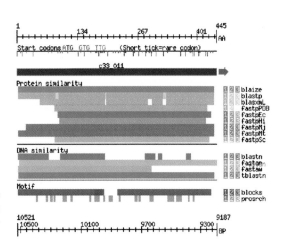

Fig. 8. Evidence summary. Evidence is grouped by type, and displayed as one line for all hits from a tool. Better evidence is darker and closer in the foreground.

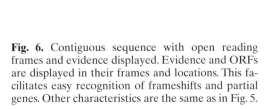

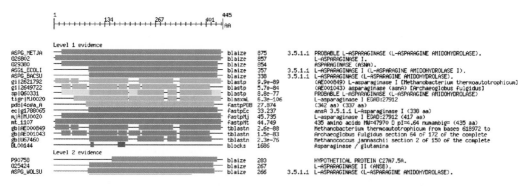

Fig. 9. Expanded evidence. Evidence is sorted by level, tool, score, and length. The first column links to the database ID of the similar sequence. The second displays the similarity location. It is linked to the original tool report. The third names the tool used. The fourth displays the tool's scoring of the match. The fifth displays the Enzyme Commission Number, hyperlinked to further enzyme information. The last column displays the matching sequence description.

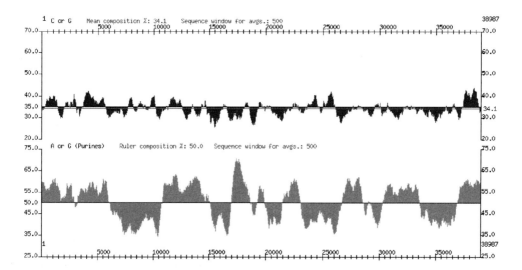

Fig. 10. Base compositions. Average A + G and G + C compositions are calculated using a sliding window of 500 bases. The moving average is displayed as a filled graph, both above and below the centerline average. Unusual G + C may indicate the presence of transposable elements. Majority A + G indicates coding strand in many organisms.

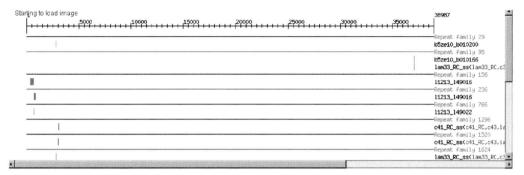

Fig. 11. Sequence repeats. The image has scrollbars (as part of an applet) because of its large dimensions. Repeats are sorted into families of matching subsequences, which are further sorted by size in descending order. The matching sequence name is in the left part of the right hand column, while the location and size of the match are displayed as filled boxes under the ruler. Red, blue and green boxes represent forward, reverse complement, and complement matches.

Fig. 12. Ambiguity information. This graph starts at base 50,001 because it is the second pane for a 71,519 base sequence limited to 50,000 bases per image. If the ambiguous base has been sequences on the forward, reverse, or both strands, the tick is displayed on the top, bottom, or center line respectively. The displayed ambiguities total is for the pane, not the sequence as a whole.

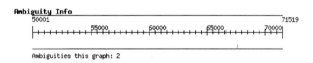

Fig. 13. Assembly information. Histograms of average positive and negative strand assembly coverage are above and below the centerline. Breaks in the blue and red bars indicate gaps in the positive and negative strand coverage. Genomic neighbors are indicated by background shading as in Fig. 5.

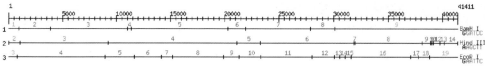

Codes are 5' to 3', with differentiation after cut.

Fig. 14. Restriction sites. Ticks correspond to the location of cut sites for the restriction enzymes listed in the right hand margin. The fragments produced are labeled 5' to 3', with the undisplayed cloning vector on the 5' end. Fragments containing vector have yellow labels, otherwise they are green. The HindIII line has restriction sites at the ends of the sequence because the clone library was HindIII restricted. By consequence, it has no yellow labeled fragments.

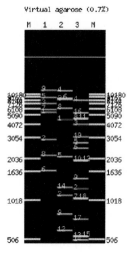

Fig. 15. Agarose gel simulation. Fragment lanes and labels correspond to those in Fig. 14. The fragment migration is calculated using the specified standard marker "M" lane migrations. Fragments containing vector are colored yellow. Fragments composed entirely of vector are colored red. Where fragments are very close, luminescence is more intense and labels are offset for readability.